Lecture Notes in Physics

For information about Vols. 1–131, please contact your bookseller or Springer-Verlag.

Vol. 132: Systems Far from Equilibrium. Proceedings, 1980. Edited by L. Garrido. XV, 403 pages. 1980.

Vol. 133: Narrow Gap Semiconductors Physics and Applications. Proceedings, 1979. Edited by W. Zawadzki. X, 572 pages. 1980.

Vol. 134: γγ Collisions. Proceedings, 1980. Edited by G. Cochard and P. Kessler. XIII, 400 pages. 1980.

Vol. 135: Group Theoretical Methods in Physics. Proceedings, 1980. Edited by K. B. Wolf. XXVI, 629 pages. 1980.

Vol. 136: The Role of Coherent Structures in Modelling Turbulence and Mixing. Proceedings 1980. Edited by J. Jimenez. XIII, 393 pages. 1981.

Vol. 137: From Collective States to Quarks in Nuclei. Edited by H. Arenhövel and A. M. Saruis. VII, 414 pages. 1981.

Vol. 138: The Many-Body Problem. Proceedings 1980. Edited by R. Guardiola and J. Ros. V, 374 pages. 1981.

Vol. 139: H. D. Doebner, Differential Geometric Methods in Mathematical Physics. Proceedings 1981. VII, 329 pages. 1981.

Vol. 140: P. Kramer, M. Saraceno, Geometry of the Time-Dependent Variational Principle in Quantum Mechanics. IV, 98 pages. 1981.

Vol. 141: Seventh International Conference on Numerical Methods in Fluid Dynamics. Proceedings. Edited by W. C. Reynolds and R. W. MacCormack. VIII, 485 pages. 1981.

Vol. 142: Recent Progress in Many-Body Theories. Proceedings. Edited by J. G. Zabolitzky, M. de Llano, M. Fortes and J. W. Clark. VIII, 479 pages. 1981.

Vol. 143: Present Status and Aims of Quantum Electrodynamics. Proceedings, 1980. Edited by G. Gräff, E. Klempt and G. Werth. VI, 302 pages. 1981.

Vol. 144: Topics in Nuclear Physics I. A Comprehensive Review of Recent Developments. Edited by T.T.S. Kuo and S.S.M. Wong. XX, 567 pages. 1981.

Vol. 145: Topics in Nuclear Physics II. A Comprehensive Review of Recent Developments. Proceedings 1980/81. Edited by T. T. S. Kuo and S. S. M. Wong. VIII, 571-1.082 pages. 1981.

Vol. 146: B. J. West, On the Simpler Aspects of Nonlinear Fluctuating. Deep Gravity Waves. VI, 341 pages. 1981.

Vol. 147: J. Messer, Temperature Dependent Thomas-Fermi Theory. IX, 131 pages. 1981.

Vol. 148: Advances in Fluid Mechanics. Proceedings, 1980. Edited by E. Krause. VII, 361 pages. 1981.

Vol. 149: Disordered Systems and Localization. Proceedings, 1981. Edited by C. Castellani, C. Castro, and L. Peliti. XII, 308 pages. 1981.

Vol. 150: N. Straumann, Allgemeine Relativitätstheorie und relativistische Astrophysik. VII, 418 Seiten. 1981.

Vol. 151: Integrable Quantum Field Theory. Proceedings, 1981. Edited by J. Hietarinta and C. Montonen. V, 251 pages. 1982.

Vol. 152: Physics of Narrow Gap Semiconductors. Proceedings, 1981. Edited by E. Gornik, H. Heinrich and L. Palmetshofer. XIII, 485 pages. 1982.

Vol. 153: Mathematical Problems in Theoretical Physics. Proceedings, 1981. Edited by R. Schrader, R. Seiler, and D.A. Uhlenbrock. XII, 429 pages. 1982.

Vol. 154: Macroscopic Properties of Disordered Media. Proceedings, 1981. Edited by R. Burridge, S. Childress, and G. Papanicolaou. VII, 307 pages. 1982.

Vol. 155: Quantum Optics. Proceedings, 1981. Edited by C.A. Engelbrecht. VIII, 329 pages. 1982.

Vol. 156: Resonances in Heavy Ion Reactions. Proceedings, 1981. Edited by K.A. Eberhard. XII, 448 pages. 1982.

Vol. 157: P. Niyogi, Integral Equation Method in Transonic Flow. XI, 189 pages. 1982.

Vol. 158: Dynamics of Nuclear Fission and Related Collective Phenomena. Proceedings, 1981. Edited by P. David, T. Mayer-Kuckuk, and A. van der Woude. X, 462 pages. 1982.

Vol. 159: E. Seiler, Gauge Theories as a Problem of Constructive Quantum Field Theory and Statistical Mechanics. V, 192 pages. 1982.

Vol. 160: Unified Theories of Elementary Particles. Critical Assessment and Prospects. Proceedings, 1981. Edited by P. Breitenlohner and H.P. Dürr. VI, 217 pages. 1982.

Vol. 161: Interacting Bosons in Nuclei. Proceedings, 1981. Edited by J.S. Dehesa, J.M.G. Gomez, and J. Ros. V, 209 pages. 1982.

Vol. 162: Relativistic Action at a Distance: Classical and Quantum Aspects. Proceedings, 1981. Edited by J. Llosa. X, 263 pages. 1982.

Vol. 163: J. S. Darrozes, C. Francois, Mécanique des Fluides Incompressibles. XIX, 459 pages. 1982.

Vol. 164: Stability of Thermodynamic Systems. Proceedings, 1981. Edited by J. Casas-Vázquez and G. Lebon. VII, 321 pages. 1982.

Vol. 165: N. Mukunda, H. van Dam, L.C. Biedenharn, Relativistic Models of Extended Hadrons Obeying a Mass-Spin Trajectory Constraint. Edited by A. Böhm and J.D. Dollard. VI, 163 pages. 1982.

Vol. 166: Computer Simulation of Solids. Edited by C.R.A. Catlow and W.C. Mackrodt. XII, 320 pages. 1982.

Vol. 167: G. Fieck, Symmetry of Polycentric Systems. VI, 137 pages, 1982.

Vol. 168: Heavy-Ion Collisions. Proceedings, 1982. Edited by G. Madurga and M. Lozano. VI, 429 pages. 1982.

Vol. 169: K. Sundermeyer, Constrained Dynamics. IV, 318 pages. 1982.

Vol. 170: Eighth International Conference on Numerical Methods in Fluid Dynamics. Proceedings, 1982. Edited by E. Krause. X, 569 pages. 1982.

Vol. 171: Time-Dependent Hartree-Fock and Beyond. Proceedings, 1982. Edited by K. Goeke and P.-G. Reinhard. VIII, 426 pages. 1982.

Vol. 172: Ionic Liquids, Molten Salts and Polyelectrolytes. Proceedings, 1982. Edited by K.-H. Bennemann, F. Brouers, and D. Quitmann. VII, 253 pages. 1982.

Lecture Notes in Physics

Edited by H. Araki, Kyoto, J. Ehlers, München, K. Hepp, Zürich
R. Kippenhahn, München, H. A. Weidenmüller, Heidelberg
and J. Zittartz, Köln

201

Group Theoretical Methods in Physics

Proceedings of the XIIth International Colloquium
Held at the International Centre for Theoretical Physics,
Trieste, Italy, September 5–11, 1983

Edited by G. Denardo, G. Ghirardi and T. Weber

Springer-Verlag
Berlin Heidelberg GmbH 1984

Editors

G. Denardo
G. Ghirardi
T. Weber
Istituto di Fisica Teorica dell'Università di Trieste
Trieste, Italy

ISBN 978-3-540-13335-3

Library of Congress Cataloging in Publication Data. Main entry under title: Group theoretical methods in physics. (Lecture notes in physics; 201) Includes index. 1. Groups, Theory of–Congresses. 2. Mathematical physics–Congresses. I. Denardo, G. (Gallieno), 1935-. II. Ghirardi, G. C. III. Weber, T. (Tullio), 1937-. IV. International Colloquium on Group Theoretical Methods in Physics (12th: 1983: International Centre for Theoretical Physics) V. Series.
QC20.7.G76G78 1984 530.1'5222 84-5597

ISBN 978-3-540-13335-3 ISBN 978-3-540-38859-3 (eBook)
DOI 10.1007/978-3-540-38859-3

Originally published by Springer-Verlag Berlin Heidelberg New York Tokyo in 1984

2153/3140-543210

PREFACE

The XII International Colloquium on Group Theoretical Methods in Physics took place at the International Centre for Theoretical Physics in Trieste, Italy, September 5 - 11, 1983.

The purpose of the Colloquium has been to bring together scientists active in the several fields of theoretical physics in which group theoretical methods are used. The Colloquium included invited general talks aimed at giving all participants an outline of the uses of group theoretical techniques in the various branches of physics, and specialized contributions presented as seminars or posters. The matter has been divided into seven sessions according to the following scheme:

1. Group Representations, Group Extensions, Contractions and Bifurcations
2. Completely Integrable Systems
3. Elementary Particles and Gauge Theories
4. Supersymmetry and Supergravity
5. Atomic and Nuclear Physics
6. Symmetries in Condensed Matter Physics and Statistical Mechanics
7. Canonical Transformations and Quantum Mechanics

We had the great pleasure of having in a special session a general talk given by Professor E.P. Wigner.

The proceedings are structured according to the previous proceedings by listed sessions, and within each session we have followed the alphabetical order of the authors. Obviously, due to the possible overlapping of some topics in various sessions, in some cases the choice of fitting a contribution in a given session has been to a certain extent arbitrary and has been done, for organizational purposes, in such a way as to balance the time devoted to the various topics.

We wish to express our gratitude to

- the Standing Committee, which decided to have the XII Colloquium in Trieste, expressing in this way confidence in the capacities of the Trieste scientific environment and of the local organizers;
- the International Advisory Committee, which was of great help in choosing the appropriate speakers for each subject;
- the International Centre for Theoretical Physics, Trieste, the International School for Advanced Studies, Trieste, and the Istituto Nazionale di Fisica Nucleare, Italy, for having allowed through their generous sponsorship the organization of the Colloquium. In particular the funds we had at our disposal have allowed us to partly support, in the spirit of the policy of ICTP, the participation of various scientists from developing countries;

- the secretarial staff of ICTP and the Institute for Theoretical Physics of the University of Trieste, for their efficient assistance in solving practical problems.

Trieste, December 1983

The Editors:

G. Denardo
G.C. Ghirardi
T. Weber

The map of Palmanova, a small town near Trieste, founded in 1593 by the Venetian government, which the participants to the Colloquium had the opportunity to visit during an organized trip. It is appropriate to use it as the symbol for the Colloquium due to its wonderfully symmetric structure.

The participants of the XII Colloquium, in Trieste, Italy

INTERNATIONAL ADVISORY COMMITTEE

Abdus Salam
ICTP
Strada Costiera 11
Miramare, Trieste
Italy

M.F. Atiyah
Mathematical Institute
University of Oxford
24-29 St. Giles
Oxford OX1 3, LB
U.K.

H. Bacry
Centre de Physique Théorique
CNRS
Case 907
13288 Marseille Cedex
France

L.C. Biedenharn
Department of Physics
Duke University
Durham, N. Carolina 27706
U.S.A.

K. Bleuler
Institut für Theoretische Kernphysik
Universität Bonn
Nussallee 14-16
5300 Bonn
Fed.Rep. Germany

A.P. Cracknell
Carnegie Lab. of Physics
University of Dundee
Dundee DD1 4HN, Scotland
U.K.

H.D. Doebner
Institut für Theoretische Physik
Technische Universität Clausthal
3392 Clausthal-Zellerfeld
Fed. Rep. Germany

M. Hamermesh
School of Physics and Astronomy
Tate Laboratory of Physics
University of Minnesota
116 Church Street S.E.
Minneapolis, Minnesota 55455
U.S.A.

A. Janner
Instituut voor Theoretische Fysica
Katholieke Universiteit
Nijmegen
The Netherlands

M. Lax
Department of Physics
The City College of
The City University of New York
New York, N.Y. 10031
U.S.A.

M. Moshinsky
Instituto de Fisica
UNAM- Universidad Nacional Autonoma
de Mexico
Apdo. Postal 20-364
Mexico 20, D.F.
Mexico.

Y. Ne'Eman
Sackler Institute of Advanced Studies
Tel Aviv University
Tel Aviv
Israel

I. Prigogine
Service de Chimie Physique II
Campus Olaine
U.L.B.- Blvd. du Triomphe
1050 Bruxelles
Belgium

T. Regge
Istituto di Fisica
Università degli Studi
Corso d'Azeglio 46
10125 Torino
Italy

I.E. Segal
Department of Mathematics
MIT, Room 2-244
Cambridge, Massachusetts 02139
U.S.A.

S. Stenberg
Department of Mathematics
Harvard University
1, Oxford Street
Cambridge, Massachusetts 02139
U.S.A.

G. 't Hooft
Instituut voor Theoretische Fysica
Princetonplein 5, Pb. 80.006
3508 TA Utrecht
The Netherlands

J. Wess
Institut für Theoretische Physik
Universität Karlsruhe (TH)
Kaiserstrasse 12
D-7500 Karlsruhe 1
Fed.Rep. Germany

E.P. Wigner
Department of Physics
Princeton University
P.O. Box 708
Princeton , New Yersey 08544
U.S.A.

P. Winternitz
Centre de Recherches de Matématiques
Appliquées
Université de Montreal
C.P. 6128, "succ. A"
Montreal, Quebec
Canada

B.C. Wybourne
Department of Physics
University of Canterbury
Christchurch 1
New Zeland

CONFERENCE ORGANIZERS

G. Denardo
Istituto di Fisica Teorica
Università di Trieste
Trieste
Italy
and ICTP - Trieste, Italy.

G. Ghirardi
Istituto di Fisica Teorica
Università di Trieste
and
ICTP
Trieste
Italy

T. Weber
Istituto di Fisica Teorica
Università di Trieste
Italy

LIST OF PARTICIPANTS

Abdus Salam
ICTP
Strada Costiera 11
Miramare, Trieste
Italy

S. Adjei
Obafemi Awalowo University
P.M.B. 5363
Ado-Ekiti
Ohdo State
Nigeria

S. Twareque Ali
Department of Mathematics
Concordia University
7141 Sherbrooke St. West
Montreal PQ H4B 1R6
Canada

A. Al-Lahham
Department of Mathematics
University of Damascus
Damascus
Syria

R.L. Anderson
Department of Physics and Astronomy
University of Georgia
Athens, Georgia 30602
U.S.A.

M.N. Angelova
Faculty of Physics
Sofia University
Blvd. A. Ivanov 5
1126 Sofia
Bulgaria

J.P. Antoine
Institut de Physique Théorique
Université Catholique de Louvain
Chemin du Cyclotron 2
1348 Louvain - la -Neuve
Belgium

F. Arickx
Dienst Teoretische en Wiskundige
Natuurkunde
Rijksuniversitaire Centrum Antwerpen
Groenenborgerlaan 171
2020 Antwerpen
Belgium

M.I. Arroyo
Faculty of Physics
University of Sofia
Blvd. A. Ivanov 5
1126 Sofia
Bulgaria

R. Arshansky
Department of Physics and Astronomy
Tel Aviv University
Tel Aviv
Israel

J.A. de Azcarraga
Dpto. de Fisica Teorica
Facultad de Ciencias Fisicas
Universidad de Valencia
Buriasot (Valencia)
Spain

H. Bacry
Centre de Physique Théorique
CNRS
Case 907
13288 Marseille Cedex
France

L. Balloomal
Department of Physics
Faculty of Science
Cairo University
cairo
Egypt

L.M. Benn
Department of Physics
University of Lancaster
Lancaster LA1 4YB
U.K.

G. Bhamathi
Department of Physics
University of Madras
Madras 25
India

L.C. Biedenharn
Department of Physics
Duke University
Durham, N. Carolina 27706
U.S.A.

K. Bleuler
Institut for Theoretische Kernphysik
Universität Bonn
Nussallee 14-16
5300 Bonn
Fed. Rep. Germany

M. Bozic
Institute of Physics
P.O. Box 57
11001 Belgrade
Yugoslavia

M. Bregola
Istituto di Fisica
Via Paradiso 12
44100 Ferrara
Italy

P. Budinich
SISSA
Strada Costiera 11
Miramare, Trieste
Italy

F. Calogero
Istituto di Fisica "G. Marconi"
Università degli Studi
Piazzale Aldo Moro 2
00185 Roma
Italy

J.F. Carinena
Dpto. de Fisica Teorica
Facultad de Ciencias
Univesridad de Zaragoza
Zaragoza
Spain

L. Castell
Max-Planck Institut
Matildstrs. 16
813 Starnberg
Fed.Rep. Germany

E. Chacon Esponda
Instituto de Fisica
UNAM- Univ.Nac.Autonoma de Mexico
Apdo. Postal 20-364
Delegacion A. Obregon
01000 Mexico City
Mexico

Chau Ling-Lie
Department of Physics
Brookhaven National Laboratories
Upton, New York
U.S.A.

F.J. Chinea
Dpto. de Metods Matematicos
de la Fisica
Facultad de Ciencias Fisicas
Universidad de Madrid
Madrid 3
Spain

G.C. Chukwumah
Department of Mathematics
University of Nigeria
Nsukka
Nigeria

J.F. Cornwell
Department of Theoretical Physics
University of St. Andrews
St. Andrews, Fife
Scotland KY16 9SS
U.K.

J.P. Dahl
Department of Chemical Physics
Technical University of Denmark
DTH 301
2800 Lyngby
Denmark

G. D'Ariano
Dipartimento Fisico
A. Volta
27100 Pavia
Italy

E. D'Emilio
Scuola Normale Superiore di Pisa
Pisa
Italy

H. De Meyer
Seminarie voor Wiskundige Natuurkunde
Rijksuniversiteit- Gent
Krijgslaan 281-S9
B-9000 Gent
Belgium

G. Denardo
Istituto di Fisica Teorica
Università di Trieste
Trieste
Italy
and
ICTP - Trieste, Italy.

Y. De Rop
Facultés Universitaires
Notre Dame de la Paix
61 Rue de Bruxelles
B-5000 Namur
Belgium

P. De Wilde
Seminarie voor Wiskundige Natuurkunde
Rijksuniversiteit-Gent
Krijgslaan 281-S9
B-9000 Gent
Belgium

R. Dirl
Institut für Theoretische Physik
TU Wien
A-1040 Wien
Karlsplatz 13
Austria

H.D. Doebner
Institut für Theoretische Physik
Technische Universität Clausthal
3392 Clausthal-Zellerfeld
Fed.Rep. Germany

H. Eichenherr
ETH - Hönggerberg
Theoretical Physics
CH-8093 Zurich
Switzerland

M. Elfazani
Department of Physics
Faculty of Science
P.O. Box 9480
Benghazi
Libya

A.R. Engelmann
Quantum Chemistry Group
University of Uppsala
Box 518
S 751 20 Uppsala
Sweden

F. Englert
Pool de Physique
Universite Libre de Bruxelles
C.P. 225
1050 Bruxelles
Belgium

W.M. Fairbairn
Department of Physics
University of Lancaster
Lancaster LA1 4YB
U.K.

P. Fortini
Istituto di Fisica
Via Paradiso
44100 Ferrara
Italy

J. Fuksa
Institute of Physics .
Czechoslovak Academy of Sciences
Na Slovance 2
180 40 Praha 8 - Liben
Czechoslovakia

L.C. Garcia de Andrade
Instituto di Fisica
Univ. Federal Rio de Janeiro
Cidade Universitaria
Rio de Janeiro 68528 (R.J.)
Brazil

A.M. Gebara
Tajoura Nuclear research Centre
P.O. Box 30878
Tajoura (Tripoli)
Libya

G.C. Ghirardi
Istituto di Fisica Teorica
Università di Trieste
and
ICTP
Trieste
Italy

A. Giovannini
Istituto Nazionale Fisica Nucleare
Sezione di Torino
Corso M. D'Azeglio 46
Torino
Italy

N. Goumry
Department of Physics
University of Jordan
P.O. Box 1682
Amman
Jordan

B. Gruber
Southern Illinois University
At Carbondale
carbondale
Illinois 62901
U.S.A.

C. Guiot
Istituto Nazionale Fisica Nucleare
Corso M. D'Azeglio 46
10125 Torino
Italy

Y. Guler
Department of Physics
Middle East Technical University
Ankara
Turkey

M. Hage Hassan
Department of Physics
Lebanese University
Hadeth - Beirut
Lebanon

J. Hainzl
Gesamthochschule Kassel
Universität Landes Hessen
Fachbereich 17 - Mathematik
D- 3500 Kassel
Fed.Rep. Germany

L. Halpern
Department of Physics
Florida State University
Tallahsse, Florida 32306
U.S.A.

M. Hamermesh
School of Phsyics and Astronomy
Tate Laboratory of Physics
University of Minnesota
116 Church Street S.E.
Minneapolis, Minnesota 55455
U.S.A.

D. Harding
Faculty of Mathematical Studies
The University of Southampton
Highfield, Southampton SO9 5NH
U.K.

D.M. Hatch
277 ESC
Brigham Young University
Provo, Utah 84057
U.S.A.

W.F. Heidenreich
Physik-Abteilung
Max-Planck Institut
Postfach 1529
D-8130 Starnberg
Fed.Rep. Germany

J. Hennig
Ist. Theoretical Phsyics
Clausthal
Leibniz str. 10
Fed.Rep. Germany

F. Herbut
Department of Physics
Faculty of Science
P.O.Box 550
11001 Belgrade
Yugoslavia

B. Jancewicz
Institute of Theoretical Physics
University of Wroclaw
Cybulskiego 36
50-205
Polan

A. Janner
Instituut voor Theoretische Fysika
Katholieke Universiteit
Nijmegen
The Netherlands

M.V. Jaric
Department of Physics
Montana State University
Bozeman, Montana 59717
U.S.A.

G. Jona Lasinio
Istituto di Fisica "G. Marconi"
Università degli Studi
Piazzale Aldo Moro 2
00185 Roma
Italy

B.R. Judd
Department of Physics
The Johns Hopkins University
Homewood Campus
Baltimore, Maryland 21218
U.S.A.

Jue Changkeun
Department of Physics
Kyungpook National University
Taegu 635
Korea

V. Kac
Department of Mathematics
Massachusetts Institute of Technology
Cambridge, Massachusetts 02139
U.S.A.

A.J. Kalnay
Centro de Fisica
IVIC-Ist.Venezolano de Investigaciones
Cientificas
Ap. Postal 1827
Caracas 1010 A
Venezuela

B. Kendirli
Department of Mathematics
Middle East Technical University
Ankara
Turkey

Y.S. Kim
Department of Physics and Astronomy
University of Maryland
College Park, Maryland 20742
U.S.A.

M. Koca
Department of Physics
Cukurova University
Adana
Turkey

V. Kolsky
Institute of Physics
Czechoslovak Academy of Sciences
No Slovance 2
180 40 Prague 8 - Liben
Czechoslovakia

J.N. Kotzev
Faculty of Physics
Sofia University
Blvd. A. Ivanov 5
1126 Sofia
Bulgaria

P. Kramer
Institut f. Theoretische Physik
Universität Tübingen
Auf der Morgenstelle 14
D-7400 Tübingen 1
Fed.Rep. Germany

J. Krause
Pontificia Univ.Catolica de Chile
Dpto. de Fisica
Casilla 114-D
Santiago
Chile

H.W. Kunert
Institute of Physics
Technical University
Plotrowo 3
60965 Poznan
Poland

M. Kupczynski
Ecole Normale Supérieure
Takkadoum - Rabat
B.P. 5118
Morocco

A.K. Kwasniewski
Institute of Theoretical Physics
University of Wroclaw
Cybulskiego 36
Wroclaw
Poland

J. Lorenc
Institute of Theoretical Physics
University of Wroclaw
Cybulskiego 36
50-205 Wroclaw
Poland

M. Lorente
Dpto. de Metodos Matematicos
de la Fisica
Facultad Ciencias Fisicas
Universidad Complutense
Madrid 3
Spain

J. Lukierski
Laboratoire de Physics Théorique
Université de Bordeaux I
Domaine du Haut-Vigneau
33170 Gradignan
France

A. Maduemezia
department of Physics
University of Ibadan
Ibadan
Nigeria

V. Man'ko
Lebedev Institute of Physics
USSR Academy of Sciences
Leninsky Prospekt 53
Moscow
USSR

S. Marculescu
Institut f. Theoretische Physik
Universitat Karlsruhe
Physikhochhaus
Kaiserstrasse 12
D-7500 Karlsruhe
Fed.Rep. Germany

H. Mavromatis
Department of Physics
American University of Beirut
Beirut
Lebanon

L. Michel
Institut des Hautes Etudes Scientifiques
91440 Bures- sur-Yvette
France

B. Milewski
Institute of Theoretical Physics
University of Wroclaw
Cybulskiego 36
50-205 Wroclaw
Poland

M. Mintchev
Istituto NF Nucleare
Sezione di Pisa
Pisa
Italy

M. Moshinsky
Instituto de Fisica
UNAM - Universidad Nacional Autonoma
de Mexico
Apdo. Postal 20-364
Mexico 20, D.F.
Mexico

B.A. Mughrabi
Atomic Energy Commission
P.O. Box 6091
Damascus
Syria

A. Mondragon
Instituto de Fisica
UNAM- Univ. Nacional Autonoma de
Mexico
Apdo. Postal 20-364
Delegacion A. Obregon
01000 Mexico D.F.
Mexico

V. Mujica
Quantum Chemistry Group
Uppsala University
Box 518
S 751 20 Uppsala
Sweden

W. Nahm
Max-Planck-Institut f. Mathematik
Gottfied-Clarenstrasse 26
D-5300 Bonn 3
Fed.Rep. Germany

A. Nduka
School of Natural and Applied Sciences
Federal University of Technology
P.M.B. 1526
Owerri
Nigeria

H. Nencka-Ficek
Institute of Molecular Physics
Polish Academy of Sciences
Smoluchowskiego 17/19
Poznan
Poland

R. Neri
Institut f. Theoretische Physik
Philosophenweg 47
Auf der Morgenstelle
D-7400 Tubingen
Fed.Rep. Germany

A. Nowicki
Institute of Teachers' Training
ODN
Dawida la
50-527 Wroclaw
Poland

M.E. Noz
New York University Medical Centre
University Hospital
560 First Avenue
New York, N.Y. 10016
U.S.A.

J. Nuyts
Physique Theorique et Mathematique
Faculte des Sciences
Universite de l'Etat a Mons
19 Av. Maistriau
7000 Mons
Belgium

C.G. Oakley
Department of Theoretical Physics
University of Oxford
1 Keble Road
Oxford OX1 3NP
U.K.

O.A.A. Odundun
Physics Department
University of IFE
Ile-Ife
Nigeria

A.A.H. Omar
Mathematics Department
University of El-Minya
P.O.Box 2807
Cairo 11511
Egypt

S. Pallua
Department of Theoretical Physics
Prirodoslovno-matematicki facultet
Marulicev trg 19/1
41000 Zagreb
Yugoslavia

Z. Papadopolos
Institut f. Theoretische Physik
Universitat Tubingen
Auf der Morgenstelle 14
D- 7400 Tubingen
Fed.Rep. Germany

D. Parashar
3116 Roop Nagar
Delhi University
Delhi
India

F. Pasemann
Institut f. Theoretische Physik
Technische Universität Clausthal
D-3392 Clausthal-Zellerfeld
Fed.Rep. Germany

M. Perroud
Institut des Hautes Etudes
Scientifiques
91440 Bures-sur-Yvette
France

A. Pramudita
Perumahan Batan CU-25
Babarsari
Yogyakarta
Indonesia

C. Quesne
Service Physique Théorique
et Mathématique, CP 229
Université Libre de Bruxelles
Blvd. de Triomphe
1050 Bruxelles
Belgium

M. Rahman
Mathematics Department
Rajshahi University
Rajshahi
Bangladesh

K. Rama Mohana Rao
Applied Mathematics
A.U.P.G. Extension Centre
Nuzvid 521202
Andhra Pradesh
India

M. Rasetti
Istituto di Fisica
Politecnico di Torino
Corso Duca degli Abruzzi 24
10129 Torino
Italy

T. Regge
Istituto di Fisica
Università degli Studi
Corso d'Azeglio 46
10125 Torino
Italy

C. reina
Istituto di Fisica
Via Celoria 16
20133 Milano
Italy

C. Ripamonti
Istituto di Scienze Fisiche
Università di Genova
Viale Benedetto XV, 5
16132 Genova
Italy

M.A. Rodriguez
Dpto. de Metodos Matematicos
de la Fisica
Facultad Ciencias Fisicas
Universidad Complutense
Madrid 3
Spain

L. Roszkowski
Institute for Theoretical Physics
Hoza 69
Warsaw
Poland

R.K. Roychoudhry
Electronics Unit
Indian Statistical Institute
Calcutta 700035
India

A. Roy Chowdhury
High Energy Phsyics Division
Department of Physics
Jadavpur University
Calcutta 32
India

P. Rudra
University of Kalyani
Kalyani
West Bengal
India

H. Ruegg
Department de Physique Theorique
Universitè de Geneve
32 Blvd. d'Ivoy
CH-1211 Geneve 4
Switzerland

M. Saeed-ul-Islam
Bahauddin Zakariya University
Multan
Pakistan

M.H. Sarmadi
Department of Physics and Astronomy
University of Pittsburg
Pittsburg, PA 15260
U.S.A.

N. Sanchez
Observatoire de Meudon
DAPHE
92190 Meudon
France

T.S. Santhanam
Institute of Mathematical Sciences
MATSCIENCE
Madras 600 113
India

M. Saraceno
Dpto. de Fisica
Comision Nacional de Energia Atomica
Av. del Libertador 8250
Buenos Aires
Argentina

M. Sato
Research Institute for Mathematical Sciences
Kyoto University
Kitashirakawa, Sakyo-ku
Kyoto 606
Japan

B. Sazdovic
Institute of Physics
Studentski trg 12/V
P.O. Box 57
11001 Belgrad
Yugoslavia

R. Scholze
Institute of Theoretical Physics
3392 Clausthal-Zellerfeld
Technische Universitat Clausthal
Fed.Rep. Germany

W. Schweizer
Institut f. Theoretische Physik
Universitat Tubingen
Auf der Morgenstelle 14
7400 Tubingen
Fed.Rep. Germany

A. Sciarrino
Istituto di Fisica Teorica
Università di Napoli
Mostra d'Oltremare, Pad. 19
80125 Napoli
Italy

I.E. Segal
Department of Mathematics
MIT, Room 2-244
Cambridge, Massachusetts
U.S.A.

J. Shabani
Institut de Physique Théorique
Université Catholique de Louvain
Chemin du Cyclotroin 2
1348 Louvain - la Neuve
Belgium

R. Sharp
McGill University
Department of Physics
Montreal
Canada

A.G. Shayeb
Najah National University
Nablus
West Bank

B.C. Sidharth
Birla Planetarium and Astronomical
Research Centre
96 Jawaharlal Road
Calcutta 700 071
India

D. Sijacki
Institute of Physics
P.O. Box 57
11001 Beograd
Yugoslavia

S. Sinzinkayo
Institut de Phsyique au Sart Tilman
Service Physique Théorique et
Mathematique
Université de Liege
Batiment B.5
4000 Liege 1
Belgium

L. Solombrino
Dipartimento di Fisica
Università di Lecce
Via Arnesano
73100 Lecce
Italy

A.I. Solomon
Faculty of Mathematics
The Open University
Walton Hall
Milton Keynes MK7 6AA
U.K.

J.M. Souriau
Centre de Physique Theorique
Luminy - Case 907
F-13288 Marseille Cedex 9
France

E. Spallucci
Istituto di Fisica Teorica
Università di Trieste
34100 Trieste
Italy

W. Stebb
Universitat Gesamthochschule Paderborn
Fach. 6, Naturwissensch. 1 - Physik
Warburgerstrasse 100
4790 Paderborn
Fed.Rep. Germany

A. Stella
Istituto di Fisica "G. Galilei"
Università di Padova
Via Marzolo 8
35100 Padova
Italy

G.E. Tanyi
Department of Mathematics
The National University of Lesotho
P.O. Roma 180
Lesotho

R. Tello-Llanos
Instituto Venezolano de Investigaciones
Cientificas - IVIC
Centro de Fisica
Apdo. 1827
Caracas 1010
Venezuela

J. Thierry-Mieg
GAR
Section d'Astrophysique Relativiste
Observatoire de Paris
92190 Meudon
France

I. Todorov
Institute of Physics
Bulgarian Institute of Sciences
Boulevard Lenin 72
1113 Sofia
Bulgaria

G. Vanden Berghe
Seminaire voor Wiskundige Natuurkunde
Rijksunversiteit -Gent
Krijgslaan 281-S9
B-9000 gent
Belgium

P.M. Van der Broek
Twente University of Technology
Dept. Applied Mathematics
P.O.Box 217
7500 AE Enschede
The Netherlands

J. Van der Jeugt
Seminaire voor Wiskundige Natuurkunde
Rijksuniversiteit -Gent
Krijgslaan 281-S9
B-9000 Gent
Belgium

A. Ventura
E.N.E.A.
Via G. Mazzini 2
40138 Bologna
Italy

M. Vujicic
Department of Physics
Faculty of Science
Universoty of Belgrade
P.O. Box 550
11001 Beograd
Yugoslavia

W. Wadia
Dept. of Mathematics and Theoretical Physics
Nuclear Research Centre
Atomic Energy Establishment
Cairo
Egypt

T. Weber
Istituto di Fisica Teorica
Università di Trieste
Italy

F. Wegner
Institut f. Theoretische Phsyik
Universitat Heidelberg
Abt. Vielteilchen Physik
Philosophenweg 19
6900 Heidelberg 1
Fed.Rep. Germany

E.P. Wigner
Department of Physics
Princeton University
P.O. Box 708
Princeton, New yersey 08544
U.S.A.

P. Winternitz
Centre de Recherches de Matématiques Appliquées
Université de Montreal
C.P. 6128 "succ A"
Montreal, Quebec
Canada

K.B. Wolf
Instituto de Investigaciones en Matematicas Aplicadas y en Sistemas
UNAM - Universidad Nacional Autonoma de Mexico
Apso. Postal 20-726
01000 Mexico D.F.
Mexico

E.S. Zaak
Tajoura Nuclear Research Centre
P.O. Box 30878
Tajoura (Tripoli)
Libya

J. Zak
Department of Physics
TECHNION - Israel Institut of Technology
Technion City
32000 haifa
Israel

S. Zerbini
Dipartimento di Fisica
Università degli Studi
di Trento
38050 Povo (Trento)
Italy

L. Ziemczonek
Department of Phsyics
Pedagogical Univesrity of Slupsk
Arciszewskiego 22B
76 -200 Slupsk
Poland

CONTENTS

SPECIAL SESSION

GROUP REPRESENTATIONS, GROUP EXTENSIONS, CONTRACTIONS AND BIFURCATIONS

COMPLETELY INTEGRABLE SYSTEMS

ELEMENTARY PARTICLES AND GAUGE THEORIES

SUPERSYMMETRY AND SUPERGRAVITY

ATOMIC AND NUCLEAR PHYSICS

SYMMETRIES IN CONDENSED MATTER PHYSICS AND STATISTICAL MECHANICS

CANONICAL TRANSFORMATIONS AND QUANTUM MECHANICS

SPECIAL SESSION

The Use and Ultimate Validity of Invariance Principles

E.P. Wigner
Department of Physics, Princeton University, P.O.Box 708,New Yersey 08544,USA

Introductory Remark

Let me first mention the three fundamental concepts of physics of which the invariance principle is one: initial conditions, laws of nature, invariance principles or symmetries. All of them underwent enormous changes in our century, the invariance principles perhaps least. The separation of initial conditions and laws of nature was made with great clarity by Newton and I consider this his greatest accomplishment, perhaps even greater than the discovery of his gravitational law. The symmetries seem to have been discovered and well formulated first by Galileo, perhaps independently by Newton. Some of them are obvious from every day life, such as the laws of both space and time displacement invariances. If it would require a different kind of effort to pick up this pencil in this room from that of picking it up in the next room, or if it had to be done in a different way tomorrow from the way it can be done today, our whole life would be different - if at all possible. But the invariance with respect to a uniform motion along a straight line does not manifest itself in everyday life - Aristotles' laws of physics surely deny it and the lack of its recognition led to much of the opposition to railroads. Perhaps I mention also that the assumption of the existence of ether also led to questioning this invariance even by rather recent outstanding scientists. Since the days of Galileo and Einstein the only significant change in the invariance principles was Einstein's introduction of the relativity theories.

As far as initial conditions and laws of nature are concerned the change was almost infinitely greater. Newton defined the initial state of his systems by specifying the three position and momentum coordinates of the objects which form the system - in his most important case the planets of our solar system. These are 6n numbers if the number of objects is n. The next very basic change came from Maxwell's theory of the electromagnetic fields - these are described by 6 functions of the three space variables, giving the three components of the electric and magnetic fields. Six functions of three variables constitute an infinitely more complex mathematical structure than Newton's 6n numbers. The following basic change was introduced by quantum mechanics which proposed a variety of descriptions of the state of a system: initially wave functions, that is complex functions in 3n dimensional space where n is again the number of objects, in this case particles which form the system. The next description by vectors in an infinite dimensional so-called "Hilbert space" is equivalent herewith. However, whether the state vectors of field theories are also quite equivalent is not truly clear, but they are interesting and useful.

The laws of nature have undergone, of course, changes which correspond to the changes in the specification of the initial conditions. The present ones give, in every case, the time derivative of the description of the state of the system, i.e., the time derivative of the initial conditions. I do not need to enumerate the invariance transformations. Perhaps I should define them.

Invariance Principles

There are numbers or functions which determine the state of a system to such an extent that, together with the laws of nature (in our case laws of physics) they determine all the properties of the system for all times as long as the system remains free from outside influence (this obvious condition will not be repeated subsequently). These numbers or functions are supposed to be independent of each other - if any of them can be obtained as a function of the others, it will be omitted and its value attributed to the laws of nature. The initial conditions, together with the time at which they are valid, are such numbers or functions - to be called determinants of state of the system - but there are much more general such sets of information. For instance, in classical mechanics not all information on the position of the objects has to refer to the same time. But totally, the determinants of state are minimal amounts of information which, together with the laws of nature, can furnish all the information on the properties of the system to which they refer.

There is, obviously, an infinite set of "determinants of state" which are equivalent. Surely, the initial conditions which apply at two different times, coupled with these times as determinants of state, are equivalent. The laws of nature produce the equivalences of the many different determinants of the same state. It is clear, for instance, that the determinant of state for a single particle which is free consists of six numbers giving the position and velocity components at the times specified. Denoting these with $r_x(t_x)$, $r_y(t_y)$, $r_z(t_z)$, $v_x(t_x')$, $v_y(t_y')$ $v_z(t_z')$, it is easy to see that two such determinants of state, each consisting of the above six numbers (and of course the times to which these refer) represent the same state if they satisfy the relations

$$r_\kappa(t_\kappa) = a_\kappa + b_\kappa t_\kappa \qquad v_\kappa(t_\kappa') = b_\kappa$$

for all three κ with the same a and b but for any of the value of the t variables. For more complicated systems the determination of the equivalence of determinants of state is, of course, more complicated since it depends on more complicated laws of nature.

We will now consider operators which transform the determinants of state. Let F and G represent two different determinants of the same state and Q be an operator which transforms determinants of state. This Q will be called a valid operator if it transforms two equivalent determinants of state into equivalent determinants of state, i.e. if it follows from $F \simeq G$ that $Q\,F \simeq Q\,G$. Evidently, the Q form a group which contains as an invariant subgroup the transformations of determinants of state into equivalent determinants of state, i.e. determinants of the same state. The symmetry operators form the factor group of this invariant subgroup - with the exception of the unit element of this factor group they transform determinants of state in general into determinants of other states.

The preceding is a perhaps unnecessarily general definition of the symmetry operators. In practice there are only two types of such operators, and their products - they commute. The first type consists of the members of what I call the Poincaré group - displacements in space and time, Lorentz transformations and reflections, and the combinations of these. We believe that the laws of physics are invariant under the first two types of transofrmations and at least approximately invariant under some reflections. The systems connected by these transformations have the same appearance if viewed from coordinate systems related to each other by the aforementioned transformations. The second type of transformations replace particles by particles of a different nature. The one most generally recognized such transformation replaces electrons by positrons, protons by antiprotons, etc. and conversely, and surely has high accuracy. The other transformations of this nature are very approximate - the best known one replaces protons by neutrons and conversely and represent only a crude approximation but in many cases a useful one.

Let me admit that there is a certain lack of logical simplicity in the definition of the states of the systems and hence of their determinants. As the example given already shows, the determinants' values depend on the positions and times of events. These are defined by measurements with macroscopic instruments which are considered to be "at rest" (or in uniform motion) that is the functioning of which is supposed to be governed by the laws which they are expected to define. To verify laws of nature for small systems their validity for large systems, consisting of small parts, has to be assumed so that what we verify is only the possible consistency of the laws. But these seem to be there!

The reservation just presented is hardly valid for the general relativity theory but the confirmation of this, as I hope we'll see, is even more difficult.

The Use of the Invariance Principles

The principal and most natural use of the invariance principles is the checking of any proposed law of nature - whether it obeys it. One can well say that the recognized invariance principles are laws which the laws of nature obey - if the invariance principle is valid. This is an important use - it also helps the proposer of a new law to formulate it, because he realizes that his proposal should conform with the accepted invariance principles.

The next oldest use of the invariance principles is the derivation, and hence justification, of conservation laws: those of momentum, energy, angular momentum, and the motion of the center of mass. The derivation of these conservation laws from the invariance principles is due, I believe, to Hamel (1905, but it may have been obtained even earlier) but is usually attributed to Emmy Noether. In classical theory the derivation is not easy but the conservation laws do follow easily from the principles of quantum mechanics. I propose, therefore, that in order to obtain them for classical mechanics, they be derived quantum mechanically and the classical theory then be considered as an approximation to the quantum theory.

Actually, the invariance principles have much wider applications in quantum mechanics than in the earlier theories. The main reason for this is the linear nature of the equations and of the description of states. There is an infinity of states the time dependence of which is completely determined by the postulate of the invariance of the time development equations given by the special theory of relativity, i.e. by the postulate of the invariance of the equations with respect to the Poincaré group - as was mentioned earlier.

Furthermore, it can be shown that the quantum description, i.e. the state vector of every state of every system, is in principle decomposable into a linear manifold of such states and if this were done, the time development of the system, that is of its state vector, would be determined. The trouble is that, as a rule, the decomposition of a state vector, given by the usually observable properties, into the state vectors the development of which is fully given by the Poincaré invariance, is as a rule not possible practically unless the system consists only of a single elementary particle. Thus, in practice, the time development is obtainable from the postulate of the invariance of the equations with respect to the special theory of relativity only for elementary particles, such as electrons, light quanta, etc. but in these cases the quantum mechanical descroption of the properties of the particle goes very much further than in classical theory.

But quantum theory provides also in practice interesting and important, though not complete, information also for many complex systems. The state vector of each stationary state of the electrons surrounding a nucleus can be written as a linear combination of a finite number of state vectors - of state vectors representing states with a definite angular momentum component in a given direction. The components of a state with a given angular momentum component in another direction can be written then as a linear combination of these state vectors and the coefficients are fully, though not obviously, given by the rotational invariance of the whole system. And there are many other consequences of the invariances on the quantum mechanical properties of simple systems - in the case just considered of the rotational invariance.

Uses of the invariance principles in quantum mechanics to give approximate regularities are even more frequent, and perhaps equally interesting. The oldest such application is based on the so-called

Russell-Saunders (or L-S coupling) approximation for the study of atomic spectra. In first approximation this neglects all magnetic interaction - both that due to the spin of the electron and also to the orbital motion of these. This is, as a rule, a good approximation but the result obtained by taking these effects into account in first approximation are even more interesting.

And there are many other approximate applications of the symmetry principles. The Bloch theory of the electron motion in crystals is one, the theory of molecules is another - perhaps the approximation assuming equal masses for, and equal interactions between proton-proton, neutron-neutron, and proton-neutron pairs, is another. The latter one leads to the so-called supermultiplet theory. And there are many other similar very interesting approximations - we heard a good deal about the theory of quarks. On the whole it is clear that many interesting and important consequences of the quantum mechanical theory can be easily obtained by using the invariance principles and their consequences - consequences which would be difficult to obtain by a straight application of the basic equations.

Criticisms

The first point of criticism was made before: how are the coordinate systems defined for which the basic laws of nature are fundamentally valid? But in practice they are defined and so are the Poincaré (or Lorentz) transformations which leave them invariant. It was mentioned also that this criticism does not apply to the general relativity but the application of this to microscopic, that is atomic or nuclear or particle systems, can be seriously questioned. But this will not be done here.

We must also admit that almost all useful applications of the symmetry principles are approximate. The derivation of the Russell-Saunders coupling, for instance, neglects in addition to those effects which it acknowledges, also the interaction with the radiation field which causes a decay of all excited states and thus interferes with time-displacement invariance. The same is true for nuclei and the β-decay is to be added to this - this violates another (the reflection) invariance.

The solid state applications are also very approximate and it is perturbing that we do not have a real explanation of the immense life-time of the superconductivity currents, once induced.

Next, we must admit that quantum theory is not truly valid for macroscopic systems - just as, in this writer's opinion, general relativity is not valid for microscopic ones. The former point was made, first, by H. D. Zeh but it is not quite generally recognized.

Finally, we must admit that the present theories of physics do not describe the wonderful phenomenon of life. This may be a basic weakness of the present physical theories, affecting also the linear nature of the quantum mechanical equations - as does, in a way, Zeh's remark - and this is another limitation on the validity of all that I have discussed.

But, in spite of all these criticisms, we can maintain that our physics is not only a very interesting but also very useful discipline and we owe very much to its development - not only to its help in everyday life but also to its role as an intellectual inspiration.

Ultimate Validity of the Invariance Principles

If man continues to live and to be interested in science, it is likely that all scientific doctrins will be changed - hence it may not be right to speak about the "ultimate validity" of any part of our science. What may be alright, however, is to try to find the parts of science which surely need changes and also to try to guess the areas in which change is likely to occur in the foreseeable future.

As to the former question, the desirability to extend physics to the phenomenon of life was already mentioned. But there is another problem which one may consider. Heisenberg, in his late years, said that the ultimate purpose of quantum mechanics is to provide us with the collision matrices. This raises the question whether the elements of this are observable. I tried to prove that they are - and the absolute values of the matrix elements surely seem to be. But I could give prescription for the observation of the complex phases of these elements only by assuming that the gravitational effects of the measuring apparata on the colliding particles can be neglected. This also indicates that our microscopic and macroscopic theories, quantum and general relativity theories, are not in harmony with each other.

This manifests itself in another way. General relativity's basic concept is the metric tensor, the g_{ik}. This gives distances between space-time points. But are space-time points valid concepts in the microscopic domain? In the general relativity theory they are thought of as crossing points of two world lines. But the point of collision is not defined in quantum theory - the collision matrix surely does not give it. I believe, therefore, that the true unification of quantum and general relativity theories will require fundamental changes. And this belief is indicated in several recent articles and also by the need for renormalization in quantum field theories. This, the true unification of the macroscopic and microscopic theories, appear to be a necessary objective. This is my view on the first question I raised.

As to the second question, the indications of violations of our symmetry principles, I like to remind of Dirac's suggestion of the long-time change of the ratio of electromagnetic and gravitational interactions. This would violate the principle of time displacement invariance. Other tiny violations of our invariances may exist - there is a coordinate system at rest with respect to the center of mass of the Universe and the direction toward it may play a special role. These may give departures from the displacement and rotational invariances, surely tiny ones, but still departures.

And very finally, phenomena may be discovered which violate some of the invariances wildly - as the β-decay violates the reflection invariance. That was, in fact, a shock to me.

After all these pessimistic forecasts let me remind you how useful the invariance principles were in the past, and are now. Without their existence we could not have recognized laws of nature and all our physics, in fact our existence, would be different. They were, and are now, useful even before actually formulated and their usefulness increased enormously since they were recognized and used as described very incompletely above.

With these remarks I want to close my address. I also want to ask you to forgive me for having so much engaged in speculations. It is, in some ways, and in my opinion, good to engage in that - though only infrequently.

GROUP REPRESENTATIONS
GROUP EXTENSIONS
CONTRACTIONS AND BIFURCATIONS

COHOMOLOGY AND CONTRACTION: THE "NON-RELATIVISTIC" LIMIT REVISITED

V. Aldaya and J.A. de Azcárraga

Departamento de Física Teórica, Facultad de Ciencias Físicas
Universidad de Valencia, Burjasot (Valencia)

Abstract

In this note we reconsider the transition from $P\otimes U(1)$ to the extended Galilei group $\tilde{G}_{(m)}$, first discussed by Saletan. To this aim, we first analyse the relations between the groups $G\otimes U(1)$ and $\tilde{G}_c$, where G is a Lie group of trivial $H_0^2(G,U(1))$ cohomology and $\tilde{G}_c$ is a central extension of G_c (obtained from G by contraction) by U(1).

1. Introduction and results

The problem of obtaining a non trivial U(1) extension $\tilde{G}_{(m)}$ [1] of the Galilei group from the -necessarily direct product- U(1) extensions of the Poincaré group was first considered by Saletan [2]. In modern language, one may say that the cohomological constructions on both (Poincaré and Galilei) groups do not commute with the contraction process, and thus it is possible to obtain a true Galilei cocycle from a Poincaré coboundary in the contraction limit. In this note we wish to examine this problem with generality in a geometrical language and to pay some attention to the contraction limit behaviour of the canonical structures which can be defined on a Lie group.

Let G, G_c be defined as in the Abstract, and $\xi(g',g)=\delta(g'*g)-\delta(g')-\delta(g)$ be the 2-coboundary generated by $\delta(g)$ $(g',g\in G)$. Let us call g_i $(i=1,\ldots r)$ the parameters of G, and consider the action of the direct product law of $G\otimes U(1)$ on itself as given by the group law

$$\tilde{g}'*\tilde{g}\equiv(g',\zeta')*(g,\zeta)=(g'*g,\zeta'\zeta\exp\xi(g',g))$$
$$\tilde{g}=(g_1,\ldots g_r;\zeta),\ \tilde{g}'\in G\otimes U(1)\quad;\ g=(g_1,\ldots g_r),\ g'\in G\quad. \tag{1}$$

Taking into account that $\left.\frac{\partial\delta(g*g)}{\partial g_i}\right|_{g_i=0}=L_{X_{g_i}}(g)$, the left invariant generators $\tilde{X}^L$ of $G\otimes U(1)$ are given by the r+1 vector fields

$$\tilde{X}^L_{(g_i)}=X^L_{(g_i)}+\left[L_{X_{(g_i)}}\delta(g)-\left[\frac{\partial\delta(g)}{\partial g_i}\right]_{g_i=0}\right]\Xi\equiv X^{Lj}_{(g_i)}\frac{\partial}{\partial g^j}+X^{L\zeta}_{(g_i)}\Xi \tag{2}$$
$$\tilde{X}^L_{\zeta}\equiv\Xi=i\zeta\frac{\partial}{\partial\zeta}\qquad\left(X^L_{(g_i)}\in\mathcal{G}(G)\,;\ \tilde{X}^L_{(g_i)},\Xi\in\mathcal{G}(G\otimes U(1))\right)$$

and the Lie brackets are given by $(\tilde{X},\tilde{Y},\tilde{Z}\in\mathcal{G}(G\otimes U(1)))$

$$[\tilde{X},\tilde{Y}] = Z + (L_Z \delta(g))\Xi \equiv \tilde{Z} + \left[\frac{\partial\delta(g)}{\partial g_z}\right]_{g=e} \cdot \Xi \quad ; \tag{3}$$
$$[\Xi,\tilde{X}] = 0 \quad \forall \tilde{X} \ .$$

When

$$\left.\frac{\partial\delta(g)}{\partial g_z}\right|_{g=e} \tag{4}$$

is zero we see that there is no modification of the structure constants, which are the same as those of the explicit direct product law (obtained for $\delta(g)=0$; this of course may always be obtained by absorbing $\delta(g)$ in a redefinition of ξ).

Let us now look at the component of the left invariant canonical 1-form on $G\otimes U(1)$ dual to Ξ ($\Theta(\tilde{X}_{g_i})=0$, $\Theta(\Xi)=1$). This component is particularly interesting because it may be used to define a left-invariant connection [3] on the principal U(1)-bundle $G\otimes U(1)(G,U(1))$ with respect which the vector fields $\tilde{X}_{g_i}$ are horizontal. Writing

$$\Theta = \Theta_{g_i} dg^i + d\varsigma/i\varsigma \tag{5}$$

we obtain

$$\Theta_{g^i} X^j_{(g^i)} + X^\varsigma_{(g^i)} = 0 \ . \tag{6}$$

Thus, when $\delta(g)=0$ Θ reduces to $d\varsigma/i\varsigma$,and when $\left.\frac{\partial\delta(g)}{\partial g_z}\right|_{g_i=0} = 0$ we get

$$\Theta = \frac{d\varsigma}{i\varsigma} - d\,(\delta(g)) \ . \tag{7}$$

In both cases, the curvature associated with Θ is zero and the horizontal vector fields generate a subalgebra, the Lie algebra of G. It is clear that in these cases the contraction process [4] cannot alter the coboundary character of ξ and so it will not lead to $\tilde{G}_c$, but to the trivial direct product $G_c\otimes U(1)$. (In this reasoning $\delta(g)$ is to be understood as $\delta(g,\lambda)$ where λ is the contraction parameter; we shall not write λ explicitly).

Thus, a necessary condition for the contraction of $G\otimes U(1)$ to give a non direct product central extension $\tilde{G}_c$ is that (4) be non zero. We may thus restrict ourselves to $\delta(g)$ linear in the group parameters.But not all such $\delta(g)$ will give rise to a true cocycle in the contraction process; it will be necessary that (although in general the $\delta(g)$'s may not have a limit) $\xi(g',g)$ be defined in that limit. In this case, $\xi_c(g',g)$ does not have a $\delta_c(g)$ to proceed from and is not a trivial cocycle (is not a coboundary).

2. The case of the Poincaré group ($G=P, G_c=$Galilei group, $r=1,\ldots 10$)

Let us parametrize P by $\vec{\alpha}=\gamma[2(1+\gamma)]^{-\frac{1}{2}}\vec{V}/c$ (or by $\vec{p}=2mc\vec{\alpha}\sqrt{1+\alpha^2}$, $p^0=mc(1+2\alpha^2)$), $\vec{\epsilon}$ ($|\vec{\epsilon}|=\sin\varphi/2$) and x^μ. As a result of the previous analysis, the possible coboundaries $\delta(g)$ (all with dimensions of $\hbar$) may be exemplified by the following examples: a) x^0p^0 b) $\vec{p}\cdot\vec{x}$ c) mcx^1. The first one leads to a cocycle of the Galilei group; the second does not fulfill the condition (4)$\neq 0$ and the third leads to a ξ which has not a limit ξ_c. With the above parametrization, and with the Poincaré left invariant vector fields given by

$$X^L_{(x^0)}=(1+2\alpha^2)\frac{\partial}{\partial x^0}+2\sqrt{1+\alpha^2}\,\vec{\alpha}\,\frac{\partial}{\partial\vec{x}}\;;\quad X^L_{(\epsilon^i)}=\left(\sqrt{1-\epsilon^2}\,\delta^j_i+\eta^j_{\cdot ik}\,\epsilon^k\right)\frac{\partial}{\partial\epsilon^j}\;;$$

$$X^L_{(x^k)}=2\sqrt{1+\alpha^2}\,R^j_k\,\alpha_j\frac{\partial}{\partial x^0}+\left(R^i_k+2\alpha^i\alpha_j R^j_k\right)\frac{\partial}{\partial x^i}\;; \tag{8}$$

$$X^L_{(\alpha^i)}=(1+\alpha^2)^{-\frac{1}{2}}\left\{\left[R^j_i+\alpha_\ell R^\ell_i\,\alpha^j\right]\frac{\partial}{\partial\alpha^j}+\left[\sqrt{1-\epsilon^2}\,\eta^j_{\cdot ki}-\eta^j_{\cdot sn}\,\eta^n_{\cdot ki}\,\epsilon^s\right]R^{-1\,k}_{\ \ m}\,\alpha^m\frac{\partial}{\partial\epsilon^j}\right\}$$

we obtain, for P⊗U(1) a) $\delta(g)=x^0p^0$

$$\tilde{X}^L_{(x^0)}=X^L_{(x^0)}+4m\alpha^2(1+\alpha^2)\,\Xi\;;\quad \tilde{X}^L_{(x^k)}=X^L_{(x^k)}+2m\sqrt{1+\alpha^2}\,(1+2\alpha^2)\,\alpha_j R^j_k\,\Xi\;;$$

$$\tilde{X}^L_{(\epsilon^i)}=X^L_{(\epsilon^i)}\;;\quad \tilde{X}^L_{(\alpha^i)}=X^L_{(\alpha^i)}+4m\,x^0\sqrt{1+\alpha^2}\,\alpha_j R^j_i\,\Xi\;; \tag{9}$$

$$\left[\tilde{X}^L_{(\alpha^i)},\tilde{X}^L_{(x^k)}\right]=2\,\delta_{ik}\left[\tilde{X}^L_{(x^0)}+mc\,\Xi\right],\ \text{etc.}\;;$$

$$\Theta=\frac{d\zeta}{i\zeta}-x^0dp^0-\vec{p}\cdot d\vec{x}$$

b) $\delta(g)=\vec{p}\cdot\vec{x}$

$$\tilde{X}^L_{(x^0)}=X^L_{(x^0)}+4m\alpha^2(1+\alpha^2)\,\Xi\;;\quad \tilde{X}^L_{(x^k)}=X^L_{(x^k)}+2m\sqrt{1+\alpha^2}\,(1+2\alpha^2)\,\alpha_j R^j_k\,\Xi\;;$$

$$\tilde{X}^L_{(\epsilon^i)}=X^L_{(\epsilon^i)}\;;\quad \tilde{X}^L_{(\alpha^i)}=X^L_{(\alpha^i)}+2m\,x_\ell\left[R^\ell_i+2\alpha^\ell\alpha_j R^j_i\right]\Xi\;; \tag{10}$$

$$\left[\tilde{X}^L_{(\alpha^i)},\tilde{X}^L_{(x^k)}\right]=2\,\delta_{ij}\,\tilde{X}^L_{x^0},\ \text{etc.}\qquad;\quad \Theta=\frac{d\zeta}{i\zeta}-d(\vec{x}\cdot\vec{p})\,.$$

It is evident from (9) that the limit $c\to\infty$ will lead to the extended Galilei group $\tilde{G}_{(m)}$ for which $[K_i,P_j]=m\delta_{ij}$ (where $\vec{K}$ is the boosts generator). In fact, it is simple to check that the limit ξ_c of the Poincaré coboundary generated by x^0p^0 is given by

$$\xi_c(g',g) = m\left\{\frac{1}{2} B'V^2 + B'\vec{V}'\cdot R'\vec{V} + BV'^2 + \vec{V}'\cdot R'\vec{A} + B\vec{V}'\cdot R'\vec{V}\right\} \quad (11)$$

which is indeed a non-trivial Galilean cocycle ($\vec{A}$,B now parametrize the space-time translations). In contrast, $\delta(g)=\vec{p}\cdot\vec{x}$ generates a Poincaré coboundary whose limit is the Galilean coboundary generated by $\delta_c(g) = m\vec{A}\cdot\vec{V}$. It is finally interesting to notice that in this second case $d\Theta = 0$; in the first example we obtain a presymplectic form. The role played by the connection form Θ is relevant in connection with a quantization programme based on a group manifold recently introduced [5] and will be discussed elsewhere.

References

[1] V. Bargmann, Ann.Math. 59,1 (1954)

[2] E. J. Saletan, J. Math. Phys. 2,1 (1961)

[3] S. Kobayashi and K. Nomizu, Foundations of Differential Geometry vol.1, Interscience Pub. N.Y. (1969),p.103

[4] E. Inönü and E. P. Wigner, Proc. Nat. Acad. Sci. U.S. 39,518 (1953) E. Inönü in Istanbul Summer School of Theoretical Physics (1962),F. F. Gürsey Ed. Gordon and Breach (1964),p. 391.

[5] V. Aldaya and J.A. de Azcárraga, J. Math. Phys. 23, 1297 (1982). For further extensions, see the lectures in the International School on Supersymmetry and Supergravity (Karpaçz,1983) B. Milewski Ed.,to be published by World Sci.Co.(Singapore) and references therein.

Acknowledgments

This work has been partially supported by the Comisión Asesora para la Investigación Científica y Técnica through a grant #1261-82.

LINEARIZATION - A UNIFIED APPROACH

R.L. Anderson
Dept of Physics and Astronomy
University of Georgia
Athens, GA 30622, USA

E. Taflin
Dept de Physique Théorique
Université de Genève
1211 Genève 4, Switzerland

ABSTRACT

We show, by using examples, that the IST method, Hirota τ_N formalism, and the Kac-Moody constructions of Date et.al. all yield the same inverse linearization operator in the sense of Flato et.al. Hence, they solve the Cauchy problem for the same set of initial conditions.

1. NON-LINEAR GROUP REPRESENTATIONS

The notion of non-linear repr.s of Lie groups was introduced in [3] as a tool to study the Cauchy problem for covariant non-linear evolution (NLE) Eq.s. Heuristically, a (non-linear) analytic repr. (U,E) of a real Lie group G in a Fréchet space E is a local analytic action U of G in E, with a fixed point (let us say the origin). Two analytic repr.s are equivalent if they are intertwinned by an invertible map, analytic around the origin. A repr. (U,E) equivalent to its linear part is said to be linearizable. Linearizability is characterized by the solvability of equations in Chevally-Eilenberg cohomology. If a linearization exists, it is obtained solving these equations iteratively. As in the case of linear group repr.s, one can, for analytic group repr.s, canonically associate non-linear analytic repr.s with the corresponding Lie algebra. By definition covariant NLE Eq.s belong to such a Lie algebra repr. [4], [5]. The existence of a linearization operator gives the evolution operator for the NLE Eq. in terms of the linear evolution operator and implies non-linear super position principles, hierarchies of NLE Eq.s and infinite numbers of constants of motions [1], [13], [14], [15]. Important examples of linearizable equations are given by massive relativistic NLE Eq.s in 1 + 3 dim. [11], [12]. The Cauchy problem for the KdV [14] Eq, and for the Benjamin-Ono Eq. (small initial conditions) [1] has been solved by linearization.

2. INVERSE SCATTERING TRANSFORM (IST)

Thus far, the IST method of constructing solutions, when realized for NLE Eq.s, has only produced solutions for equations which are linearizable to their linear parts. Specifically, given a NLE Eq. solvable by IST, iteration of the associated Gelfand-Levitan-Marchenko (G-L-M) Eq.(s) yields the Taylor series representation of the inverse linearization map A^{-1}. For example, following Taflin [14], formulas (3.11), (4.02), and (4.14) of Rosales [9] can be given a rigorous mathematical meaning and hence represent analytic solutions to the cohomological equations for A^{-1} for the KdV, MKdV - pot.s-G, and NLS Eq.s, respectively. These maps are then identical to those obtained by iterating the G-L-M Eq.s corresponding to these NLE Eq.s. In order to illustrate this point here, consider the example of the K-P Eq.. Following Manakov et.al. [8], we take the K-P Eq. in the form

$$(\partial_t u + 6u\partial_x u+\partial_x^3 u)(x,y,t) = -3\alpha^2\partial_y^2 \int_{-\infty}^x d\xi u(\xi,y,t), \alpha\in\mathbb{C}. \tag{1}$$

Then, for appropriate distributions F,

$$v(x,y,t) = \frac{1}{2\pi}\int_{\mathbb{C}^2} dpdq\, F(p,q)\, \exp\{i(p-q)\, x+\alpha^{-1}(p^2-q^2)y+4i(p^3-q^3)t\},$$

is a solution of the linear part of (1). With these definitions $u(t) = A^{-1}(v(t))$, where

$$[u(t)](x,y) = \frac{\partial}{\partial x}\frac{1}{i}\sum_{n=i}^{\infty} (2\pi)^{-n} \int_{\mathbb{C}^{2n}} dp_1dq_1\dots dp_ndq_n \frac{F(p_1,q_1)\dots F(p_n,q_n)}{(p_1-q_1)\qquad(p_n-q_n)}$$

$$\cdot\frac{\exp\{i(p_1-q_1+\dots+p_n-q_n)x+\alpha^{-1}(p_1^2-q_1^2+\dots+p_n^2-q_n^2)y+4i(p_1^3-q_1^3+\dots+p_n^3-q_n^3)t\}}{(p_1-q_2-i\epsilon)(p_2-q_3-i\epsilon)\dots(p_{n-1}-q_n-i\epsilon)} \tag{2}$$

Equation (2) is identical to the map obtained by iterating the corresponding G-L-M Eq. $K(x,x',y,t) + F(x,x',y,t) + \int_{-\infty}^{x} dx'' K(x,x'',y,t)F(x'',x',y,t) = 0$, $x \geq x'$, where $[u(t)](x,y) = -\partial_x K(x,x,y,t)$, and $F(x,x',y,t) = (2\pi)^{-1}\int dp\int dq F(p,q)\exp\{ipx-iqx' + \alpha^{-1}(p^2-q^2)y + 4i(p^3-q^3)t\}$. Conversely, given (2), the G-L-M Eq. follows as in the case of the KdV [9].

3. HIROTAS τ_N FORMALISM

The new result here is that Hirota's dependent variable substitution method or τ_N formalism contains the essential information for the construction of inverse linearization maps. We illustrate this with the KP equation. Formula (2) gives the following expression for the linearization map A^{-1} :

$$u(t) = + 2\partial_x^2 \ell n \{\tau(v(t))\} = A^{-1}(v(t)), \tag{3}$$

where we have introduced

$$[\tau(v(t))](x,y) = \sum_{n=o}^{\infty} \frac{1}{n!} \int_{\mathbb{C}^{2n}} dp_1 dq_1 \ldots dp_n dq_n \left(\prod_{1\leq\mu<\nu\leq n,} C(p_\mu, q_\mu; p_\nu, q_\nu) \right)$$

$$\prod_{j=1}^{n} \frac{F(p_j,q_j)}{i(p_j-q_j)} e^{\eta(p_j,q_j;x,y,t)} , \tag{4}$$

with $C(p_i,q_i;p_j,q_j) = \dfrac{(p_i-p_j)(q_i-q_j)}{(p_i-q_j)(q_i-p_j)}$ and $\eta(p,q;x,y,t) =$

$= i(p-q)x+\alpha^{-1}(p^2-q^2)y+4i(p^3-q^3)t$. If now the distribution $F(p-q)/(p-q)$ has support at a finite number of points (P_i,Q_i), $1\leq i\leq N$, P_i and Q_i positive imaginary and $P_i-Q_j \neq 0, \forall i,j$, then (4) is the well known expression for the Hirota τ_N-function.[10] for Eq. (1).

This construction is valid for all cases known to us and is independent of whether or not the τ_N leads to solitons or not. Conversely, given the τ_N functions for a NLE Eq. it is not difficult to find an expression analogous to (4) and then the construct the inverse linearization operator A^{-1} by the corresponding dependent variable substitution.

4. LINEAR HIERARCHIES AND KAC-MOODY ALGEBRAS

In the construction of Data et.al. [2], a commuting subset of symmetries of the linear part of a given NLE Eq., which contains the linear part of the original Eq., fixes a realization of a fundamental representation of a Kac-Moody algebra. Once fixed, the realization determines the c_{ij}'s appearing in the expression for τ_N. Then as discussed in § 3, we find that $\tau(v)$ (see e.q.(4)) when composed with Hirota's independent variable substitution (see e.g. (3)) yields an inverse linearization map A^{-1} for the original NLE Eq. Further the image of the linear hierarchy, under A^{-1}, yields a nonlinear hierarchy of commuting symmetries of the original Eq.

We shall illustrate this by considering the following construction for the s-G Eq. The linear hierarchy

$$\partial_{\rho_k} v = w_k(\partial_\xi^{-1})v,\ w_k(\partial_\xi^{-1}) = \partial_\xi^{-k},\ k = 1,3,5,\dots, \tag{5}$$

is a subset of the commuting symmetries of the linear part of the s-G Eq. (in "light-cone" coordinates)

$$\partial_{\rho_1} u = \partial_\xi^{-1} \sin u, \tag{6}$$

and it includes the linear part of (6). The linear hierarchy (5) sets the "scale" in the exponential generation function

$$X(p) = -\frac{1}{2} \exp\{\sum_{k_{odd}} w_k(p^{-1})\rho_k\} \exp\{-\sum_{k_{odd}} \frac{4}{k w_k(p^{-1})} \partial_{\rho_k}\}$$

in the realization of the basic $A_1^{(1)}$-module as a Lie algebra of linear differential operators in infinitely many variables acting on the space of polynomials in these variables [7],[6]. This in turn fixes the c_{ij}'s in the expression for the counterpart of $\tau(v(\rho))$ introduced in § 3. In particular, here $\tau(v(\rho))$ is given by $\tau(v(\rho)) = \sum_{n=o}^{\infty} i^n \tau^n(v(\rho))$, where $\tau^1=1$, $[\tau^n(v(\rho))](\xi) = \frac{1}{n!} \int_{\mathbb{C}^n} dp_1 \dots \int dp_n (\frac{1}{4})^n \prod_{1\le\nu<\mu\le n} C_{\nu\mu} \prod_{j=1}^{n} \tilde{v}(p_j)\, e^{\eta_j}$,

$[v(\rho)](\xi) = \int_{\mathbb{C}} dp\, e^{\eta(p)} \tilde{v}(\rho)$, and $\eta_j = ip_j\xi + \sum_{k_{odd}} (ip_j)^{-k} \rho_k$. Then the map

$$[u(\rho)](\xi) = 2i\ \{\ln(\sum_{n=o}^{\infty}(-i)^n[\tau^n(v(\rho))][\xi)) - \ln(\sum_{n=o}^{\infty}(i)^n[\tau^n(v(\rho))](\xi))\},$$

defines an inverse linearization operator for a commutative s-G hierarchy formulation. The hierarchy (5) is its linear part, and (6) is included in the image of (5) under (6). Similarly, the linearization of the s-G Eq. in "space-time" coordinates can be obtained. (Date, Jimbo and Miwo's work on the s-G Eq. was not available to the authors at the time this manuscript was prepared.)

[1] R.L. Anderson and E. Taflin : "Explicit non-soliton solutions of the Benjamin-Ono equation" Lett. Math. Phys. 7 243-248 (1983).

[2] E. Date, M. Jimbo, M. Kashiwara and T.Miwa : "Transformation groups for soliton equations - euclidean Lie algebras and reduction of the KP hierarchy" Publ. RIMS, Kyoto University 18, 1077-1110 (1982).

[3] M. Flato, G. Pinczon and J. Simon : "Non-linear representations of Lie groups" Ann. Sci. Ec. Norm. Sup. 10, 405-418 (1977).

[4] M. Flato and J. Simon : "Non-linear equations and covariance" Lett. Math. Phys. 3, 279-283 (1979).

[5] M. Flato and J. Simon : "On a linearization program of non-linear field equations" Phys. Lett. 94B, 518-522 (1980).

[6] V.G. Kac, D.A. Kazhdan, J. Lepowsky and R.L. Wilson : "Realization of the basic representations of the Euclidean Lie algebras" Adv. Math. 42, 83-112 (1981).

[7] J. Lepowsky and R.L. Wilson : "Construction of the affine Lie algebra $A_1^{(1)}$" Comm. Math. Phys. 62, 43-53 (1978).

[8] S.V. Manakov, V.E. Zakharov, L.A. Bordag, A.R. Its and V.B. Matveev: "Two-dimensional solitons of the K-P equation and their interaction" Phys. Lett. 63A, 205-206 (1977).

[9] R.R. Rosales : "Exact solutions of some nonlinear evolution equations" Stud. Appl. Math. 59, 117-151 (1978).

[10] J. Satsuma : "N-soliton solution of the two-dimensional KdV-equation" J. Phys. Soc. Jpn 286-290 (1976).

[11] J.C.H. Simon : "Wave operators for non-linear Klein-Gordon equations" Preprint, University of Dijon, France (1983).

[12] J. Simon : "Non-linear representations of the Poincaré group and global solutions of relativistic wave equations", Pac. Jour. Math. (to appear).

[13] E. Taflin : "Analytical linearization, Hamiltonian Formalism, and Infinite Sequences of Constants of Motion for the Burgers Equation" Phys. Rev. Lett. 47, 1425-1428 (1981).

[14] E. Taflin : "Analytical linearization of the KdV equation" Pac. Jour. Math. (to appear).

[15] E. Taflin : "Dynamical symmetries and conservation laws for the KdV equation" Rep. Math. Phys. (Delayed).

WEYL KINEMATICAL GROUPS OF ELECTROMAGNETIC
AND ENERGY-MOMENTUM TENSORS

J. BECKERS and S. SINZINKAYO,
Université de Liège,
Physique théorique et mathématique,
Institut de Physique au Sart Tilman, Bâtiment B.5,
B-4000 LIEGE 1, Belgique.

Abstract : We determine the "kinematical" groups of electromagnetic fields F with respect to Weyl transformations. We find different structures for parallel and perpendicular (constant and uniform) fields. We compare the results with those obtained in the Poincaré context. As a physical application, we consider the invariance conditions on associated energy-momentum tensors and get meaningful results in connection with the description of zero rest mass particles.

1. Introduction

The Weyl group (Poincaré and scale transformations) has an obvious physical interest, especially in connection with zero rest mass particle description[(1)-(3)]. We want here to determine the largest Weyl subgroups leaving invariant a non-trivial constant and uniform electromagnetic field F . We show that for $F_{//}$, according to Bacry-Combe-Richard (B.C.R.)[(4)] notations, the kinematical group remains six-dimensional as in the Poincaré context, but for $F_{\perp}$, we obtain a seven-dimensional kinematical group (§ 2). The same considerations are applied to the electromagnetic energy-momentum tensor T (§ 3).

2. F-kinematical groups

Let us consider a Weyl infinitesimal coordinate transformation

$$x'^{\mu} = x^{\mu} + a^{\mu} + \omega^{\mu}{}_{\nu}x^{\nu} + \rho x^{\mu} \quad , \quad (\mu = 0,1,2,3) \quad , \qquad (2.1)$$

where the infinitesimal parameters $a^{\mu},\{\omega^{\mu\nu}\} \equiv (\vec{\phi},\vec{\theta})$ and ρ are respectively associated with the infinitesimal generators $P^{\mu},\{M^{\mu\nu}\} \equiv (\vec{K},\vec{J})$ and D corresponding to translations, restricted homogeneous Lorentz transformations (boosts $\vec{K}$ and pure rotations $\vec{J}$) and dilatations. From the tensorial character of $F^{\mu\nu}$:

$$F'^{\mu\nu}(x') = \frac{\partial x'^{\mu}}{\partial x^{\alpha}} \frac{\partial x'^{\nu}}{\partial x^{\beta}} F^{\alpha\beta}(x) \qquad (2.2)$$

we deduce the invariance conditions for a constant and uniform field :

$$\begin{aligned} &\vec{\phi}\wedge\vec{B} + \vec{\theta}\wedge\vec{E} + 2\rho\vec{E} = 0 \quad , \\ &\vec{\phi}\wedge\vec{E} - \vec{\theta}\wedge\vec{B} - 2\rho\vec{B} = 0 \quad . \end{aligned} \qquad (2.3)$$

We directly notice that the two Lorentz invariants

$$\Delta(F) = \frac{1}{2} F_{\mu\nu}F^{\mu\nu} = \vec{B}^2-\vec{E}^2 \quad ,$$
$$\phi(F) = \frac{1}{2} F_{\mu\nu}{}^{*}F^{\mu\nu} = 2\vec{E}.\vec{B} \quad , \qquad (2.4)$$

where ${}^{*}F$ is the dual of F, remain scale invariant in the Weyl context so that the B.C.R. method(4) can be used.

The stability group of F, i.e. the subgroup of the homogeneous part $O(3,1) \otimes O(1,1) \equiv \{\vec{K},\vec{J},D\}$ of the Weyl group, leaving invariant a constant and uniform electromagnetic field F, can easily be determined using Eqs. (2.3) and unitary operators such that :

$$U(\omega,\rho) = \exp i(\vec{\phi}.\vec{K}-\vec{\theta}.\vec{J}+\rho D) \quad . \qquad (2.5)$$

We find two different stability groups : for $F_{//}$ we get a 2-dimensional group generated by

$$G_1 = \vec{B}.\vec{J}-\vec{E}.\vec{K} \quad , \quad G_2 = \vec{E}.\vec{J}+\vec{B}.\vec{K} \qquad (2.6)$$

and for $F_{\perp}$ we get a 3-dimensional group generated by

$$G_1 \quad , \quad G_2 \quad , \quad G_3 = (\vec{E}\wedge\vec{B}).\vec{K} + \frac{1}{2}\vec{E}^2 D \quad . \qquad (2.7)$$

Our stability group (2.7) contains evidently the corresponding B.C.R. stability group as a subgroup. The supplementary generator G_3 is a linear combination of Lorentz boosts and dilatations. For $F_{//}$ we recover the results of B.C.R., showing that dilatations do not interfere in this case.

Then with the B.C.R. conventions by choosing $F_{//}$: $\vec{E} = (0,0,E)$; $\vec{B} = (0,0,B)$ and $F_{\perp}$: $\vec{E} = (E,0,0)$; $\vec{B} = (0,E,0)$, we obtain two different kinematical groups :

$$G^W_{F_{//}} \equiv \{J^3,K^3,P^\mu\} \quad , \quad n = 6 \quad , \qquad (2.8)$$

and

$$G^W_{F_{\perp}} \equiv \{A^1,A^2,2K^3+D,P^\mu\} \quad , \quad n = 7 \quad , \qquad (2.9)$$

where $A^1 \equiv J^1+K^2$, $A^2 = J^2-K^1$,

the superscript W refering to the Weyl context.

3. T-kinematical groups

The electromagnetic energy-momentum tensor T is defined in general in Minkowski space by :

$$T^{\mu\nu} = F^{\mu\lambda}F_{\lambda}{}^{\nu} + \frac{1}{4}\eta^{\mu\nu}F_{\alpha\beta}F^{\alpha\beta} \qquad (3.1)$$

with $G_M \equiv \{\eta_{\mu\nu}\} = \text{diag}\,(+,-,-,-)$. The invariance conditions on T for the constant and uniform case are given by

$$\omega^\mu{}_\sigma T^{\sigma\nu} + \omega^\nu{}_\sigma T^{\sigma\mu} + 2\rho T^{\mu\nu} = 0 \quad . \qquad (3.2)$$

Let $T_{//}$ and $T_{\perp}$ be the electromagnetic energy-momentum tensors associated with the particular $F_{//}$ and $F_{\perp}$ respectively. Then a study of Eq. (3.2) in a parallel way with that of Section 2, shows that the kinematical groups of $T_{//}$ and $T_{\perp}$ are in the Poincaré (P) and Weyl (W) cases :

$$\text{i)}\quad G^P_{T_{//}} \equiv \{J^3,K^3,P^\mu\} \; ; \; G^P_{T_\perp} \equiv \{A^1,A^2,J^3,P^\mu\} \quad , \tag{3.3}$$

$$\text{ii)}\quad G^W_{T_{//}} \equiv \{J^3,K^3,P^\mu\} \equiv G^P_{T_{//}} \quad , \quad n_{//} = 6 \quad , \tag{3.4}$$

and

$$G^W_{T_\perp} \equiv \{A^1,A^2,J^3,K^3+D,P^\mu\} \supset G^P_{T_\perp} \quad , \quad n_\perp = 8 \quad . \tag{3.5}$$

It is interesting to remark that

$$G^W_{T_\perp} \equiv G^W_A \quad , \quad A \equiv (A^\mu) \; : \; A^\mu A_\mu = 0 \tag{3.6}$$

where G^W_A is the Weyl kinematical group of a lightlike fourvector A , given by the Weyl extended little group[(1)(2)] of A , supplemented by the four translation generators P^μ .

With the B.C.R. conventions for $F_\perp$, the associated $T_\perp$ takes the form

$$T^{\mu\nu} = \begin{pmatrix} E^2 & 0 & 0 & E^2 \\ 0 & 0 & 0 & 0 \\ 0 & 0 & 0 & 0 \\ E^2 & 0 & 0 & E^2 \end{pmatrix} \quad . \tag{3.7}$$

We notice that in this case

$$T^{\mu\nu} = A^\mu A^\nu \tag{3.8}$$

with

$$A \equiv (A^\mu) = (|\vec{E}|,0,0,|\vec{E}|) \tag{3.9}$$

i.e. a lightlike fourvector whose components are given by the concerned electromagnetic field. The tensor $T^{\mu\nu} \equiv (3.8)$ is invariant if and only if A^μ is invariant, a fact which explains the relation (3.6).

These results are typical of perpendicular considerations and show how they are correlated with physical reality through zero rest mass particle descriptions.

References

1) V. Hussin, Lett.Nuovo Cim. 35(1982)305
2) J. Beckers and V. Hussin, J.Math.Phys. (to appear 1984)
3) Yu.P. Stepanovkii, Theor.Math.Phys. 47(1981)501
4) H. Bacry, Ph. Combe and J.L. Richard, Il Nuovo Cim. 67A(1970)267.

FROM SPINORS TO PROBABILITY AMPLITUDES OF EXTERNAL AND INTERNAL VARIABLES FOR SPINNING PARTICLES

M. Božić

Institute of Physics
11001 Beograd, P.O.B. 57, Yugoslavia

Dahl[1] has recently suceeded to incorporate the Bopp and Haag[2] results on quantum-mechanical rotor into Dirac theory of electron. Eigenfunctions of Dahl's Hamiltonian (whose matrix representation is Dirac Hamiltonian) are probability amplitudes of internal and external degrees of freedom. In the nonrelativistic limit those eigenfunctions tend to the simple product of orbital functions and spin functions which are:

$$X_{1/2}(\alpha\beta\gamma) = (8\pi^2)^{-1/2}\cos(\beta/2)\exp(i\alpha/2)\exp(i\gamma/2)$$

$$X_{-1/2}(\alpha\beta\gamma) = (8\pi^2)^{-1/2}\sin(\beta/2)\exp(-i\alpha/2)\exp(i\gamma/2)$$

In connection with those results we consider that it might be of interest to answer the following questions:

1. Is there a quantum-mechanical concept/physical phenomena related to spin for whose interpretation/explanation the informations contained in spinors are not sufficient but probability densities of additional variables in the states with fixed values of m are necessary?

2. Is it possible to verify experimentally the reality of probability densities $X_m(\alpha\beta\gamma)$?

It is well known that the explanation of the basic experiments with independent spin -1/2 particles, like Stern-Gerlach type experiments or the experiments on coherent spinor rotation[3,4] are based on: the quantized quantity m, the spinors$\binom{u}{v}$, u and v being probability densities of m=1/2 and m=-1/2, respectively, and on transformation properties of spinors under rotation[5]. This set of information on the states of individual particles has also been used in the study of phenomena related to the systems of coupled spins. That was, in our opinion, the reason that for the simplest system of coupled spins (two-spins system) the expressions "parallel spins" and "antiparallel spins" are introduced to denote the triplet and singlet states, respectively. This vocabulary has later been enlarged by the expressions "totally aligned spins" and "the states of parallel spins" referring to the ground state of Heisenberg ferromagnet. Those latter expressions although correct when applied to the state $|S_t=n/2, S_t^z=n/2\rangle$ of n 1/2-spins are not appropriate when associated with the states $|S_t=n/2, S_t^z\neq n/2\rangle$ because they implicitly contain the assumption that the state $|S_t=n/2, S_t^z\neq n/2\rangle$ may be obtained by the rotation of the state $|S_t=n/2, S_t^z=n/2\rangle$ through a certain angle, which is not true. By S_t and S_t^z we denote the values of the total spin and of the z-projection of the total spin, respectively.

The above mentioned expressions and the reasoning implied by the assumptions incorporated into those expressions had, in our opinion, important consequence in the theory of magnetism. The main one is related to the fact that the difficulty which led Weiss to assume the existence of domains remained in the quantum theory of magnetism based on the Heisenberg discovery that Weiss molecular field is due to exchange interaction. The difficulty consists in reconciling the assumption of the large molecular field (to account for the large values of critical temperatures) with the phenomena consisting in the possibility to change the over-all magnetization from an initial value of zero (in the absence of an applied magnetic field) to a saturation value, by the application of a very weak magnetic field.

If the state $|S_t=n/2, S_t^z=n/2\rangle$, and especially the state $|S_t=n/2, S_t^z=0\rangle$ would be interpreted directly on the basis of their expressions which are linear combinations of products $X_{m_1}(\alpha_1\beta_1\gamma_1)\ldots X_{m_n}(\alpha_n\beta_n\gamma_n)$ then one should not have principial difficulties to understand the change of overall magnetization from zero to the saturated value by the application of weak magnetic field, since this corresponds to the process of increase of the component S_t^z without the change of the quantity S_t, and therefore without the change of exchange energy, which is large compared to the energies responsible for the splitting of energy levels having the same value of S_t.

So, instead of double triumph of quantum mechanics through the explanation of two Weiss assumptions: the existence of the molecular field and the existence of domain structure, quantum mechanics triumphed only through the Heisenberg explanation that the molecular field is due to exchange interaction whereas the domain theory[6] is still a sophisticated mixture of the elements of classical and quantum theory of magnetism.

The purely quantum theory of domains should be based, in our opinion, on the richness of the subspaces of states characterised by the fixed value of the total spin. But, this richness may emerge in a "physically interpretable way" only if one passes from the vector i.e. spinor description of the states of individual particles to the description based on the probability amplitudes of some internal variables, like Euler angles α,β,γ of the quantum mechanical rotor.

This affirmation may clearly be seen in the two-particle case. Probability amplitudes of two sets of Euler angles in the coupled states are given by:

$$X_{SM}(\alpha_1\beta_1\gamma_1\alpha_2\beta_2\gamma_2)=\sum_m(\tfrac{1}{2}\, m\, \tfrac{1}{2}\, M-m|SM)X_m(\alpha_1\beta_1\gamma_1)X_{M-m}(\alpha_2\beta_2\gamma_2)$$

The joint probability density is:

$$P_{SM}(\alpha_1\beta_1\gamma_1\alpha_2\beta_2\gamma_2)=|X(\alpha_1\beta_1\gamma_1\alpha_2\beta_2\gamma_2)|^2$$

By giving to the quantum numbers S and M all possible values we find:

$$P_{11}(\alpha_1\beta_1\gamma_1\alpha_2\beta_2\gamma_2) = (8\pi^2)^{-2}\cos^2(\beta_1/2)\cos^2(\beta_2/2)$$

$$P_{1-1}(\alpha_1\beta_1\gamma_1\alpha_2\beta_2\gamma_2) = (8\pi^2)\sin^2(\beta_1/2)\sin^2(\beta_2/2)$$

$$P_{10}(\alpha_1\beta_1\gamma_1\alpha_2\beta_2\gamma_2) = \frac{1}{2}(8\pi^2)^{-2}\Big[\cos^2(\beta_1/2)\sin^2(\beta_2/2)+\cos^2(\beta_2/2)\sin^2(\beta_1/2)+ + \frac{1}{2}\cos(\alpha_2-\alpha_1)\sin\beta_1\sin\beta_2\Big]$$

$$P_{00}(\alpha_1\beta_1\gamma_1\alpha_2\beta_2\gamma_2) = \frac{1}{2}(8\ {}^2)^{-2}\Big[\cos^2(\beta_1/2)\sin^2(\beta_2/2)+\cos^2(\beta_2/2)\sin^2(\beta_1/2)- - \frac{1}{2}\cos(\alpha_2-\alpha_1)\sin\beta_1\sin\beta_2\Big]$$

From the above we see that:

a) the density P_{10} is not obtainable by rotation from the density P_{11};

b) by comparing the densities P_{00} and P_{10} one sees that the difference of signs in the linear combinations has for the consequence the difference of the joint probability densities;

c) the following identity is valid

$$P_{10}(\alpha_1\beta_1\gamma_1\alpha_2\beta_2\gamma_2) = P_{00}(\alpha_1+\pi,\beta_1\gamma_1\alpha_2\beta_2\gamma_2)$$

By generalizing to $N>2$ we would come to the analogous conclusions. All that means that the number of up and down spins in a state $|S_t, S_t^z=S_t\rangle$ is not an appropriate characteristic of the subspace of states with the fixed value of S_t and should not be taken to be the basis for the visual presentation of the spin coupling in this subspace.

To conclude, the answer to our question from the beginning is: There exist quantum mechanical concepts/physical phenomena related to spin for the interpretation/explanation of which the informations contained in spinors are not sufficient. On the systems of many interacting spinning particles the reality of probability amplitudes $X_m(\alpha\beta\gamma)$ should be, in principle, verifiable.

REFERENCES

1. J.P. Dahl, K.Dan.Vidensk.Selsk.Ma.Fys.Medd. 39, 12 (1977).

2. F. Bopp and R. Haag, Z.Naturforsh 5a, 644 (1950).

3. H. Rauch, A. Zeillinger, G. Badurker, A. Wilfing, W. Bauspass and U. Bonse, Phys.Lett. 54A, 425 (1975).

4. S.A. Werner, R. Collela, A.W. Overhauser and C.F. Eagen, Phys.Rev.Lett. 35, 1053 (1975).

5. Z. Marić and M. Božić, Proc. XI Int.Coll.Gr.Theor.Meth.Phys., edited by M. Serdaroglu and E. Inonu (Springer-Verlag, Berlin, 1983) p. 486.

6. Ch. Kittel and J. Galt, Sol.St.Phys. 3, 437 (1956).

A CHARACTERIZATION OF FACTOR SYSTEMS OF LOCALLY-OPERATING REPRESENTATIONS

José F. Cariñena
Departamento de Física Teórica. Facultad de Ciencias. Zaragoza. SPAIN.
Mariano A. del Olmo and Mariano Santander
Departamento de Física Teórica. Facultad de Ciencias. Valladolid. SPAIN.

1.- The relevance of multiplier locally-operating representations of transitive transformations groups in quantum mechanics was pointed out by Hoogland(1). In a recent paper(2) we have shown that such representations can be obtained from the linear locally-operating representations of any splitting group $\tilde{G}$ for G in a similar way to what happens for general multiplier representations(3,4) The concept of superequivalence of factor systems finds then its natural setting. A first step in the determination of a "minimal" splitting group solving our problem is the search of possible factor systems arising in the locally-operating multiplier representations we are considering . Once this matter has been solved, a similar process to the one developed in Ref 3 permits us to find a representation group $\bar{G}$ for multiplier locally-operating representations of G. Every linear locally-operating representation of G is gauge equivalent to an induced representation of G as has been shown in a recent paper(5). We aim then to look for a criterion for deciding whether a factor system of G can arise in a locally-operating representation or not. It will be given in next section.

2.- Factor systems of locally-operating representations are defined by means of gauge-functions $A(g,x)$ satisfying $A(g_1,g_2x)\ A(g_2,x) = \omega(g_1,g_2)\ A(g_1g_2,x)$. Let x_0 be an arbitrarily chosen point in the homogeneous space for G, and Γ the corresponding isotopy group. The defining relation for ω when restricted to $\Gamma \times \{x_0\}$ leads to a multiplier matrix representation of Γ namely $A(\gamma', x_0)\ A(\gamma, x_0) = \tau(\gamma',\gamma)\ A(\gamma'\gamma, x_0)$ where the factor system $\tau \in H^2(\Gamma,T)$ is but the restriction of ω to $\Gamma\times\Gamma$. In the case of Γ being a connected and simply connected Lie group such factor systems τ are to be equivalent to the trivial factor system and only factor system ω whose restrictions to Γ are equivalent to the trivial one could arise in a multiplier locally-operating representation of G.

The more general case of Γ being a connected but not simply connected Lie group can be analyzed by making use of the inflation-restriction sequence(4) corresponding to the covering morphism $p: \Gamma^* \to \Gamma$. We denote Γ^* the universal covering group of Γ; the kernel of p is the first homotopy group $\pi_1(\Gamma)$. The restriction, inflation and transgression homomorphisms will be denoted ρ, Λ and δ respectively. The fundamental point is the exactness of the inflation-restriction sequence (Mackey-Moore cohomology is considered. See e.g. Ref 6)

$$1 \to H^1(\Gamma,T) \to H^1(\Gamma^*,T) \overset{\rho}{\to} H^1(\pi_1(\Gamma),T)^{\Gamma} \overset{\delta}{\to} H^2(\Gamma,T) \overset{\Lambda}{\to} H^2(\Gamma^*,T)$$

The action of Γ on T is trivial because Γ is connected and Γ^* is a central extension of Γ by $\pi_1(\Gamma)$ and therefore $H^1(\pi_1(\Gamma),T)^{\Gamma} = H^1(\pi_1(\Gamma),T) = \widehat{\pi_1(\Gamma)}$ where $\hat{A}$ deno notes the dual group of the Abelian group A. Moreover, the first cohomology groups $H^1(\Gamma,T)$ and $H^1(\Gamma^*,T)$ are $\hat{\Gamma}$ and $\hat{\Gamma}^*$ respectively and then the inflation-restriction sequence reduces to

$$1 \to \hat{\Gamma} \to \hat{\Gamma}^* \overset{\rho}{\to} \widehat{\pi_1(\Gamma)} \overset{\delta}{\to} H^2(\Gamma,T) \overset{\Lambda}{\to} H^2(\Gamma^*,T)$$

Exactness of this sequence means that $\ker \Lambda = \delta(\widehat{\pi_1(\Gamma)})$. This is an important fact because any finite-dimensional multiplier τ-representation of Γ, P, gives rise to one $P\circ p$ of Γ^* with a factor system in the class of $\Lambda(\bar{\tau})$ which has to be trivial because of the simply-connectedness of Γ^*. We have found our first result:

Theorem 1. Factor systems arising in locally-operating multiplier representations of G have their restrictions to $\Gamma \times \Gamma$ in the kernel of the inflation map Λ.

The set of (equivalence class of) such factor systems will be denoted $H^2_{loc}(G,T)$ and it is easy to check that it is a subgroup of $H^2(G,T)$. A straightforward consequence of the continuity of the inflation map when $H^2(\Gamma,T)$ and $H^2(\Gamma^*,T)$ are endowed with appropriate topologies(7) is that $H^2_{loc}(G,T)$ is closed in $H^2(G,T)$. So that, if $H^2(G,T)$ is a Lie group, $H^2_{loc}(G,T)$ is a Lie group too.

3.- In order to characterize the subgroup $\ker \Lambda$ we recall that the Abelian group $H^2(\Gamma,T)$ is isomorphic to $\mathbb{R}^r \otimes \mathbb{Z}^s \otimes C$ (with $r,s \in \mathbb{N}$) where C is a compact group ($C \simeq T^1 \otimes F$ with F a finite group) while $H^2(\Gamma^*,T)$ is isomorphic to $\mathbb{R}^n$ ($n \in \mathbb{N}$). If $\widehat{\pi_1(\Gamma)}$ is endowed with the compact-open topology of $\pi_1(\Gamma)$ and we consider the usual topologies(7) on $H^2(\Gamma,T)$ and $H^2(\Gamma^*,T)$, the inflation and transgression morphisms are continuous. The condition $\ker \Lambda = \delta(\widehat{\pi_1(\Gamma)})$ implies that $\ker \Lambda$ is compact i.e. $\ker \Lambda \subset C$. On the other hand the group $\mathbb{R}^n$ has no proper compact subgroups and therefore $C \subset \ker \Lambda$. This leads to the following theorem:

Theorem 2. The subgroup $H^2_{loc}(G,T)$ is made up by factor systems $\bar{\omega} \in H^2(G,T)$ such that their restrictions to $\Gamma \times \Gamma$ belong to the maximal compact subgroup of $H^2(\Gamma,T)$.

4.- Examples: i) It is well known that if G is the 3 + 1 Galilei group, $H^2(G,T) = Z_2 \otimes \mathbb{R}$ and Γ being isomorphic to the threedimensional Euclidean group, $H^2(\Gamma,T) = Z_2$. Theorem 2 shows that $H^2_{loc}(G,T) = H^2(G,T)$.

ii) Another interesting example is that of G being the 2 + 1 Galilei group. In this case $H^2(G,T) = \mathbb{R}^2$, corresponding to the new commutators $[K_1,K_2] = kI$ and $[K_i,P_j] = m\delta_{ij}I$ (we remark that $H^2(G^*,T) = \mathbb{R}^2$). The isotopy subgroup is the twodimensional Euclidean group E(2) in which case $H^2(\Gamma,T) = \mathbb{R}$. The factor system $[k,m]$ of G when restricted to $\Gamma \times \Gamma$ gives the factor system $[k]$ of Γ. The criterion given by Theorem 2 shows that only $[m,o]$ factor systems can appear in locally-operating multiplier representations of G.

iii) In the case of G being the 1 + 1 Galilei group a similar analysis leads to $H^2(G,T) = H^2_{loc}(G,T) = \mathbb{R}^2$ which would be reduced to $\mathbb{R}$ if a spatial or temporal inversion is included.

iv) Two other cases worth of mention are those of symmetry groups of an uniform and constant non relativistic magnetic or electric field pointing in the OZ direction. In the first case the symmetry group is the 6-dimensional subgroup of the Galilei group generated by P_0, $\vec{P}$, J_3 and K_3. The isotopy subgroup will be $SO(2) \otimes \mathbb{R}$ (it is generated by J_3 and K_3) and therefore $H^2(\Gamma, T) = 1$, i.e, $H^2_{loc}(G,T) = H^2(G,T)$. In the case of an electric field the isotopy subgroup includes K_2 and K_3 too and it is isomorphic to $E(2) \otimes \mathbb{R}$. The second cohomology group is $H^2(G,T) = \mathbb{R}^4$ (see Ref 1 where it is proved that $H^2(G^*,T) = \mathbb{R}^6$). Standard techniques quickly lead to $H^2(\Gamma,T) = \mathbb{R}$ corresponding to the new commutator $[K_1, K_2] = k\, I$. Consequently $H^2_{loc}(G,T) = \mathbb{R}^3$, i.e. $H^2_{loc}(G,T) \neq H^2(G,T)$.

V) Hereafter we will assume that the isotopy group Γ is such that there exists a finite-dimensional σ-representation of Γ for any $\bar{\sigma}$ in the kernel of the inflation map Λ. In this case $H^2_{loc}(G,T)$ coincides with $\ker \Lambda$ and is closed in $H^2(G,T)$. Infact, for any ω $Z^2(G,T)$ such that $\bar{\omega}_{|\Gamma\times\Gamma} \in \ker \Lambda$ and any Borel section $s: X \to G$, we can choose a representative ω such that $\omega_{|s(X)\times\Gamma}=1$ and if Σ is a finite-dimensional σ-representation of Γ with $\sigma = \omega_{|\Gamma\times\Gamma}$, $A(g,x)=\omega(g,s(x))\,\Sigma\,(s^{-1}(gx)gs(x))$ is the gauge matrix of a L.O.M.R. of G with factor system ω.

We recall that the existence of a representation group for G depends on the existance of a topology τ on $H^2(G,T)$ making continuous the maps $W(g,h): \bar{\omega} \to [\eta(\bar{\omega})](g,h)$, with η being a homomorphic section $\eta: H^2(G,T) \to Z^2(G,T)$. A similar theory can be developed for local representation groups with the substitution of $H^2_{loc}(G,T)$ for $H^2(G,T)$. The point is that as $H^2_{loc}(G,T)$ is closed, then if it exists a representation group for G there is also a local representation group. Moreover if the representation group is a Lie group, the local representation is a Lie group too.

REFERENCES

1. H.Hoogland, Nuovo Cimento B 32 427(1976), Ph.D.Thesis, University of Nijmegen (1977); J.Phys. A 11 1557 (1978).
2. J.F. Carinena, M.A. del Olmo and M.Santander, Physica 114 A 420 (1982).
3. J.F. Carinena and M. Santander J.Math. Phys., 20 2168 (1979).
4. U. Cattaneo; J.Math.Phys. 19 452 (1978).
5. M. Asorey, J.F. Carinena and M.A. del Olmo , J.Phys. A 16 1603 (1983).
6. U. Cattaneo and A. Janner , J. Math.Phys. 15 1155 (1974).
7. C.C. Moore Trans. Am. Math. Soc. 113 40 (1964), 113 64 (1964).

RECENT DEVELOPMENTS ON SHIFT OPERATORS

P. De Wilde(°), J. Van der Jeugt(†), H. De Meyer(‡)
and G. Vanden Berghe

Seminarie voor Wiskundige Natuurkunde, Rijksuniversiteit-Gent
Krijgslaan 281 - S9, B-9000 Gent, BELGIUM

The shift operator technique, developed by Hughes and Yadegar [1)] for an A_1 subalgebra of a Lie algebra G, is extended to other Lie algebra-subalgebra chains.

1) $B_3 \supset A_1^1 \oplus A_1^1 \oplus A_1^2$

For the basis elements of B_3 we choose the basis generators of the subalgebra, denoted by $s_{0,\pm1}$, $t_{0,\pm1}$ and $u_{0,\pm1}$, and the components of a tensor operator $T^{[1/2\ 1/2\ 1]}$ of the bispinor-vector type which was introduced by Vanden Berghe et al [2)]. The shift operators can be expressed in terms of these generators. As in the case of A_1 [1)], relations between the matrix elements of the $A_1 \oplus A_1 \oplus A_1$ shift operators between subalgebra states and the reduced matrix elements of the tensor can be established. In this way relations between quadratic products of shift operators lead to relations between quadratic products of the reduced matrix elements of the bispinor-vector. We have calculated 11 such relations, and obtained from them a set of coupled recursion relations for the matrix elements of the tensor. For three particular classes of B_3 irreducible representations (irreps), namely $[v,0,0]$, $[v,0,1]$ and $[0,0,v]$ $(v \in \mathbb{N}_0)$, we solved the recursion relations explicitly, taking into account the $B_3 \to A_1 \oplus A_1 \oplus A_1$ branching rules. The three cases considered have the common property that no degeneracy occurs in the reduction to the subalgebra. As an example, we give the results for the $[0,0,v]$ irreps (the results for $[v,0,0]$ and $[v,0,1]$ are published elsewhere [3)]). The meaning of the labels s, t and u in the following expressions is obvious from the notation of the subalgebra generators.

$$|\langle s+1/2,t+1/2,u+1 \| T^{[1/2\ 1/2\ 1]} \| s,t,u\rangle|^2 = 1/4(s+t+u+2)(s+t+u+3) \times (s-t+u+1)(-s+t+u+1)\frac{1}{(u+1)}(v/2+s+t+3)(v/2-s-t)\ , \tag{1}$$

$$|\langle s+1/2,t+1/2,u \| T^{[1/2\ 1/2\ 1]} \| s,t,u\rangle|^2 = 1/4(s-t)^2(s+t+u+2) \times (s+t-u+1)\frac{(2u+1)}{u(u+1)}(v/2+s+t+3)(v/2-s-t)\ , \tag{2}$$

(°) Research assistant I.W.O.N.L.
(†) Research assistant N.F.W.O.
(‡) Research associate N.F.W.O.

$$|\langle s+1/2,t+1/2,u-1 \| T^{[1/2\ 1/2\ 1]} \| s,t,u\rangle|^2 = 1/4(s+t-u+2)(s+t-u+1)$$
$$\times (s-t+u)(-s+t+u)\frac{1}{u}(v/2+s+t+3)(v/2-s-t). \qquad (3)$$

The remaining reduced matrix elements follow from (1)-(3) by means of a simple substitution.

Considering the subalgebra scalars, a pair of commuting operators in the enveloping algebra of B_3 was found : $[C^{(1102)},C^{(0022)}] = 0$. Hereby, the superscripts denote the degrees of the homogeneous operator in the sets s_k, t_k, u_k and $T^{[1/2\ 1/2\ 1]}_{i\ \ j\ \ k}$ $(i,j=\pm 1/2\ ;\ k=0,\pm 1)$ respectively. On account of expressions like (1)-(3), eigenvalues of the commuting operators were determined in the three cases mentioned above. It was proved that the commuting pair is a special case of a general result concerning mutually commuting subalgebra scalars in the case $G \supset A_1 \oplus \ldots \oplus A_1$ [4].

2) $G_2 \supset A_2$

In the reduction chain $G_2 \to A_2$, the states contained in any G_2 irrep can be labelled by the A_2 labels (λ,μ), a set of internal labels of A_2, e.g. the "hypercharge" Y and the isospin labels T and T_0, and a "missing label" s :

$$\left|\begin{matrix}\lambda\ ,\ \mu\ ;\ s\\ Y\ ,\ T\ ,\ T_0\end{matrix}\right\rangle . \qquad (4)$$

Operators shifting the (λ,μ)-values by certain integral values are constructed and denoted as $A^{i,j,k}_{\lambda,\mu,Y}$, where $(i,j,k) \in \{(\pm 1,0,\mp 2/3),(0,\mp 1,\mp 2/3),(\mp 1,\pm 1,\mp 2/3)\}$. Their action upon the states (4) of a certain G_2 irrep reads as follows :

$$A^{1,0,-2/3}_{\lambda,\mu,Y}\left|\begin{matrix}\lambda,\mu;s\\ Y,T,T_0\end{matrix}\right\rangle \sim \sum_{s'}\left|\begin{matrix}\lambda+1\ \ ,\mu;s'\\ Y-2/3,T,T_0\end{matrix}\right\rangle . \qquad (5)$$

Relations connecting quadratic product operators of shift operators are determined, e.g. :

$$A^{-1\ ,0,\ 2/3}_{\lambda+1,\mu,Y-2/3}A^{1,0,-2/3}_{\lambda,\mu,Y} + A^{1\ \ ,-1\ ,\ 2/3}_{\lambda-1,\mu+1,Y-2/3}A^{-1,+1,-2/3}_{\lambda\ ,\ \mu\ ,Y} - \frac{1}{\sqrt{6}}(\lambda+1)S$$
$$+\frac{1}{9}(\lambda+1)(\lambda+2\mu+6)C^{*} - \frac{1}{2}(\lambda+1)I_2 = 0 . \qquad (6)$$

Herein, I_2 is the second order Casimir operator of A_2, C^{*} is proportional to the Casimir operator of G_2, and S is a subalgebra invariant in the enveloping algebra of G_2.

On account of relations like (6), the following formal commutation relations of shift operators are deduced (H_1 and H_2 are some well-defined combinations of S, C^{*} and I_2) :

$$[A^{\pm 1,0}, A^{\mp 1,\pm 1}] = \pm \frac{1}{\sqrt{6}} A^{0,\pm 1} \quad , \quad [A^{0,\pm 1}, A^{\mp 1,0}] = \mp \frac{1}{\sqrt{6}} A^{\mp 1,\pm 1} \; ,$$

$$[A^{\mp 1,\pm 1}, A^{0,\mp 1}] = \pm \frac{1}{\sqrt{6}} A^{\mp 1,0} \quad , \quad [A^{1,0}, A^{-1,0}] = \frac{1}{\sqrt{3}} H_1 \; , \tag{7}$$

$$[A^{0,1}, A^{0,-1}] = \frac{1}{2\sqrt{3}} H_1 + \frac{1}{2} H_2 \quad , \quad [A^{1,-1}, A^{-1,1}] = \frac{1}{2\sqrt{3}} H_1 - \frac{1}{2} H_2 \; ,$$

and all other $[A^{i,j}, A^{i',j'}]$ are zero. This obviously implies that the shift operators form an A_2 Lie algebra. But all the $|\lambda,\mu\rangle$-states of a G_2 irrep are connected to one another by repeated actions of the shift operators, and the multiplicity of the states is determined by the relations between the shift operators. Consequently, the branching multiplicity diagrams for $G_2 \to A_2$ must coincide with A_2 weight diagrams. In this way, the relationship between G_2 and A_2 — in literature often described as "strange and peculiar" [5] — is generated in a very natural way.

As a second application on the shift operator technique, the eigenvalues of the scalar operator S, appearing in relation (6), are determined. The operator S is an SU(3) invariant in the enveloping algebra of G_2. The set of six operators I_2, I_3 (the second and third order Casimir operators of A_2), S, Y, T^2 and T_0 form a complete set of mutually commuting operators which label uniquely the basis states of any irrep of G_2. Hence, the determination of the eigenvalues of S solve the one missing label problem $G_2 \to A_2$ completely. Moreover they give rise to an orthonormal basis of the states.

---

1) J.W.B. Hughes and J. Yadegar : J. Math. Phys. 19, 2068 (1978)

2) G. Vanden Berghe, H. De Meyer, P. De Wilde : J. Phys. A : Math. Gen. 15, 2677 (1982)

3) P. De Wilde and J. Van der Jeugt : J. Math. Phys. (submitted)

4) J. Van der Jeugt : J. Math. Phys. (submitted)

5) C. Fronsdal, "Elementary Particle Physics and Field Theory" Brandeis Summer Institute 1962, Benjamin, N.Y., p. 532 (1963).

UNITARY AND NON-UNITARY, MULTIPLICITY FREE IRREDUCIBLE REPRESENTATIONS OF $\overline{SL(3,R)}$*

Y.Güler, Physics Department
Middle East Technical University, Ankara, Turkey

ABSTRACT: Irreducible representations, which are multiplicity free with respect to SU(2) subgroup, are obtained by a constructive method. It is observed that finite dimensional representations are labeled by a positive integer or semi-integer number J_0 and three complex numbers. A <u>new</u> unitary, multiplicity free representation, with J content $J = j_0, j_0+1, j_0+2$. is determined.

I. INTRODUCTION: The problem of determination of all representations of semi-simple real, non-compact groups has not been fully solved yet. There are mainly two approaches in this respect. First one is a method which is used by Gel'fand and Graev[1]. They use the functions defined on the coset spaces of the group as the representation space and determine principal series of unitary, irreducible representations of SL(n,R).

The second method is initiated by Harish-Chandra[2] who determines all irreducible, unitary representations using the representations of the maximal compact subgroup.

In this work, all representations, unitary and non-unitary, of $\overline{SL(3,R)}$ are obtained by a method initiated by Naimark[3]. The advantage of the method is that it makes it possible to obtain finite and infinite dimensional representations at once. Besides, it predicts new representations which can not be obtained by other methods[4].

2. THE DETERMINATION OF REPRESENTATION SPACE

Any representation of the SU(2) subalgebra is the direct sum of irreducible representation given by an integer or half integer J. Since any representation of $\overline{SL(3,R)}$ contains a representation of SU(2) subalgebra, one should consider the space

*Research partially supported by the Scientific and Technical Research Council of Turkey.

$$R = \oplus M_j \qquad \text{II.1}$$

as the multiplicity free representation space of $\overline{SL(3,R)}$ where M_j is the $(2j+1)$ dimensional vector space corresponding to an irreducible representation of SU(2). Hence, the problem of determination of irreducible representations of $\overline{SL(3,R)}$ is reduced to the problem of determination of possible j values contained in R. In general it is proved as a theorem[3] that any eigenvector f of the Hermitean infinitesimal operator J_3 of SU(2) subalgebra, corresponding to the eigenvalue m ($m = -j,-j+1,\ldots j$) and satisfying $J_3^p f = 0$ can be written as a linear combination of the eigenstates f_m^j, $j = m,m+1,\ldots,m+p-1$ where p is a non-zero positive integer.

$\overline{SL(3,R)}$ commutation relations[5] give the following equations:

$$jA_{j+3}B_{j+1} - (j+4)A_{j+2}B_{j+3} = 0 \qquad \text{II.2}$$

$$(2j-1)A_{j+2}C_j - 3B_{j+1}B_{j+2} - (2j+7)A_{j+2}C_{j+2} = 0 \qquad \text{II.3}$$

$$3(j-1)A_{j+1}B_j \ (j-3)B_{j+1}C_j - (j+5)B_{j+1}C_{j+1} - 3(j+3)A_{j+2}B_{j+2} = 0 \qquad \text{II.4}$$

$$(-2j+3)A_j^2 \ (-j+2)B_j^2 + 2C_j^2 + (j-3)B_{j+1}^2 + (2j+5)A_{j+2}^2 = 0 \qquad \text{II.5}$$

where A_j, B_j and C_j are complex numbers appearing in the linear combinations. Simultaneous solution of Eqns.(II.2-5) determine the following series of representations:

1. Finite dimensional representations are labeled by three complex numbers and j_0. The j content is

$$j = j_0, j_0+1, j_0+2, \ldots \qquad j_0 = 0, \tfrac{1}{2}, 1, \ldots \qquad \text{II.6}$$

2. There are two series of unitary representations:

a) Unitary representations labeled by one imaginary, one real number and j_0. The j content is

$$j = 0,2,4,6 \ldots \qquad \text{for } j_0 = 0 \qquad \text{II.7}$$

$$j = \tfrac{1}{2}, \tfrac{5}{2}, \tfrac{9}{2}, \ldots \qquad \text{for } j_0 = \tfrac{1}{2} \qquad \text{II.8}$$

$$j = 1,3,5,7, \qquad \text{for } j_0 = 1 \qquad \text{II.9}$$

b) Unitary representations labeled by a real number and j_0. The j content is

$$j = j_0, j_0+1, j_0+2, \ldots \qquad j_0 \geq 3 \qquad \text{II.10}$$

REFERENCES:

1) I.M.Gel'fand ard M.I.Graev, Am.Math.Soc.Transl., Ser.2,2, 147(1956).

2) H.Chandra, Proc.Nat.Acad.Sci.,USA,37,363(1951).

3) M.A.Naimark, Linear Representations of the Lorentz Group, Pergamon Press, 1964

4) Dj.Sijacki, J.Math.Phys., 16, 298 (1975).

5) Y.Güler, J.Math.Phys.,19,508(1978).

THE SYMMETRY GROUP OF A DIFFERENTIAL EQUATION

M. Hamermesh
University of Minnesota
Minneapolis, MN 55416/USA

The use of group theoretical methods for the study of differential equations goes back to Sophus Lie, who developed methods that are still the essential ones used today. The effective use of such techniques was shown by Bluman and Cole and by Ovsyannikov. The recent book by Ovsyannikov gives a complete description of his methods, but I find the notation difficult to understand and use. A clear presentation of modern methods and algorithms can be found in the thesis and papers of P.J.Olver: Ol-3 is a brief but clear description of the algorithms for finding the most general symmetry group of a system of ordinary or partial differential equations, even for the case of nonlinearity, Ol-2 is a rigorous presentation of the mathematics, and Ol-4 gives a large number of applications. The lectures of Sattinger are also a good source of examples. I shall give here only a brief outline of the subject, and most of the material will be taken from the papers and thesis of Olver.

For ordinary differential equations Lie showed that a first order equation can be invariant under an infinite number of one-parameter groups. For second and higher order equations one can have only a finite number of independent one-parameter groups: at most 8 for n=2, and at most n + 4 for $n>2$.. Furthermore, these groups can be determined systematically.

A local group of transformations acting on the space $\mathbb{R}^n$ consists of a Lie group G and a smooth map $\Phi: V \to \mathbb{R}^n$, where $V \subset G \times \mathbb{R}^n$, and

$$\Phi(e,x) = x \text{ for } x \in \mathbb{R}^n; \quad \Phi(g,\Phi(h,x)) = \Phi(g\cdot h,x)$$

when the arguments make sense. Associated with the local group are its generators. Suppose that $\exp(t\alpha)$ is the one-parameter group generated by the element α. Then the corresponding vector field on $\mathbb{R}^n$ is $\varphi(\alpha)$, where

$$\varphi(\alpha)\rfloor_x = \frac{d}{dt}\Big|_{t=0} \Phi(\exp(t\alpha)x).$$

If the coordinates on $\mathbb{R}^n$ are $x=(x_1,\dots,x_n)$, we write the vector field as

$$\vec{v} = \sum_{i=1}^{p} \xi^i \, \partial/\partial x_i \, .$$

So the coordinate functions for the vector field are

$$\xi^i(x) = \frac{d}{dt}\Big|_{t=0} (\Phi^i(t,x))$$

if

$$\Phi(\exp(t\alpha)x) = (\Phi^1(t,x), \dots \Phi^n(t,x)).$$

The vector fields also act as operators on functions $F: \mathbb{R}^n \to \mathbb{R}$:

$$\vec{v}\,F(x) = \sum_i \xi^i(x)\, \partial F/\partial x_i \, .$$

A function $F: \mathbb{R}^n \to \mathbb{R}^m$ is said to be G-invariant if $F(gx)=F(x)$ whenever gx is defined. A set S is said to be a G-invariant subset if, for every $x \in S$, $gx \in S$ whenever gx is defined. Note that the vanishing of a function F gives a subvariety $S=\{x: F(x)=0\}$ which may be G-invariant even though the function F does not have invariant level sets for other values.

Theorem: F is a G-invariant function iff $\alpha F(x)=0$ for every infinitesimal generator α of G and every $x \in \mathbb{R}^n$. The subvariety $S=\{x: F(x)=0\}$ is G-invariant iff this same condition holds.

Now let us consider the case of partial differential equations. To simplify the notation we assume a single dependent variable, and p independent variables x_i (i=1...p) and spaces $X=\mathbb{R}^p$ (coords.x), $U=\mathbb{R}$ (coord. u). The solution of the system S is a function u=f(x), whose graph is a p-dimensional submanifold in the space X x U.

A symmetry group of S is a local Lie group of transformations G acting on X x U that transforms solutions u=f(x) of S into solutions $u= g\cdot f(x)$ of S. How does the symmetry group transform functions?. It takes the graph of u=f(x) in X x U and changes it to a new graph (for g sufficiently close to the identity e) $\tilde{u} = g\cdot f(\tilde{x})$, where

$$\tilde{x} = X_g(x,f(x)) = X_g \circ (I\times f)(x)$$

$$\tilde{u} = \mathcal{U}_g(x,f(x)) = \mathcal{U}_g \circ (I\times f)(x).$$

Solving,

$$g\cdot f = [\mathcal{U}_g \circ (I\times f)] \circ [X_g \circ (I\times f)]^{-1}.$$

We want to find the largest local symmetry group of the partial differential equation. To do this we first construct a space that represents all the derivatives that appear in the pde. For a function of p independent variables, the number of k'th order derivatives is

$$p_k = \binom{p+k-1}{k}.$$

We denote the derivatives by

$$\partial_J = \frac{\partial^{|J|}}{\partial x_1^{j_1}\cdots\partial x_p^{j_p}},$$

where $J=(j_1,\dots,j_p)$, and the j's are nonnegative integers with $|J|=\sum j_\nu = k$.

If f: X→U (i.e., u = f(x)), these derivatives of order k are $u_J=\partial_J f(x)$, giving a space $U_k=\mathbb{R}^{p}$.

Let $U^{(k)}= U \times U_1 \times U_2 \times \dots \times U_k$. This is the space that represents all the derivatives of order $\leq k$. We denote a point in $U^{(k)}$ by $u^{(k)}$, where $u^{(k)}$ has $(1+p_1+\dots+p_k)$ components u_J, $0\leq|J|\leq k$, and $u_0\equiv u$. Given a smooth function u=f(x), there is induced a function $u^{(k)}= pr^{(k)}f: X\to U^{(k)}$, the k'th prolongation of f (the erweiterung, extension, continuation, prodolzhenie), defined by the equations

$$u_J = \partial_J f(x).$$

For example, $pr^{(2)} f(x,y) = (f(x,y), f_x, f_y, f_{xx}, f_{xy}, f_{yy})$.

$X \times U^{(k)}$ is the k-jet space. (For k=1, we get the one-jet, or tangent bundle.) The function $pr^{(k)}f(x)$ represents the Taylor polynomial of degree k of f at x.

Our partial differential equation is $\Delta(x,u^{(k)})=0$. A solution of the equation is the subvariety $S\subset X \times U^{(k)}$ given by the vanishing of the smooth function $\Delta: X \times U^{(k)}\to\mathbb{R}$, i.e.,

$$\Delta(x, pr^{(k)}f(x)) = 0.$$

In other words, a transformation on x,u induces a prolongation to $X \times U^{(k)}$.

Given a one-parameter group with generator α, we get a prolongation of the one-parameter group $\exp(t\alpha)$ to $X \times U^{(k)}$, $pr^{(k)}\exp(t\alpha)$, and its infinitesimal generator is

$$pr^{(k)}\alpha = \frac{d}{dt}\Big|_{t=0} pr^{(k)}\left[\exp(t\alpha)\right].$$

We then have the Theorem:

If, for every infinitesimal generator α of G, $pr^{(k)}_{\alpha}\Delta(x,u^{(k)}) = 0$ whenever $\Delta(x,u^{(k)}) = 0$, then G is a symmetry group of the equation $\Delta = 0$.

Now all we need is a formula for finding the prolongation of a vector field. The algorithm is as follows:

Given the vector field α on $X \times U$,

$$\alpha = \sum_{i=1}^{p} \xi^i(x,u)\frac{\partial}{\partial x_i} + \varphi(x,u)\frac{\partial}{\partial u},$$

the k'th prolongation of α is

$$pr^{(k)}\alpha = \alpha + \sum_J \varphi^J(x,u^{(k)})\frac{\partial}{\partial u_J},$$

a vector field on $X \times U^{(k)}$, where the sum goes over all J with $|J| = \sum j_\nu = k$.

The functions φ^J are given by

$$\varphi^J = D^J\Big(\varphi - \sum_{i=1}^{p} u_i\xi^i\Big) + \sum_{i=1}^{p} u_{J,i}\,\xi^i,$$

where

$$u_i = \partial u/\partial x_i\ ; \quad J,i = (j_1,\dots j_{i-1}, j_i+1, j_{i+1},\dots j_p);$$

$$D^J = D_1^{j_1}D_2^{j_2}\cdots D_p^{j_p}\ ; \quad D_i = \frac{\partial}{\partial x_i} + \sum_J u_{J,i}\frac{\partial}{\partial u_J}.$$

For example, suppose that p=2,k=2,i.e., we have one second order pde in two independent variables x,y. Then

$$D_x = \frac{\partial}{\partial x} + u_x\frac{\partial}{\partial u} + u_{xx}\frac{\partial}{\partial u_x} + u_{xy}\frac{\partial}{\partial u_y} + u_{xxx}\frac{\partial}{\partial u_{xx}} + u_{xxy}\frac{\partial}{\partial u_{xy}} + u_{xyy}\frac{\partial}{\partial u_{yy}}.$$

Thus

$$D_x(xuu_{xy}) = uu_{xy} + xu_xu_{xy} + xuu_{xxy}.$$

Suppose that $v = x\partial_u - u\partial_x = \xi\partial_x + \varphi\partial_u$, so $\xi = -u$, $\varphi = x$.

$$pr^{(1)}\vec{v} = \vec{v} + \varphi^x\partial_{u_x}\ ; \quad \varphi^x = D_x(\varphi - u_x\xi) + u_{xx}\xi$$
$$= D_x(x + uu_x) - uu_{xx} = 1 + u_x^2.$$

For $pr^{(2)}\vec{v}$ we need $\varphi^{xx}\partial_{u_{xx}} = 3u_xu_{xx}$.

Thus $pr^{(2)}\vec{v} = \vec{v} + (1 + u_x^2)\partial_{u_x} + 3u_xu_{xx}\partial_{u_{xx}}$. If this acts on u_{xx} we get

$$pr^{(2)}\vec{v}(u_{xx}) = 3u_xu_{xx} = 0 \quad \text{if } u_{xx} = 0.$$

We give some examples:

1. The one-dimensional heat equation, $u_t = u_{xx}$, with p=2,k=2. The heat equation solution set is the subvariety of $X \times U^{(2)}$ given by $\Delta \equiv u_t - u_{xx} = 0$. Suppose that there is a local symmetry group with vector field $\alpha = \xi\partial_x + \tau\partial_t + \varphi\partial_u$ on $X \times U$. Then $pr^{(2)}_{\alpha}\Delta = \varphi^t - \varphi^{xx} = 0$, whenever $u_t = u_{xx}$, and ξ, τ, φ are functions of x,t,u.

$$\begin{aligned}
\varphi^t &= D_t(\varphi - u_x\xi - u_t\tau) + u_{xt}\xi + u_{tt}\tau \\
&= D_t\varphi - u_xD_t\xi - u_tD_t\tau \\
&= \varphi_t + u_t\varphi_u - u_x\xi_t - u_x(u_t\xi_u) - u_t\tau_t - u_t(u_t\tau_u) \\
&= \varphi_t + u_t(\varphi_u - \tau_t) - u_x\xi_t - u_xu_t\xi_u - u_t^2\tau_u.
\end{aligned}$$

$$\begin{aligned}
\varphi^{xx} &= D_x^2(\varphi - u_x\xi - u_t\tau) + u_{xxx}\xi + u_{xxt}\tau \\
&= D_x^2\varphi - u_xD_x^2\xi - u_tD_x^2\tau - 2(D_x\xi)(D_xu_x) - 2(D_x\tau)(D_xu_t).
\end{aligned}$$

$$D_x\varphi = \varphi_x + u_x\varphi_u\ ;\quad D_x\varphi_x = \varphi_{xx} + u_x\varphi_{xu};$$

$$D_x(u_x\varphi_u) = u_x D_x\varphi_u + u_{xx}\varphi_u = u_x\varphi_{ux} + u_x^2\varphi_{uu} + u_{xx}\varphi_u.$$

$$D_x^2\varphi = \varphi_{xx} + 2u_x\varphi_{ux} + u_{xx}\varphi_u + u_x^2\varphi_{uu}\ ,\ \text{etc.}$$

$$\begin{aligned}\varphi^{xx} = {} & \varphi_{xx} + u_x(2\varphi_{xu} - \xi_{xx}) - u_t\tau_{xx} + u_x^2(\varphi_{uu} - 2\xi_{xu}) \\ & - 2u_xu_t\tau_{xu} - u_x^3\xi_{uu} - u_x^2u_t\tau_{uu} + u_{xx}(\varphi_u - 2\xi_x) \\ & - 2u_{xt}\tau_x - 3u_{xx}u_x\xi_u - u_{xx}u_t\tau_u - 2u_{xt}u_x\tau_u .\end{aligned}$$

We set $\varphi^{xx} - \varphi^t = 0$ subject to the condition that $u_{xx} - u_t = 0$. Our procedure is to replace u_t by u_{xx} everywhere in the equation $\varphi^{xx} - \varphi^t = 0$. Then we set the coefficients of the independent coordinates equal to zero. For example, the term not containing derivatives of u gives $\varphi_{xx} - \varphi_t = 0$. We get a system of equations (the symmetry equations ,or defining equations) which must be solved. Their most general solution gives the maximal symmetry group.

2. We carry out the complete calculation for the Burgers equation:

$$u_t + uu_x + u_{xx} = 0.$$

Here we get $\varphi^t + u\varphi^x + u_x\varphi + \varphi^{xx} = 0$, subject to $u_t + uu_x + u_{xx} = 0$. We can use the results of Example 1 for φ^t and φ^{xx}, and

$$\begin{aligned}\varphi^x &= D_x(\varphi - u_x\xi - u_t\tau) + u_{xx}\xi + u_{xt}\tau \\ &= \varphi_x + u_x(\varphi_u - \xi_x) - u_xu_t\tau_u - u_x^2\xi_u - u_t\tau_x .\end{aligned}$$

We set $u_t = -uu_x - u_{xx}$.

$$\begin{aligned}\varphi^{xx} = {} & \varphi_{xx} + u_x(2\varphi_{xu} - \xi_{xx} + u\tau_{xx}) + u_x^2(\varphi_{uu} - 2\xi_{xu} + 2u\tau_{xu}) \\ & - u_x^3\xi_{uu} + u_{xx}(\varphi_u - 2\xi_x + \tau_{xx}) - 3u_{xx}u_x\xi_u \\ & + u_x^3u\tau_{uu} + u_x^2u_{xx}\tau_{uu} + 2u_xu_{xx}\tau_{xu} + u_xu_{xx}u\tau_u \\ & + 2\tau_x(u_x^2 + uu_{xx} + u_{xxx}) + u_{xx}^2\tau_u \\ & + 2u_x\tau_u(\ ''\quad ''\quad ''\ );\end{aligned}$$

$$\begin{aligned}\varphi^t = {} & \varphi_t - (\varphi_u - \tau_t)uu_x - u_{xx}(\varphi_u - \tau_t) - u_x\xi_t + u_x^2u\xi_u \\ & + u_xu_{xx}\xi_u - (uu_x + u_{xx})^2\tau_u;\end{aligned}$$

$$\begin{aligned}u\varphi^x = {} & u\varphi_x + u_xu(\varphi_u - \xi_x) - u_x^2u\xi_u \\ & + \tau_xu(uu_x + u_{xx}) + u_xu\tau_u(uu_x + u_{xx});\end{aligned}$$

$$\varphi u_x = \varphi u_x .$$

Equating the coefficients of the independent coordinates to zero gives:

$$u_xu_{xxx}:\ \tau_u = 0\quad ;\quad u_{xxx}:\ \tau_x = 0,$$

so $\tau=\tau(t)$ only, and we can cancel all the terms involving τ_x and τ_u. Then

$$\varphi_t + u\varphi_x + \varphi_{xx} = 0$$

$$u_x:\ 2\varphi_{xu} - \xi_{xx} + u\tau_t - u\xi_x - \xi_t + \varphi = 0$$

$$u_{xx}:\ \tau_t - 2\xi_x = 0$$

$$u_x^2:\ \varphi_{uu} - 2\xi_{xu} = 0$$

$$u_x u_{xx}:\ \xi_u = 0 \Rightarrow \xi = \xi(x,t) \Rightarrow \xi_{xu} = 0 \Rightarrow \varphi_{uu} = 0$$

$$u_x^3:\ \xi_{uu} = 0 \qquad \Downarrow$$

$$\varphi = \alpha(x,t) + u\beta(x,t).$$

The equation for u_x gives

$$\tau_t - \xi_x + \beta = 0\ ;\ -\xi_t + \alpha + 2\beta_x - \xi_{xx} = 0.$$

We find

$$\xi_{xx} = 0,\ \beta_x = 0,\ \alpha = \xi_t,\ \varphi_x = \alpha_x.$$

The equation $\varphi_t + u\varphi_x + \varphi_{xx} = 0$ becomes

$$\alpha_t + u\beta_t + u\alpha_x + \alpha_{xx} = 0.$$

Equating coefficients of u^0 and u^1 to zero, we find

$$\beta_t + \alpha_x = 0,\ \alpha_t + \alpha_{xx} = 0.$$

Since $\alpha = \xi_t$, $\alpha_{xx} = \xi_{xxt} = 0$, so α is linear in x. Then $\alpha_t = 0$, so that $\xi_{tt} = 0$, and ξ is linear in t. We also found $\xi_{xx} = 0$ so that ξ is linear in x. Using all the relations obtained, we find

$$\alpha = c_4 + c_5 x$$
$$\beta = -(c_3 + c_5 t)$$
$$\varphi = c_4 + c_5 x - (c_3 + c_5 t)u$$
$$\xi = c_1 + c_3 x + (c_4 + c_5 x)t$$
$$\tau = c_2 + 2c_3 t + c_5 t^2.$$

Thus the infinitesimal symmetry algebra of Burgers' equation consists of the five vector fields:

$$\vec{v}_1 = \partial_x\ ;\ \vec{v}_2 = \partial_t\ ;\ \vec{v}_3 = x\partial_x + 2t\partial_t - u\partial_u;$$
$$\vec{v}_4 = t\partial_x + \partial_u\ ;\ \vec{v}_5 = xt\partial_x + t^2\partial_t + (x - tu)\partial_u.$$

The structure constants are given by the Table:

	v_1	v_2	v_3	v_4	v_5
v_1	0	0	v_1	0	v_4
v_2	0	0	$2v_2$	v_1	v_3
v_3	$-v_1$	$-2v_2$	0	v_4	$2v_5$
v_4	0	$-v_1$	$-v_4$	0	0
v_5	$-v_4$	$-v_3$	$-2v_5$	0	0

G_1 and G_2 are translation groups; G_3 gives scale transformations; G_4 is a Galilean group; G_5: $(x,t,u) \longrightarrow \left(\frac{t}{1-\lambda t}, \frac{x}{1-\lambda t}, u + \lambda(x - tu)\right)$; $\lambda \in \mathbb{R}$.

3. The Korteweg-deVries equation is

$$u_t + uu_x + u_{xxx} = 0.$$

So $\varphi^t + u\varphi^x + \varphi u_x + \varphi^{xxx} = 0$ if $u_t + uu_x + u_{xxx} = 0$. Proceeding as in Example 2, we find four vector fields: $\vec{v}_1 = \partial_x$; $\vec{v}_2 = \partial_t$; $\vec{v}_3 = t\partial_x + \partial_u$; $\vec{v}_4 = x\partial_x + 3t\partial_t + 2u\partial_u$,

with the Table :

	v_1	v_2	v_3	v_4
v_1	0	0	0	v_1
v_2	0	0	v_1	$3v_2$
v_3	0	$-v_1$	0	$-2v_3$
v_4	$-v_1$	$-3v_2$	$-2v_3$	0

We note that for both the Burgers and Korteweg-deVries equations ξ and τ do not depend on u, i.e., the independent variables are transformed among themselves, independently of the dependent variables. A symmetry group for which

$$\vec{v} = \sum_{i=1}^{p} \xi^i(x)\partial/\partial x_i + \varphi(x,u)\partial/\partial u$$

is said to be projectable. One finds similarly that the one-dimensional heat equation and the two-dimensional Laplace equation are projectable.

4. An example where the symmetry group is nonprojectable occurs for p=2, k=2, and the equation $u_{xx}=0$, with $\vec{v} = \xi\partial_x + \eta\partial_y + \varphi\partial_u$. One finds that the symmetry group is given by all vector fields of the form

$$(c_1 + c_4 x + c_5 u)\partial_x + c_2\partial_y + (c_3 + c_6 x + (c_7 + 2c_4 x)u)\partial_u,$$

where the c_1 are arbitrary functions of y.

For linear pde's in one dependent variable one can prove that for order $n \geqslant 3$, all symmetries are projectable. An excellent discussion is contained in the thesis O1-1.

References

P.J.Olver. 1. Symmetry Groups of Partial Differential Equations; Thesis, Harvard University, 1976.

2. Symmetry Groups and Group Invariant Solutions of Partial Differential Equations; J.Diff.Geom. 14,497-542 (1979).

3. How to Find the Symmetry Group of a Differential Equation; Appendix to D.H.Sattinger, Group-Theoretic Methods in Bifurcation Theory, Springer,1979.

4. Applications of Lie Groups to Differential Equations, Lecture Notes, Math. Inst.,Oxford,1980.

D.H.Sattinger,Les Symetries des Equations et Leurs Applications dans la Mecanique et la Physique, Lecture Notes, Orsay,1980.

L.V.Ovsyannikov, Group Analysis of Differential Equations, Academic Press,1982 (Russian edition, Nauka, Moscow,1978).

J.M.Hill, Solution of Differential Equations by Means of One-parameter Groups, Research Notes in Math., Pitman,1982.

Bluman and Cole, Similarity Methods for Differential Equations, Springer Applied Math. Sci. #13, 1974.

Group Contractions and the E(2)-like Little Group for Massless Particles as an Infinite-momentum/zero-mass Limit of the O(3)-like Little Group for Massive Particles.

D. Han
Systems and Applied Sciences Corporation, Riverdale, Maryland 20737

Y. S. Kim
Department of Physics and Astronomy, University of Maryland, College Park, Maryland 20742

Marilyn E. Noz
Department of Radiology, New York University, New York, New York 10016

D. Son
Department of Physics, Columbia University, New York, New York 10027

One of the beauties of Einstein's special relativity is the unified description of the energy-momentum relation for massive and massless particles through $E = [P^2 + M^2]^{1/2}$. In addition to mass, energy and momentum, relativistic particles have internal space-time symmetries. Is there then a unified way to describe internal space-time symmetries for both massive and massless paticles? More specifically, can the internal symmetry of massless particles be obtained as a limiting case of that of massive particles?

In 1939, Wigner formulated a method of studying the internal space-time symmetries of massive and massless particles based on the little groups.[1] The little group is a subgroup of the Lorentz group which leaves the four-momentum of a given particle invariant. The little groups for massive and massless particles are locally isomorphic to the three-dimensional rotation group and the two-dimensional Euclidean group respectively, as summarized in Table I.

Table I. Little groups for massive and massless particles.

P: four-momentum	Subgroup of O(3,1)	Subgroup of SL(2,c)
Massive: $P^2 > 0$	O(3)-like subgroup of O(3,1): <u>hadrons</u>	SU(2)-like subgroup of SU(2,c): <u>electrons</u>
Massless: $P^2 = 0$	E(2)-like subgroup of O(3,1): <u>photons</u>	E(2)-like subgroup of SL(2,c): <u>neutrinos</u>

The first step in attacking this problem is to gain a thorough understanding of each of the four cases listed in Table I. The representation suitable for electrons and positrons was discussed in Wigner's original paper. The representations for relativistic extended hadrons in the quark model have also been worked out.[2] The representation suitable for neutrinos has also been worked out.[3] It has been shown in Ref. 3 that the invariance under transformations of the E(2)-like little group is responsible for the polarization of neutrinos.

As for photons, the history is somewhat complicated. The E(2) group is generated by one rotation and two translation operators applicable to functions defined on the two-dimensional xy plane:

$$L_3 = -i\left(x \frac{\partial}{\partial y} - y \frac{\partial}{\partial x}\right), \quad P_1 = -i\frac{\partial}{\partial x}, \quad P_2 = -i\frac{\partial}{\partial y}. \tag{1}$$

These operators satisfy the commutation relation

$$[P_1, P_2] = 0, \quad [L_3, P_1] = iP_2, \quad [L_3, P_2] = -iP_1. \tag{2}$$

We can construct repesentations of this group diagonal in P_1 and P_2, or in L_3. These possibilities have been considered in the past as summarized in Table II.

Table II. Representations of the E(2) group.

Diagonal in	Unitary infinite-dimensional	Non-unitary finite-dimensional
P_1 and P_2	Wigner in 1939[1]	Trivial
L_3	Inonu and Wigner in 1953[4]	Han et al. in 1982[3]

It has been shown in Ref. 3 that the four-vector representation for photons corresponds to a finite-dimensional non-unitary representation of the E(2) group.

In order to obtain a unified description of the little groups for massive and massless particles, we are led to consider the possibility of obtaining E(2) as a limiting case of O(3). This idea is not new. Inonu and Wigner in 1953 introduced the method of group contraction to the physics world and worked out in detail how the infinite-dimensional unitary representation of the E(2) group can be obtained as a limiting case of the spherical harmonics for large values of angular momentum. However, in view of the result obtained in Ref. 3, we have to obtain a finite-dimensional representation of the E(2) group as a contraction of the spherical harmonics of finite angular momentum.

With this point in mind, let us consider the coordinate transformation of the E(2) group. The transformation in which a rotation is followed by a translation takes the form

$$\begin{pmatrix} x' \\ y' \\ 1 \end{pmatrix} = \begin{pmatrix} \cos\phi & -\sin\phi & u \\ \sin\phi & \cos\phi & v \\ 0 & 0 & 1 \end{pmatrix} \begin{pmatrix} x \\ y \\ 1 \end{pmatrix} . \tag{3}$$

The way in which the above three-parameter matrix describes the helicity and gauge degrees of freedom in the four-vector representation of the electromagnetic field has been discussed in detail in Ref. 3.

The three-by-three matrix in Eq.(1) can be exponentiated as

$$D(\phi, u, v) = \exp[-i(uP_1 + vP_2)] \exp(-i\phi L_3) . \tag{4}$$

The generators in this case are

$$L_3 = \begin{pmatrix} 0 & -i & 0 \\ i & 0 & 0 \\ 0 & 0 & 0 \end{pmatrix}, \quad P_1 = \begin{pmatrix} 0 & 0 & i \\ 0 & 0 & 0 \\ 0 & 0 & 0 \end{pmatrix}, \quad P_2 = \begin{pmatrix} 0 & 0 & 0 \\ 0 & 0 & i \\ 0 & 0 & 0 \end{pmatrix} . \tag{5}$$

These generators satisfy the commutation relations given in Eq.(2).

On the other hand, in O(3), three-by-three rotation matrices applicable to coordinate variables (x, y, z) are generated by L_3 of Eq.(5) and

$$L_1 = \begin{pmatrix} 0 & 0 & 0 \\ 0 & 0 & -i \\ 0 & i & 0 \end{pmatrix}, \qquad L_2 = \begin{pmatrix} 0 & 0 & i \\ 0 & 0 & 0 \\ -i & 0 & 0 \end{pmatrix} . \tag{6}$$

Both E(2) and O(3) share the same L_3. The question is how P_1 and P_2 can be obtained from L_1 and L_2. For this purpose, let us consider the surface of a sphere with a large radius, and a small area near the north pole.[5] Then z is very large and is approximately equal to the radius of the sphere R. We can then write

$$\begin{pmatrix} x \\ y \\ z \end{pmatrix} = \begin{pmatrix} 1 & 0 & 0 \\ 0 & 1 & 0 \\ 0 & 0 & R \end{pmatrix} \begin{pmatrix} x \\ y \\ 1 \end{pmatrix} . \tag{7}$$

The column vectors on the left- and right-hand sides are respectively the coordinate vectors on which O(3) and E(2) transformations are applicable.

We shall use A for the three-by-three matrix on the right hand side. Then, in the limit of large R,

$$L_3 = A\, L_3\, A^{-1} \;, \quad P_1 = (1/R)\, A\, L_2\, A^{-1} \;, \quad P_2 = -(1/R)\, A\, L_1\, A^{-1} \;, \tag{8}$$

where L_3, P_1, and P_2 are given in Eq.(5). This limiting procedure is the contraction of the spherical harmonics of $\ell = 1$ to a finite-dimensional representation of E(2).

Let us return to physics. As was noted before, the internal space-time symmetries of free particles are governed by the little groups. If a massive particle is at rest, the symmetry group is generated by the angular momentum operators J_1, J_2 and J_3. If this particle moves along the z direction, J_3 remains invariant, and its eigenvalue is the helicity. However, we have been avoiding in the past the question of what happens to J_1 and J_2, particularly in the infinite-momentum limit where the particle appears massless. Do they transform themselves to accommodate the E(2)-like symmetry for massless particles?

The purpose of this paper is to show that J_1 and J_2 become proportional to the generators of gauge transformations in the infinite-momentum/zero-mass limit. Let us start with a massive particle at rest. The generators of the little group satisfy the O(3)-like commutation relation

$$[J_i, J_j] = i\varepsilon_{ijk}\, J_k \;. \tag{9}$$

If we boost this massive particle along the z direction, its momentum and energy will become P and $E = [P^2 + M^2]^{1/2}$ respectively, and J_3 remains invariant:

$$BJ_3B^{-1} = J_3 \;, \tag{10}$$

where B is the boost operator. However, for J_2 and J_1, we are led to consider

$$\begin{aligned} G_1 &= -\,(M/E)\; BJ_2B^{-1} = -\,J_2 + (P/E)K_1 \;, \\ G_2 &= (M/E)\; BJ_1B^{-1} = J_1 + (P/E)K_2 \;, \end{aligned} \tag{11}$$

where K_1 and K_2 are the boost generators along the x and y directions respectively.

Then the forms given in Eq.(11) are exactly like those in Eq.(8). The B matrix in Eq.(11) is like the A matrix in Eq.(8), and the ratio (M/E) is like (1/R) in Eq.(8) measured in a suitable unit. Now, in terms of the operators J_3, G_1 and G_2, the O(3)-like commutation relations for J_i can be written as

$$[J_3, G_1] = iG_2 \;, \quad [J_3, G_2] = -iG_1 \;, \quad [G_1, G_2] = (M/E)^2\, J_3 \;. \tag{12}$$

In the infinite-momentum/zero-mass limit, the quantity $(M/E)^2$ vanishes, and the G operators become

$$G_1 \rightarrow N_1 , \quad \text{and} \quad G_2 \rightarrow N_2 , \tag{13}$$

where

$$N_1 = K_1 - J_2 , \quad N_2 = K_2 + J_1 , \tag{14}$$

and the O(3)-like commutation relations of Eq.(12) become

$$[N_1, N_2] = 0 , \quad [J_3, N_1] = iN_2 , \quad [J_3, N_2] = -iN_1 . \tag{15}$$

The matrices N_1 and N_2 defined in Eq.(14) together with J_3 form the generators of the E(2)-like little group for massless particles.[1] They satisfy the above commutation relations which are identical to those for the generators of E(2) given in Eq.(2). J_3 is like the generator of rotation while N_1 and N_2 are like the generators of translations. These N operators are known to generate gauge transformations.[3,6]

We have thus shown that rotations around the axes perpendicular to the momentum become gauge transformations in the infinite-momentum/zero-mass limit. An explicit calculation leading to the above-mentioned result has been carried out in the O(3,1) regime where both the O(3)-like little group and the E(2)-like little group are subgroups of O(3,1).[7] It appears to be straightforward to show that the E(2)-like subgroup of SL(2,c) applicable to neutrinos[3] is the same limiting case of SU(2) within the framework of the SL(2,c) formalism of Lorentz transformations.

REFERENCES

1. E. P. Wigner, Ann. Math. 149, 40 (1939); V. Bargmann and E. P. Wigner, Proc. Nat. Acad. Scie. 34, 211 (1948).
2. Y. S. Kim, M. E. Noz, and S. H. Oh, J. Math. Phys. 20, 1341 (1979); D. Han, M. E. Noz, Y. S. Kim, and D. Son, Phys. Rev. D 25, 1740 (1982).
3. D. Han and Y. S. Kim, Am. J. Phys. 49, 348 (1981); D. Han, Y. S. Kim, and D. Son, Phys. Rev. D 26, 3717 (1982).
4. E. Inonu and E. P. Wigner, Proc. Nat. Acad. Sci. 39, 510 (1953).
5. R. Gilmore, Lie Groups and Lie Algebras, Some of Their Applications (John Wiley, New York, 1974).
6. S. Weinberg, Phys. Rev. 134, B882 (1964); 135, B1049 (1964).
7. D. Han, Y. S. Kim, and D. Son, Phys. Lett. B (to be published).

REPRESENTATION APPROACH TO LATTICES OF SUBGROUPS OF SPACE GROUPS

Karel Hršel and Vojtěch Kopský

Institute of Physics, Czechoslovak Academy of Sciences
Na Slovance 2, POB 24, 180 40 Praha 8, Czechoslovakia

The ground theorem of the theory of subgroups of the space groups is the well known theorem by Hermann /1/, a direct consequence of the "diamond isomorphism theorem" /2/, which claims in this context that any subgroup F of a (generally n-dimensional) space group G is an equiclass subgroup of a group H, uniquely defined by $H = F \cup T_G$ as an equitranslational subgroup of G. Lattices of equitranslational subgroups /3/ and tables of maximal equiclass subgroups and of minimal supergroups /4/ provide very useful information about relations between space groups. However, the contemporary stage of crystal physics requires to look for the connection of subgroups of space groups with their representations. The evident areas of application are the phase transition theory, theory of domains and domain walls and, perharps, the theory of incommensurate structures.

The main idea of the representation approach is to determine the epikernels of irreducible representations (ireps) of space groups. These important subgroups, known originally inder the name "co-kernels" /5/, were redefined by Ascher /6/ as stabilizers of vectors of irreducible representation spaces. The well known criterion of phase transition theory - the "chain subduction criterion" /7/ - is just the criterion for a subgroup being an epikernel. The importance of epikernels in the theory of subgroups is based on an important, though somewhat trivial observation, that epikernels generate the whole lattice of subgroups in the sense that any subgroup is an intersection of epikernels /8/.

The investigation of epikernels and of subgroups of space groups should be divided into two parts. We have to distinguish subgroups of the same dimension as the original group from subgroups with dimension deficiency. The first ones are connected with ireps, corresponding to rational vectors of Brillouin zones, the others are connected with irrational k-vectors. Under the dimension of the group we mean here the dimension of its translation subgroup.

The first part actually reduces to investigation of finite groups. Indeed, if G and its subgroup F are of the same dimension, then the index $[T_G:T_F]$ as well as the index of corresponding point groups are finite, so that $[G:F]$ is finite as the product of the first two. Moreover, the group $U = \text{core } F = \bigcap_i F_i$ - the intersection of conjugates F_i in G - is also of finite index in G. Hence the subgroups without dimension deficiency form a sublattice $\mathcal{L}_o(G)$ in the whole lattice $\mathcal{L}(G)$. This sublattice itself is infinite, but every group $F \in \mathcal{L}_o(G)$ can be embedded into a finite fractional sublattice $\mathcal{L}(G/U)$, which is mapped by a homomorphism σ with $\ker \sigma = U = \text{core } F$ onto a finite lattice of a group $\mathcal{U}$, isomorphic with the factor group G/U. Only those ireps of G play the role in the fractional sublattice $\mathcal{L}(G/U)$ which are engendered by ireps of $\mathcal{U}$.

An irep of a space group G is determined by a star $\{k\}$ and by a label α of an allowable irep of the little group G_k (which in turn is a label of a projective representation of the corresponding point group G_k with corresponding factor system). The space $L_k^{(\alpha)}$ splits into direct sum of primitivity spaces $L_{k_i}^{(\alpha)}$. We start the investigation from one of these primitivity spaces for each $\{k\}$, α. The little group G_k contains all elements of G which leave the space $L_k^{(\alpha)}$ invariant. These elements act on $L_k^{(\alpha)}$ according to corresponding allowable ireps of G_k. From this follows a useful

Theorem: Epikernels of allowable ireps of the little group G_k are also epikernels of the corresponding induced ireps of G. All such epikernels have the translation subgroup $T_k = \ker \chi_k(T_G)$.

Corollary: If G_k is symmorphic, then the equiclass subgroup of G_k with translation subgroup T_k is an epikernel, corresponding to the identity irep of the point group G_k.

Generally, we investigate first the allowable ireps of G_k. Then we must investigate pairs, triplets, etc. of primitivity spaces, i.e. such sets of these spaces which transform among themselves under some point group which lies between the point group G_k and the whole G. In this way the epikernels are divided into sets which have the same translation subgroup.

This program has been so far carried to end for the cubic space groups of the class O_h, for the special points of the Brillouin zone. The results include epikernels, corresponding subduction coefficients and stability spaces.

References:

/1/ C. Hermann, Zs. Kristallogr. 69 (1929), 533

/2/ C. McLane, G. Birkhoff, Algebra. McMillan, New York. 1967

/3/ E. Ascher, Battelle Inst. Report. Lattices of Equitranslation Subgroups of Space Groups. (1968)
L.L. Boyle, J.E. Lawrenson, Acta Cryst., A 28 (1972), 485

/4/ J. Neubüser, H. Wondratschek, Minimal subgroups of space groups. Report (1969). Maximal supergroups of space groups. Report (1970)
L.L. Boyle, J.E. Lawrenson, Acta Cryst., A 28 (1972), 489

/5/ R.S. McDowell, J. Molec. Spectr., 17 (1965), 365

/6/ E. Ascher, J. Phys. C.: Solid St. Phys., 10 (1977), 1365

/7/ F.E. Goldrich, J.L. Birman, Phys. Rev., 167 (1968), 528

/8/ V. Kopský, Group Lattices, Subduction of Bases and Fine Domain Structures for the Magnetic Point Groups. Academia, Praha. 1982.
V. Kopský, Czech. J. Phys. B 33 (1983), 485 and in print

YOUNG TABLEAUX FOR THE LIE SUPERALGEBRA OSP(M/N)

J.-P. Hunri, B. Morel, H. Ruegg
Département de Physique Théorique
Université de Genève
1211 Genève 4, Switzerland

A. Sciarrino
Istituto di Fisica Teorica
Napoli, Italy
INFN, Sezione di Napoli, Italy

A. Sorba
Laboratoire de Physique des Particules (LAPP)
Annecy-le-Vieux, France

ABSTRACT

Starting from the Kac-Dynkin diagrams for irreducible representations, we introduce supertableaux.

Irreducible representations (IRs) of superalgebras can be characterized by their highest weight in the root space (Ref.1). The explicitations of the content of the representations in terms of IRs of the bosonic part has been given in Ref. 2 and 3 for, respectively, the SU(M/N) and Osp(M/N) algebras. A description of representations of SU(M/N) in terms of Young (super) tableaux (YST) has been proposed in Ref. 4 and in Ref. 5 the relation between the two approaches has been clarified. The generalization of the YST to the orthosymplectic case is sketched here. An IR of Osp(M/N) looks like a reducible representation of Sp(2n) x O(M). In the following, due to lack of space, we will discuss only the superalgebra B(m,n) (M = 2m+1, m 1) with vectorial representations of O(M). The YTs can characterize the IRs of O(M) and Sp(2n) using the following correspondance between the Dynkin indices (a_i, i = 1,..,n for Sp(2n); a_j, j =1,.. .,m for O(2m+1)) and the labels λ_i (λ_j) - λ_i number of boxes in the i-th row of the YT.

$$(1) \qquad \begin{aligned} a_i &= \lambda_i - \lambda_{i+1} , \quad a_n = \lambda_n \\ a_j &= \lambda_j - \lambda_{j+1} , \quad a_m = 2\lambda_m \end{aligned}$$

The Kac-Dynkin diagram for B(m,n) is:

(2)

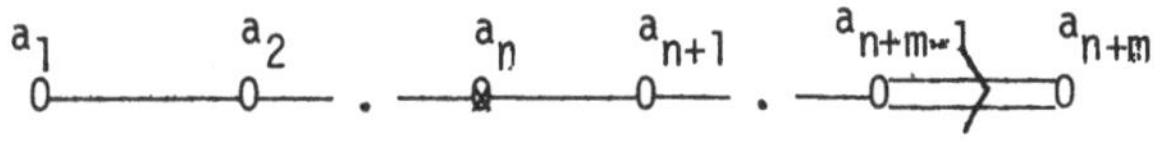

The coordinates of the highest weight of the IR of O(M) can be directly read in the diagram (2), while one of the simple root of Sp(2n) is hidden by the odd root. The value of this component can be computed :

$$(3) \qquad b = a_n - a_{n+1} - \ldots\ldots - a_{n-m+1} - 1/2\, a_{n+m}$$

This implies that a_n has to be integer (a_{n+m} even, vectorial representations) or half-integer (a_{n+m} odd, spinorial representations) and some consistency relations:

$$(4) \qquad b < m \; ; \quad a_{n+b+j} = 0 \qquad (j = 1,\dots,m-b)$$

The remaining IRs of $Sp(2n)\times O(M)$ contained in the IR of $B(m,n)$ can be obtained by the h.w. Λ by repeated application of the negative odd roots β_j^{i-} (see ref. 3). As $B(m,n)$ is a class II superalgebra the distinction between positive and negative roots can be made introducing a $SU(n)\times O(M)$ gradation and, consequently the negative (positive) roots belong to a $IR(n,m)$ $((\bar{n}, m))$ of $SU(n)\times O(M)$.

It may happen that the operator $\prod \beta_j^{i-}$ when applied to Λ has no inverse; in this case the representation is called atypical. The atypicality conditions are given in Ref. (1) and (3). Otherwise the representation is called typical and its dimension can be computed by general formula (see Refs. 1 and 3).

Given a Kac-Dynkin diagram a YST can be defined as it follows:

$$(5) \qquad \begin{aligned} \lambda_i &= b + \sum_{t=i}^{n-1} a_t && (i = 1,\dots,m) \\ \rho_j &= n + 1/2\, a_{n+m} + \sum_{t=j}^{m-1} a_{n+t} && (j = 1,\dots,m) \end{aligned}$$

where λ_i (ρ_j) are the number of boxes in the i-th (j-th) row (column) of the YST. Such a tableau is legal if b is bigger than the highest non vanishing index a_{n+k} $(k \geq 1)$. If $b < m$ one recovers consistency relations (4). The content of IRs of $Sp(2n)\times O(M)$ of the YST $\{\mu\}$ is given by:

$$(6) \quad \{\mu\} = \sum_\sigma \left(L_\sigma \times [\mu], \sum_i P^{2i} \times \tilde{L}_\sigma \right)$$

where: L_σ is a negative generalized YT (see Ref. 6), the sum is over all the possible partitions σ defining YTs and in the product $L_\sigma \times [\mu]$ one has to keep only the positive YTs (considered as $Sp(2n)$ YTs); $\tilde{L}_\sigma$ is the positive trasposed YT of L_σ and it has to be considered as $O(M)$ YT. P^{2i} are the negative GYTs introduced in Ref.7 and in the product $P^{2i} \times \tilde{L}_\sigma$ only positive YTs have to be retained. If a YT in the product $P^{2i} \times \tilde{L}_\sigma$ has columns with $p > m$ boxes one has to replace the column with $M - p$ boxes in order to give a meaning to the YT. YTs obtained with rule and equal YTs obtained by $P^{2j} \times \tilde{L}_\sigma$ $(j > i)$ has to be considered only once. In this decomposition one has to keep only the terms in which they appear legal $Sp(2n)$ and $O(M)$ YTs.
If $\rho_1 > n$ or the IR is atypical not all the terms of the $O(M)$ sector of r.h.s. have to be retained. Details and full discussion will be given elsewhere. Using YST one can perform the direct product of two IRs and the rules are similar to those for the product of $Sp(2n)$ IRs (Ref. 7).
Final remarks:

a) In contrast with $SU(M/N)$ case there is no need to introduce controvariant tensors and, therefore, "mixed YSTs".
b) For $Osp(M/N)$ the use of YSTs seems to provide an useful tool to read off the explicit content of IRs of the bosomic part.

References:

1) V. Kac; Adv. Math. 26, 8 (1977); Commun. Math. Phys. 53, 33 (1977) Lect. notes in Math. Vol. 676 (Springer-Verlag, N.Y., 1978)

2) J.P. Hurni, B.Morel; Jour. Math. Phys. 24, 157 (1983)

3) J.P. Hurni, B.Morel; Jour. Math. Phys. 23, 2236 (1982)

4) B. Balantekin, I.Bars; Jour. Math. Phys. 22, 1149, 1810 (1981)

5) I. Bars, B. Morel, H. Ruegg; Jour. Math. Phys. (1983)

6) G. Girardi, A. Sciarrino, P. Sorba; J. Phys. A 15, 1119 (1982)

7) G. Girardi, A. Sciarrino, P. Sorba; J. Phys. A 16, 2609 (1983)

THE ASSOCIATED LIE ALGEBRA OF $\ddot{x} + f_2\dot{x} + f_1x = f_0$

J. Krause
Universidad Católica de Chile, Casilla 114-D, Santiago, Chile

M. Aguirre
Universidad Católica de Valparaíso, Casilla 4059, Valparaíso, Chile

Recently there has been an increasing interest in studying the symmetry principles involved in classical mechanics[1]. This interest is motivated by the identity of symmetry groups operating in classical and quantum mechanics. Indeed, there is the feeling that this common symmetry structure should be the guide for having a complete quantization procedure ("geometric quantization")[2]. As a consequence, a fundamental problem arises: that of finding the symmetry group associated with a given mechanical system.

In this work we examine some features of this problem by means of the similarity analysis of a single particle's motion in Newtonian mechanics. In similarity analysis one usually tryes to determine the general form of the differential equations (of some required order) which admit a given group as a symmetry group[3]. Here we tackle the converse (and more interesting) problem of finding the Lie group symmetries of any given inhomogeneous ordinary differential equation of the second order: $\ddot{x} + f_2(t)\dot{x} + f_1(t)x = f_0(t)$. To this effect, we consider the symmetries generated by the infinitesimal point transformations: $t' = t + \varepsilon\eta(t,x)$, and $x' = x + \varepsilon\theta(t,x)$, as usual. Thus, we obtain the generators in the following forms:

$$\eta(t,x) = \phi_1(t)x + \phi_2(t) \quad ,$$

$$\theta(t,x) = \{\dot{\phi}_1(t) - f_2(t)\phi_1(t)\}x^2 + \phi_3(t)x + \phi_4(t) \quad ,$$

where $\phi_1,\ldots,\phi_4$ are functions of t, which are determined from a system of linear homogeneous differential equations; i.e.,

$$\ddot{\phi}_1 - f_2\dot{\phi}_1 + (f_1 - \dot{f}_2)\phi_1 = 0 \quad ,$$

$$\dddot{\phi}_2 + (4f_1 - f_2^2 - 2\dot{f}_2)\dot{\phi}_2 + (2\dot{f}_1 - f_2\dot{f}_2 - \ddot{f}_2)\phi_2 = (f_0f_2 - \dot{f}_0)\phi_1 - 3f_0\dot{\phi}_1 \quad ,$$

$$2\dot{\phi}_3 = 3f_0\phi_1 + \ddot{\phi}_2 - f_2\dot{\phi}_2 - \dot{f}_2\phi_2 \quad ,$$

$$\ddot{\phi}_4 + f_2\dot{\phi}_4 + f_1\phi_4 = 2f_0\dot{\phi}_2 + \dot{f}_0\phi_2 - f_0\phi_3 \quad ,$$

wherefrom,

$$\eta(t,x) = q^a\{\phi_{1.a}(t)x + \phi_{2.a}(t)\} ,$$

$$\theta(t,x) = q^a\{(\dot\phi_{1.a}(t) - f_2(t)\phi_{1.a}(t))x^2 + \phi_{3.a}(t)x + \phi_{4.a}(t)\} .$$

The constants of integration q^a, $a=1,\ldots,8$, behave as a set of eight essential parameters of the Lie group. Interesting enough, we have already sufficient information to formally obtain the associated Lie algebra. Indeed, we easily get the expressions

$$f^c_{ab}\eta_c = \left[\eta_a , \eta_{bt}\right] + \left[\theta_a , \eta_{bx}\right] , \quad f^c_{ab}\theta_c = \left[\eta_a , \theta_{bt}\right] + \left[\theta_a , \theta_{bx}\right] ,$$

where the f^c_{ab} denote the structure constants. (The antisymmetrization corresponds to the indices "a" and "b" only.) These are identities which hold for all t; thus let us consider them at t=0. So, in order to represent the algebra, we adopt the following parametrization:

$$q^1 = \eta(0,0) , \quad q^2 = \theta(0,0) , \quad q^3 = \eta_t(0,0) , \quad q^4 = \theta_x(0,0) ,$$

$$q^5 = \eta_x(0,0) , \quad q^6 = \theta_t(0,0) , \quad q^7 = \tfrac{1}{2}\eta_{tt}(0,0) , \quad q^8 = \tfrac{1}{2}\theta_{xx}(0,0) .$$

Therefore, upon substituting from these initial data into the equations above, we obtain the following set of non-zeroth structure constants of the Lie algebra associated with the linear differential equation $\ddot x + f_2(t)\dot x + f_1(t)x = f_0(t)$:

$$f^1_{13} = 1 , \quad f^1_{25} = 1 ; \quad f^2_{16} = 1 , \quad f^2_{24} = 1 ;$$

$$f^3_{17} = 2 , \quad f^3_{25} = f_2(0) , \quad f^3_{28} = 1 , \quad f^3_{56} = -1 ;$$

$$f^4_{13} = -\tfrac{1}{2}f_2(0) , \quad f^4_{15} = \tfrac{3}{2}f_0(0) , \quad f^4_{17} = 1 , \quad f^4_{28} = 2 , \quad f^4_{56} = 1 ;$$

$$f^5_{15} = f_2(0) , \quad f^5_{18} = 1 , \quad f^5_{35} = -1 , \quad f^5_{45} = 1 ;$$

$$f^6_{12} = -f_1(0) + \tfrac{1}{2}\dot f_2(0) , \quad f^6_{13} = 2f_0(0) , \quad f^6_{14} = -f_0(0) , \quad f^6_{16} = -f_2(0) ,$$

$$f^6_{23} = -\tfrac{1}{2}f_2(0) , \quad f^6_{25} = \tfrac{3}{2}f_0(0) , \quad f^6_{27} = 1 , \quad f^6_{36} = 1 , \quad f^6_{46} = -1 ;$$

$$f^7_{13} = -2f_1(0) + \dot f_2(0) + \tfrac{1}{2}f_2^2(0) , \quad f^7_{15} = -f_0(0)f_2(0) , \quad f^7_{18} = -\tfrac{3}{2}f_0(0) ,$$

$$f^7_{25} = -f_1(0) + \tfrac{1}{2}\dot f_2(0) + \tfrac{1}{2}f_2^2(0) , \quad f^7_{28} = \tfrac{1}{2}f_2(0) , \quad f^7_{35} = f_0(0) ,$$

$$f^7_{37} = 1 , \quad f^7_{45} = -\tfrac{1}{2}f_0(0) , \quad f^7_{56} = -\tfrac{1}{2}f_2(0) , \quad f^7_{68} = 1 ;$$

$$f^8_{15} = -f_1(0) + \tfrac{1}{2}\dot f_2(0) , \quad f^8_{35} = \tfrac{1}{2}f_2(0) , \quad f^8_{48} = 1 , \quad f^8_{57} = 1 .$$

It must be observed that the chosen parametrization is precisely the one that brings the infinitesimal operators of the projection group (i.e., $\ddot{x} = 0$) in the standard basis adopted in the current literature[3].

Finally, we wish to mention here the obvious fact that the one-dimensional time-independent Schrödinger equation comes quite directly under the scope of the similarity techniques as presented in this note. Details of the present work (with some miscellaneous applications to Newtonian mechanics)[4], as well as a detailed analysis of the associated Lie algebra, will be published elsewhere.

This work was supported in part (M.A.) by DGI (UCV), through Grant N°123.726/82, and in part (J.K.) by DIUC, through Grant N°46/83.

REFERENCES

[1] K. Mariwalla, Phys. Rep. 20C, 287-362 (1975).

[2] Cf. R.T. Prosser, J. Math. Phys. 24, 548 (1983), and references quoted therein.

[3] G.W. Bluman and J.D. Cole, Similarity methods for differential equations, Springer-Verlag, New York (1974).

[4] M. Aguirre and J. Krause, "Infinitesimal symmetry transformations of some one-dimensional linear systems", preprint, UCV (1983).

THREE-DIMENSIONAL COMMUTATIVE DIAGRAM OF GROUP HOMOMORPHISMS

Marko V. JARIĆ

Department of Physics, Montana State University,
Bozeman, Montana 59717, U.S.A.

It will be shown in this short contribution how Michel's commutative diagram of group homomorphisms (short exact sequences) can be enlarged and imbedded in three dimensions (e.g. on a tetrahedron).

Some twenty years ago Michel[1] presented a planar commutative diagram of short exact sequences of groups applicable to any group and any two of its invariant subgroups. This diagram is very useful in summarizing the information it contains. However, the diagram does not contain all the available[1,2] information and, besides not being completely symmetrical, it contains some obscuring crossings. These shortcomings may be overcome by adding five more short exact sequences to Michel's diagram and by imbedding the whole diagram in three dimensions, on a tetrahedron.

Let G be a group (e.g. a space group) and T and K two of its invariant subgroups (e.g. T is the translation group of a space group and K is the kernel of a representation of G). Then $K\cap T$ and $K\circ T=\{k\circ t, k\in K, t\in T\}$ are also invariant subgroups of G. Furthermore, K and T are invariant subgroups of $K\circ T$ while $K\cap T$ is the invariant subgroup of K,T and $K\circ T$. This leads to the following quotient groups and isomorphisms:

$$Z \equiv T/(K\cap T) = (K\circ T)/K; \tag{1}$$

$$Y \equiv K/(K\cap T) = (K\circ T)/T; \tag{2}$$

$$G/T = (G/(K\cap T))/Z; \tag{3}$$

$$G/K = (G/(K\cap T))/Y; \tag{4}$$

$$(K\circ T)/(K\cap T) = Y\times Z; \tag{5}$$

$$(G/(K\cap T))/(Y\times Z) = (G/K)/Z = (G/T)/Y = G/(K\circ T). \tag{6}$$

Therefore, the following short exact sequences are added to Michel's diagram:

$$1 \to K\cap T \to K\circ T \to Y\times Z \to 1; \tag{7}$$

$$1 \to Y\times Z \to G/(K\cap T) \to G/(K\circ T) \to 1; \tag{8}$$

$$1 \to K \to K\circ T \to Z \to 1; \tag{9}$$

$$1 \to T \to K\circ T \to Y \to 1; \tag{10}$$

$$1 \leftrightarrow Y \leftrightarrow Y\times Z \leftrightarrow Z \leftrightarrow 1. \tag{11}$$

All this information can be summarized in a condensed form by placing Y, Z, $K\cap T$ and $G/(K\circ T)$ at the vertices of a tetrahedron. K is placed half-way along the $K\cap T$,Y edge; T is half-way along the $K\cap T$,Z edge; G/K is half-way along the Z,$G/(K\circ T)$ edge; G/T is half-way along the Y,$G/(K\circ T)$ edge; $Y\times Z$ is half-way along the Y,Z edge. $K\circ T$ is placed at the center of the face Y,$K\cap T$,Z while $G/(K\cap T)$ is at the center of the face Y,$G/(K\circ T)$,Z. The group G is placed at the center of the tetrahedron. Finally, by "drawing" the arrows indicating the exact sequences one arrives at the diagram

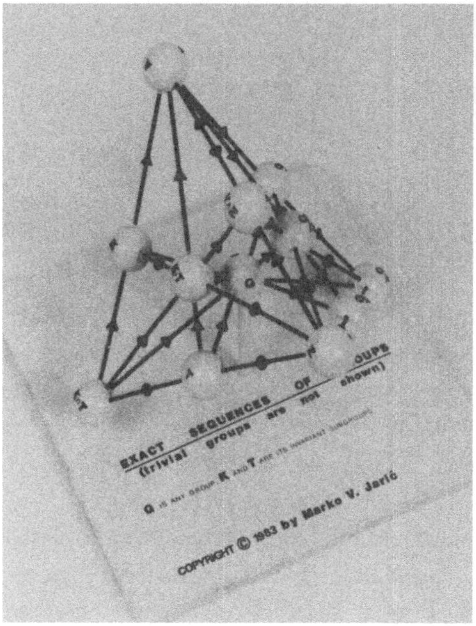

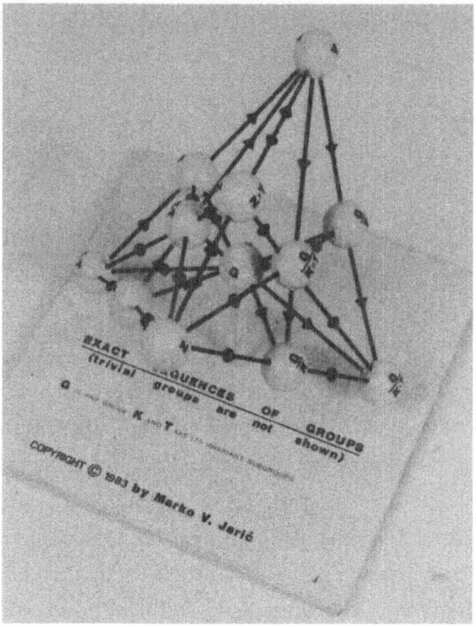

shown in the photographs.

Such a diagram represents a simple way of summarizing the results, eqs.(1) to (6). Thus, it can be used as a mnemonic as well as a pedagogic tool.

The author is grateful to Prof. Rugheimer of the Montana State University who made the holograms of the diagram which were presented at this conference. He also acknowledges the hospitality and financial support from the Einstein Center for Theoretical Physics at the Weizmann Institute of Science, Rehovot, Israel.

References

1. L. Michel, course presented at *l'Ecole d'Eté de Physique Théorique de Cargese - 1965*, preprint no. A81.366(39); L. Michel and J. Mozrzymas, Lec. Notes Phys. 180 (1983).
2. L. Michel, lectures given at *l'Ecole Normale Supérieure* in the spring of 1983 (unpublished).

INDECOMPOSABLE REPRESENTATIONS OF VERMA TYPE

M. Lorente Páramo
Departamento de Métodos Matemáticos de la Física
Facultad de Ciencias Físicas
Universidad Complutense. Madrid-3. (Spain)

1. Universal envelopping algebras of Lie Algebras

For the sake of simplicity, let us take the Lie Algebra A_2 of the unitary group SU(3), the generators of which satisfy the canonical commutation relations:

$$\left.\begin{array}{l} [h_i, h_j] = 0 \\ [h_i, f_{jk}] = \delta_{ij} f_{jk} - \delta_{ik} f_{jk} \\ [f_{ij}, f_{kl}] = \delta_{jk} f_{il} - \delta_{il} \delta_{kj} \end{array}\right\} \qquad i,j,k,l = 1,2,3 \qquad [1]$$

where the h_i form the complete set of commuting generators and f_{ij} are the root vector corresponding to the root $(e_i - e_j)$.

According to the Poincaré-Birchoff-Witt theorem, a basis for the universal envelopping algebra of A_2 can be chosen as the following set of ordered tensor products

$$\Omega = \{ \mathbb{I} \, , \, f_{32}^m \, f_{31}^n \, f_{21}^p \, f_{12}^r \, f_{13}^s \, f_{23}^t \, h_1^u \, h_2^v \, h_3^w \} \qquad [2]$$

where m,n,p,r,s,t,u,v,w are non-negatives integers. This example can be generalized to any Lie Algebra in which the universal envelopping Algebra is defined by ordered tensor product of its generators.

2. Verma modules for semisimple Lie Algebras

We define a basis as the ordered tensor products of negative root vectors

$$\Omega_- = \left\{ \begin{array}{l} f_{32}^r \, f_{31}^s \, f_{21}^t \equiv X(r,s,t) \\ \qquad \mathbb{I} \equiv X(0,0,0) \end{array}\right. \qquad [3]$$

Obviously the space generated by this basis Ω_- is a subspace of Ω.

It can be proved that a representation ρ can be defined on Ω_- with the following conditions:

$$\left.\begin{array}{l} \rho(h_i)\mathbb{I} = \Lambda_i \, \mathbb{I} \quad , \; i = 1,2,3 \, , \quad \Lambda_i \in \mathbb{C} \\ \rho(f_{32})\mathbb{I} = \rho(f_{31})\mathbb{I} = \rho(f_{21})\mathbb{I} = 0 \end{array}\right\} \qquad [4]$$

where the action of the generators on the basis is given as follows:

$$\left.\begin{aligned}
\rho(h_1)\ X\ (r,s,t) &= (\Lambda_1-s-t)\ X\ (r,s,t)\\
\rho(h_2)\ X\ (r,s,t) &= (\Lambda_2-r+t)\ X\ (r,s,t)\\
\rho(h_3)\ X\ (r,s,t) &= (\Lambda_3+r+t)\ X\ (r,s,t)\\
\rho(f_{12})\ X\ (r,s,t) &= t(\Lambda_1-\Lambda_2+1-t)\ X\ (1,s,t-1) - sX(r+1,s-1,t)\\
\rho(f_{23})\ X\ (r,s,t) &= r(\Lambda_2-\Lambda_3+1-r-s+t)\ X\ (r-1,s,t)
\end{aligned}\right\} \qquad [5]$$

and similar expresions for the negative root generators can be calculated with the help of the canonical commutation relations. This type of representations, called "Verma modules" can be generalized to any semisimple Lie Algebra, and it will be denoted by ρ_Λ, where $\Lambda = (\Lambda_1,\Lambda_2,..,\Lambda_l)$ is the highest weight of the representation.

3. Extremal vectors.

An element of the space Ω_- is called "extremal vector" if the following conditions are satisfied:

$$\left.\begin{aligned}
&\rho(f_{12})Y = \rho(f_{23})Y = 0\\
&\rho(h_i)Y = M_iY \qquad (M_1+M_2+M_3 = 0)\quad M_i \in \mathbb{C}
\end{aligned}\right\} \qquad [6]$$

From [5] the simplest solutions are:

i) $t = \Lambda_1-\Lambda_2+1 \quad , r = s = 0 \qquad \Longrightarrow M = (\Lambda_2-1,\Lambda_1+1,\Lambda_3)$ [7]

ii) $r = \Lambda_2-\Lambda_3+1 \quad , s = t = 0 \qquad \Longrightarrow M = (\Lambda_1,\Lambda_3-1,\Lambda_2+1)$ [8]

provided that r,s,t are non negatives.

The vectors Y and the corresponding weight M generate a subspace.

We apply to Y all the basis vectors [3] and take M as the highest weight. The new representation ρ_M is obviously a Verma module which is contained in ρ_Λ, and it is called a Verma submodule. We can construct extremal vectors as before, if we substitue Λ by M. In order that an extremal vector exsists, it is necessary that the weight M should satisfy

$$M = \Lambda + r(e_3-e_2) + s(e_3-e_1) + t(e_2-e_1). \qquad [9]$$

This method can be iterated until all possible "extremal vectors" are exhausted.

4. Indecompossable representations

In order to make clear this construction, let us take the weight $\Lambda = (1,0,-1)$ as the highest weight of a Verma module in A_2. All the extremal vectors and the corresponding weights are given in the following diagram:

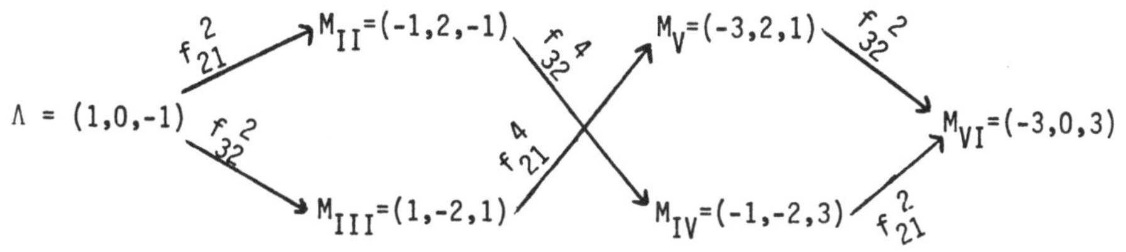

In other words, the vector f_{21}^{2} generate a Verma submodule of highest weight M_{II}, which is spanned by the basis $X(r,s,t)f_{21}^{2}$ (in standard order); f_{32}^{2} generate a Verma submodule of highest weight M_{III} and basis $X(r,s,t)f_{32}^{2}$; f_{32}^{4} generate a Verma submodule of weight M_{IV} and basis $X(r,s,t)f_{32}^{4}\ f_{21}^{2}$;$f_{21}^{4}$ generate a Verma submodule of weight M_V and basis $X(r,s,t)f_{21}^{4}\ f_{32}^{2}$; and finally we have the extremal vector $f_{21}^{2}f_{32}^{4}\ f_{21}^{2} =$ $f_{32}^{2}\ f_{21}^{4}\ f_{32}^{2}$ which generate a Verma submodule of highest weight M_{VI}. Each of the aforementioned basis span an subspace which is invariant under the representation ρ_Λ, therefore this representation is indecompossable.

5. Construction of extremal vector in the general case.

Let L be a semisimple Lie Algebra of rank l, the generators of which H_i, $E_{\pm\alpha}$ satisfy the C.R. A basis can be constructed with the help of negative root generators

$$X(m,n,\dots,r) \equiv \{E_{-\alpha}^{m}\ E_{-p}^{n}\ \dots\ E_{-\delta}^{r}\} \qquad [10]$$

and a Verma module is defined, as before,

$$\left.\begin{aligned} \rho(H_i)\ \mathbb{1} &= \Lambda_i\ \mathbb{1} & i &= 1,2,\dots,l \\ \rho(E_\alpha)\ \mathbb{1} &= 0 & &\alpha,\ \text{simple root of } L \end{aligned}\right\} \qquad [11]$$

The extremal vectors are defined, as before,

$$\left.\begin{aligned} \rho(H_i)Y &= M_i Y & i &= 1,2,\dots,l \\ \rho(E_\alpha)Y &= 0 & &\forall\alpha\ ,\ \text{simple root of } L \end{aligned}\right\} \qquad [12]$$

for any simple root α, it can be proved that

$$Y = E_{-\alpha}^{m} \quad , \quad (m = \frac{2(\Lambda,\alpha)}{(\alpha,\alpha)} + 1) \qquad [13]$$

is extremal vector provided m is non negative. It is easy to proved that the weight $M = (M_1,M_2,\ldots,M_l)$ is obtained from Λ by

$$M = \Lambda - S_{\alpha}(\Lambda+R)-R \qquad [14]$$

where R is the half sum of ale positive roots and S_{α} a Weyl reflection on the root α. Y generates a Verma subsmodule ρ_M with basis $X(m,n,\ldots,r)Y$, which span an invariant subspace under the representation ρ_{Λ}. On each Verma submodule ρ_M we can construct new extremal vectors for all simple roots, and repeat the process until all extremal vectors are exhausted.

6. Is this construction complet?

According to Bernshtein, Gel'fand and Gel'fand[1], given two Verma modules: $\rho_M \subset \rho_{\Lambda}$ iff there exsist positive root $\gamma_1,\gamma_2,\ldots,\gamma_k$, such that:

i) $M + R = S_{\gamma k}\ldots S_{\gamma 1}(\Lambda+R)$, S_{γ} being a Weyl reflection,

ii) $2(S_{\gamma i-i}\ldots S_{\gamma 1}(\Lambda+R), \gamma_i)/(\gamma_i,\gamma_i)$ is a non-negative integer, for $i = 1,2,\ldots k$.

For γ_i a simple root, this theorem gives all the extremal vector describe before. For γ_i non simple root, we can reduce the construction to a product of simple root generators. Suppose, for the sake of simplicity, that there exsist a positive root γ such that (i) $M+R = S_{\gamma}(\Lambda+R)$, (ii) $2(\Lambda+R,\gamma)/(\gamma,\gamma)$ is a non negative integer. Since S_{γ} is always the product of odd number of Weyl reflections on simple roots, say, $S_{\gamma} = S_3S_2S_1$, then, if we define

$$m_1 = \frac{2(\Lambda+R,\alpha_1)}{(\alpha_1,\alpha_1)} , \quad m_2 = \frac{2(S_1(\Lambda+R),\alpha_2)}{(\alpha_2,\alpha_2)} , \quad m_3 = \frac{2(S_2S_1(\Lambda+R),\alpha_3)}{(\alpha_3,\alpha_3)}$$

we can prove that

$$Y = E_{-\alpha_3}^{m_3} E_{-\alpha_2}^{m_2} E_{-\alpha_1}^{m_1} \qquad [15]$$

is an extremal vector with m_1, m_2, m_3 integer numbers and that the corresponding weight

$$M = \Lambda - m_1\alpha_1 - m_2\alpha_2 - m_3\alpha_3 \qquad [16]$$

satisfies conditions i) and ii).

[1] I.N. Bernshtein, I.M. Gel'fand and S.I. Gel'fand, Funk. Anal. Prilozh., 5 (1971) 1-19

SOME RECENT RESULTS ON THE SU(3)$\supset$SO(3) STATE LABELLING PROBLEM

C. Quesne*
Physique Théorique et Mathématique CP 229, Université Libre de Bruxelles,
B 1050 Brussels, Belgium.

Introduction

The specification of a complete, nonredundant set of SU(3)$\supset$ SO(3) basis states is the prototype of labelling problems in group theory. There is one missing label, and hence, a degree of arbitrariness in the definition of states. Since the pioneering work of Elliott [1] , Bargmann and Moshinsky [2] , and Racah [3] , several different methods [4] have been proposed to solve this problem.

The purpose of the present contribution is to report on a recently proposed solution [5] , which clearly exhibits the operation of Littlewood's branching rule for U(3)$\supset$ O(3) [6] , and hence may be termed canonical. Moreover it can be generalized in a straightforward way to provide a solution to the SU(n)$\supset$SO(n) state labelling problem for arbitrary n [7, 8] .

SU(3)$\supset$ SO(3) Branching Rule

Littlewood's branching rule for U(3)$\supset$ O(3) [6] , supplemented with Newell's modification rules [9] , states that the multiplicity of the SO(3) irrep L in the SU(3) irrep $[h_1 h_2]$ is equal to the total multiplicity of the SU(2) irrep $j=(h_1-h_2)/2$ in the product representations $j_L \times j_s$, where j_L is equal to $L/2$ or $(L-1)/2$ according to whether h_1+h_2-L is even or odd, and $j_s=(h_1^s-h_2^s)/2$ (with h_1^s, h_2^s even integers) runs over all integers having the parity of $h_s/2 = [(h_1+h_2-L)/2]$ and satisfying the following inequalities

$$|j-j_L| \leq j_s \leq \min(j+j_L, h_s/2) . \tag{1}$$

We now proceed to show that the highest weight states (HWS) of the equivalent irreps L belonging to $[h_1 h_2]$ can be further specified by j_s and constructed by coupling to total angular momentum j two states respectively characterized by the angular momenta j_L and j_s . For such purpose, it is first necessary to identify the SU(2) group, whose irreps are denoted by j_L, j_s, and j .

The U(3)$\supset$ O(3) and Sp(4,R)$\supset$ U(2) Complementary Chains

It is well known [10] that the bases for the irrep $[h_1 h_2]$ of SU(3) (or of U(3)) can be built as polynomials in six boson creation operators η_{im}, i = 1,2, m = 1,0, -1, acting on the boson vacuum state $|0\rangle$. In terms of these creation operators and the corresponding annihilation operators $\xi^{im} = (\eta_{im})^+$, the generators of U(3) and of its O(3) subgroup are given by

$$C_m^{m'} = \sum_{i=1}^{2} \eta_{im} \xi^{im'}, \qquad m, m' = 1, 0, -1, \tag{2}$$

and

$$L_1 = -(C_1^0 + C_0^{-1}), \quad L_0 = C_1^1 - C_{-1}^{-1}, \quad L_{-1} = C_0^1 + C_{-1}^0, \tag{3}$$

respectively. Let us also consider the chain Sp(4,R) ⊃ U(2), where U(2) is generated by

$$C_{ij} = \sum_{m=-1}^{1} \eta_{im} \xi^{jm}, \tag{4}$$

and Sp(4,R) by

$$D_{ij}^+ = \sum_m (-1)^m \eta_{im} \eta_{i-m}, \quad D_{ij} = \sum_m (-1)^m \xi^{im} \xi^{j-m}, \quad E_{ij} = C_{ij} + \tfrac{3}{2}\delta_{ij}. \tag{5}$$

The groups U(3) and U(2) [10], as well as O(3) and Sp(4,R) [11], are complementary within an irrep of a larger group Sp(12,R) [12]. This means that all the boson states transforming according to a given irrep of one group, and of highest weight with respect to it, belong to a single irrep of the complementary group and that both irreps are in one-to-one correspondence within the irrep of the larger group. Hence we obtain two complementary chains

$$\begin{array}{ccc} U(3) & \supset & O(3) \\ [h_1 h_2] & & L \end{array} \qquad \text{and} \qquad \begin{array}{ccc} Sp(4,R) & \supset & U(2), \\ \langle 3/2, L+3/2 \rangle & & [h_1 h_2] \end{array} \tag{6}$$

where below each one of the groups we have indicated the quantum numbers characterizing its irreps. It is then clear that the state labelling problems for the irreps of the chains given in Eq. (6) are entirely equivalent. In each case there is one missing label which can be chosen to be the same.

Let us now consider the SU(2) subgroup of U(2), whose generators are given by

$$J_+ = C_{12}, \quad J_0 = (C_{11} - C_{22})/2, \quad J_- = C_{21}. \tag{7}$$

Eq. (6) shows that the irreps of this subgroup are characterized by $j = (h_1 - h_2)/2$, so that we have actually identified the SU(2) group we need to derive the canonical solution to the U(3) ⊃ O(3) - or the SU(3) ⊃ SO(3) - state labelling problem. All the HWS $P(\eta_{im})|0\rangle$ of the equivalent O(3) irreps L belonging to the U(3) irrep $[h_1 h_2]$ are solutions of the following system of equations

$$L_0 \, P|0\rangle = L \, P|0\rangle \,, \qquad L_1 \, P|0\rangle = 0 \,, \tag{8a}$$

$$N \, P|0\rangle = h \, P|0\rangle \,, \qquad J_0 \, P|0\rangle = j \, P|0\rangle \,, \qquad J_+ \, P|0\rangle = 0 \,, \tag{8b}$$

where $N = C_{11} + C_{22}$, and $h = h_1 + h_2$.

The L = o Case

The L = o irreps are contained with multiplicity one in all the U(3) irreps characterized by even integers h_1^s, h_2^s (where s stands for scalar), and do not appear in the remaining irreps. In this case, the solution of Eq. (8), where we put L = o and replace h and j by h_s and j_s , is obtained very easily.

The complementarity relationship between O(3) and Sp(4,R) implies that the solutions of Eq. (8a) can be written as $P(D_{ij}^+)|0\rangle$, where $P(D_{ij}^+)$ is some polynomial in the D_{ij}^+ generators. We next impose Eq. (8b) to these solutions. Since with respect to SU(2), the D_{ij}^+ operators form a vector $\underset{\sim}{D}^+$, whose spherical components are equal to $D_1^+ = D_{11}^+$, $D_0^+ = \sqrt{2}\, D_{12}^+$, $D_{-1}^+ = D_{22}^+$, and we have the relation $[N, \underset{\sim}{D}^+] = 2\underset{\sim}{D}^+$, the solution of Eqs. (8a) and (8b) is given, up to some normalization constant, by

$$|h_s j_s m_s ; 0\rangle = P_{h_s j_s m_s}(D_{ij}^+)|0\rangle \propto (\underset{\sim}{D}^+ \cdot \underset{\sim}{D}^+)^{(\frac{1}{2}h_s - j_s)/2} \, \mathcal{Y}_{j_s m_s}(\underset{\sim}{D}^+)|0\rangle \,, \tag{9}$$

where $\mathcal{Y}_{j_s m_s}$ is a solid spherical harmonic, and m_s must be set equal to j_s . For $m_s < j_s$, the states (9) are still scalar under O(3), but they are no more of highest weight with respect to U(2).

The L ≠ o Case for the Irrep [L]

When L ≠ o and $[h_1 h_2] = [L]$,there is no missing label, and it is straightforward to see that the HWS is given by

$$|L j_L m_L ; L\rangle = P_{L j_L m_L}(\eta_{im})|0\rangle \propto (\eta_{11})^{j_L + m_L} (\eta_{21})^{j_L - m_L} |0\rangle \,, \tag{10}$$

where $j_L = L/2$ and m_L must be set equal to j_L . When $m_L < j_L$, the states (10) are still of highest weight with respect to O(3), though not with respect to U(2).

Canonical Basis

Let us now consider the general case corresponding to arbitrary irreps $[h_1 h_2]$ and L . By coupling the polynomials $P_{L j_L m_L}(\eta_{im})$ and $P_{h_s j_s m_s}(D_{ij}^+)$ to total angular momentum $j = (h_1 - h_2)/2$ and projection $m = j$ by means of an ordinary SU(2) Wigner coefficient, and by applying the result to the vacuum state $|0\rangle$, we clearly obtain states characterized by the irreps j of SU(2) and L of O(3), and of highest weight with respect to both groups. They are also specified by the irrep $[h_1 h_2]$ of U(2) - and therefore of U(3) - provided $h_1+h_2=h_s+L$. It was shown in Ref. 5 that when h_1+h_2-L is even, we get in this way a complete set of independent solutions of Eq. (8), which can be distinguished by the additional label j_s :

$$|[h_1h_2]j_sL\rangle \propto \sum_{m_Lm_s} \langle j_L m_L, j_s m_s | j j\rangle P_{Lj_Lm_L}(\eta_{im}) P_{h_1+h_2-L\, j_s m_s}(D^+_{ij})|0\rangle. \quad (11)$$

When h_1+h_2-L is odd, the solutions of Eq. (8) cannot be obtained in this simple way, but can be written as

$$|[h_1h_2]j_sL\rangle \propto \eta_{12,10}\ |[h_1-1,h_2-1]j_s\,L-1\rangle\ , \quad (12)$$

where $\eta_{12,10}=\eta_{11}\eta_{20}-\eta_{10}\eta_{21}$ is the polynomial corresponding to the HWS of the irrep L = 1 contained in $[1^2]$, and $|[h_1-1,h_2-1]j_s\,L-1\rangle$ is given by an expression similar to Eq. (11) with h_1, h_2, and L respectively replaced by h_1-1, h_2-1, and $L-1$.

In both Eqs. (11) and (12), the additional label j_s has the parity of $h_s/2 = [(h_1+h_2-L)/2]$ and satisfies the inequalities (1). We have therefore obtained solutions of Eq. (8) which exhibit the operation of Littlewood's modified branching rule in a transparent way. It is clear from their definition that they are not orthogonal with respect to j_s.

Relation With Bargmann-Moshinsky Basis

Among the various analytic, nonorthogonal SU(3)⊃ SO(3) basis [4], Bargmann-Moshinsky one [2] plays an important part because it is based upon Littlewood's branching rule through the use of the elementary permissible diagram (epd) method [13]. It is therefore interesting to relate it to the canonical basis.

In Bargmann-Moshinsky basis, the additional label is the power q of the HWS polynomial associated with the epd characterized by the irreps $[2^2]$ and L = o [13]. This quantum number is linked to the intermediate angular momentum j_s, used as missing label in the canonical basis, by the relation $j_s=h_s/2-2q$. The Bargmann-Moshinsky HWS can therefore be alternatively specified by j_s and denoted by $|[h_1h_2]j_sL)$, where we use a round bracket to distinguish them from the canonical HWS.

In Ref. 5, we have determined the expansion of the canonical HWS in terms of the Bargmann-Moshinsky ones. We have shown that the transformation matrix is triangular, i.e.,

$$([h_1h_2]j'_s L\,|\,[h_1h_2]j_s L\rangle = 0 \qquad \text{unless} \quad j'_s=(j_s)_{min},(j_s)_{min}+2,\dots,j_s\,,$$

where $(j_s)_{min}=|j-j_L|$ or $|j-j_L|+1$ according to whether h_1-L is even or odd. Moreover, for the nonzero matrix elements, we have obtained a simple expression, including no summation.

Generalization to SU(n) ⊃ SO(n)

The canonical basis corresponding to the decomposition of the U(n) irrep $[h_1 h_2 \ldots h_d]$ into O(n) irreps $(\lambda_1 \lambda_2 \ldots \lambda_d)$ is obtained by considering the chain Sp(2d,R) ⊃ U(d) [7, 8]. The building blocks are now the HWS polynomials associated with the irreps $[h_1^s h_2^s \ldots h_d^s]$ (0) and $[\lambda_1 \lambda_2 \ldots \lambda_d]$ $(\lambda_1 \lambda_2 \ldots \lambda_d)$ respectively. When $d \leq [n/2]$, a complete, nonredundant set of basis states is built [7] by coupling both polynomials to a definite irrep $[h_1 h_2 \ldots h_d]$ by means of a U(d) Wigner coefficient. The operator pattern characterizing the latter [14] provides the right number $d(d-1)/2$ of additional independent labels. When $d > [n/2]$, the modification rules make the construction of the canonical basis more difficult. The procedure to be followed in this case has been detailed in Ref. 8.

References

1. J.P. Elliott, Proc. Roy. Soc. A 245 (1958) 128, 562
2. V. Bargmann and M. Moshinsky, Nucl. Phys. 23 (1961) 177
3. G. Racah, in "Group Theoretical Concepts and Methods in Elementary Particle Physics" ed. F. Gürsey (Gordon and Breach, New York, 1964)
4. M. Moshinsky, J. Patera, R.T. Sharp,and P. Winternitz, Ann. Phys. (N.Y.) 95 (1975) 139
5. C. Quesne, submitted to J. Phys. A : Math. Gen.
6. D.E. Littlewood, "The Theory of Group Characters" (Clarendon, Oxford, 1950)
7. J. Deenen and C. Quesne, J. Phys. A : Math. Gen. 16 (1983) 2095
8. C. Quesne, submitted to J. Phys. A : Math. Gen.
9. M.J. Newell, Proc. Roy. Irish Acad. 54A (1951) 143, 153
10. M. Moshinsky, J. Math. Phys. 4 (1963) 1128
11. M. Moshinsky and C. Quesne, J. Math. Phys. 12 (1971) 1772
12. M. Moshinsky and C. Quesne, J. Math. Phys. 11 (1970) 1631
13. M. Moshinsky and V. Syamala Devi, J. Math. Phys. 10 (1969) 455
14. L.C. Biedenharn, A. Giovannini, and J.D. Louck, J. Math. Phys. 8 (1967) 691

IRREDUCIBLE PROJECTIVE REPRESENTATIONS OF THE GENERALIZED SYMMETRIC GROUPS B_n^m

M.Saeed-ul-Islam,
Department of Mathematics,
Bahauddin Zakariya University,
Multan, Pakistan.

A set of generators and relations for the generalized symmetric group B_n^m is given by:

$$r_1,\dots,r_n : r_i^2 = 1 = r_n^m, i=1,\dots,n-1; (r_i r_{i+1})^3 = 1, i=1,\dots,n-2;$$

$$r_{n-1}r_n)^2 = (r_n r_{n-1})^2 (r_i r_j)^2 = 1, i,j=1,\dots,n, j\neq i,i+1. \text{(see [1])}.$$

If m=2, B_n^2 is the Weyl group of type B_n. Inequivalent irreducible projective representations (henceforth i.p.r.'s) of B_n have been determined by Read [9] with factor set α in [-1,-1,-1]. In this paper, we determine the i.p.r.'s of B_n^m with factor set α in [-1,-1,-1] and $m>2$. Also m must be even (see [2]).

Let P be a projective representation of B_n^m with factor set $\alpha \varepsilon [-1,-1,-1]$ and $P(r_i) = T_i, i=1,\dots,n-1$, then $T_1,T_2,\dots,T_n$ satisfy the following:

$$\text{(A)}\dots\begin{cases} T_i^2 = I, i=1,\dots,n-1; T_n^m = I; T_i T_{i+1} T_i = T_{i+1} T_i T_{i+1}, i=1,\dots,n-2; \\ T_i T_j = -T_j T_i, i,j=1,\dots,n, j\neq i,i+1; (T_{n-1}T_n)^2 = -(T_n T_{n-1})^2 \end{cases}$$

Conversely, if a set of n matrices $T_1,\dots,T_n$ satisfy equations [A] then they generate a projective representation of B_n^m with factor set belonging to the class [-1,-1,-1] (see [2]).

An arbitrary element $\sigma \varepsilon B_n^m$ may be expressed uniquely as the product of disjoint cycles $= \theta_1,\dots,\theta_t$,

where
$$\theta_i = \begin{pmatrix} b_{i1} & \dots & b_{it_i} \\ \xi^{k_{i1}} b_{i2} & \dots & \xi^{k_{it_i}} b_{it_i} \end{pmatrix}$$

$b_{ij} \varepsilon \{1,\dots,n\}$, $k_{ij} \varepsilon \{1,\dots,m\}$ and t_i is the length of the cycles $\theta_i, i=1,\dots,t$.

Define $f(\theta_i) = \sum_{j=1}^{t_i} k_{ij}$ and put $f(\sigma) = \sum_{i=1}^{t} f(\theta_i)$. σ is said to be positive if $f(\sigma)$ is even and negative otherwise.

Let $a_{rs}(\sigma)$ denote the number of cycles θ_i of σ lenght s such that $f(\theta_i) \equiv r$ (mod m), $1\leq r\leq m$, $1\leq s\leq n$. Then mxn matrix $a_{rs}(\sigma)$ is called the type of σ written as type (σ). $\sigma,\sigma_1 \varepsilon B_n^m$ are conjugate if and only if type (σ) = type (σ_1) (see [4])

<u>Definition</u>: Let $(t_1,\dots,t_k)$ be a k-tuple where $t_i \varepsilon \{0,1,\dots,n\}$ and $t_1+\dots+t_k = n$. We shall call $(t_1,\dots,t_k)$ a permissible k-tuple. Define $P_0=0$ and $P_i = \sum_{j=1}^{i} t_i$ $i=1,\dots,k$. Let B_n^m be the generalized symmetric group on the t_i symbols

$$P_i = \{p_{i-1}+1,\dots,p_i\}, \; i = 1,\dots,k.$$

Denote the direct product $B_{t_1} \times B_{t_2} \times \dots \times B_{t_k}$ by $B_{(t_1,\dots t_k)}$. $B^m_{(t_1,\dots,t_k)}$ is called a generalized Young subgroup on the symbols $1,\dots,n$ (see [6]).

<u>Lemma:</u> (See [7]). Let $\alpha \varepsilon [-1,-1,-1]$. Then $\sigma \varepsilon B_n^m$ is α-regular if and only if

(i) $\sigma=\tau_1\ldots\tau_r\nu_1\ldots\nu_s$ where all ν_i are disjoint even and positive cycles and all τ_j are disjoint odd and negative cycles,or

(ii) (only if n is odd), $\sigma=\tau_1\ldots\tau_r$ where all τ_j are disjoint negative cycles.

Lemma: Let $\alpha\ \varepsilon[-1,-1,-1]$. The number of α-regular classes of B_n^m is given by : $p(t_1)\ldots p(t_{\frac{1}{2}m})$ if n is even and $2p(t_1)\ldots p(t_{\frac{1}{2}m})$ if n is odd, where the summation is is taken over all permissible $\frac{1}{2}$m-tuples $(t_1,\ldots,t_{\frac{1}{2}m})$ and $p(t_i)$ denotes the number of partitions of t_i, $i=1,\ldots,\frac{1}{2}m$.

Proof: The proof may be established by a one-to-one correspondence $(a_{rs})\leftrightarrow(\pi(t_1),\ldots,\pi(t_{\frac{1}{2}m}))$ if n is even and a two-to-one correspondence $\begin{matrix}(a_{rs})\\(a'_{rs})\end{matrix}\Big\rangle\ (\pi(t_1),\ldots,\pi(t_{\frac{1}{2}m}))$ if n is odd; where

$$a_{rs}=\begin{cases}b_{\frac{r+1}{2},s} & \text{if r is odd and s is even}\\ b_{\frac{r}{2},s} & \text{if r is even and s is odd}\\ 0 & \text{otherwise}\end{cases}\qquad a'_{rs}=\begin{cases}b_{\frac{r+1}{2},s} & \text{if r is odd}\\ 0 & \text{otherwise}\end{cases}$$

and $(\pi(t_1),\ldots,\pi(t_{\frac{1}{2}m}))\equiv((1^{b_{11}}\,2^{b_{12}}\ldots),\ldots,(1^{b_{\frac{1}{2}m,1}}\,2^{b_{\frac{1}{2}m,2}}\ldots))$ is a $\frac{1}{2}$m-partition of n.

Definition: Let $\{N_1,\ldots,N_{2k+1}\}$ be the set of matrices of degree 2^k, as defined in Read[8],for $k=[\frac{1}{2}n]$. Define $T_i=\frac{1}{2}(N_i-N_{i+1}),i=1,\ldots,n-1$, and $T_n=N_n$. Then $\{T_i:i=1,\ldots,n\}$ satisfy equations(A) and therefore generate a projective representation U of B_n^m with factor set $\alpha\ \varepsilon[-1,-1,-1]$. We call U the <u>basic projective representation</u> of B_n^m.

Definition:Let k≤m and $\sigma=\sigma_1\ldots\sigma_k\varepsilon B^m_{(t_1,\ldots,t_k)}$ where $\sigma_i\ \varepsilon B^m_{t_i}$,$i=1,\ldots,k$. Define

$$\chi_{(t_1,\ldots,t_k)}(\sigma)=\xi^z,z=\sum_{i=1}^{k} i\times f(\sigma_i)$$

where ξ is some primitive mth root of unity. Then $\chi_{(t_1,\ldots,t_k)}$ is an irreducible linear representation(i.l.r.) of $B^m_{(t_1,\ldots,t_k)}$ and we call it the basic linear representation of $B^m_{(t_1,\ldots,t_k)}$.

We now give our main result:

Theorem: Let U be the basic projective representation of B_n^m with factor set in [-1,-1, -1]. If $(t_1,\ldots,t_{\frac{1}{2}m})$ is a permissible $\frac{1}{2}$m-tuple, let $\chi_{(t_1,\ldots,t_{\frac{1}{2}m})}$ be the basic linear representation of $B^m_{(t_1,\ldots,t_{\frac{1}{2}m})}$ as above. Let $\mathbb{P}$ be an i.l.r. of $B^m_{(t_1,\ldots,t_{\frac{1}{2}m})}$ lifted from an i.l.r. P of $S_{(t_1,\ldots,t_{\frac{1}{2}m})}$. Then a full set of i.p.r.s' of B_n^m is given by:

$$\{[\chi_{(t_1,\ldots,t_{\frac{1}{2}m})}\otimes(U\downarrow B^m_{(t_1,\ldots,t_{\frac{1}{2}m})})\otimes\mathbb{P}]\uparrow B_n^m\}\quad\text{if n is even, and}$$

$$\{[\chi_{(t_1,\ldots,t_{\frac{1}{2}m})}\hat{\otimes}(U\downarrow B^m_{(t_1,\ldots,t_{\frac{1}{2}m})})\otimes\mathbb{P}]\uparrow B_n^m\}\ \bigcup\{(-1)^{g(\sigma)}\otimes[(\chi_{(t_1,\ldots,t_{\frac{1}{2}m})}\otimes(U\downarrow B^m_{(t_1,\ldots,t_{\frac{1}{2}m})})\otimes\mathbb{P})\uparrow B_n^m]\}\quad\text{if n is odd; where}$$

where P ranges over all inequivalent i.l.r.s' of $S_{(t_1,\ldots,t_{\frac{1}{2}m})}$, $(t_1,\ldots,t_{\frac{1}{2}m})$ ranges over all permissible $\frac{1}{2}$m-tuples and g is defined by:

$$g(\sigma) = \begin{cases} 0 \text{ if } \sigma \text{ is the product of an even number of generators} \\ \quad \{ r_1,\ldots,r_n \} \\ 1 \text{ otherwise.} \end{cases}$$

Proof: Since the number of i.p.r.s' in the set is equal to the number of α-regular classes in B_n^m ,we need only prove that these are inequivalent and irreducible representations of B_n^m.This can be easily acheived by considering the inner products of different members of the set.

References:

[1] H.S.M.Coxeter,The abstract groups $R^m = S^m = (R^j S^j)^{pj} = 1$, $S^m = T^2 = (S^{-j}T)^{2pj} = 1$ and $S^m = T^2 = (S^{-j}TS^jT)^{pj} = 1$,Proc. Lond. Math. Soc.,41 (1938) 278-301.

[2] M.Saeed-ul-Islam,On the second cohomology group of the generalized symmetric group B_n^m,J. of pure and applied sciences,Vol.1,(1982,53-56.

[3] M.Saeed-ul-Islam, The irreducible projective representations of finite imprimitive unitary reflection groups G(m,p,n),Ph.D.Thesis(1980),U.C.W.Aberystwyth,U.K.

[4] A.Kerber, Representations of Permutation Groups I, Lecture Notes in Mathematics, No.240,Springer-Verlag,Berlin/N.Y.,1971.

[5] A.O.Morris, Projective representations of finite groups,Proc. of the Conference on Clifford Algebra and its Generalizations and applications,Matscience,Madras, 1971,(1972),43-86.

[6] M.B.Puttaswamaiah,Unitary representations of generalized symmetric groups,Canad. J.of Mathematics,21(1969) 28-38.

[7] E.W.Read,The α-regular classes of the generalized symmetric group, Glasgow Math. J.Vol.17, part 2, (1976) 144-150.

[8] E.W.Read,On projective representations of the finite reflection groups of type B_ℓ D_ℓ,J. of Lond. Math. Soc.(2), 10(1975,129-142.

INDECOMPOSABLE REPRESENTATIONS OF SOME GRADED LIE ALGEBRAS*

T. S. Santhanam**
Physikalisches Institut
der Universität Bonn,
Federal Republic of Germany

Indecomposable representations of some Graded Lie algebras like the algebra of parabose oscillators and $Gs\ell(2)$ are studied.

1. Introduction

The representation theory of graded Lie algebras (also called Lie superalgebras) has been widely used in supersymmetry. A complete classification of simple Lie superalgebras has already been given [1]. Many results on finite dimensional representations have been derived, but relatively few on infinite dimensional representations have been worked out. We present here two such studies, one on the algebra of parabose oscillators [2] and the $Gs\ell(2)$, Z_2-graded Lie algebra of $s\ell(2)$ [3]. We explicity obtain the indecomposable (not fully reducible) representations in these cases. The merit of this study lies in the fact that the usual irreducible representations are intimately tied up with these indecomposable representations in the sense that they are either induced on quotient spaces or subduced on invariant subspaces of these indecomposable representations.

*Invited seminar at XII International colloquium on Group Theoretical Methods in Physics, Sep 5-10, 1983

**Permanent Address: MATSCIENCE, Madras - 600 113, India

2. Parabose oscillators

The algebra of parabose oscillators is a prototype version of a graded Lie algebra. It is defined [4] as by

$$[a, N] = a,$$
$$[N, a^+] = a^+,$$

and

$$N = \frac{1}{2}\{a, a^+\} \equiv \frac{1}{2}(aa^+ + a^+a) \qquad (2.1)$$

where the operator N has a nonnegative infinite dimensional spectrum. The creation operator a^+ and the annihilation operator a do not satisfy the commutation relation of the usual harmonic oscillator. The normal Fock representation has been obtained earlier [5] by recognizing that $\frac{1}{2}a^{+2}$, $\frac{1}{2}a^2$ and $\frac{1}{2}$ N close under commutation and the algebra is that of the Lorentz group SO(2,1). By using the standard representation of this algebra and extracting the square roots, one obtains for a and a^+ the following representation

$$N|n\rangle = \frac{1}{2}(aa^+ + a^+a)|n\rangle = n + \frac{p}{2},$$
$$[a,a^+] = 1 + (p-1)(-1)^n, \quad p = 0,1,2,\ldots$$
$$\langle 2n|a|2n+1\rangle = \sqrt{2n+p},$$
$$\langle 2n-1|a|2n\rangle = \sqrt{2n} \qquad (2.2)$$

As is seen, for p (referred to as the order of the statistics) = 1, the distinction between the odd and even states disappear and we recover the standard harmonic oscillator.

Now we study the algebra Eq.(2.1) in all generality. We follow the procedure used by Gruber and Klimyk for the Lie algebras [6]. We choose a basis Ω for the universal enveloping algebra of the parabose oscillators as

$$\Omega : \{a^{+n} a^m N^r, \ n, m, r, = 0, 1, 2, \ldots\} \equiv X(n, m, r),$$
$$X(0, 0, 0) = I \qquad (2.3)$$

where X(n,m,r) is an ordered (tensor) product of the elements a^+, a and N. An element y ε Ω is called an extremal vector for the representation ρ on Ω if

$$\rho(\beta)y = 0, \quad \beta = a \text{ or } a^+ \ldots \qquad (2.4)$$

Actually, the degeneracy of the solutions of Eq.(2.4) is responsible for indecomposable representations. By repeated use of the relations Eq.(2.1) we get the following basic master representation

$$\rho(a^+)\, X(n, m, r) = X(n+1, m, r), \tag{2.5}$$

$$\rho(a)\, X(2k, m, r) = 2k\, X(2k-1, m, r) + X(2k, m+1, r), \tag{2.6}$$

$$\rho(a)\, X(2k+1, m, r) = 2(k-m)\, X(2k, m, r) + 2X(2k, m, r+1) - X(2k+1, m+1, r), \tag{2.7}$$

$$\rho(N)\, X(n, m, r) = (n-m)\, X(n, m, r) + X(n, m\ r+1) \tag{2.8}$$

This representaion is in general infinite dimensional, with neither a highest nor a lowest weight. It is seen that the action of ρ is to keep m and r the same or to increase them. Thus, each of the subspaces $V(m,r) \in \Omega$,

$$V(m, r) : \{\ X(n, m+k, r+k_2),\ \ k_1, k_2 \text{ are non-negative integers } \} \tag{2.9}$$

is an invariant subspace with respect to the action of ρ on Ω, which subduces representations on these unvariant subspaces and unduces representations on the quotient spaces. The basis Ω can be written as [7]

$$\Omega = \Omega_+ \Omega_- \Omega_N , \tag{2.10}$$

where $\Omega_+ = a^{+n}$, $\Omega_- = a^m$ and $\Omega_N = N^r$.

The representation on $\Omega_+\Omega_-$ can be obtained from Eqs.(2.5) - (2.8) by setting (N-diagonal)

$$\rho(N)\, I = \lambda I\ \ , \lambda \in C \tag{2.11}$$

In fact, for m=0, (i.e. $\Omega \sim \Omega_+$), they reduce to the known ones of Eq.(2.2). Notice, however, that we neither take any square root nor explicitly use SO(2,1). These are automatically defined on the enveloping algebra. To exhibit the indecomposable representations, let us consider the very special case for which m= 0 and 1 and r=0. The master representation for this case becomes [8]

$$\begin{aligned}
\rho(a^+)\ X(n, i) &= X(n+1, i), \quad i = 0,1 , \\
\rho(a)\ X(2k, 0) &= 2k\ X(2k-1, 0) + X(2k, 1) , \\
\rho(a)\ X(2k, 1) &= 2k\ X(2k-1, 1) , \\
\rho(a)\ X(2k+1,0) &= 2(k+\lambda)\ X(2k, 0) - X(2k+1, 1) , \\
\rho(a)\ X(2k+1,1) &= 2(k+\lambda-1)\ X(2k, 1) , \\
\rho(N)\ X(n, 0) &= (\lambda+n)\ X(n, 0) , \\
\rho(N)\ X(n, 1) &= (\lambda+n-1)\ X(n, 1)
\end{aligned} \tag{2.12}$$

It follows that besides

$$\rho(a)\ X(0, 1) = 0 \tag{2.13}$$

we have

$$\rho(a)\ Z \equiv \rho(a)\ \{ X(0,0) - \frac{1}{2(\lambda-1)}\ X(1,1) \} = 0 \tag{2.14}$$

The representation is reducible to

$$a^{+^n}\ \{ X(0, 1) \} + a^{+^n} Z \tag{2.15}$$

However, for $\lambda = 1$, the representation becomes truely indecomposable, since Z is undefined, i.e. X(0,1) can not be reached from X(1,1) while X(1,1) can be reached from X(0,1). Similar representations can be worked out for arbitrary m, which amounts to giving an additional degree of freedom (taking values 0,1,2,...m) besides n. This fact is of great interest in gauge theories where the vacuum becomes degenerate with respect to another degree of freedom. The method is quite general that one can discuss the representations of coherent states for which

$\rho(a) I = \mu I \ , \ \mu \varepsilon C.$

3. Graded Lie algebra $Gs\ell(2)$

This algebra is defined by

$$[\ell_3, \ell_\pm] = \pm\ell_\pm , \quad [\ell_+, \ell_-] = \ell_3$$

and

$$[\ell_\pm, V_\pm] = 0, \quad [\ell_\pm , V_\mp] = \frac{1}{\sqrt{2}} V_\pm , \quad [\ell_3, V_\pm] = \pm\frac{1}{2} V_\pm \tag{3.1}$$

where the odd elements $(V_\pm)$ are defined by

$$V_\pm^2 = \frac{1}{\sqrt{2}} \ell_\pm \tag{3.2}$$

A basis for the universal enveloping algebra Ω of Gsℓ(2) can be chosen as

$$\Omega \ : \ \{ I \ , \ V_+^i \ V_-^k \ \ell_+^r \ \ell_-^s \ \ell_3^t ; \quad i,k = 0,1 \quad r,s,t = \text{non-negativ integers} \} \tag{3.3.}$$

By straightforward induction one can get the master representation. We shall just quote the results for the special case for which

$$\rho(\ell_3) \ I = \Lambda I \ , \quad \Lambda \varepsilon C$$

and when $s = 0$.
It can be shown that the extremal vectors which satisfy

$$\rho(V_-)Y = 0, \quad \rho\,(\ell_-)Y = 0 \tag{3.4}$$

exist for $r = 0$ and $r = -2\Lambda+1$. In the former case, two extremal vectors

$$Y_1 = (1 + \frac{1}{\Lambda - \frac{1}{2}} V_+V_-), \quad \Lambda \neq \frac{1}{2}$$

and

$$\bar{Y}_1 = V_- \ , \tag{3.5}$$

with the corresponding weights Λ and $\Lambda - \frac{1}{2}$ respectively are obtained. For the latter case, we get two additional extremal vectors

$$Y_2 = (1 - \frac{1}{\Lambda - \frac{1}{2}} V_+V_-)\ell_+^{-2\Lambda+1}$$

$$\bar{Y}_2 = \bar{Y}_1 \ \ell_+^{-2\Lambda+1} \ , \tag{3.6}$$

which correspond to the weights $-\Lambda+1$ and $-\Lambda+\frac{1}{2}$ respectively. In the former case ($r=0$) the representaion decomposes into a direct sum of two infinite dimensional irreducible representations. If, $-2\Lambda+1 = r$, then the representation decomposes into the direct sum of two indecomposable representations which are infinite dimensional. These have $(Y_1,\bar{Y}_2)$ and $(Y_2,\bar{Y}_1)$ as extremal vectors respectively. They induce in the quotient space a finite dimensional irreducible representation of dimension $(-4\Lambda+1)$ in the former case and $(-4\Lambda+3)$ in the latter case. The two cases are related by the substitution $\Lambda \to \Lambda - \frac{1}{2}$.

I am grateful to Professor Bruno Gruber from whom I learnt about the indecomposable representations. I had profitable discussions with Prof. E. C. G. Sudarshan during my brief visit to Austin, Texas. The material presented here has been worked during my stay in the Physikalisches Institut, Bonn and I am very grateful to Professors G. von Gehlen and K. Dietz for their gracious hospitality.

References

1. V. Kac, Advances in Math. 26, 8 (1977);
P. G. O. Freund and I. Kaplansky, J. Math. Phys. 17, 228 (1976);
M. Scheunert, W. Nahm and V. Rittenberg, J. Math. Phys. 17, 1626 (1976).

2. B. Gruber and T. S. Santhanam, J. Math. Phys. 24, 1032 (1983).

3. B. Gruber, T. S. Santhanam and R. Wilson, (preprint).

4. H. S. Green, Phys. Rev. 90, 270 (1953).

5. T. F. Jordan, N. Mukunda and S. V. Papper, J. Math. Phys. 4, 1089 (1963);
L. O. Raifeartaigh and C. Ryan, Proc. Roy. Ir. Acad. Sec A 62, 93 (1963).

6. B. Gruber and A. V. Klimyk, J. Math. Phys. 16, 1816 (1975).

7. N. Jacobson, Lie Algebras (Interscience, New York, 1962).

8. This has also been worked out by E. C. G. Sudarshan (unpublished) and private communication.

STEPHEN PANEITZ: A brief appreciation

I. E. Segal

1. He left us suddenly, not quite 4 days ago, so it is too early to try to sum up his life and work in any definitive way. He was on the program for today, and always gave the highest priority to keeping his commitments, and so would want to have his work presented as scheduled. I am moreover rather sure that he would consider the best memorial to be our dedication to the high ideals, intellectual and social, that he held and actively exemplified, in his remarkably short but germinal life.

Thus I will try to give his lecture for him on the material he designated. It will however not be nearly as informative and precise as it would have been, had he been here to do it. He was unsurpassed among my 40 odd Ph. D. students, a number of whom now have outstanding international reputations, on the comparative basis of their early scientific works. He had a combination of cultural breadth, a steel-trap mind, and a dedication to fundamental physical applications of the higher mathematics that was truly exceptional and most effective.

His productivity in the 28 years he had was remarkable for finished work and even more for new directions set forth. It involves much more than group theory and its applications,- for example, nonlinear wave equations, infinite-dimensional differential geometry, stability and operator theory; briefly, whatever the physical context required. He hardly had time to write up, even in several unpublished but careful working papers, many of his interesting ideas or results. It will have to suffice here to describe some very interesting fragments of his group-theoretical work that he presented with so little fanfare that they could be easily overlooked.

2. He developed a quite explicit relation between causality and invariance in wave equations that provides a kind of theoretical explanation of the apparent primacy of fields of spins 0, $\frac{1}{2}$, and 1 in Nature.

Let **C** denote the totality of all conformal vector fields on Minkowski space $\mathbf{M}_0$ that displace every point into a future direction. Then **C** is a closed convex cone in the Lie algebra **L** of all conformal vector fields on $\mathbf{M}_0$. Moreover, **C** is conformally invariant, i.e. invariant under the adjoint representation of the conformal group **G**. There are singularities in the action of **G** on $\mathbf{M}_0$, but they are not present in its infinitesimal action, so that the vector fields in **C** are quite smooth. Let **C'** denote the smallest closed invariant subcone of **C** that contains the Minkowski energy operator $\partial/\partial x_0$ (where x_0, x_1, x_2 x_3 denote the usual coordinates in $\mathbf{M}_0$). Then

1°. **C'** is the minimal conformally invariant closed convex cone in **C**.

2°. For each spin s, there is a closed convex cone **C**(s) of positivity for the energies corresponding to vectors in the cone.

In other terms, to each generator X of **G** there corresponds, relative to a given conformally invariant wave equation, a function E(X) from the solution variety of the equation to the real numbers by virtue of Noether's theorem. (Alternatively, there is a corresponding self-adjoint operator in the Hilbert space of normalizable solutions of the conformally invariant wave equation of spin s as formulated by Bargmann and Wigner.) **C**(s) is then defined as the totality of those X for which E(X) takes on only non-negative values (or alternatively, for which the corresponding self-adjoint operator is non-negative).

Conclusion: **C**(s) is all of **C** if and only if $s = 0, \frac{1}{2}$, or 1. Moreover, as s increases, **C**(s) shrinks to the minimal cone **C'**.

A future direction in **C** that is represented by an indefinite energy is physically interpretable as a type of tachyon: 'observers' moving in these directions would experience instability in accordance with their indefinite energy. Since every future direction,- as represented by a future-directed conformal vector field,- is a conceptually reasonable one for a physical observer, and since the essence of relativity is that different observers should see nature in coherent ways, this theorem carries the philosophical implication that higher spin fields may be inappropriate models for fundamental empirical entities.

A counter argument to this interpretation is that the fundamental dynamical equations will be nonlinear, rather than linear as in the case of the Bargmann-Wigner higher spin equations. However, essentially the same applies to conformally invariant nonlinear equations. Thus, for the Yang-Mills equation, or the related scalar equation $\Box\phi + \phi^3 = 0$, there are classically no tachyons, in the sense of evolutionary directions, or rays in **C**, for which the corresponding classical energy is indefinite. The corresponding quantized result can only be valid in the present partially unsettled state of quantum field theory on a formal basis. It is reasonable on this basis to the extent at least of formal semi-boundedness of the corresponding energies.

More mathematically speaking, the association of positive functionals having non-trivial invariance features has played a fundamental role in the global treatment of evolutionary partial differential equations. However, apart from the energy itself, such functionals have been found or developed largely on an ad hoc or trial basis. Paneitz' theory provides a beginning to the rational classification of some of the most important of such functionals.

To show a different facet of Paneitz' group-theoretical range, I will describe work having to do with nonlinear Lie groups, illustrative of his technical power. As little as 50 years ago, nonlinear Lie groups apepared as somewhat exotic objects; in a paper by Cartan around that time he cited the universal cover of SL(2,R). This is now a familiar object in principle, but as an infinite covering and lacking any faithful finite-dimensional representation, it is constructed by a topological procedure that is far from explicit. Physically important groups having similar features are SO(2,3) and SO(2,4).

It puzzled me for a long time that there was no analytically explicit expression for the universal cover in even the simplest nontrivial cases. Thus SL(2,R) can be considered as the projective group, of which the real projective line S^1 is a homogeneous space. Its universal cover therefore acts correspondingly on the universal cover of S^1, i.e. R^1. It therefore follows from general theory that if X,Y, Z form a basis for the Lie algebra of SL(2,R), and if q is any real number, there exists a well-defined analytic function F(q;a,b,c) giving the action on the element q of R^1 of the element of the universal cover: $e^{aX}e^{bY}e^{cZ}$, where a, b, and c are arbitrary real numbers.

I had brought this puzzlement up in earlier years with various by now well known group theorists, to no avail. Recently I mentioned it again in a casual conversation at which Paneitz was present. Less than a week later, he mentioned, in a somewhat incidental way, that he had found a closed analytic expression for F(q;a,b,c). Moreover, quite contrary to my expectation, which was that a non-elementary transcendental function was involved in this function, only the standard elementary functions were required in his expression. As always, there was no visible pride or strain in his presentation of his results; unfailingly considerate and modest as he was in manner, he was probably too intelligent not to be aware of his remarkable talent. In the cases of the groups SO(2,3) and SO(2,4) the same method applied, and he determined the most interesting part of the action explicitly, but it would have been arduous to give the action for the entire universal covers.

3.Paneitz never sought to excuse his work from the strictest canons of mathematics on the basis of its potential or actual physical relevance. His work was invariably in accordance with contemporary mathematical ideas and standards, and presented in a quite direct and complete although succinct manner.

On the other hand, he was quite deliberate in seeking to develope mathematics that would eventually impact on fundamental physical theory. Thus the work that he planned to present here originated from chronometric physics, or more specifically, its application at the ultra-microscopic level. However the chronometric theory may ultimately relate to experiment and observations, Paneitz' work may appear more generally intelligible when viewed from the vantage point of this conceptually simple and natural theory, and may also be thereby potentially more adaptable. Accordingly, before giving his scheduled lecture per se,- which for completeness and brevity would have stressed detailed mathematical results,- I shall sketch briefly the main ideas of chronometric theory.

The chronometric theory asserts, briefly, that physical space-time is this 4-dimensional Lorentzian analogue to the 2-sphere, say **M**, which is in the small is causally identical to Minkowski space $\mathbf{M}_0$, and which in particle observations is seen as if it were group-theoretically $\mathbf{M}_0$. This space-time **M** is the maximal one to which the Maxwell, Yang-Mills, and similar conformally covariant equations canonically extend, without singularities in their conformal transformation properties. It is conformally equivalent to the Einstin Universe $R^1 \times S^3$ but has no distinguished metric; rather it has an 8-parameter family of conformally equivalent metrics, each of which is an Einstein Universe metric for a corresponding separation of space-time into time and space components.

In group-invariant terms, **M** is describable as a homogeneous space of the universal cover of the local group of causality-preserving transformations at any given point of $\mathbf{M}_0$; more specifically this group **G** (the universal cover of SU(2,2)) modulo the subgroup **P**, describable as the scale-extended universal cover of the Poincare group. Canonically contained in **G** is the universal cover $\mathbf{P}_0$ of the Poincare group, which modulo its homogeneous subgroup is $\mathbf{M}_0$, from which the canonical imbedding of $\mathbf{M}_0$ into **M** is invariantly definable.

M is endowed ab initio only with a causal structure, not a metric one, and is to be thought of as empty space-time. The Einstein metric (of which as noted there are an 8-parameter family on **M**) reflects in addition the large-scale gravitational structure of the energetic contents of **M**, i.e. the universe. In other terms the Einstein metric depends on the actual state of the universe, and not only on the fundamental geometry and nonlinear partial differential equations, which are naturally presumed invariant under the full group of causality-preserving transformations in **M**.

To develope some empirical consequences from the hypothesis that space-time in the large is representable as **M**, one may observe that given any point of **M** and Einstein frame (or choice of Einstein metric within the 8-parameter family), there is a canonical, Lorentz-covariant decomposition of the Einstein energy,- i.e. the representative of $\partial/\partial t$, where t is the Einstein time,- into a scale-covariant component, and a scale contravariant component. It is then natural to make the physical interpretation that the first component, which is simply $\partial/\partial x_0$, where x_0 is the Minkowski time coordinate osculating t, is the energy observed in microscopic particle experiments (e.g. the frequency of a photon), while the second component represents diffuse large-scale processes(including conventional gravity and definable as generalized gravity, including the redshift process). For a positive energy field, such as Maxwell, all 3 components are positive, and for a localized statethe expected value of the Einstein energy H differs by unmeasurably little from the expectation value of the Minkowski energy H_0, so that the gravitational energy H_1, defined by the equation $H = H_0 + H_1$, is minimal for a maximally localized state. Thus it represents an attractive force, and to the extent that it may be approximated by an effective potential, this potential must be of the form $-g/r$, where r is the distance and g is a constant, by virtue of the scale contravariance and euclidean invariance it inherits from the presumptive quantum field and state of which it represents an expectation value.

To the extent that H_0 can be approximated by an effective potential, such as presumably for an assembly of subnuclear particles, it correspondingly is represented by a potential of the form ℓr, serving to explain asymptotic freedom, etc., and to clarify it by its parallel with gravity. Mathematically, H_1 is simply the transform of H_0 under conformal inversion; i.e. the gravitational force is supposed to be just the transform of the totality of the conventional (and possibly as yet unobserved) microscopic forces, and not really a separate force, except by an analytical (well-defined, but man-made) definition. It is the very long time scale on which gravitational forces are implemented that largely account for their apparent difference from microscopic forces. The natural chronometric time unit is the eon, the time required for light to encircle space; this conformally invariant unit is of the order of perhaps a billion years, on the basis of redshift measurements. The many eons required for mass and inertia to develope explain the action at a distance appearance of gravity, quite consistently with causal conformally invariant fundamental dynamical equations.

The theory is natural , essentially non-parametric, and explanatory of otherwise puzzling general matters such as the mechanism by which Mach's principle operates, the origin of the seemingly strange subnuclear potential that varies as r, and the relation of the Einstein Equivalence principle to particle considerations (i.e., conformal invariance underlying both). The hypothesis that cosmologically redshifted energy is represented by H_1 gives rise to predictions that have been in excellent quantitative agreement with systematic observations on galaxies and quasars, and without any objective counterindications, however intense the subjective reactions of those who have long espoused a Doppler explanation of the cosmological redshift. In essence the chronometric theory providesa conceptually simple and unique framework for the physics of extreme distances, which has been well substantiated in the large scale, but remains largely to be developed at the microscopic scale.

It is in connection with this latter development that the spannor representation of the scale-extended Poincare group, and corresponding spannor fields on **M**, were investigated by Paneitz. The term arises from the conception of a spannor as a twisted or wrenched (as if by a spanner) spinor. Spannors provide the simplest and most interesting example of the indecomposable representations with which the mathematical part of Paneitz' report was concerned, and to this I now turn.

REFERENCES

The references given below include relevant bibliographies

1. S. M. Paneitz, Global solutions of the hyperbolic Yang-Mills equations and their sharp asymptotics. In press, Proc. Amer. Math. Soc. Institute, Berkeley, Calif., July 1983, on Nonlinear Functional Analysis.

2. S. M. Paneitz, Parametrization of causal actions of universal covering groups and global hyperbolicity. Submitted for publication.

3. I. E. Segal, Chronometric cosmology and fundamental fermions. Proc. Natl. Acad. Sci. USA 79, 7961-7922 (1982).

INDECOMPOSABLE REPRESENTATIONS OF THE POINCARE GROUP AND ASSOCIATED FIELDS

Stephen M. Paneitz*

I. Introduction

Conventionally, physical fields are assumed to transform trivially under space-time translations. This is natural enough in Minkowski space $\mathbf{M}_0$, but is questionable in the case of a curved space-time.

For example, in the case of the universal cosmos **M**, which is the maximal spatially and temporally isotropic globally causal space-time, the relativistic space-time translations act in a natural global manner, but have no preferred role relative to other causal transformations.

On the other hand, thecounter question may be raised of whether such a non-trivial action would make any physical difference. Would not the effect be simply of the order of the space curvature, and hence too small to affect microscopic considerations?

The answer is yes and no: yes, energetically the effect should be small, but no, like the inclusion of spin from a particle classification point of view, non-relativistic theory having been greatly clarified by the introduction of spin although the energetic effects were typically slight. In a theory associated with the Yang-Mills equations, the gauge group could be affected.

But even granting the possibility of new selectionrules, would not non-trivial transformation properties under space-time translations involve the introduction of a new parameter,- the curvature,- simply adding to many existing uncertainties?

No; there is a natural way to introduce the curvature and relate it to empirical results. Moreover the curvature is conformally invariant and provides a natural third unit together with $\hbar = c = 1$, fully specifying physical units in a causally invariant way.

In fact, there is a very natural modification of spinor fields, which is quite simple, that provides an interesting example, and the lowest-dimensional one.

II. Example: The Spannors

The basic finite-dimensional representation. As earlier, let **P** denote the universal (2-fold) cover of the 11-dimensional scale-extended Poincare group. Then **P** is isomorphic to the semi-direct product of $GL(2,C)_r$ and $H(2)$, where $GL(2,C)_r$ denotes the 7-dimensional subgroup of $GL(2,C)$ having real determinant, and $H(2)$ denotes the set of all 2 x 2 complex hermitian matrices, as a 4-dimensional additive group, and the action of $GL(2,C)_r$ on $H(2)$ is as follows: the element T of $GL(2,C)_r$ carries the element H of $H(2)$ into THT^* (where * denotes ther hermitian conjugate).

Now define the representation $\mathbf{S}^+$ of **P** as follows: $\mathbf{S}^+(L \# F) = \begin{pmatrix} L & (i/2)FL^{*-1} \\ 0 & L^{*-1} \end{pmatrix} (\det L)^{w/2}$, where # denotes the semi-direct

*Deceased 1 September 1983. Posthumous presentation by I. E. Segal.

product, and L and F are arbitrary in $GL(2,C)_r$ and H(2) respectively.

Then the following hold:

1° S^+ is a representation of **P**.

2° S^+ has an irreducible 2-dimensional subspace, and an irreducible quotient modulo this subspace, but is indecomposable.

3° It admits CP, but admits neither C nor P alone.

4° The direct sum $\mathbf{S} = \mathbf{S}^+ \oplus \mathbf{S}^-$, where $\mathbf{S}^-$ denotes the C-conjugate (or equivalently, P-conjugate) of $\mathbf{S}^+$ admits substantially unique C, P, and T.

5° As the distance scale, say R (interpretable as the 'radius of the universe', a conformally invariant unit), becomes infinite(in laboratory units), $\mathbf{S}^+$ converges to the spin representation,

$$L \# F \longrightarrow \begin{pmatrix} L & 0 \\ 0 & L^{*-1} \end{pmatrix} (\det L)^{w/2}$$

In connection with 5° it should be noted that R is not fixed by the commutation relations of **P**, due to its 1-parameter family of outer automorphisms, in any natural way. However the commutation relations of **G** do serve to fix it, by virtue of the simplicity of **G**, which implies that apart from discrete possibilities, the automorphisms are inner. Choosing R = 1 = ℏ = c corresponds to a basis for the Lie algebra of **G** such that the constants of structure are all ±1 or 0.

To see 5°, observe that the introduction of laboratory units of length, apart from the units ℏ = c = 1, alters $\mathbf{S}^+$ only by replacing F by F/R. Accordingly as R becomes infinite, the off-diagonal term in $\mathbf{S}^+$ tends to zero and the conventional spin representation results. Group-theoretically, there is an obvious analogy with the limit as c becomes infinite, in which case the Poincare group deforms into the Galilean group, as emphasized by Minkowski in his classic adress in 1908. In the latter case, however, the only generators affected are the boosts, which are not directly observed, since they do not commute with the energy, and serve only to express the Lorentz invariance. In the present case, the boosts are unaffected as R becomes infinite, but the energy itself is changed.

$\mathbf{S}^{\pm}$ are defined as the <u>half-spannor</u> representations of **P** of weight w. They are mutually contragredient apart from the weight.

The direct sum $\mathbf{S}^+ \oplus \mathbf{S}^-$ defines the <u>full-spannor</u> representation **S** of **P**. It is a rather striking representation from both physical and mathematical standpoints in view of the foregoing.

Mathematically, however, there is an alternative definition of the spannors that exhibits most clearly the natural character of the representation they define. The causal group of **M**, which locally is identical to the local causal group SU(2,2) of $\mathbf{M}_0$, naturally contains the scale-extended Poincare group as a subgroup; it is simply the subgroup of all local causal transformations that may be extended globally to all of $\mathbf{M}_0$ without singularities. The defining representation of SU(2,2) is a 4-dimensional linear one. Thus there is a natural local imbedding of **P** into GL(4,C), defined by its imbedding in **G** and the local representation of **G** as SU(2,2). It is natural to consider also the complex conjugate representation, which is equivalent to $\mathbf{S}^-$; and there are no other 4-dimensional indecomposable linear representations of **G**, nor such of lower dimension, which are non-trivial on the translation subgroup of the Poincare group.

Higher-dimensional representations having these properties were studied by Paneitz, who found that below a certain dimension there are only a finite number of inequivalent such representations (apart from weight). For a sufficiently high dimension, it was however possible to construct an infinite number of inequivalent indecomposable representations of the Poincare group each of which was non-trivial on the translation subgroup. In the lower dimensions the situation may be summarized as follows:

Dimension	Situation
5	1 real representation and its dual
6	1 real self-contragredient representation
7	1 representation, its complex conjugate, dual and antidual, all inequivalent
8	8 representations.

Of these, all admit T; only those of dimensions 5 and 6 admit P; that of dimension 6 also admits C. These representations are for the most part related to the tensor product of the half-spannor representations, which appear as the most interesting and basic. For theoretical physics, it is the fields induced from the representations that are ultimately relevant, and to this matter we now turn.

III. Fields induced from indecomposable representations of **P**

According to the general theory of induced representations, to induce from a representation of **P** requires a group, say **G'**, of which P is a subgroup. If **G'**/**P** is to have a **G'**-invariant causal (or conformal) structure and be 4-dimensional, there is only one possibility: **G'** is the universal cover of SU(2,2), which may be denoted simply **G** here. This follows from the fact that **P** must be (isomorphic to) the isotropy subgroup of the causal group of the putative space-time **G**/**P**. The resulting homogeneous space **G**/**P** is then the universal cosmos **M**, and is conformal to the Einstein Universe $R^1 \times S^3$, with the metric $dt^2 - ds^2$, where ds is the element of arc length on S^3 in radians, and t is the Einstein time, also in radians.

For orientation purposes it should be noted that the simply connected versions (i.e. universal covers) of the de Sitter, anti-de Sitter, and Minkowski spacesare canonical subspaces of **M**. Relative to any given point (physically, of observation) of **M**, and Einstein frame at the point, there is a canonical imbedding of each of these spaces in **M**. The isometry groups of these spaces-times are subgroups of **G**, and the imbedding is obtainable as the orbit under this subgroup of a suitable point of **M**. Fields on **M** therefore determine canonically fields on each of these symmetrical subspaces.

In order for physical fields to represent stable physical states, they must satisfy appropriate positive energy constraints. In general, the action of **G** on the section space of a bundle induced from **P** (or any other subgroup) will not be irreducible, but will have a composition series of invariant subspaces, and corresponding quotient spaces; and in general these associated irreducible actions of **G** will fail to satisfy the positive-energy constraint. Indeed, for an arbitrary representation of **P** there will in general be no positive-energy 'subquotient' for the section space of the induced bundle. The question of the existence and structure of a positive energy subspace of the section space, and the character of its composition series (or invariant subspace chain, together with the corresponding quotient representations, or 'factors') is therefore essential for the correlation of the mathematics with the physics. The familiar technique involving the Fourier transform that is effective in the case of Minkowski space does not apply to other than flat spaces. Even in the simpler cases of spinor and other fields on **M** transforming trivially under Minkowski space-time translations, these problems are difficult, and the composition series are non-trivial. That is to say, as regards the latter, the entire space does not decompose into a direct sum or integral of invariant subspaces, as in the familiar cases of the Poincare and de Sitter groups, but is in part indecomposable. Thus there is a 'massless' invariant subspace, which is small in the sense that vectors in the subspace are determined by their Cauchy data at a fixed time, which has no invariant complement in the total positive-energy subspace; however, both this massless representation, and the quotient modulo the subspace in which it acts, which is 'massive' (lacking the possibility of determination of vectors by their Cauchy data at a fixed time), are unitary relative to an appropriate inner product.

In the case of spannor fields the situation is still more complicated. Technically, the treatment is complicated by the higher multilicities of the K-types, i.e. the representations of the maximal essentially compact subgroup K that are contained in the representation in question, which may be 2 rather than 1 as in the spinor case. In spite of this the generic composition series is similar to that for the spinor fields. However, for the value $w = 4$, it is exceptional. There is a 10-step composition series. Restricting consideration to the one-sided energy subspace, there are two massive factors in the composition series, and one massless factor, which is situated quite differently from that in the case of spinor fields: it is defined by a higher quotient space, rather than by a subspace.

Physically speaking, it could make a significant difference if the fundamental fermions,- leptons, neutrinos, and subnuclear particles,- are spannors rather than spinors, as commonly supposed, especially if they cohere together in a single indecomposable representation.

REFERENCES

1. S. M. Paneitz and I. E. Segal, Analysis in space-time bundles. I, General considerations and the scalar bundle, J. Funct. Anal. 47, 78-142 (1982); II, The spinor and form bundles, ibid. 49, 335-414 (1982).

2. S. M. Paneitz, Analysis in space-time bundles, III: Higher spin bundles. J. Funct. Anal. 1983, in press.

3. S. M. Paneitz, Indecomposable finite-dimensional representations of the Poincare group. Ann. Inst. H. Poincare, 1983, in press.

SL(n,R)/SO(n) UNIRREPS AND GROUP DECONTRACTION

Dj. Šijački

Institute of Physics
P.O.Box 57, Belgrade, Yugoslavia

Unitary irreducible representations (unirreps) of the SL(n,R) group are studied in the basis of its maximal compact subgroup SO(n). These representations are obtained from the corresponding unirreps of the Wigner-Inönü contracted semidirect-product group $T_r \circledS SO(n)$, $r=\frac{1}{2}(n-1)(n+2)$ by making use of the decontraction formula. Matrix elements of the SL(n,R) generators are explicitely constructed in the Gel'fand-Tsetlin basis of the SO(n) subgroup representations.

The theory of unitary irreducible representations (unirreps) of real noncompact semisimple Lie groups is not devoid still of certain subtle points. The most complete results concerning the general SL(n,R) case representations were obtained by Gel'fand and Graev[1]; though certain interesting explicit expressions are still unknown. We single out especially those problems related to the construction of spinorial (double valued) unirreps of the SL(n,R) groups.

We will advocate here the group decontraction approach to the explicite SL(n,R) unirreps construction. It follows from the work of Harish-Chandra[2] that the all real-noncompact-group unirreps can be obtained only if one utilizes the space of functions over the corresponding maximal compact subgroup. Thus, we start from the semidirect-product group $T_r \circledS SO(n)$, $r=\frac{1}{2}(n-1)(n+2)$, which is a Wigner-Inönü contraction of the SL(n,R) group w.r.t. its maximal compact subgroup SO(n).

Let L_{ab} and Q_{ab}, a,b=1,2,...,n generate respectively the SO(n) and T_r groups; $L_{ab}=-L_{ba}$, $Q_{ab}=Q_{ba}$ and $\sum Q_{aa}=0$. The corresponding Lie algebra is as follows

$$[L_{ab}, L_{cd}] = i(\delta_{ac}L_{bd} - \delta_{ad}L_{bc} - \delta_{bc}L_{ad} + \delta_{bd}L_{ac}),$$

$$[L_{ab}, Q_{cd}] = i(\delta_{ac}Q_{bd} + \delta_{ad}Q_{bc} - \delta_{bc}Q_{ad} - \delta_{bd}Q_{ac}),$$

$$[Q_{ab}, Q_{cd}] = 0.$$

The SL(n,R) generators L_{ab} and T_{ab}, a,b=1,2,...,n, where $L_{ab}=-L_{ba}$, $T_{ab}=T_{ba}$, $\sum T_{aa}=0$ have the following commutation relations

$$[L_{ab}, L_{cd}] = i(\delta_{ac}L_{bd} - \delta_{ad}L_{bc} - \delta_{bc}L_{ad} + \delta_{bd}L_{ac}),$$

$$[L_{ab}, T_{cd}] = i(\delta_{ac}T_{bd} + \delta_{ad}T_{bc} - \delta_{bc}T_{ad} - \delta_{bd}T_{ac}),$$

$$[T_{ab}, T_{cd}] = -i(\delta_{ac}L_{bd} + \delta_{ad}L_{bc} + \delta_{bc}L_{ad} + \delta_{bd}L_{ac}).$$

Now, the T_{ab} generators can be obtained from the Q_{ab} ones by making use of the decontraction formula[3], viz.

$$T_{ab} = pQ_{ab} + \frac{i}{2}(Q_{cd}Q^{cd})^{-1/2}\left[L_{cd}L^{cd}, Q_{ab}\right],$$

where p is an arbitrary real parameter. It is clear now that at least some representations of the T_{ab} generators follow from the Q_{ab} ones.

The SL(n,R) unirreps construction is carried out in two steps. In the first step one makes use of the group decontraction formula to evaluate the form of the T_{ab} representation expressions from the Q_{ab} ones. In the second step one finds all possible irreducible Hilbert spaces; their scalar products are in one-to-one correspondence to the SL(n,R) unirreps. In both steps of the construction one works in the spaces of functions of the parameters of the SO(n) group. The basis vectors for the SO(n) unirreps, determined by the group chain $SO(n) \supset SO(n-1) \supset \ldots \supset SO(2)$, are given by the Gel'fand and Tsetlin labels as follows

$$|m_{jk}\rangle = |M_m, M_{n-1}, \ldots, M_2\rangle,$$

where M_j stands for $(m_{j1}, m_{j2}, \ldots, m_{j[j/2]})$. The representation D matrix elements of SO(n) are given by

$$D^{(J)}_{\{K\}\{M\}}(\{\theta_n\}) = \left\langle \begin{matrix}(J)\\ \{K\}\end{matrix} \right| D(g(\{\theta_n\})) \left| \begin{matrix}(J)\\ \{M\}\end{matrix} \right\rangle,$$

where J and $\{K\}$, $\{M\}$ stand for M_n and $(K_{n-1}, K_{n-2}, \ldots, K_2)$, $(M_{n-1}, M_{n-2}, \ldots, M_2)$ respectively, and $g(\{\theta_n\}) \in SO(n)$. The most general Hilbert space to start with is the (separable) Hilbert space of functions on SO(n) which are square integrable w.r.t. the invariant measure on SO(n), and the scalar product of two such functions is in the general case given by

$$(f, h) = \iint dg_1 dg_2 f^*(g_1)\varkappa(g_1,g_2)h(g_2), \qquad g_1,g_2 \in SO(n)$$

where $\varkappa(g_1,g_2)$ is a kernel function and dg_1 and dg_2 are invariant measures over SO(n). The most convenient basis for these functions is provided by the $D^{(J)}_{\{K\}\{M\}}$ matrix elements.

We denote the Q_{ab} generators in the spherical basis by Q_A, and they are proportional to the D-functions of the r-dimensional SO(n) representation which we denote by $M_n=R$. In the simplest case we write

$$Q_{AB} = D^{(R)}_{\{0\}\{A\}}.$$

The matrix elements of the corresponding T_A generators are given, in terms of the generalized Wigner's functions and the reduced matrix elements, by the following expression

$$\left\langle \begin{matrix}(J')\\ \{K'\}\,\{M'\}\end{matrix} \right| T_A \left| \begin{matrix}(J)\\ \{K\}\{M\}\end{matrix} \right\rangle = \begin{pmatrix}(J') & (R) & (J)\\ \{M'\} & \{A\} & \{M\}\end{pmatrix} \left\langle \begin{matrix}(J')\\ \{K'\}\end{matrix} \right\| T \left\| \begin{matrix}(J)\\ \{K\}\end{matrix} \right\rangle ,$$

where

$$\left\langle \begin{matrix}(J')\\ \{K'\}\end{matrix} \right\| T \left\| \begin{matrix}(J)\\ \{K\}\end{matrix} \right\rangle = (N(J')N(J))^{1/2}\left[p + \frac{i}{2}\left(C_2(J') - C_2(J)\right)\right] \begin{pmatrix}(J') & (R) & (J)\\ \{K'\} & \{0\} & \{K\}\end{pmatrix} ,$$

and where $C_2(J)$ is the SO(n) Casimir operator value for the (J) representation.

All SL(n,R) unirreps can be obtained from the expressions provided by the unitarity and positive definitness of the representation space scalar product, i.e.

$$\varkappa((J'),(J)) = \varkappa^*((J'),(J)),$$

and

$$\varkappa((J'),(J)) \geqslant 0,$$

and the hermiticity condition in terms of the quantities

$$\varkappa((J'),(J)) \left\langle \begin{matrix}(J')\\ \{K'\}\end{matrix} \right\| T \left\| \begin{matrix}(J)\\ \{K\}\end{matrix} \right\rangle .$$

References

1. I.M. Gel'fand and M.I. Graev, Am.Math.Soc.Transl. 2(1956)147.
2. Harish-Chandra, Proc.Natl.Acad.Sci. USA 37(1951)170,362,366,691.
3. Y. Dothan and Y. Ne'eman, in "Symmetry Groups in Nuclear and Particle Physics", F.J. Dyson, ed. (Benjamin, N.Y., 1966).

HYSTERESIS & UNIVERSAL BIFURCATION
IN NATURAL PROCESSES.

G.E.Tanyi*
The National University
P.O.Roma 180, Lesotho.

1.ABSTRACT

Let $H_{p-1} = \{(a,u) \in R^k \times R^n$ s.t. $D_2^k f(a,u)h^k = 0$ has a non-trivial solution vector h for $k = 0,1,\dots,p-1\}$, where f is a p-determined k-parameter family of smooth maps. Then for any perturbation $f(a,u) \pm \varepsilon$ for which (a,u) belongs to H_{p-1}, the equilibrium configuration of f moves into a hysteresis loop with p-1 thresholds. Various examples in theoretical biology are discussed.

2.THE STRUCTURE OF H_{p-1}

2.1 Proposition (Compatibility Conditions)

Let N(m,n) denote the number of basis monomials of the homogeneous m - degree polynomials $D_2^m f(a,u)h^m$ and set $D_2^m f^i(a,u) = (c_1^i, c_2^i, \dots, c_N^i)(a,u)$, $i = 1,2,\dots,n$. Then a necessary and sufficient condition for the system $D_2^m f(a,u)h^m = 0$ to have common roots is that the coefficients c_A^i lie on the algebraic conical locus,

$$c_A^i c_B^j = c_B^i c_A^j , \quad (i,j = 1,2,\dots,n \;\&\; A,B = 1,2,\dots,N) \tag{2.1}$$

2.2 Theorem (Sufficient Conditions for H_{p-1} to be nonempty)

i.For $m = 3,4,\dots,p-1$, $i = 1,2,\dots,n$ and $A = 1,2,\dots,N(m,n)$ the common vertex $c_A^i = 0$ of the algebraic conical locus (2.1) is equivalent to the conditions,

$$\partial^m f^i(a,u)/\partial u_1^{i_1} \dots \partial u_n^{i_n} = 0 \tag{2.2}$$

where $i_1 + i_2 + \dots + i_n = m$.

ii. A sufficient condition for an equilibrium configuration $u = u_o$ to lie on H_{p-1} is that in addition to (2.2) it also satisfy the equations,

$$f(a,u_o) = 0 , \quad \text{Det } D_2 f(a,u_o) = 0 \;\&\; \text{Det } D_2^2 f^i(a,u_o) = 0 . \tag{2.3}$$

$(i = 1,\dots,n)$.

2.3 Distributive Systems.

For a distributive system defined on a domain M of R^q, $u_1,\dots,u_n$ are smooth functions of the local coordinates $x_1,\dots,x_q$ of M and f(a,u) is a p-determined k-parameter family of Fréchet differentiable operators on a Banch space U subject to the condition that the linear operator $\partial f(a,u)/\partial u$ is Fredholm at every point (a,u_o) on the zero set of f. Conditions analogous to the proposition (2.1) and the theorem (2.2) exist.

3. APPLICATIONS

3.1 A 3-determined 2-component Ecosystem.

Evolution Mechanism :

$$E_1 \xrightarrow{\delta} p_1 \xrightarrow{d_1} P_1 \quad ; \quad p_1 \underset{r_1}{\overset{R_1}{\rightleftharpoons}} 2p_1 \underset{r_{32}}{\overset{r_{23}}{\rightleftharpoons}} 3p_1$$

$$p_2 \underset{r_2}{\overset{R_2}{\rightleftharpoons}} 2p_2 \quad ; \quad p_1 \xrightarrow{\alpha} p_2 \xrightarrow{d_2} P_2$$

$(R_i = r_i E_i,\ i = 1,2\ ;\ E_i = \text{Constant energy supply}).$

Equilibrium Configuration :

$$(x^3 + \lambda xy - ax^2 + bx - c\ ,\ y^2 - \lambda xy - dy) = 0$$

$$x = p_1/E_1\ ,\ y = p_2(r_2/r_{32}E_1)^{1/2}/E_1, \qquad \lambda = \alpha(r_{32}E_1/r_2)^{1/2}/r_{32}E_1$$

$$(a,b,c) = ((r_{32} - r_1)/r_{32}E_1\ ,\ (d_1 - r_1E_1)/r_{32}E_1^2\ ,\ \delta/r_{32}E_1^2\ )$$

$$d = (r_2E_2 - d_2)(r_{32}E_1/r_2)^{1/2}/r_{32}E_1\ .$$

H_2 : Subject to the compatibility condition $\lambda^4 - a^2 + 3b = 0$,

H_2 is the hypersurface, $(u,v) \qquad (C_1,C_2,C_3,x_o,y_o)$

where $C_1(u,v) = u^2 - v^4$, $C_2(u,v) = u^3 - u^2v^2 - uv^4 + v^6$

$C_3(u,v) = v^3 - uv$, $x_o = u - v^2$

and $y_o = (v^2 - 3uv)/3^{3/2}$.

3.2 A 3-determined 1-component reaction-diffusion operator.

Evolution Mechanism :

$$E_1 \xrightarrow{\delta} p_1 \quad , \quad p_1 + E_1 \underset{r_1}{\overset{r_1}{\rightleftharpoons}} 2p_1 \underset{r_{32}}{\overset{r_{23}}{\rightleftharpoons}} 3\,p_1 \quad , \quad p_1 \xrightarrow{d_1} P_1$$

$(r_{32} \neq 0)$.

Equilibrium Configuration :

$$d^2u(x)/dx^2 + f(a, b,c,u(x)) = 0 , 0 \leq x \leq \pi, \ u'(0)= 0 = u'(\pi)).$$

$$a = (r_{23} - r_1)/r_{32}E_1, \quad b = (d_1 - r_1E_1)/r_{32}E_1^2 , \quad c = \delta/r_{32}E_1^2 .$$

$$f(a,b,c,u) = c - bu + au^2 - u^3 .$$

$$H_2 = \left\{(a,b,c,u_0) \text{ s.t. } b = 3(a/3)^2 - n^2, \ c = (a/3)^2 - an^2/3 \ \& \ u_0 = a/3\right\} .$$

3.3 Group Action on the Energy Space.

Each bifurcation diagram corresponding to a given integer n is the result of the action of a 1 - parameter group of translations on the 2 - dimensional energy plane with coordinates (a,b) given by,

$$\tau(n^2)(a,b) = (a, b+n^2) = (a,b').$$

---

* Work supported by a visiting Maths. grant of the I.C.T.P.
P.O.B.586, Miramare - 34100 Trieste (ITALY).

IRREDUCIBLE REPRESENTATIONS OF

THE BASIC CLASSICAL LIE SUPERALGEBRAS

SU(m/n) ; SU(n/n)/U(1) ; OSp(m/2n) ;

D(2/1 ; α) ; G(3) ; F(4).

Jean THIERRY-MIEG
Groupe d'Astrophysique Relativiste
Observatoire
92190 MEUDON - FRANCE

ABSTRACT :

Extending the results of Victor Kac, we construct and tabulate exhaustively the irreducible representations, typical and atypical, tensorial and spinorial, of the basic classical Lie superalgebras.

INTRODUCTION :

Irreducible representations of the simple Lie superalgebras have not completely been constructed. Kac (1978) |1| has established the complete set of representations of OSp(1/2n). Also, he has enumerated the representations of the other basic classical Lie superalgebras OSp(m/2n), SU(m/n), D(2/1;α), G(3) and F(4). He has sorted them into typical and atypical, constructed the typical representations, and shown that they encompass an equal number of even and odd states (Bosons and Fermions). However, he has not described the atypical representations, which are the most important for physical applications, since they are usually of lower dimension and seem to break supersymmetry, as they often encompass an unequal number of Bosons and Fermions.

Other attempts have yielded few results. In an early paper, Scheunert, Nahm and Rittenberg (1977) |2| have given the representations of OSp(1/2) and of SU(1/2). Later, the representations of SU(1/n) for all n were found by several methods - partial results by Ne'eman and Sternberg (1980) |3|,- full construction by ThierryMieg and Morel (1981) |4| and by Sun and Han (1981) |5|. Jarvis and Green (1979) |6| and Scheunert (1982) |7| have studied the spectrum of the Casimir operators of Gl(m/n), but this method does not separate the atypical representations. Dondi and Jarvis (1980) |8| and independently Balantekin and Bars (1981-1982) |9| have tried to extend the method of Young tableaux to the case of SU(m/n), $m \neq n$, OSp(m/2n)/Z_2 and P(n). Unfortunately, their method has been shown last year by Bars himself, Morel and Ruegg (1982) |10| to be unreliable in the general case.

In this note, elaborating on the results of Kac |1|, we present an explicit construction of the irreducible representations of the basic classical Lie superalgebras. For the mathematician, our main result is a set of lemmas which characterizes the atypical representations. For the physicist, our main result is the prosaic numerical tabulation of the typical and atypical representations of the superalgebras SU(m/n), $mn \leqslant 8$; SU(n/n)/U(1), $n \leqslant 3$; OSp(1/2n) $n \leqslant 3$; OSp(2n/2), $n \leqslant 3$; D(2/1 ; α), $\forall \alpha$; G(3) and F(4).

The concrete knowledge of those numbers should promote new phenomenological applications of superalgebras.

TYPICAL REPRESENTATIONS :

This section summarizes Kac's construction of the typical representations. Let G denote a superalgebra over $\mathbb{C}$, G_0 and G_1 its even and odd subspaces. Let H be a Cartan subalgebra (maximal commuting). The dimension r of H is the rank of G.

Let (α_i^+ : i = 0,1,... r-1) be a system of positive simple roots. We choose the α such that one and only one simple root $\beta^+ = \alpha_0^+$ is odd. Let C_{ij} be the symmetrized Cartan matrix. The superalgebra is characterized by the graded Jacobi identity, and the relations :

$$[\alpha_i^+, \alpha_j^-] = \delta_{ij}\, h_j \quad ; \quad [h_i, h_j] = 0 \quad ; \quad [h_i, \alpha_j^{\pm}] = \pm C_{ij}\, \alpha_j^{\pm}$$

We normalize C_{ij} by a) $C_{ij} = C_{ji}$; b) either $C_{00} = 2$ or ($C_{00} = 0$ and $C_{01} = 1$)

Let $\Delta \subset H^*$ denote the set of all roots, Δ^+ and Δ^- the positive and negative roots, Δ_0 and Δ_1 the even and odd roots ; $\bar{\Delta}$ the regular roots and Δ' the exceptional ones defined by

$$\Delta_0' = \{ \alpha \in \Delta_0 \ / \exists \beta \in \Delta_1 \ , \ \alpha = 2\beta \ \} \ ,$$

$$\Delta_1' = \{ \beta \in \Delta_1 \ / \exists \alpha \in \Delta_0 \ , \ \beta = \alpha/2 \ \} \ ; \qquad \Delta = \Delta' + \bar{\Delta} \ .$$

The existence of exact squares in G is a peculiarity of superalgebras, however by Jacobi :

$$\forall\, g \in G, \quad [g, [g,g]] = 0 \quad \Longleftrightarrow \quad \forall\, \alpha \in \Delta \ , \ 3\alpha \notin \Delta$$

Let ρ denote the half supersum of all positive roots :

$$2\rho = \sum \alpha - \sum \beta \ , \qquad \alpha \in \Delta_0^+ \ , \ \beta \in \Delta_1^+ \ .$$

The matrix C_{ij} defines a Weyl invariant metric on the root space H^* which is the restriction of the Killing metric on G. By definition, in a "basic classical" Lie superalgebra, the even roots, therefore also the exceptional odd roots, have positive definite norm. However, the regular odd roots are of norm zero and do not correspond to any Weyl symmetry : the Weyl group of G is that of G_0.
In OSp(2/2n) and SU(m/n) $m \neq n$ (type I superalgebras), G_0 does contain a U(1) factor, and we call k the corresponding Cartan operator. In the other superalgebras (type II), the odd root β^+ hides a simple positive root δ^+ of G_0, and we call k the operator $k = [\ \delta^+, \delta^-]$.

Theorem 1 : (Kac proposition 2.2)

Every finite dimensional irreducible representation V_Λ contains a unique vector $|\Lambda\rangle$, the highest weight, which is annihilated by the positive roots and is an eigen vector of H.

$$\alpha_i^+ \ |\Lambda\rangle = 0, \quad h_i \ |\Lambda\rangle = \Lambda(h_i) \ |\Lambda\rangle = \lambda_i \ |\Lambda\rangle \ ,$$

$$|\Lambda\rangle \in V_\Lambda \ , \qquad \Lambda \in H^* \ , \qquad \lambda_i \in \mathbb{C} \ .$$

Then V_Λ is included in the linear span S_Λ of the action on $|\Lambda\rangle$ of supersymmetric products of negative roots. However, as in the case of Lie algebras, one must consider the orbit of $\Lambda+\rho$ under the Weyl group W of G, i.e. of G_0, and add all contributions with weight ± 1, according to the parity $\varepsilon(W)$ of the Weyl element.

Lemma 1 : $\quad V_\Lambda \subset S_\Lambda^W = S_\Lambda \cap \left(\sum_{w \in W} \varepsilon(w) \ S_{w(\Lambda+\rho)} - \rho \right) .$

Lemma 2 : The G_o orbit of $|\Lambda>$ is finite dimensional if and only if the Dynkin weights

$$a_i = 2<\Lambda+\rho|\alpha_i> / <\alpha_i|\alpha_i> \quad , i = 1, \dots r-1$$

$$k = 2<\Lambda+\rho|\delta^+> / <\delta^+|\delta^+> \quad \text{for type II}$$

are non negative integers.
This lemma is trivial, but together with theorem 1, it provides an enumeration of all finite dimensional representations.

Let a "principal weight" be the highest weight μ of a G_o submodule, and M_Λ be the set of principal weights of V_Λ. Let the "susy crystal" be the convex hall C_o generated by the exterior product of the negative odd roots. Let C_Λ be the same crystal translated by Λ.

Lemma 3 : $M_\Lambda \subset C_\Lambda$

The lemma generalizes to superalgebras the finite development of superfunctions :

$$\Phi(x,\theta) = f_0(x) + \theta^i f_i(x) + \dots$$

$\Phi(o,o)$ corresponds to Λ, $f_n(x)$ to the G_o submodules and θ^i to β_i^+

Using lemmas 1 and 3, we derive

Lemma 4 : $M_\Lambda \subset C_\Lambda^W = C_\Lambda \cap (\sum_{w\in W} \varepsilon(w)\, C_{w(\Lambda+\rho)} - \rho)$

Theorem 2 : (Kac proposition 2.11) : $B(\Lambda) = \prod_{\beta\in\bar{\Delta}_1} <\Lambda+\rho|\beta> = 0 \Rightarrow M_\Lambda = C_\Lambda^W$

In this case, the representation is called typical, the superdimension vanishes (except in OSp(1/2n)), and the dimension is :

$$\dim(V_\Lambda) = 2^F \prod_{\alpha\in\Delta_0^+} <\Lambda+\rho|\alpha> / <\rho_0|\alpha> ,$$

$$F = \mathrm{card}(\Delta_1^+) , \quad 2\rho_0 = \Sigma\, \alpha , \quad \alpha \in \Delta_0^+ .$$

This theorem achieves the description of the typical finite dimensional representations. In particular, it gives the spinorial representations of OSp(m/2n), which a priori cannot be constructed using supertableaux.

ATYPICAL REPRESENTATIONS :

When Λ is a root of the polynome $B(\Lambda) = 0$, some principal weights of C_Λ^W decouple from M_Λ. Consider the example $b = <\Lambda+\rho|\beta> = 0$. The principal weight $|\mu> = \beta^-|\Lambda>$ is the highest weight of a submodule since $\beta^+|\Lambda> = b|\Lambda> = 0$. This module must be quotiented out, and we have :

$$b = o \Longrightarrow M_\Lambda \subset (C_\Lambda \cap C_{\Lambda+\beta^+})$$

In general, we shall prove the following theorem : Let $\beta_i^+ \in \bar{\Delta}_1^+$,

Theorem 3 : If $|\Lambda + \rho> \neq 0$ and $<\Lambda+\rho|\beta_i^+> = 0$, Λ is called atypical of type i, C_Λ is reducible, and its principal weight $|\mu_i> = |\Lambda - \beta_i^+>$ is a highest weight.

Proof : Let P_i be the H eigen subspace of S_Λ which includes $\beta_i^- | \Lambda>$. Let Q_i be the principal subspace of P_i.

$$Q_i = \{ q_i, q_i \in P_i, \alpha_j^+ | q_i> = 0 \}$$

Let A^+ be the ring of raising operators generated by α_j^+ and β^+, and call $R_i(\Lambda)$ the rank of the system.

$$A^+ | q_i> = | \Lambda>$$

a) No two negative odd roots coincide in H*. Therefore, by lemma 3, Q_i is at most dimension 1, the rank is at most 1, and $R_i(\Lambda)$ is linear in b.

b) If the Dynkin weights of P_i are negative, by lemma 4, the rank $R_i(\Lambda)$ is zero and the theorem is proven.

c) If the Dynkin weights are positive, and if the representation is typical, then by theorem 2 the rank is 1. Therefore $R_i(\Lambda)$ divides $B(\Lambda)$, and there is a j such that : $R_i(\Lambda) = <\Lambda+\rho | \beta_j>$

d)If $|\mu> = |\Lambda - \beta_j>$ decouples from V_Λ, then $V_\mu \subset (C_\Lambda \cap C_\mu)$, and the principal weight $|\Lambda - 2\beta_j>$ decouples from V_μ. Therefore the equation $R_i(\Lambda) = 0$ is invariant under β_i translation :

$$R_i(\Lambda) = 0 \quad \Rightarrow \quad R_i(\Lambda - \beta_i) = 0$$

e) Together, equations (c) and (d) imply :

$$<\beta_j | \beta_i> = 0$$

In SU(1/n), OSp(2/2n), OSp(3/4), OSp(n/2), D(2/1 ; α), G(3) and F(4), no two different odd roots are orthogonal, and i = j. In SU (m/n), and OSp(m/2n), pairs of orthogonal negative odd roots exist, but one of them is also a root of a subalgebra with lower m or n. Therefore by recurrence in m, n, i = j in all cases. QED.

Lemma 5 : If $<\Lambda + \rho | \beta_i> = 0$ and $|\Lambda + \rho> \neq 0$ then $M_\Lambda \subset C_\Lambda \cap C_{\Lambda+\beta_i^+}$

Indeed, when the principal weight $|\mu> = |\Lambda - \beta_i>$ decouples from V_Λ, it carries away the whole of V_μ, and a fortiori the submodules of S_μ which are decoupled from V_μ.

Lemma 6 : If the Dynkin weights are positive but not integers, and if Λ is atypical of type i,j... exclusively, then :

$$M_\Lambda = \tilde{C}_\Lambda = C_\Lambda \cap C_{\Lambda+\beta_i} \cap C_{\Lambda+\beta_j} \cap \quad \ldots$$

We call $\tilde{C}_\Lambda$ the cleaved crystal.

Proof a contrario : If a weight μ in $\tilde{C}_\Lambda$ is a highest weight, then μ is of the form : $|\mu> = |\Lambda - \beta_m - \beta_n>$, with m,n different from i, j.... ,or contains more β. Necessarily, $V_\mu \subset (\tilde{C}_\Lambda \cap C_\mu)$, therefore μ must be atypical of type i,j..m,n , whereas Λ is typical in m and n.

Therefore
$$\begin{cases} <\beta_m + \beta_n | \beta_i> = 0 \,, & <\Lambda + \rho | \beta_i> = 0 \,, \\ <\Lambda + \rho | \beta_m> = <\Lambda + \rho | \beta_n> = -<\beta_m | \beta_n> \neq 0 \,. \end{cases}$$

In SU(1/n), B(n), C(n), D(2/1 ; α), G(3), F(4), the first equation has no solution. In SU(m/n) and OSp(m/2n) the system has no solution.

To study the case where Λ is atypical and the weights are integer we follow 2 trajectories $\Lambda_1(t)$ and $\Lambda_2(t)$ in the root space. $\Lambda_1(t)$ is atypical $\forall$ t. $\Lambda_2(t)$, obtained by varying k while keeping the a_i fixed, is typical for $t \neq 0$.
By lemmas 4 and 6 : $M_{\Lambda_1} \subset \tilde{C}_{\Lambda_1}$ and $M_{\Lambda_2} \subset C^W_{\Lambda_2}$.
In the limit :

Lemma 7 : $M_\Lambda \subseteq \widetilde{C^W_\Lambda} \cap \lim_{t\to 0}(C^W_{\Lambda_2})$.

In type I superalgebra, equality holds. In type II superalgebras, there is a complication. When k is non integer, the K-Weyl symmetry is not active. In G(3) for k = 3 and F(4) for k = 4 only, some weights occur with multiplicity n_1 in $\widetilde{C^W_\Lambda}$, n_3 in C^W_Λ and n_2 in lim $C^W_{\Lambda_2}$ with $n_1 = n_2 = 2$, $n_3 = 0$. In this case, we have found that the correct multiplicity is 1. In all other cases min $(n_1, n_2) \leq n_3$ and equality holds.

SUPERDIMENSION :

Let S_Λ be the superdimension of a representation atypic of type i. Let

$$\Delta_i = \{ \alpha , \alpha \in \Delta_0^+ , \langle\alpha|\beta\rangle = 0 \}$$

be the positive even roots which commute with β_i^+ . Let ρ_i be their half sum. Let

$$d^i = \prod_{\alpha\in\Delta_i} \langle\Lambda+\rho|\alpha\rangle / \langle\rho_i|\alpha\rangle .$$

Lemma 8 : In SU(1/n) and C(n+1)=OSp(2/2n), $S_\Lambda = d^i$.

In OSp(m/2), D(2/1,α), G(3) and F(4), we have found numerically that $S_\Lambda = \nu\, d^i$ with ν = 1 or 2.

RESULTS :

Using lemma 7 and a mini-computer, we have tabulated the irreducible representations of the superalgebras listed in the introduction. We recover the results of ref |1,5|, but disagree with those given by the supertableaux method |9|.

The author is grateful to Victor Kac for help and comments.

1. V Kac, Lect. Notes in Math. **676** (1978), 597-626.
1. M. Schneunert, W. Nahm & V. Rittenberg, J.M.P. **18** (1977) 155.
3. Y. Ne'eman, S. Sternberg PNAS USA **77** (1980) 3127.
4. J. Thierry-Mieg & B. Morel in Superspace and Supergravity, S. Hawking, M. Rocek ed., Cambridge Univ. Press (1981).
5. Sun Hong Zhou, Han Qi Zhi, Sc. Sinica **24** (1981) 914-923.
6. P.D. Jarvis, H.S. Green J Math. Phys. **20** (1979) 2115.
7. M. Scheunert, Bonn Univ preprints 1982-83.
8. P.H. Dondi, P.D. Jarvis J. Phys. A. **14** (1981) 547.
9. A. Balantekin, I. Bars, J.M.P. **22** (1981) 1149, 1810 **23** (1982) 1239.
10. I. Bars, B. Morel, H. Ruegg, J.M.P. in press.

GROUP REPRESENTATIONS IN INDEFINITE METRIC SPACES

P.M. van den Broek
Department of Applied Mathematics,
Twente University of Technology,
P.O. Box 217, 7500 AE Enschede,
The Netherlands.

Let V be a n-dimensional complex vector space with scalar product (,) and let η be a Hermitian non-singular linear operator on V. The indefinite metric of V is given by $\langle\phi,\psi\rangle = (\phi,\eta\psi)$. Let G be a finite group of symmetry transformations of the indefinite metric space V. According to a generalisation of Wigners theorem [1,2] one then comes to deal with a n-dimensional projective linear-antilinear (PLA) representation of G; each $g \in G$ is represented by an operator D(g) on V which is either η-unitary (ηU), or η-antiunitary (ηAU), or η-pseudounitary (ηPU), or η-pseudoantiunitary (ηPAU). In terms of matrices, taken with respect to an orthonormal basis of V we then have, instead of unitarity,

$$D^{\dagger}(g)\ \eta\ D(g) = (-)_g\ \eta^g. \tag{1}$$

Here η is now a Hermitian non-singular matrix, $(-)_g$ is equal to -1 if g is represented by a ηPU or a ηPAU operator and equal to +1 otherwise and the superscript g denotes complex conjugation if and only if g is represented by a ηAU or a ηPAU operator. Let G_0 be the normal subgroup of G consisting of those $g \in G$ which are represented by ηU operators and let a,b and c be elements of G (if any exist) which are represented by ηAU, ηPU and ηPAU operators respectively. Then aG_0, bG_0 and cG_0 denote the cosets of G with respect to G_0 which are represented by ηAU, ηPU and ηPAU operators respectively. We will only consider here the case that G_0 has index 1 or 2. This leaves the following 4 possibilities: $G = G_0$ (case I), $G = G_0 + aG_0$ (case II), $G = G_0 + bG_0$ (case III) and $G = G_0 + cG_0$ (case IV).
For any non-singular n×n-matrix A an equivalence transformation of D is given by $D'(g) = A^{-1}D(g)A^g$ and $\eta' = A^{\dagger}\eta A$. D is said to be decomposable if there exists an equivalence transformation such that

$$D'(g) = \begin{pmatrix} D_1'(g) & 0 \\ 0 & D_2'(g) \end{pmatrix} \quad \forall g \in G; \qquad \eta' = \begin{pmatrix} \eta_1' & 0 \\ 0 & \eta_2' \end{pmatrix} \tag{2}$$

A PLA representation D satisfying (1) is decomposable into a direct sum of undecomposable PLA representations. If D is irreducible it is undecomposable but if D is undecomposable it is not necessarily irreducible. We have investigated which are the

undecomposable PLA representations, with the following results.
In case I and II the undecomposable PLA representations are just the irreducible PLA representations.
In the cases III and IV we distinguish between two types of irreducible PLA representations D : those for which $D \downarrow G_0$ is irreducible (type A) and those for which $D \downarrow G_0$ is reducible (type B).
Consider case III. If D is an irreducible PLA representation of type B then the irreducible PLA representation D', defined by $D'(g) = (-)_g D(g)$ is not equivalent with D. Then D and D' are said to be related. The undecomposable PLA representations of G are the irreducible PLA representations of type A and the direct sums of two related irreducible PLA representations of type B. It follows that in each PLA representation of case III related irreducible PLA representations have equal multiplicity.
In case IV the undecomposable PLA representations of G are the irreducible PLA representations of type B and the direct sums of two irreducible PLA representations of type A. It follows that in each PLA representation of case IV the irreducile PLA representations of type A have even multiplicity. More details and full proofs will be published in [3].

[1] L. Bracci, G. Morchio and F. Strocchi : Commun. Math. Phys. 41, 289-299 (1975).

[2] P.M. van den Broek : "Symmetry transformations in indefinite metric spaces", Memorandum nr. 428, Department of Applied Mathematics, Twente University of Technology, The Netherlands (1983).

[3] P.M. van den Broek : "Group representations in indefinite metric spaces", Journ. Math. Phys. (to be published).

TENSOR OPERATOR REALISATIONS OF THE CLASSICAL LIE ALGEBRAS AND NON-TRIVIAL ZEROS OF THE 6J-SYMBOL

J. Van der Jeugt(°), H. De Meyer(†), G. Van den Berghe and
P. De Wilde(‡)

Seminarie voor Wiskundige Natuurkunde, Rijksuniversiteit-Gent
Krijgslaan 281-S9, B-9000 Gent, BELGIUM

The existence of an infinity of zeros of Racah's 6j-symbol which are non trivial in the sense that they do not result from triangle condition violation has been discussed recently by Biedenharn and Louck [1]. In their book the topic is illustrated by means of an extensive table containing more than 1400 structural zeros. Clearly, in remaining within the framework of the SO(3) Lie algebra A_1 in which the 6j-symbol naturally arises, the structural zeros coincide with the zeros of a function depending on six non-negative integer or half-odd integer variables of which the domain of definition is restricted to all entries which satisfy the triangle conditions. As an example, it is easy to deduce from Racah's well-known algebraic formula [2] of the 6j-symbol that the one-parameter family of 6j-coefficients $\begin{Bmatrix} 3a-4 & a & 2a-1 \\ a & a & 2a-3 \end{Bmatrix}$, where $2a \in \mathbb{Z}_+$ and $a \geqslant 2$, yields an infinity of such zeros.

However, yet another approach to these zeros emerges in the larger framework of the classical semi-simple Lie algebras which all contain a variety of distinct A_1 subalgebras. Indeed, it was already recognized by Racah [3] that the non-trivial vanishing of the particular 6j-coefficient $\begin{Bmatrix} 5 & 5 & 3 \\ 3 & 3 & 3 \end{Bmatrix}$ elucidates the embedding of the exceptional Lie algebra G_2 into the SO(7) Lie algebra B_3. A convincing proof relies upon the standard realisation of the SO(7) generators as tensor operators with respect to the generators of the principal SO(3) subgroup in the chain SO(7) $\supset G_2$ $\supset$ SO(3). Moreover, the argument can be inverted in the sense that the restriction of this SO(3) tensor operator realisation to the maximal subalgebra G_2 on its own already necessitates the mentioned 6j-coefficient to vanish. In this perspective the property that G_2 can be inbedded in B_3 is not even essential for explaining the structural zero.

From the above observations the question arises whether similar arguments can be repeated for other Lie algebras too. As it was already suggested by Biedenharn

(°) Research assistant N.F.W.O.
(†) Research associate N.F.W.O.
(‡) Research assistant I.W.O.N.L.

and Louck [1] it is preferable to make first the exceptional Lie algebras object of an investigation. In what follows we shall demonstrate on F_4 and E_6 that the question can be answered affirmatively. However, for the sake of comprehension we first go into some details of the definition and properties of SO(3) tensor operators and tensor realisations. How structural zeros can be explained from a realisation will be discussed on one illustrative example. Finally, all results obtained at present are assembled in a table at the end.

SO(3) tensor operators are defined by means of reduced matrix elements [4,5] :

$$<\tau_2'\ell_2' \| v^k(\tau_2\ell_2,\tau_1\ell_1) \| \tau_1'\ell_1'> = [k]^{1/2}\delta_{\tau_2'\tau_2}\delta_{\ell_2'\ell_2}\delta_{\tau_1'\tau_1}\delta_{\ell_1'\ell_1} , \qquad (1)$$

where ℓ and k are SO(3) representation labels, [k] = 2k+1 is the dimension of the SO(3) tensor representation and τ is an additional label to distinguish irreps with the same ℓ. These operators obey the following commutation relations :

$$\left[v_{q_1}^{k_1}(\tau_1\ell_1,\tau_2\ell_2),v_{q_2}^{k_2}(\tau_3\ell_3,\tau_4\ell_4)\right] = \sum_{k_3,q_3} \{ [k_1]\,[k_2]\,[k_3] \}^{1/2} \begin{pmatrix} k_1 & k_2 & k_3 \\ q_1 & q_2 & q_3 \end{pmatrix}$$

$$\times (-1)^{2\ell_4+\ell_3-\ell_2-q_3}\Big(\delta_{\tau_2\tau_3}\delta_{\ell_2\ell_3}(-1)^{k_1+k_2+k_3+\ell_1+\ell_2+\ell_3+\ell_4}\begin{Bmatrix} k_1 & k_2 & k_3 \\ \ell_4 & \ell_1 & \ell_3 \end{Bmatrix} v_{q_3}^{k_3}(\tau_1\ell_1,\tau_4\ell_4)$$

$$- \delta_{\tau_1\tau_4}\delta_{\ell_1\ell_4}\begin{Bmatrix} k_1 & k_2 & k_3 \\ \ell_3 & \ell_2 & \ell_1 \end{Bmatrix} v_{q_3}^{k_3}(\tau_3\ell_3,\tau_2\ell_2)\Big). \qquad (2)$$

If A denotes an irrep of a classical Lie group G we learn from tables [6] how A decomposes into irreps (γk) of a particular SO(3) subgroup, γ being used to distinguish again between similar irreps. Hence, the SO(3) tensor operators which constitute an operator realisation of A are labelled by (γkq) where q runs in unit steps from -k to k. In order to define a space spanned by states $|\tau\ell m>$ on which these operators act, we select an irrep B of G for which B occurs in the decomposition of the Kronecker product $A\times B$ [4]. If B decomposes into irreps ($\tau\ell$) the set $\{|\tau\ \ell\ m> : m=-\ell,-\ell+1,\dots,\ell\}$ is an acceptable basis. Since it is the aim to construct realisations of Lie algebras, A should be identified with the adjoint irrep of G. Moreover, it is preferable to choose the dimension of B as low as possible in which case a so-called minimal realisation will be obtained. Without making the choice of B explicit, the G-generators can be written as :

$$G_q^{\gamma k} = \sum_{\tau_1\ell_1\tau_2\ell_2} g[\gamma k;\tau_1\ell_1,\tau_2\ell_2]\, v_q^k(\tau_1\ell_1,\tau_2\ell_2) , \qquad (3)$$

whereby the g-coefficients remain to be determined. To do so, we use the property that the set of generators is closed under commutation which allows to write :

$$[G^{\gamma_2 k_2}_{q_2}, G^{\gamma_1 k_1}_{q_1}] = (-1)^{q_1+q_2} \sum_{k,\gamma} \{[k][k_1][k_2]\}^{1/2} \begin{pmatrix} k_2 & k_1 & k \\ q_2 & q_1 & -q_1-q_2 \end{pmatrix} C^{\gamma k}_{\gamma_2 k_2, \gamma_1 k_1} G^{\gamma k}_{q_1+q_2}, \tag{4}$$

where the C-coefficients remain to be determined. Substitution of (3) in both sides of (4) leads with the application of (2) to the following equations w.r.t. the unknown g's and C's :

$$C^{\gamma k}_{\gamma_2 k_2, \gamma_1 k_1} \, g[\gamma k; \tau_2 \ell_2, \tau_1 \ell_1] = (-1)^{\ell_2-\ell_1} \sum_{\tau,j} \Big\{ (-1)^{k+k_1+k_2+2j} \begin{Bmatrix} k_2 & k_1 & k \\ \ell_1 & \ell_2 & j \end{Bmatrix}$$
$$\times\, g[\gamma_2 k_2; \tau_2 \ell_2, \tau j]\, g[\gamma_1 k_1; \tau j, \tau_1 \ell_1] - \begin{Bmatrix} k_1 & k_2 & k \\ \ell_1 & \ell_2 & j \end{Bmatrix} g[\gamma_1 k_1; \tau_2 \ell_2, \tau j]\, g[\gamma_2 k_2; \tau j, \tau_1 \ell_1] \Big\} . \tag{5}$$

If we let all parameters herein vary, a usually overcomplete system of equations linear in the C's and quadratic in the g's is generated. Substituting the solution for the latter coefficients in (3) an operator realisation of the group generators is established.

As an example let us consider the chain $F_4 \supset SO(3)$ where SO(3) is the principal subgroup. The corresponding branching rules for F_4 irreps are found in tables [6]. From them we learn that (1000), the adjoint irrep of F_4 decomposes into the SO(3) irreps (11),(7),(5) and (1), whereas (0001), the 26-dimensional irrep of F_4 reduces into (8) and (4). The method outlined above leads to the following SO(3) tensor operator realisation of F_4 :

$$G^1_q = v^1_q(4,4) + \sqrt{\tfrac{34}{5}}\, v^1_q(8,8)$$

$$G^5_q = v^5_q(4,4) + \sqrt{\tfrac{38}{51}}\, v^5_q(8,8) + (-1)^\alpha 2\sqrt{\tfrac{10}{51}}[v^5_q(4,8) + v^5_q(8,4)]$$

$$G^7_q = v^7_q(4,4) - \sqrt{\tfrac{391}{209}}\, v^7_q(8,8) + (-1)^\alpha 3\sqrt{\tfrac{238}{627}}[v^7_q(4,8) + v^7_q(8,4)]$$

$$G^{11}_q = v^{11}_q(8,8) - (-1)^\alpha \sqrt{\tfrac{11}{30}}[v^{11}_q(4,8) + v^{11}_q(8,4)] \tag{6}$$

wherein α may be freely chosen.

The fact that G^3_q and G^9_q are missing in (6) is a source for the explanation of certain non-trivial zeros of the 6j-symbol. Indeed, let us first consider the commutator $[G^{11}, G^{11}]$ which could give rise to a term proportional to $v^3(4,4)$. Since moreover this term would be generated with a coefficient proportional to $\begin{Bmatrix} 11 & 11 & 3 \\ 4 & 4 & 8 \end{Bmatrix}$ its absense from (6) explains the non-trivial vanishing of the 6j-symbol. It should be noticed that for instance the term $v^3(8,8)$ is not generated from the commutator $[G^{11}, G^{11}]$ proportional to a single 6j-coefficient. Instead of a structural zero a relation between distinct 6j-coefficients follows from it. The second structural zero which can be explained from (6) is $\left\{\begin{smallmatrix} 11 & 11 & 9 \\ 8 & 4 & 8 \end{smallmatrix}\right\} = 0$ as a consequence of the absense of a term of the form $v^9(8,4)$ in the algebra (6) which on the other hand could have been generated from the commutator $[G^{11}, G^{11}]$. For more details on SO(3) tensor operator realisations of F_4 the reader is referred to an earlier paper [7].

In the table below are assembled the structural zeros $\left\{\begin{matrix} j_1 & j_2 & j_3 \\ \ell_1 & \ell_2 & \ell_3 \end{matrix}\right\}$ which we succeeded to explain from minimal realisations of F_4 and E_6. One should notice that also direct products of SO(3) subgroups have been considered at the tail of the chains. This requires an extension of the tensor operator formalism which has been developed elsewhere [8]. Also, only the zeros marked with an ° are directly explained within the chains mentioned; the other zeros follow from them on account of Regge symmetries.

j_1	j_2	j_3	ℓ_1	ℓ_2	ℓ_3		chain
5	5	3	3	3	3	°	$G_2 \supset A_1$
5	4	4	3	4	2		
11	11	3	4	4	8	°	$F_4 \supset A_1$
11	10	2	4	5	9		
11	11	9	8	4	8	°	
11	10	10	4	9	7		
12	11	8	5	8	7		
13	10	8	6	7	7		
13	9	9	6	8	6		
11	8	6	4	4	8	°	$E_6 \supset F_4 \supset A_1$
11	9	5	4	5	7		
10.5	9.5	5	3.5	3.5	7		
9.5	8.5	4	3.5	6.5	8		
9.5	9.5	6	5.5	2.5	8		
9.5	9	6.5	2.5	5	8.5		

j_1	j_2	j_3	ℓ_1	ℓ_2	ℓ_3		chain
3	2	2	1	2	2	°	$F_4 \supset A_1 \oplus G_2 \supset A_1 \oplus A_1$
7	4.5	4.5	2.5	4	4	°	$F_4 \supset A_1 \oplus C_3 \supset A_1 \oplus A_1$
6.5	4.5	2	2	4.5	4		
7	6	5	4	6	4	°	$E_6 \supset C_4 \supset A_1$
7	6.5	4.5	4	5.5	4.5		
7.5	5.5	5	4.5	5.5	4		
7.5	6	4.5	4.5	5	4.5		
6.5	6	5.5	5.5	3	5.5		
6	6	6	6	5	3	°	
6	6	6	5	4	3		
7	6	5	4	4	4		
7	5.5	5.5	4	4.5	3.5		
6.5	6.5	5	4.5	3.5	4		
6.5	6	5.5	3.5	5	3.5		
6.5	6	5.5	4.5	3	4.5		
9	6	4	2	5	5	°	
8	6	5	1	5	6		

By simply looking through branching rule tables [6] it is possible to predict also preliminary results concerning the exceptional algebras E_7 and E_8. Furthermore there is an indication that a non-minimal realisation of F_4 is a good framework for explaining certain zeros too. All these results which will still have to be confirmed by explicit calculations are listed in the next table.

j_1	j_2	j_3	ℓ_1	ℓ_2	ℓ_3		chain	j_1	j_2	j_3	ℓ_1	ℓ_2	ℓ_3		chain
29	21	11	11	19	11	o	$E_8 \supset A_1$	17	17	3	4.5	4.5	13.5	o	$E_7 \supset A_1$
21	20	20	19	20	2			17	16	2	4.5	5.5	14.5		
29	20	12	11	20	10			17	17	15	4.5	13.5	13.5	o	
								17	16	16	4.5	12.5	14.5		
11	9	5	5	7	5	o	$F_4 \supset A_1$	20.5	17	11.5	21.5	12.5	15		
11	8	6	5	8	4		(n.-m.)	8	13.5	10	9	9	13.5		
9	8	8	7	8	2			21.5	16	11.5	9	12.5	10		
								20.5	16	12.5	8	14.5	9		

Notice that many of these zeros are not contained in the tables of Biedenharn and Louck [1]. A FORTRAN programme which we developed to generate 6j-coefficients analytically permitted us to verify that the coefficients above become zero indeed.

Clearly we have not been directly concerned with the embeddings of the exceptional algebras in higher dimensional algebras, which could have been clarified by the explicit tensor operator realisations. Although in general this problem is far from trivial, it can be solved for the example $F_4 \supset SO(3)$ which we treated before. To that aim one can make use of a theorem [9] which corrects a previously established result (see e.g. ref. 5) and which learns that the set of SO(3) tensors $\{v^k(\ell\ell), v^{k'}(\ell'\ell'), v^{k''}(\ell\ell')+\varepsilon(-1)^{k''}v^{k''}(\ell'\ell) \mid k,k',k''\in \mathbb{Z}_+,\ k \text{ and } k' \text{ odd},\ \varepsilon\in\{-1,+1\}\}$ realises the Lie algebra $D_{\ell+\ell'+1}$. Hence, it is readily verified that the realisation (6) elucidates the embedding of F_4 into the Lie algebra D_{13}.

As a conclusion we can claim that certainly not all structural zeros of the 6j-symbol can be explained in the way exhibited here. It remain at present open questions to trace out the precise subset of zeros that can be explained, whether this subset is finite or infinite and whether its elements can be grouped in one or more parametrized families of zeros, such as the one mentioned at the beginning of this paper.

1. Biedenharn L. and Louck J., *The Racah-Wigner Algebra in Quantum Theory*, London : Addison-Wesley (1981)
2. Racah G., Phys. Rev. 62, 438 (1942)
3. Racah G., Phys. Rev. 76, 1352 (1949)
4. Wadzinski H., Il Nuovo Cim. 62B, 247 (1969)
5. Judd B., *Operator techniques in Atomic Spectroscopy*, New York : McGraw-Hill (1963)
6. McKay M. and Patera J., *Tables of dimensions, indices and branching rules for representations of simple Lie algebras*, New York : Marcel Dekker (1981)
7. Van der Jeugt J., Vanden Berghe G. and De Meyer H., J. Phys. A 16, 1377 (1983)
8. De Meyer H., Vanden Berghe G. and Van der Jeugt J., J. Math. Phys. (in press)
9. Vanden Berghe G. and De Meyer H., J. Math. Phys. (in press)

COMPLETELY INTEGRABLE SYSTEMS

YANG - BAXTER ALGEBRAS OF DYNAMICAL CHARGES
IN THE CHIRAL GROSS - NEVEU MODEL

H. Eichenherr
Laboratoire de Physique Théorique et Hautes Energies,
Université Pierre et Marie Curie, Tour 16 - 1er étage,
4, place Jussieu, 75230 Paris Cedex 05 (France)

A certain class of two-dimensional field theories containing the non-linear σ models on symmetric spaces and various spinor models is distinguished by the existence of an infinite series of conserved non-local charges. These dynamical charges are non-abelian with respect to Poisson brackets or quantum commutators and are therefore supposed to constitute an example of a dynamical symmetry group in field theory. In this seminar I shall derive their classical canonical as well as quantum commutator algebra for the example of the chiral Gross-Neveu model. The result[1-3] is that the classical and quantum charge algebras are quadratic Lie algebras; the matrices containing their structure constants solve the classical and quantum Yang-Baxter equations, respectively.

a) Classical charge algebra

The canonical formalism for the U(N) chiral Gross-Neveu model is given by the Hamiltonian

$$H = \int dx \left[-i\psi_a^\dagger \gamma^5 \partial_1 \psi_a - g((\bar{\psi}_a\psi_a)^2 - (\bar{\psi}_a\gamma^5\psi_a)^2)\right]$$

and the Poisson brackets

$$\{A,B\} = i\int dx \sum_{a,\alpha} A\left(\frac{\overleftarrow{\delta}}{\delta\psi_{a\alpha}^\dagger(x)}\,\frac{\overrightarrow{\delta}}{\delta\psi_{a\alpha}(x)} + \frac{\overleftarrow{\delta}}{\delta\psi_{a\alpha}(x)}\,\frac{\overrightarrow{\delta}}{\delta\psi_{a\alpha}^\dagger(x)}\right)B$$

where $\overleftarrow{\delta}/\delta\psi$ ($\overrightarrow{\delta}/\delta\psi$) denotes the left (right) derivative in the Grassmann-algebra of the anticommuting classical spinor field $\psi_{a\alpha}$ ($a=1..N$, $\alpha=1,2$). From the conserved and curvatureless currents

$$(A_\mu)_{ab} = -4ig(\bar{\psi}_a\gamma_\mu\psi_b)\,, \qquad \lim_{|x|\to\infty} A_\mu(t,x) = 0$$

$$\partial_\mu A^\mu = 0\,, \qquad \partial_0 A_1 - \partial_1 A_0 + [A_0,A_1] = 0$$

we construct the compatible linear system

$$\partial_\mu \Phi(t,x|\lambda) = -L_\mu(t,x|\lambda)\Phi(t,x|\lambda) \ , \qquad \Phi(t,+\infty|\lambda) = \mathbb{1}$$

where $\quad L_\mu = \dfrac{\lambda}{\lambda^2-1}(\lambda A_\mu - \varepsilon_{\mu\nu}A^\nu) \ , \qquad \varepsilon_{01} = 1 \quad ,$

and its monodromy matrix $T(\lambda) = \Phi(t,-\infty|\lambda)$. $T(\lambda)$ is the generating functional of the non-local charges :

$$T(\lambda) = \sum_0^\infty \lambda^n T^{(n)} \ , \quad \{H,T(\lambda)\} = \{H,T^{(n)}\} = 0 \ ,$$

in particular,

$$T^{(1)} = -\int dx \ A_0(x) \ , \quad T^{(2)} = \int dxdy \ \Theta(y-x)A_0(x)A_0(y) - \int dx \ A_1(x) \quad \text{etc.}$$

We shall employ the tensor notation

$$(A\otimes B)_{ik,jl} = A_{ij}B_{kl} \ , \quad \{A \overset{\otimes}{,} B\}_{ik,jl} = \{A_{ij},B_{kl}\} \ .$$

To compute the charge algebra, we use the chain rule to write

$$\{T(\lambda) \overset{\otimes}{,} T(\mu)\} = \int dxdy \left(\frac{\delta T(\lambda)}{\delta L_{1ab}(x,\lambda)} \otimes \frac{\delta T(\mu)}{\delta L_{1cd}(y,\mu)}\right)\{L_{1ab}(x,\lambda),L_{1cd}(y,\mu)\}, \quad (1)$$

insert the expression

$$\{L_1(x,\lambda) \overset{\otimes}{,} L_1(y,\mu)\} = \delta(x-y) \ [r(\lambda^{-1}-\mu^{-1}),L_1(x,\lambda)\otimes\mathbb{1} + \mathbb{1}\otimes L_1(y,\mu)]$$

with $\quad r_{ik,jl}(x) = 4g \ x^{-1} \ \delta_{il}\delta_{jk} \quad ,$

and obtain[2], observing that the integrand of eq.(1) is a complete derivative :

$$\{T(\lambda) \overset{\otimes}{,} T(\mu)\} = [r(\lambda^{-1}-\mu^{-1}),T(\lambda)\otimes T(\mu)] \qquad (2)$$

and hence for the charges $T^{(n)}$

$$\{T_{ij}^{(n)},T_{kl}^{(m)}\} = 4g\sum_{r=0}^{m-1}\left(T_{il}^{(r+n)}T_{kj}^{(m-r-1)} - (kj\leftrightarrow il)\right) \ .$$

The matrix $r(x)$ which solves the classical Yang-Baxter equations[4] provides the structure constants of the canonical charge algebra. The r.h.s. of eq.(2) being quadratic, the $T_{ij}^{(n)}$ do not form a closed basis of the Lie algebra : one has to include products of the $T_{ij}^{(n)}$ with any number of factors.

b) Quantum charge algebra

The charges $T_{ij}^{(n)}$ are supposed to have quantum counterparts which are obtained by renormalization of the classical expressions. The first two conserved charges in the quantum SU(N) chiral Gross-Neveu model act on, say, an outgoing k-particle state as follows[3] :

$$T^{(1)}_{ab}|\theta_1 c_1 \ldots \theta_k c_k\rangle_{out} = 4ig|\theta_1 d_1 \ldots \theta_k d_k\rangle_{out} \sum_{i=1}^{N} \left[I^{(i)}_{ab}\right]_{\{d\}\{c\}},$$

$$\left[T^{(2)}_{ab} - \frac{1}{2}T^{(1)}_{ae}T^{(1)}_{eb}\right]|\theta_1 c_1 \ldots \theta_k c_k\rangle_{out} = (-8g^2)|\theta_1 d_1 \ldots \theta_k d_k\rangle_{out} \cdot \tag{3}$$

$$\cdot \left[\sum_{i<j} (I^{(i)}_{ae} I^{(j)}_{eb} - I^{(j)}_{ae} I^{(i)}_{eb}) + \frac{iN}{\pi} \sum_{j=1}^{k} \theta_j I^{(j)}_{ab}\right]_{\{d\}\{c\}} .$$

Here $\left[I^{(i)}_{ab}\right]_{\{d\}\{c\}} = (\delta_{ad_i} \delta_{bc_i} - \frac{1}{N}\delta_{ab}\delta_{d_i c_i}) \prod_{j \neq i} \delta_{d_j c_j}$, the c_j are the SU(N) labels of the particles and the θ_j are their rapidities ($\theta_i < \theta_j$ for $i<j$ on out states). To determine the algebra of the quantum charges without explicit renormalization of infinitely many operators, we make use of a factorization principle based on the following classical property of $T(\lambda)$:

$$\text{Let} \quad \psi(t,x) = \begin{cases} \psi_1(t,x) & x<A \\ 0 & A \leq x \leq B \\ \psi_2(t,x) & B<x \end{cases}$$

$$\text{then} \quad T_{ab}(\lambda;\psi) = T_{ac}(\lambda;\psi_1) T_{cb}(\lambda;\psi_2) \tag{4}$$

The quantum analogue of eq. (4) reads[5)]

$$T_{ab}(\lambda)|\theta_1 c_1, \theta_2 c_2\rangle_{out} = T_{aa_1}(\lambda)|\theta_1 c_1\rangle T_{a_1 b}(\lambda)|\theta_2 c_2\rangle \tag{5}$$

(analogously for k>2 particles). Assuming such a factorized action of the quantum monodromy operator $T_{ab}(\lambda)$, its k-particle matrix elements in the asymptotic Fock space are determined by the one-particle matrix elements

$$\langle\theta d|T_{ab}(\lambda)|\theta c\rangle = \delta(\theta-\theta')\left[\delta_{ab}\delta_{cd} f_1(\lambda,\theta) + \delta_{ad}\delta_{cb} f_3(\lambda,\theta)\right] . \tag{6}$$

To get functional equations for the f_i, we use

i) conservation of $T_{ab}(\lambda)$ in $2 \to 2$ scattering, schematically :

$$_{out}\langle 2|T_{ab}(\lambda)|2'\rangle_{in} \begin{cases} = \sum_{2''} {}_{out}\langle 2|2''\rangle_{in}\ {}_{in}\langle 2''|T_{ab}(\lambda)|2'\rangle_{in} \\ = \sum_{2''} {}_{out}\langle 2|T_{ab}(\lambda)|2''\rangle_{out}\ {}_{out}\langle 2''|2'\rangle_{in} \end{cases} \tag{7}$$

ii) invariance under parity ($\mathcal{P}$) and time reversal ($\mathcal{T}$) :

$$T_{ac}(-\lambda)\mathcal{P}T_{cb}(\lambda)\mathcal{P}^{-1} = \delta_{ab}\,\mathbf{1} \tag{8}$$

$$T_{ac}(\lambda)(\mathcal{PT})T_{cb}(\lambda)(\mathcal{PT})^{-1} = \delta_{ab}\,\mathbf{1} \tag{9}$$

Inserting eqs.(5),(6) into eqs.(7),(8),(9) and analyzing the resulting functional equations for f_1 and f_3, one finds[3] that $\langle\theta d|T_{ab}(\lambda)|\theta' c\rangle$ can be expressed through the S-matrix of the model :

$$\langle\theta d|T_{ab}(\lambda)|\theta' c\rangle = \delta(\theta-\theta')F_{ad,bc}(\theta+\gamma(\lambda))e^{i\phi(\lambda,\theta)} \tag{10}$$

Here $\phi(\lambda,\theta)$ is a phase, $\gamma(\lambda)$ is a spectral parameter arising naturally from the factorization principle, and the particle-antiparticle scattering amplitude $F_{ad,bc}(\theta)$ is given by[7]

$$F_{ad,bc}(\theta) = \delta_{ab}\delta_{cd}t_1(\theta) + \delta_{ad}\delta_{cb}t_2(\theta)$$

$$t_1(\theta) = \frac{\Gamma(1/2+\theta/2\pi i)\Gamma(1/2-1/N-\theta/2\pi i)}{\Gamma(1/2-1/N+\theta/2\pi i)\Gamma(1/2-\theta/2\pi i)} \quad , \quad \frac{t_1(\theta)}{t_2(\theta)} = \frac{N}{2\pi i}(\theta-i\pi)$$

(up to CDD-poles). In fact, eq.(7) formally coincides with the Yang-Baxter equation ("factorization equation") for the factorized S-matrix of the chiral Gross-Neveu model[6] :

$$S_{mn,ik}(\theta-\theta')F_{ic,jd}(\theta)F_{kd,le}(\theta') = F_{nc,kd}(\theta')F_{md,ie}(\theta)S_{ik,jl}(\theta-\theta') \tag{11}$$

with $S_{ac,bd}(\theta) = F_{ad,bc}(i\pi-\theta)$ the particle-particle scattering amplitude, whereas eqs.(8) and (9) yield the real analyticity and unitarity equations for F together with the constraints $\gamma(\lambda) = -\gamma(-\lambda)$ and $\gamma(\lambda) = \gamma^*(\lambda)$, respectively. Comparing eqs.(3) and (10), one finds

$$\phi(\lambda,\theta) = \frac{i\pi}{N}\mathrm{sgn}(\gamma(\lambda)) + O(\lambda^3) \quad ,$$
$$\gamma(\lambda) = \frac{\pi}{2gN\lambda} + O(\lambda) \qquad \text{as } \lambda \to 0. \tag{12}$$

Now the whole series of quantum non-local charges is given by eqs.(5) and (10). The commutator algebra of the quantum monodromy operators $T_{ab}(\lambda)$ is determined from the observation that the k-particle matrix element of $T_{ab}(\lambda)$ (cf. eq.(5)) looks like the monodromy matrix in the statistical mechanics of an inhomogenous vertex model on a line of k sites with $\langle\theta_j c_j|T_{a_{j-1}a_j}(\lambda)|\theta'_j c'_j\rangle$ as the statistical weights :

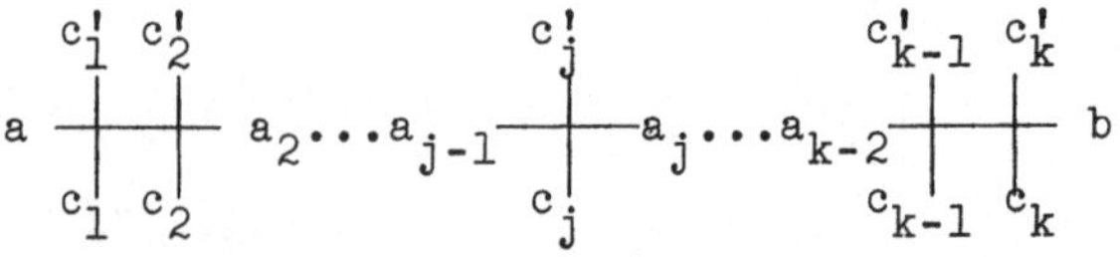

The quantum Yang-Baxter equation (11) for the weights therefore implies

$$S_{mn,ik}(\gamma(\lambda)-\gamma(\mu))\Big(T(\lambda)\bullet T(\mu)\Big)_{ik,jl} = \Big(T(\mu)\bullet T(\lambda)\Big)_{nm,ki}S_{ik,jl}(\gamma(\lambda)-\gamma(\mu))$$

where the operator product in Fock space is understood. So the structure constants of the quantum algebra are provided by the two-particle S-matrix. For the commutator we find

$$[T_{ij}(\lambda),T_{kl}(\mu)] = \frac{2\pi i}{N}\frac{1}{\gamma(\lambda)-\gamma(\mu)}\Big[T_{kj}(\lambda)T_{il}(\mu) - T_{kj}(\mu)T_{il}(\lambda)\Big] \quad . \tag{13}$$

If the expression (12) for $\gamma(\lambda)$ is exact, then the classical and quantum charge algebras (2) and (13) are isomorphic with respect to

$$\{\ ,\ \} \leftrightarrow -i[\ ,\ ] \quad .$$

It is remarkable that their behaviour under the Lorentz group is quite different : A boost

$$\Lambda(\varepsilon) = \begin{pmatrix} \cosh\varepsilon & \sinh\varepsilon \\ \sinh\varepsilon & \cosh\varepsilon \end{pmatrix}$$

leaves the classical $T(\lambda)$ invariant. However, the rapidity transforms as $\theta \to \theta+\varepsilon$, implying that the quantum spectral parameter carries a representation of the Lorentz group

$$\lambda \to \lambda(\varepsilon) \quad , \quad \gamma(\lambda) \to \gamma(\lambda)+\varepsilon \quad .$$

Again assuming $\gamma(\lambda)=\pi/2gN\lambda$ to be exact, we find

$$\lambda(\varepsilon) = \frac{\lambda}{1+\Delta\varepsilon\lambda} \quad , \quad \Delta = \frac{2gN}{\pi} \quad ,$$

and the generator of the Lorentz group $\ell = \partial_\varepsilon \Lambda(\varepsilon)\big|_{\varepsilon=0}$ obeys the commutation relations

$$[\ell,T_{ab}(\lambda)] = \Delta\lambda^2\partial_\lambda T_{ab}(\lambda)$$
$$[\ell,T^{(n)}_{ab}] = (n-1)\Delta T^{(n-1)}_{ab} \quad .$$

References

(1) H.J. de Vega, H. Eichenherr, J.M. Maillet, preprint PAR-LPTHE 83.9
(2) H.J. de Vega, H. Eichenherr, J.M. Maillet, preprint PAR-LPTHE 83.17
(3) H.J. de Vega, H. Eichenherr, J.M. Maillet, forthcoming preprint
(4) L.D. Faddeev, Les Houches lectures 1982, Saclay preprint T/82/76
(5) Al. Zamolodchikov, Dubna preprint E2-11485 (1978), unpublished
(6) B. Berg, M. Karowski, P. Weisz, V. Kurak, Nucl. Phys. B134, 125 (1978)
(7) B. Berg, P. Weisz, Nucl. Phys. B146, 2o5 (1978)

SUBGROUPS OF LIE GROUPS AND SYMMETRY REDUCTION FOR NONLINEAR PARTIAL DIFFERENTIAL EQUATIONS

A.M. Grundland[1], J. Harnad, and P. Winternitz[2]
Centre de Recherche de Mathématiques Appliquées,
Université de Montréal, Montréal, Québec, Canada H3C 3J7

Abstract

A partial differential equation in N variables that is invariant under a Lie group G can be reduced to a PDE in fewer variables, or to an ODE, by requiring that solutions depend only on the invariants of some subgroup $G_0 \subset G$. A classification of subgroups of G then provides a systematic tool for introducing such symmetry variables. A subgroup G_0 with generic orbits of codimension k $(1 \le k \le N-1)$ in the space of independent variables will reduce the number of variables in the PDE from N to k. The example of a quite general nonlinear scalar equation invariant under the Poincaré group P(n,1) is studied in detail and in particular new solutions of the n+1 dimensional sine-Gordon equation are obtained.

The theory of Lie groups was originally developed in connection with the study of ordinary and partial differential equations[1]. Not surprisingly, group theory can be applied to great advantage to study, and ultimately solve, both linear and nonlinear differential equations[2,3,4].

Among the numerous applications of the theory of Lie groups and Lie algebras to the study of nonlinear equations (the construction of Bäcklund transformations for nonlinear evolution equations, or the derivation of nonlinear superposition formulas for certain systems of nonlinear ODE's, to name just two), we concentrate, in this contribution, on one of the oldest applications, namely that of "symmetry reduction". By this we have in mind the introduction of new independent variables in a PDE, that will reduce the studied equation to a lower dimensional PDE, or in particular, to an ODE.

To be more specific, let us consider a quite general relativistically invariant second order scalar equation in an n+1 dimensional Minkowski space M(n,1)

$$H(\Box u,(\nabla u)^2,u) = 0, \tag{1}$$

where $u(x_0,x_1,\ldots,x_n)$ is a scalar function of a point $\vec{x} \in M(n,1)$, H is some given sufficiently smooth function and

$$\Box u = u_{x_0x_0}-u_{x_1x_1}-\ldots-u_{x_nx_n}, \quad (\nabla u)^2 = (u_{x_0})^2-(u_{x_1})^2-\ldots-(u_{x_n})^2 \tag{2}$$

---

1. On leave of absence from Institute of Geophysics, University of Warsaw, Warsaw, Poland.

2. Work supported in part by the Natural Sciences and Engineering Research Council of Canada and the "Fonds FCAC pour l'aide et le soutien à la recherche du Gouvernement du Québec".

(the subscripts denote derivatives). Equation (1) is invariant under the Poincaré group P(n,1) and contains, as special cases, numerous PDE's of considerable physical interest, such as the sine-Gordon equation $\Box u = \sin u$, the Liouville equation $\Box u = e^u$, the Hamilton-Jacobi equation with a nonlinear "potential" $(\nabla u)^2 + V(u) = E$, and many others.

The problem which we now pose is: How does one systematically obtain all "symmetry variables" $\xi_i(\vec{x})$ $(i=1,2,\dots,k)$, $1 \le k \le n)$, such that the Ansatz $u = u(\xi_1,\dots,\xi_k)$ reduces equation (1) to a differential equation in k variables? Let us give the answer in the form of two theorems, dropping the proofs, which are quite elementary[5].

Theorem 1

The Ansatz $u(\vec{x}) = u(\xi_1,\dots,\xi_k)$ with $1 \le k \le n$ will reduce the PDE $H(\Box u, (\nabla u)^2, u) = 0$ to a differential equation in the k variables ξ_i if and only if the ξ_i satisfy

$$\Box \xi_i = \alpha_i(\xi_1,\dots,\xi_k), \quad (\nabla\xi_i, \nabla\xi_\ell) = \beta_{i\ell}(\xi_1,\dots,\xi_k) \tag{3}$$

where α_i and $\beta_{i\ell} = \beta_{\ell i}$ are some functions of $\xi_1,\dots,\xi_k$. □

Theorem 2

Let $\xi_i(\vec{x})$ $(i=1,\dots,k)$ be a set of functionally independent invariants of a subgroup G of the Poincaré group P(n,1), having generic orbits of codimension k in M(n,1). These variables will then satisfy equations (3) for some functions α_i and $\beta_{i\ell}$. □

We shall call such variables $\xi_1,\dots,\xi_k$ "codimension k symmetry variables". They can be obtained by solving a system of linear partial differential equations:

$$X_i\phi(x_0,x_1,\dots,x_n) = 0, \quad i=1,\dots,m \tag{4}$$

where $\{X_i\}$ is some basis of the Lie algebra $L \subset p(n,1)$, corresponding to the Lie group G of Theorem 2. If G has generic orbits of codimension k, then the general solution of (4) will be an arbitrary function of precisely k "elementary" functionally independent invariants $\xi_1,\dots,\xi_k$.

As an example consider the Lie algebra $L = \{M_{01}+aP_2, P_0-P_1, P_3,\dots,P_n\}$, where $M_{\mu\nu}$ are infinitesimal O(n,1) transformations, P_μ are translations and a is a fixed real constant. Equations (4) in this case are

$$\begin{gathered}(M_{01}+aP_2)\phi = (-x_0\partial_1 - x_1\partial_0 + a\partial_2)\phi = 0, \quad (P_0-P_1)\phi = (\partial_0-\partial_1)\phi = 0 \\ P_i\phi = \partial_i\phi = 0, \quad i=3,\dots,n\end{gathered} \tag{5}$$

and their solution is easily found to be

$$\phi = \phi(\xi), \quad \xi = x_2 + a\ln(x_0+x_1). \tag{6}$$

All subgroups of P(3,1) have been classified[6] and it is hence easy to obtain all codimension 1, 2 and 3 symmetry variables in M(3,1)[5].

A complete subgroup classification of all subgroups of $M(n,1)$ for arbitrary n is a difficult task. We have however been able to find all codimension 1 symmetry variables in $M(n,1)$ [5]. Let us again present the result as a theorem, the proof of which is somewhat involved and can be found in Ref. 5.

<u>Theorem 3</u>

Let $G \subset P(n,1)$ be a Lie group with generic orbits of dimension n (hypersurfaces) in $M(n,1)$. These orbits can all be obtained by Poincaré transformations from the level sets of one of the following invariants:

$$x_0,\ x_1,\ x_0+x_1,\ \rho = x_2+\tfrac{1}{4}(x_0+x_1)^2,\ \ \sigma = x_2+a\ln(x_0+x_1),\ \ (0 \neq a \in R),$$
$$r_k = (x_1^2+x_2^2+\ldots+x_{k+1}^2)^{\frac{1}{2}}\ \ (k=1,\ldots,n-1),\ \ \tau_k = (x_0^2-x_1^2-\ldots-x_k^2)^{\frac{1}{2}}\ \ (k=1,\ldots,n). \tag{7}$$

The Ansatz $u(\vec{x}) = u(\xi)$ reduces the PDE(1) to the ODE

$$H(\varkappa[u_{\xi\xi}+\tfrac{k}{\xi}u_\xi],\ \varkappa u_\xi^2,\ u) = 0. \tag{8}$$

The values of $\varkappa$ are $\varkappa = 0$ for $\xi = x_0+x_1$, $\varkappa = 1$ for x_0 and τ_k, and $\varkappa = -1$ in all other cases. The value of k in (8) is indicated by the subscript for r_k and τ_k and we have $k = 0$ for all the remaining variables ξ. □

In other words, all codimension 1 symmetry variables are listed in (7), or can be obtained from one of these "standard" variables by a Poincaré transformation, yielding:

$$(\vec{x},\vec{A}),\ (\vec{x}+\vec{A},\vec{B})+a(\vec{x}+\vec{A},\vec{C})^2,\ (\vec{x}+\vec{A},\vec{B})+a\ln(\vec{x}+\vec{A},\vec{C})$$
$$[\sum_{a=1}^{k+1}(\vec{x}+\vec{A},\vec{B}_a)^2]^{\frac{1}{2}},\ \ [(\vec{x}+\vec{A},\vec{T})^2-\sum_{a=1}^{k}(\vec{x}+\vec{A},\vec{B}_a)^2]^{\frac{1}{2}}, \tag{9}$$

where $0 \neq a \in R$, $\vec{A},\vec{B},\vec{C},\vec{T},\vec{B}_a \in M(n,1)$, $\vec{A}$ is an arbitrary vector, $\vec{B}$ and $\vec{B}_a$ are spacelike, $\vec{C}$ is lightlike and $\vec{T}$ is timelike.

While the variables (9) are the most general codimension 1 symmetry variables for a $P(n,1)$ invariant equation in $M(n,1)$, they are not the only variables that reduce (1) to an ODE. Indeed, symmetry reduction may lead to a subspace with degenerate metric (containing at least one vector, orthogonal to the entire space). In $M(n,1)$ this occurs if we consider a subalgebra of $L = \{M_{ab},M_{0a}-M_{1a},P_a,P_0-P_1\}$ $(a,b=2,\ldots,n)$, containing P_0-P_1. The presence of this last operator places us in a subspace with coordinates $(x_0+x_1,x_2,\ldots,x_n)$ and $\eta = x_0+x_1$ is an invariant of L. Now consider a subgroup invariant of the form $\phi(\eta,x_2,\ldots,x_n)$. We have $(\nabla\phi)^2 = -(\phi_{x_2}^2+\ldots+\phi_{x_n}^2)$, $\Box\phi = -(\phi_{x_2x_2}+\ldots+\phi_{x_nx_n})$, i.e. the derivatives with respect to η drop out. This allows us to introduce "degenerate symmetry variables" as follows:

1. Consider codimension 1 symmetry variables in the Euclidean space E(n-1) spanned by $\{x_2,\dots,x_n\}$, i.e. $r_k = (x_2^2+\dots+x_{k+1}^2)^{\frac{1}{2}}$ $(k=1,\dots,n-1)$.

2. Apply a general Euclidean transformation to r_k with coefficients depending on x_0+x_1, or more generally, on $\eta = (\vec{A},\vec{x})$ with $\vec{A}^2=0$, $\vec{A}$ a constant vector in M(n,1).

The "degenerate symmetry variables" obtained in this manner are:

1. $\xi = x_2+\phi(x_0+x_1)$, or more generally,

$$\xi = (\vec{B}(\eta),\vec{x})+\phi(\eta), \quad \eta = (\vec{A},\vec{x}), \quad (\vec{B}(\eta))^2 = -1, \quad \vec{A}^2 = 0, \quad (\vec{A},\vec{B}) = 0. \tag{10}$$

We have $(\nabla\xi)^2 = -1$, $\Box\xi = 0$, so equation (1) is reduced to the ODE (8) with $\varkappa = -1$, $k = 0$.

2. $\xi_k = \{[x_2+C_2(x_0+x_1)]^2+\dots+[x_k+C_k(x_0+x_1)]^2\}^{\frac{1}{2}}$ or more generally

$$\xi_k = \{\sum_{i=1}^{k+1} (\vec{x}+\vec{C}(\eta),\vec{B}_i(\eta))^2\}^{\frac{1}{2}}, \quad (\vec{B}_i(\eta),\vec{B}_j(\eta)) = -\delta_{ij}$$
$$\eta = (\vec{A},\vec{x}), \quad \vec{A}^2 = 0, \quad (\vec{B}_i(\eta),\vec{A}) = 0. \tag{11}$$

We have $(\nabla\xi_k)^2 = -1$, $\Box\xi_k = -k/\xi_k$, so equation (1) is reduced to the ODE (8) with $\varkappa = -1$ and the value of k indicated by the subscript of ξ_k.

As an application of the method of symmetry reduction let us obtain new solutions of the n+1 dimensional sine-Gordon equation

$$\Box u = \sin u. \tag{12}$$

Putting $u = u(\xi)$, where ξ is any of the variables (7)-(11), we obtain

$$u_{\xi\xi}+ \frac{k}{\xi}u_\xi = \varepsilon \sin u, \quad \varepsilon = \pm 1, \quad k=0,1,\dots,n. \tag{13}$$

For k=0, i.e. $\xi = (\vec{A},\vec{x})$, $\vec{A}^2 = \pm 1$, or ξ as in (10), we obtain the exact pendulum equation with the classical periodic, non periodic and "kink" solutions:

$$\begin{gathered} u = 2\arccos[d\ (\xi+\alpha,m)]+ \frac{1+\varepsilon}{2}\pi, \quad 0 < m < 1 \\ u = 2\arccos[\mathrm{cn}(\frac{\xi+\alpha}{m},m)]+ \frac{1+\varepsilon}{2}\pi, \quad \alpha \in \mathbb{R} \\ u = 4\arctan\alpha e^{\varepsilon_0\xi} - \frac{1-\varepsilon}{2}\pi, \quad \varepsilon_0 = \pm 1 . \end{gathered} \tag{14}$$

These are the well known "travelling wave solutions" for $\xi = (\vec{A},\vec{x})$; as functions of a much more general variable (10), involving an arbitrary vector function $\vec{B}(\eta)$ and scalar $\phi(\eta)$, they are new.

For $k \geq 1$ in (13) we set $u = 2i\ell n y$ and obtain

$$y\ddot{y} = \dot{y}^2- \frac{k}{\xi}y\dot{y} + \frac{\varepsilon}{4}(y^4-1). \tag{15}$$

For k=1 this is a special case of P_{III}, the third Painlevé transcendent[7].

For $k \geq 2$ equation (15) does not have the "Painlevé property"; indeed the generic solution will possess moving logarithmic singularities. Expanding the solution about a singular point ξ_0, we find

$$y = 2\sqrt{\varepsilon}(\xi-\xi_0)^{-1} - \frac{k}{\xi_0}\sqrt{\varepsilon} + a(\xi-\xi_0) + \frac{2k(k-1)}{3\sqrt{\varepsilon}\,\xi_0^2}(\xi-\xi_0)\ln(\xi-\xi_0)+\ldots \qquad (16)$$

where ξ_0 and a are arbitrary constants, depending on the initial conditions. If we accept the "Painlevé conjecture"[8] we conclude from this analysis that the sine-Gordon equation is not integrable by inverse scattering techniques in more than 1+1 dimension .

To conclude we would like to remark that the analysis presented here does not depend on the form of the PDE (1), but only on its invariance properties. A generalization to arbitrary Riemannian (or pseudo-Riemannian) spaces with nontrivial isometry groups is straightforward, as is the generalization to arbitrary PDE's with nontrivial invariance groups.

Acknowledgements

One of the authors (P.W.) thanks the ICTP for its hospitality and assistance in the preparation of this manuscript.

References

1. S.Lie, Vorlesungen über Differentialgleichungen mit bekannten infinitesimalen Transformationen. Teubner, Leipzig, 1891.
2. L.V.Ovsyannikov, Gruppovoĭ analiz differentsial'nykh uravneniĭ (Group Theoretical Analysis of Differential Equations), Nauka, Moscow, 1978.
3. G.W.Bluman and J.D.Cole, Similarity Methods for Differential Equations, Springer 1974.
4. P.Winternitz, Lie Groups and Solutions of Nonlinear Differential Equations (To be published in Proceedings of School and Workshop on Nonlinear Phenomena, Lecture Notes in Physics 189, Springer 1983).
5. A.M.Grundland, J.Harnad and P.Winternitz, Preprint CRMA-1162, Montreal, 1983; and also KINAM, Rev. de Fisica 4, 333 (1982).
6. J.Patera, P.Winternitz and H.Zassenhaus, J.Math.Phys. 16, 1597 (1975).
7. E.L.Ince, Ordinary Differential Equations, Dover, 1956.
8. M.J.Ablowitz, A.Ramani and H.Segur, J.Math.Phys. 21, 715 and 1006 (1980).

SPINORIAL DESCRIPTION OF LIE SUPERALGEBRAS */

Z. Hasiewicz, A.K. Kwaśniewski
Institute of Theoretical Physics
University of Wrocław
50-205 Wrocław, ul. Cybulskiego 36

A canonical derivation of extended classical Lie superalgebras right from the metric structure of underlying space time is presented. The canonical method is that of Clifford algebras $\mathcal{C}(p,q)$ representation theory.

Due to the isomorphism between even subalgebra $\mathcal{C}^{+}(p,q)$ and $\mathcal{C}(p,q-1)$ Clifford algebra, considerations carried out for binors are easily translated into those for spinors, where binors are defined as elements of faithful $\mathcal{C}(p,q)$ representation module, while spinors are elements of $\mathcal{C}^{+}(p,q)$ irreducible representation module.
We choose to discuss our construction in "binor language".

Once a "space time" $E(p,q)$ with nondegenerate quadratic form /signature; p-pluses, q-minuses/ is chosen one canonically associates with it its $\mathcal{C}(p,q)$ Clifford algebra and hence the faithful /say-right/ A-module $S(p,q)$ of $\mathcal{C}(p,q)$ where $A = F, {}^{2}F$ and $F = \mathbb{R}, \mathbb{C}, \mathbb{H}$; A being uniquely determined by /p.q/ [1÷3]

Simultaneously one has two /also canonically defined for any /p.q/ signature/ main anti-involutions $\beta_{\pm}$ of End $(S(p,q)) \sim \mathcal{C}(p,q)$ $(\beta_{\pm}(x) = \pm x \; ; \; x \in E(p,q) \subset \mathcal{C}(p,q))$ These $\beta_{\pm}$ anti-involutions in turn, induce /symmetric or skew/ sesquilinear forms $\theta_{\pm}$ [2-4] on the representation A-module $S(p,q)$
Thus one arrives a the possibility to assign to each /p.q/ the groups $G_{\pm}(1,F)$ and $G_{\pm}(p,q)$ which are correspondingly sesquilinear form preserving groups on F-ring and on the A-module $S(p,q)$ [4].
This could be summarized in a sequence of correspondencies

$$E(p,q) \longleftrightarrow \mathcal{C}(p,q) \sim \mathrm{End}\,(S(p,q)) \longleftrightarrow \begin{cases} A = F, {}^{2}F \\ S(p,q) \end{cases} \longleftrightarrow \begin{cases} G_{\pm}(1,F) \\ G_{\pm}(p,q) \end{cases} \qquad /1/$$

Similarily one can associate with /p.q/ a Lie superalgebra. This is due to an observation, that Z_2 graded associative algebra

$$\left\{ \begin{pmatrix} f & \psi^{(\pm)} \\ \varphi & c \end{pmatrix} \; ; \; f \in A, \; \psi, \varphi \in S(p,q), \; c \in \mathcal{C}(p,q) \right\} \qquad /2/$$

where $\psi^{(\pm)}$ means: $(\pm)$ conjugated ψ ; $(\pm)$ conjugation being that induced by $\beta_{\pm}$ anti-involutions [2 - 4], has as its associated Lie superalgebra $SC(p,q)$

$$SC(p,q) = \begin{cases} \ell(1,m,F) & p-q \neq 1 \; (\mathrm{mod}\,4) \\ {}^{2}\ell(1,m,F) & p-q = 1 \; (\mathrm{mod}\,4) \end{cases} \qquad /3/$$

whrere $m = \dim_F S(p,q)$ and for notation see [5].
This canonical and rather trivial assigning of $SC(p,q)$ to (p,q) signature, enables one in turn to distinguish in $SC(p,q)$, also in a canonical way, a Lie super-subalgebra $a(p,q)$, for which its odd part $a_1(p,q)$ is the module of faithful representation of Pin (p,q) group.
Note that thus we arrive at Lie superalgebras for which the supersymmetry generators have definite transformation properties of binors

*/based on Ref. 4

/hence, spinors/. The $a(p,q)$ Lie superalgebras are projected out from $sc(p,q)$ by projectors $P_\pm = \frac{1}{2}(id \pm J)$ where J is an anti-involution defined as follows

$$sc(p,q) \ni \begin{pmatrix} f & \psi^{(t)} \\ \varphi & c \end{pmatrix} \longmapsto J\begin{pmatrix} f & \psi^{(t)} \\ \varphi & c \end{pmatrix} := \begin{pmatrix} f^{(\pm)} & \varphi^{(\pm)} \\ \psi & \beta_\pm(c) \end{pmatrix} \in sc(p,q) \qquad /4/$$

correspondingly to weather β_+ or β_- is chosen. Hence the odd part $a_1(p,q)$ is J-selfadjoint (P_+) or J-antiselfadjoint (P_-).
Both possibilities lead to the same nontrivial reduction of $sc(p,q)$ to $a(p,q)$ - if at all [4].
This is because, with the odd part $a_1(p,q)$ being, say, J-selfadjoint linear space $a(p,q)$ becomes Lie superalgebra iff the even part $a_0(p,q)$ is J-antiselfadjoint /and vice versa/, which is then equivalent to the requirement that Θ_+ or Θ_- sesquilinear forms on $sc(p,q)$ are to be skew.
This is exactly this very $a(p,q)$ Lie superalgebra that we canonically assign to /p.q/ -signature, thus adding to the sequence /1/ another object - i.e. $a(p,q)$ Lie superalgebra.
In order to identify it, note that even part

$$a_0(p,q) = g_\pm(1,F) \oplus g_\pm(p,q) \qquad /5/$$

where $g_\pm(1,F)$/1.F/ and $g_\pm(p,q)$ are the Lie algebras of $G_\pm(1,F)$ and $G_\pm(p,q)$ groups.
With use of Clifford algebras properties only one can calculate explicitely commutators and anticommutators of $a(p,q)$ algebra elements.
For example

$$\left\{\begin{pmatrix} 0 & \psi^{(t)} \\ \psi & 0 \end{pmatrix}, \begin{pmatrix} 0 & \varphi^{(t)} \\ \varphi & 0 \end{pmatrix}\right\} = \begin{pmatrix} \Theta_\pm(\psi,\varphi) & 0 \\ 0 & Z^{\alpha}(\psi,\varphi)E_\alpha \end{pmatrix} \qquad /6/$$

where $Z^\alpha(\psi,\varphi)$ coefficients are nonzero only for E_α's satisfying:
$\beta_\pm(E_\alpha) = -E_\alpha$;
$\{E_\alpha\}_{\alpha\in\Gamma}$ is the canonical basis of $\mathcal{C}(p,q)$ hence $\Gamma = Z_2 \oplus \dots \oplus Z_2$ /p+q summands/ [4].
As all sesquilinear /on A-module/ form preserving groups $G_\pm(p,q)$ are known [2,3] one is ready now to identify all $a(p,q)$ Lie superalgebras, the result being presented in the table 1.
$a(p,q)$ Lie superalgebras correspond to the so called simple $(N=1)$ supersymmetries for $E(p,q)$ "space-time/.
Corresponding $a^N(p,q)$ Lie superalgebras for extended supersymmetries are easily obtained from $a(p,q)$ ones [4]. The above Lie superalgebras are classical in language of physics as well as in that of [5].
In order to incorporate eventually also $F(4)$ and $G(3)$ Lie superalgebras [5] one is naturally led to recognize the importance of the existing relations between spinors and octonions. Surely one should also replace $(\pm)$ conjugation in definition of $sc(p,q)$ by some other, a hint for that being the specific relation between 0/8,0/ spinors and octonions [6]. We then hope to arrive at similar derivation of $F(4)$ and $G(3)$ via constructions based on $\mathcal{C}(8,0)$ Clifford algebra.

Acknowledgements

We would like to thank Prof. J. Lukierski and Prof. H. Ruegg for stimulating assistence and usefull comments - correspondingly.

a(p,q) Lie superalgebras

q = 0 (row labels β+ / β− on the left; β− / β+ on the right for p = 0)

p	0	7	6	5	4	3	2	1
β+	–	–	–	UαU(1/2n;C)	USp(1/2n;H)	^{2}USp(1/2n;H)	USp(1/2n;H)	UαU(1/2n;C)
β−	–	–	OSp(1/2n;C)	OSp(1/2n;C)	USp(1/n,n;H)	–	UαU(1/2n;H)	–

p+q (mod 8) (rows) / **p−q (mod 8)** (columns)

p+q (mod 8)	0	1	2	3	4	5	6	7
0	– –		– –		UαU(1/2n;H) UαU(1/2n;H)		UαU(1/2n;H) UαU(1/2n;H)	
1		– –		– UαU(1/n,n;C)		^{2}UαU(1/2n;H) –		– UαU(1/n,n;C)
2	– OSp(1/2n;R)		OSp(1/2n;R) OSp(1/2n;R)		USp(1/n,n;H) USp(1/n,n;H)		UαU(1/2n;H) USp(1/n,n;H)	
3		– ^{2}OSp(1/2n;R)		UαU(1/n,n;C) OSp(1/2n;C)		– ^{2}USp(1/n,n;H)		UαU(1/n,n;C) OSp(1/2n;C)
4	OSp(1/2n;R) OSp(1/2n;R)		OSp(1/2n;R) OSp(1/2n;R)		USp(1/n,n;H) USp(1/n,n;H)		USp(1/n,n;H) USp(1/n,n;H)	
5		^{2}OSp(1/2n;R) –		OSp(1/2n;C) UαU(1/n,n;C)		^{2}USp(1/n,n;H) –		OSp(1/2n;C) UαU(1/n,n;C)
6	OSp(1/2n;R) –		OSp(1/2n;R) –		USp(1/n,n;H) UαU(1/2n;H)		USp(1/n,n;H) UαU(1/2n;H)	
7		– –		UαU(1/n,n;C) –		– ^{2}UαU(1/2n;H)		UαU(1/n,n;C) –

$n = \dim_F \mathcal{S}(p,q)$

TABLE 1 (*)

(*) see also [7]

References

[1] Atiyah M.F., Bott R., Shapiro S. — Topology 3 suppl. 1 pp 3-38 /1964/

[2] Porteous I — Topological Geometry /Van Nostrand -Reinhold, London 1969/

[3] Lounesto P. — Foundations of Phys. 11 1981, 721-740

[4] Z. Hasiewicz, A.K. Kwaśniewski, P. Morawiec — Supersymmetry and Clifford Algebras, Wrocław Univ. preprint No 580 /March 83/

[5] Kac V.G. — Adv. Math. 26, 8 /1977/

[6] Z. Hasiewicz, A.K. Kwaśniewski — Spinors and octonions, Wrocł. Univ. preprint No 590 /June 83/

[7] J. Lukierski, A. Nowicki — Fort. de Phys. 30, 75 (1982)

NOETHERIAN SYMMETRIES, BÄCKLUND TRANSFORMATION AND CONSERVATION LAWS FOR A COMPLETELY INTEGRABLE THREE DIMENSIONAL SYSTEM

A. Roy Chowdhury
High Energy Physics Division
Department of Physics
Jadavpur University
Calcutta - 700 032
INDIA.

The importance of symmetry group in the analysis of completely integrable nonlinear systems has been felt from the very initiation of the subject of soliton. In this respect the frame work of classical mechanics - the Noetherian transformation laws and canonical approach have been proved to be quite successful. But uptill now such analysis has been restricted only to nonlinear equations in one space and one time dimension. Here we have made an approach of Noetherian transformation law in three dimension in conjunction with Infinitesimal Backlund transformation to deduce the infinite number of conservation laws. An interesting outcome of our approach is a new form of the Backlund transformation for the special system of Kadomstev-Petviashvili equation in three dimension.

The equation under consideration reads

$$\beta^2 W_{yy} + \left(\frac{4\alpha}{3} W_t + W_x^2 + \frac{1}{3} W_{xxx}\right)_x = 0 \quad \cdots\cdots \quad (1)$$

which is obtained through the Euler-Lagrange equation;

$$\Lambda[\omega] = -\left(\frac{\partial L}{\partial W_t}\right)_t - \left(\frac{\partial L}{\partial W_x}\right)_x - \left(\frac{\partial L}{\partial W_y}\right)_y + \left(\frac{\partial L}{\partial W_{xx}}\right)_{xx}$$

from the Lagrangian;

$$L = \frac{2\alpha}{3} W_t W_x + \frac{1}{3} W_x^3 - \frac{1}{6} W_{xx}^2 + \frac{\beta^2}{2} W_y^2 \quad \cdots\cdots (2)$$

Let W_1 and W_0 be two solutions of the $K-P$ equation. Then we find;

$$\delta L = L(W_1) - L(W_0) = \frac{4\alpha}{3} \cdot \frac{\partial}{\partial x}[dS_t] - \frac{4\alpha}{3} \frac{\partial}{\partial t}[dS_x] + 2\phi$$

With ϕ given as;

$$\phi = \frac{1}{3} \cdot \frac{\partial}{\partial x}[d . d_x^2] + \frac{4\alpha}{3} . S_x d_t - \frac{1}{3} S_{xx} d_{xx} + \chi$$

$$\mathcal{X} = S_x^2 dx + \beta^2 S_y dy - \frac{2}{3} d\, d_x d_{xx}$$

$$S = \frac{1}{2}(W_1 + W_0) \; ; \quad d = \frac{1}{2}(W_1 - W_0). \quad \cdots (4)$$

To put the variation of L that is δL in the form of a total divergence we assume a functional dependence of S and as follows :

$$S_x = \lambda(d) \quad \cdots\cdots (5)$$

so it is easily seen that

$$\delta L = \frac{4\alpha}{3} \cdot \frac{\partial}{\partial t}[\mu(d)] + \frac{\partial}{\partial x}[\nu(d)] + \frac{\partial}{\partial y}[\psi(d)] - \left[\frac{1}{3}\lambda' + \frac{2}{3}d\right] d_x d_{xx}$$

with the following choice of λ and ψ

$$\frac{1}{3}\lambda'(d) + \frac{2}{3}d = b \; ; \; \psi'(d) = 2\beta^2 S_y \; ; \; \mu'(d) = \lambda(d)$$

Now invoking the consisting; $S_{yx} = S_{xy}$ we have the variational Backlund transformatic for $K-P$ equation in the form;

$$\left.\begin{aligned} S_x &= a + \gamma d - d^2 \\ S_y &= \int [\gamma - 2d]\, d_y\, dx \end{aligned}\right\} \quad \cdots\cdots (6)$$

For explicit construction of the conservation laws; we note that

$$\delta L = \frac{\delta L}{\delta W_t}\delta W_t + \frac{\partial L}{\partial W_x}\delta W_x + \frac{\partial L}{\partial W_y}\delta W_y + \frac{\partial L}{\partial W_{xx}}\delta W_{xx} \cdots (7)$$

But by Noethers identity

$$\delta L = A_t + B_x + C_y + \Lambda \delta \omega \quad \cdots\cdots\cdots\cdots (8)$$

Since under the infinitesimal variation $W \to W + \delta W = W + \epsilon f$ (which change may also take place due to Backlund Transformation) the Lagrangian change by a divergence that is

$$\delta L = \epsilon(\theta_t + \Xi_x + \Lambda_y) \quad \cdots\cdots\cdots (9)$$

So that the conservation Law become

$$T_t + X_x + Y_y = 0$$

$$\left.\begin{aligned} T &= \epsilon^{-1} A - \theta \\ X &= \epsilon^{-1} B - \Xi \end{aligned}\right\} \; Y = \epsilon^{-1} C - \Lambda$$

For example for y-translation $\delta W = \varepsilon W_y$

$$T^{(y)} = \frac{2\alpha}{3} W_x W_y \; ; \qquad X^{(y)} = \frac{2\alpha}{3} W_t W_y + W_x^2 W_y + \frac{1}{3} W_y W_{xxx} - \frac{1}{3} W_{xx} W_{yx}$$

$$Y^{(y)} = \frac{\beta^2}{2} W_y^2 - \frac{2\alpha}{3} W_t W_x - \frac{1}{3} W_x^3 + \frac{1}{6} W_{xx}^2 \; ;$$

Lastly a few points about the infinitesimal Backlund transformation. Suppose we denote the B.T with parameter γ as;

$$W_1 = \hat{B}_\gamma(x) W_0 \quad (x\text{-component})$$

$$W_1 = \hat{B}_\gamma(y) W_0 \quad (y\text{-component}).$$

Then expanding in Laurent Series

$$W_1 = \gamma + W_0 + A_1 \gamma^{-1} + A_2 \gamma^{-2} + \cdots \qquad \text{..... we have;}$$

$$A_n = - A_{(n-1)x} - \sum_{r=1}^{n-2} A_r A_{n-r-1}$$

Similarly for y-component. So really the infinite structure of conservation laws can be exploreде. The details are to be published elsewhere. The author is very grateful to Prof. Abdus Sabam for giving him the opportunity to present these results at ICTP, Trieste. He wishes to thank Prof. Abdus Salam for his generosity and support.

References

(1) H. Steudel. - Ann. der. Physik. **32**, 205 (1975)

Ann. der. Physik. **32**, 445 (1975)

(2) Tu. Gui-Zhang - Lett. Math. Phys. **6**, 63 (1982).

(3) H. Rund - In "Backlund Transformation" - Lecture Notes in Mathematics Vol. 515 (Springer-Verlag, Berlin, New York).

(4) A.S. Fokas - J. Mth. Phys. **21**, 1318, 1980.

EINSTEIN EQUATIONS WITHOUT KILLING VECTORS,

SELF-DUAL YANG-MILLS FIELD AND NON-LINEAR SIGMA MODELS

(INTEGRABILITY PROPERTIES, LINKS, NEW SOLUTIONS).

Norma SANCHEZ

ER 176 C.N.R.S., DAPHE, Observatoire de Meudon
92190 Meudon, FRANCE

We provide an unifying pattern in which non-linear sigma models, sel-dual Yang-Mills and General Relativity theories are explicitely connected through their non-linear evolution equations, integrability properties and respective solutions. We do not assume the existence of any space-time symmetry, (i.e. of any killing vector field). We report our results on the integrability of the self-dual gravitational field : Lax pair, Bäcklund transformations, infinite number of conserved laws and an infinite dimensional Lie ("loop") algebra of symmetry transformations of the self-dual Einstein equations. This extends to the case without Killing vectors the remarquable properties of the Einstein eqs with two Killing vector fields. Gravitational instantons (signature ++++), Solitons (signature -+++) and Calorons (with finite temperature) are found. In a simple way, (the only mathematical tool is complex analicity) a wide class of solutions to Euclidean Gravity is obtained.

In the last years a number of analogies and explicit links between non linear eqs. describing physically unrelated problems have been established : self-dual Yang-Mills (SDYM) field in the Yang ("R gauge") formulation, non-linear σ models and "two dimensional reduced gravity" ie the gravitational field with two Killing vectors (Ernst formulation). Since 1968, the solution of Einstein eqs in presence of two Killing vectors (stationary axisymmetric gravitational fields, colliding plane waves, cillindrical gravitational waves, cosmological solutions) have known important developpments /1-3/ It was ten years after, that these geometric techniques have been connected to those independently developped in the domain of field theory (σ model and Sine Gordon equation). /4-5/. Conversely, the algebraic and Bäcklund techniques developped to solve the Ernst equation have been useful in the subtle problem of generating (finite energy) multi-monopole SU (2) configurations /6/. See table I.

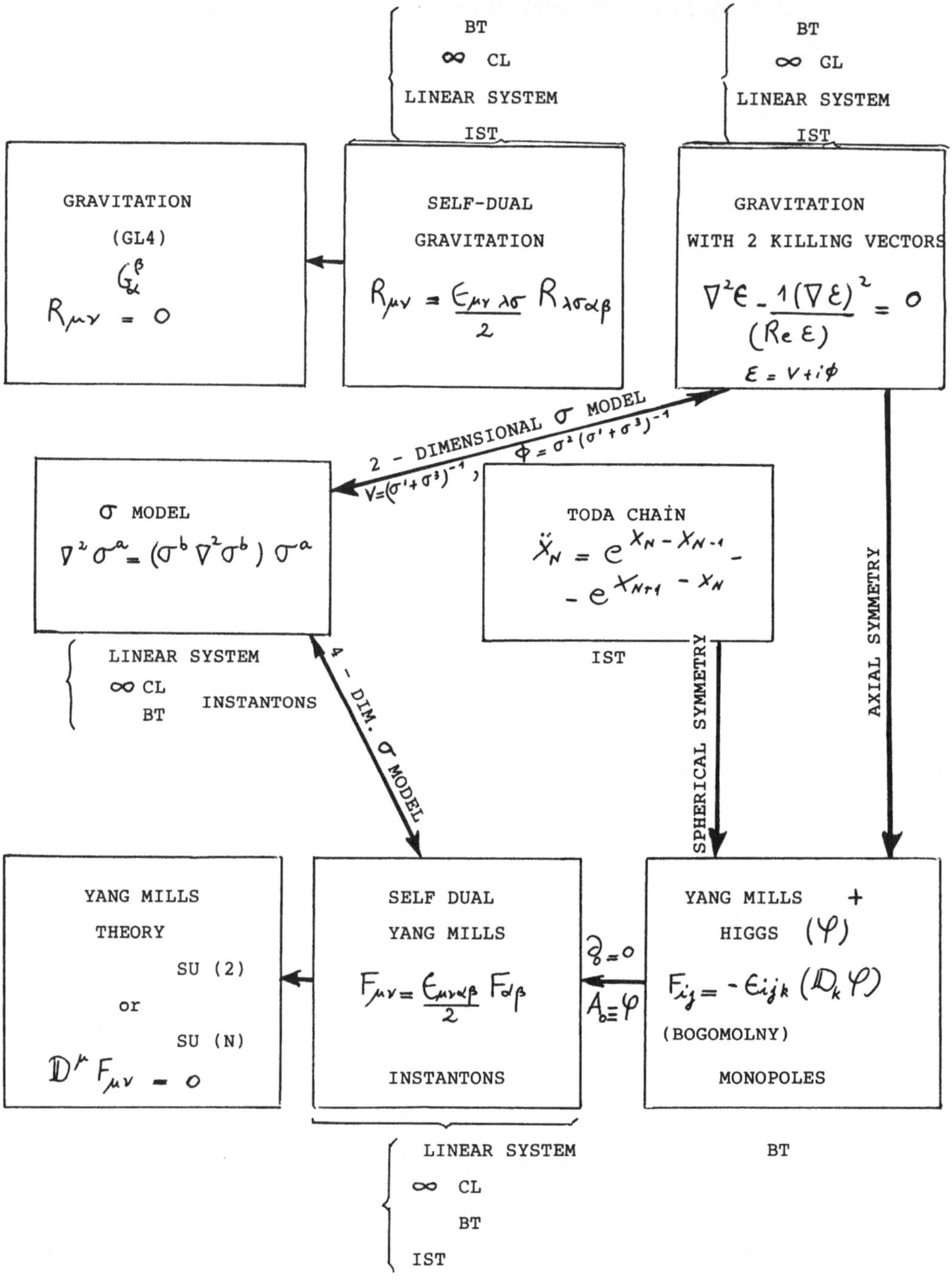

∞ CL : infinite number of conservation laws

BT : Bäcklund transformations

IST : Inverse Scattering Transform

Table I

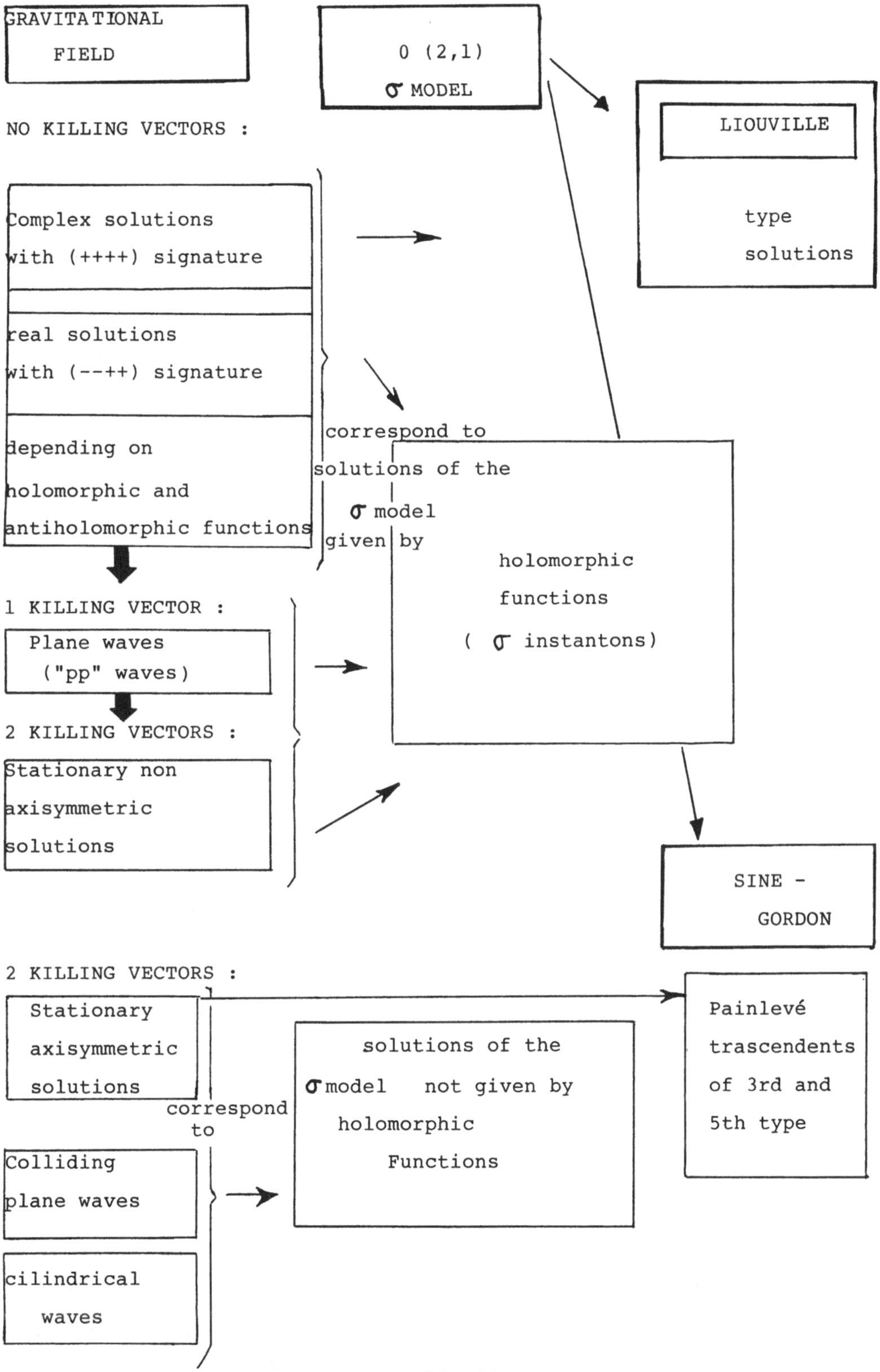
GRAVITATIONAL FIELD
0 (2,1) σ MODEL
LIOUVILLE
type solutions
NO KILLING VECTORS :
Complex solutions with (++++) signature
real solutions with (--++) signature
depending on holomorphic and antiholomorphic functions
correspond to solutions of the σ model given by
holomorphic functions
(σ instantons)
1 KILLING VECTOR :
Plane waves ("pp" waves)
2 KILLING VECTORS :
Stationary non axisymmetric solutions
SINE - GORDON
2 KILLING VECTORS :
Stationary axisymmetric solutions
correspond to
Colliding plane waves
cilindrical waves
solutions of the σ model not given by holomorphic Functions
Painlevé trascendents of 3rd and 5th type

Table II

Our aim is to go beyond the two Killing vector treatment of the gravitational field : for the general (non-selfdual) case without Killing vectors we have shown /7/ that the Einstein equations contain an O(2,1) σ model. This σ model structure emerges from a3+1 decomposition of the Einstein equations which holds irrespective of the presence of symmetries in the space-time. This includes, in particular, the stationary (one Killing vector) and the Ernst (two Killing vectors) formulations of the gravitational field. We analyse the links between the solutions of Einstein eqs and the solutions of the σ model and their reductions (Liouville eq. and Sine-Gordon) /7,8/. See table II above. For the self-dual case without Killing vectors we give a manifestly gauge invariant formulation analogous to that given by Yang for the self-dual Yang-Mills field. This formulation connects in a direct and explicit way the self-dual Yang-Mills and General Relativity theories, their respective solutions and integrability properties. In complex space $(Y,\bar{Y},Z,\bar{Z})$ where

$$\sqrt{2}\,Y = x_1 + i x_2 \qquad \sqrt{2}\,Z = x_3 + i x_4$$
$$\sqrt{2}\,\bar{Y} = x_1 - i x_2 \qquad \sqrt{2}\,\bar{Z} = x_3 - i x_4$$

the self-dual Einstein equations Rabcd = 1/2 ϵ abmn Rmncd (3)

(Rabcd stands for the Riemann tensor) can be written as

$$g^{y\bar{y}}\,\partial_{\bar{y}}\left(J^{-1}\partial_y J\right) + g^{z\bar{z}}\,\partial_{\bar{z}}\left(J^{-1}\partial_z J\right) = 0 \tag{4}$$

where $J = \begin{pmatrix} 0 & I \\ g & 0 \end{pmatrix}$ is a 4 x 4 complex, non singular matrix, element of the (GL4) group, which satisfies the condition

$$\partial_\mu J_{\alpha\beta} = \partial_\beta J_{\alpha\mu} \quad , \text{ie} \quad \partial_\mu g_{\alpha\beta} = \partial_\beta g_{\alpha\mu} \tag{5}$$

g is a 2x2 complex matrix with det g = 1. The length element is given by $dS^2 = g_{\alpha\bar{\beta}}\, dx^{\alpha}\, dx^{\bar{\beta}}$.

The (4x4) matrix eqs for J reduce to (2x2) matrix eqs. for g. If g is hermitian (ie $g^T = \bar{g}$) the self-dual Einstein eqs are connected to those of the self-dual Yang-Mills theory with the additional condition (5). In 2^n dimensions O and I are the nxn null and Identity matrices respectively and g is nxn. Even if eq (3) is meaningless for n=2, J constructed in this way provides a Kähler metric g with vanishing Ricci tensor in 2^n dimensions and the SD-Einstein eqs are connected to those of a SD-SU(N) theory with the additional condition (5). See /9,10/.

SUMMARY OF MAIN RESULTS

GRAVITATIONAL FIELD

GENERAL (NON self-dual) 4 dimensional CASE (Without any Killing vector)	. Connection with the O(2,1) model . New Solutions from Analytic Mappings
SELF-DUAL 4 dimensional CASE (without any Killing vector)	. Connection with the SDYM field . Gravitational instantons connected to instantons of Y - M . Gravitational calorons connected to calorons of Y -M

. Gravitational ("pp") waves are connected to the instantons of model

REFERENCES

1 - F.J. Ernst, Phys. Rev. 167, 1175 (1968)

2 - R. Geroch, J. Math. Phys. 12, 918 (1971) ; 13 394 (1972)
W. Kinnersley, J. Math. Phys. 18, 1529 (1977)

3 - B. Harrison, Phys. Rev. Lett. 41, 1197 (1978)

4 - D. Maison, J. Math. Phys. 20, 871 (1979)

5 - V. A. Belinski and V. E. Zakharov, Zh. Eksp. Teor. Fiz.75 1955 (1978) (Sov. Phys. JETP 48 985 (1978)

6 - P. Forgács, Z. Horvath and L. Palla, Phys. Rev. Lett. 45, 505 (1983)

7 - N. Sánchez, Phys. Rev. 26D, 2589 (1982)

8 - N. Sánchez, Phys. Lett. 94A, 125 (1983)

9 - N. Sánchez, Proc. "Journées Relativistes 1983" Acad. delle Scienze di Torino (to appear)

10 - N. Sánchez, Phys. Lett. 125B, 403 (1983).

Jet bundle technique, Lie Bäcklund vector fields and diffusion equations

W.-H. Steeb* and W. Strampp**

*Theoretische Physik, Universität Paderborn, D-4790 Paderborn, West Germany

**FB Mathematik, Gesamthochschule Kassel, D-3500 Kassel, West Germany

Recently, much attention has been focused on the classifaction of dynamical systems as integrable and nonintgerable ones. When we consider classical mechanics, the Toda lattice is a well known example of an integrable system. There are N first integrals in involution and there is a Lax representation. In field theory the Korteweg de Vries equation is an integrable system. This equation can be solved with the help of the inverse scattering transform. Moreover there is an infinite number of conservation laws. In quantum field theory the best known example of an integrable system is the quantum nonlinear Schrödinger equation. This system can be solved with the help of the Bethe ansatz or with the help of the quantum spectral method. However, most dynamical systems are non-integrable. In classical mechanics we find among the nonintegrable systems those with chaotic behaviour.

Recently, various authors have studied nonlinear reaction diffusion equations which serve as models for various phenomena with interaction and diffusion in e.g. biochemistry, developmental biology, plasma physics, population dynamics and other fields (compare [1,2] and references therein). A natural question is whether or not these nonlinear diffusion equations are integrable or not.

Nonlinear evolution equations (partial differential equations (pdes)) are usually called integrable when one of the following properties is fulfilled: (I) the initial value problem can be solved exactly with the help of the inverse scattering transform (IST), (II) they have an infinite number of conservation laws, (III) they have an auto Bäcklund transformation or a Bäcklund transformation to a linear equation, (IV) besides Lie point vector fields they admit Lie Bäcklund (L.B.) vector fields, (V) they describe pseudo-spherical surfaces, i.e. surfaces of constant negative gaussian curvature, (VI) they can be written as covariant exterior derivatives of Lie algebra valued differential forms. It is conjectured that if property (I) holds, then the property (II) through property (VI) also holds. We mention that there are evolution equations which cannot be solved by the IST. However, these equations admit Bäcklund transformations. For example the diffusion equation $\partial u/\partial t = \partial(f(u)\partial u/\partial x)/\partial x$ where $f(u) = u^{-2}$ is not a Hamiltonian system and therefore it cannot be solved with the help of IST. However, it admits a Bäcklund transformation and a hierarchy of L.B. vector fields.

In this note we study the existence of L.B. vector fields for various types of diffusion equations. We apply the jet bundle technique. Then we briefly discuss the Painlevê property. The diffusion equations under consideration are

$$\partial u/\partial t = \partial^2 u/\partial x^2 + f(u) \tag{1}$$

$$\partial u/\partial t = \partial^2 u/\partial x^2 + f_1(u)(\partial u/\partial x)^2 + f_2(u)(\partial u/\partial x) + f_3(u) \tag{2}$$

$$\partial u/\partial t = \partial(f(u)\partial u/\partial x)/\partial x \tag{3}$$

$$\partial u/\partial t = D\partial^2 u/\partial x^2 + f_1(u,v) \ , \quad \partial v/\partial t = D\partial^2 v/\partial x^2 + f_2(u,v). \tag{4}$$

Let us demonstrate the technique for the eq.(1). For the other equations we only give the results. Now we would like to find the analytic functions f where eq.(1) admits at least one L.B. vector field. It is conjectured that for an evolution equation which admits at least one L.B. vector field there is a hierarchy of L.B. fields. Within the jet bundle technique [3] we consider the submanifold

$$F \equiv u_t - u_2 - f(u) = 0 \quad , \tag{5}$$

where $u_x \equiv u_1$, $u_{xx} \equiv u_2$ and so on. Together with F = 0 we consider all its differential consequences with respect to the space coordinate. This means

$$F_1 \equiv u_{1t} - u_3 - u_1 f' = 0, \qquad F_2 \equiv u_{2t} - u_4 - u_1 f'' - u_2 f' = 0,$$

and so on. Let

$$V = g(u,u_1,u_2,u_3)\partial/\partial u \tag{6}$$

be a L.B. vector field. The assumption that the analytic function g depends also on x and t does not affect the result (i.e. the existence of a L.B. vector field). The invariance requirement is expressed as

$$L_{\bar{V}}F \hat{=} 0 \quad , \tag{7}$$

where $L_{\bar{V}}(.)$ denotes the Lie derivative and $\hat{=}$ stands for the restriction to solutions of eq.(1). $\bar{V}$ is the extended vector field of V. Due to the structure of eq.(1) we can assume, without loss of generality , that the vector field V takes the form

$$V = (g_1(u,u_1,u_2) + u_3)\partial/\partial u \quad . \tag{8}$$

Separating out term by term we find that eq.(1) admits a L.B. vector field iff f satisfies f" = 0. Consequently, f(u) = au + b ($a,b \in R$). Thus eq.(1) becomes linear. If the function f is nonlinear, then eq.(1) is not integrable.

Consider eq.(2). We take into account the vector field (8) and the condition (7). We find that eq.(2) admits a L.B. vector field iff the the functions f_1, f_2, and f_3 satisfy the following system of odes

$$f_2'f_3 = 0 \, , \quad f_2'f_1 = f_2'' \quad , \quad f_3'' + (f_1f_3)' = 0 \, . \qquad (9)$$

Furthermore eq.(2) together with eq.(9) admits a hierarchy of L.B. vector fields. Note that eq.(1) and Burgers equation are included as special cases.

Consider eq.(3). Using vector field (6) and the condition (7) we find that eq.(3) admits a L.B. vector field iff the function f satisfies the ode $2ff'' = (3f')^2$. Consequently, $f(u) = a(u + b)^{-2}$ where $a,b \in R$ (compare also [4]).

Finally, consider eq.(4). Without loss of generality, V takes the form

$$V = (h(u,v,u_1,v_1) + u_3)\partial/\partial u + (k(u,v,u_1,v_1) + v_3)\partial/\partial v \quad . \qquad (10)$$

We are only interested in the case where f or g or both are nonlinear. From condition (7) we find that eq.(4) admits a L.B. vector field iff $f(u,v)=cv^2$ $g(u,v)=0$ or $f(u,v)=0$, $g(u,v)=cu^2$, where $c \in R$. In these cases the system (4) is integrable, because the system is decoupled.

If both f and g are nonlinear, then system (4) is nonintegrable. For D=0 we obtain a system of odes. There are systems where both f and g are nonlinear and there is a first integral (integrable system). Consequently, diffusion destroys integrability.

Recently, Weiss et al [5] have introduced the Painlevé property for pdes. The quantities u, x, and t are considered in the complex domain. For the field u they make the ansatz

$$u(x,t) = \phi^n(x,t)\sum_{j=0}^{\infty} u_j(x,t)\phi^j(x,t) \qquad (11)$$

where n is a negative integer and ϕ an analytic function. If we apply this ansatz to the integrable equations given above we find that it does not work for certain cases. We are forced to introduce extensions. In some cases we must assume that n is not an integer. In other cases we must introduce logarithmic terms, i.e.

$$u(x,t) = \phi^n(x,t)\sum_{j,k=0}^{\infty} u_{jk}(x,t)\phi^j(x,t)\ln\phi^k(x,t) \quad . \qquad (12)$$

1. W. Strampp, W.-H. Steeb, and W. Erig, Prog. Theor. Phys. 68, 731 (1982).
2. W.-H. Steeb and W. Oevel, Physica Scripta (1983) (in press).
3. W.-H. Steeb and W. Strampp, Physica 114A, 95 (1982).
4. G. Bluman and S. Kumei, J. Math. Phys. 21, 1019 (1980).
5. J. Weiss, M. Tabor, and G. Carnevale, J. Math. Phys. 24, 522 (1983).

A GROUP-THEORETIC TREATMENT OF

GAUSSIAN OPTICS AND THIRD-ORDER ABERRATIONS

KURT BERNARDO WOLF
Instituto de Investigaciones
en Matemáticas Aplicadas y en Sistemas (IIMAS)
Universidad Nacional Autónoma de México
Apdo. Postal 20-726, 01000 MEXICO, D.F.

Abstract: Optical systems produce canonical transformations on the phase space of position and direction of light rays. Contractions of this pseudogrup cut the approximation order automatically to the desired terms throughout. The group $I_5 \circledS Sp(2,\mathcal{R})$ accounts for Gaussian optics and for up-to-third order aberrations. Calculations on concatenation of optical components involve rather simple 2×2-matrix plus 5-vector algebra.

1. The Hamiltonian treatment of light rays [1,2] uses for canonically conjugate coordinates the position q and the *direction* (or transverse momentum) $p = n\sin\Theta$, n being the refraction index of the medium and Θ the angle between the ray direction at q and the optical z-axis of the system. We shall work here with one transversal dimension (for reference: the q-axis is usually drawn up and the z-axis to the right). Lossless systems -homogeneous slabs and refracting surfaces- produce symplectomorphisms i.e., *canonical transformations* on the q,p-phase plane. The Hamiltonian of the system is

$$H = -\sqrt{n^2-p^2} = -n + \tfrac{1}{2}p^2/n + \tfrac{1}{8}p^4/n^3 + \tfrac{3}{48}p^6/n^5 + \cdots. \qquad (1)$$

2. Quantization replaces q and p with the familiar Schrödinger operators $\mathbb{Q}$ (argument multiplication) and $\mathbb{P}$ $(-ik_0^{-1}\partial/\partial q$ with the inverse of the wave number k_0 replacing Heisenberg's constant $\hbar$, here set to unity) acting on a Hilbert space $\mathcal{L}^2(\mathcal{R})$ of complex amplitude functions. The general z-evolution operator is the exponential of (1), $\exp(i\ell H)$, over the lenght ℓ traversed by the ray. In general the refraction index n depdns on the position (q,z); Gaussian optics deals with systems where such variations are discrete and constitute parabolic thin lenses separated by homogeneous slabs, and where the Hamiltonian (1) is cut to its first two terms, i.e., disregarding operators $\mathbb{P}^4, \mathbb{P}^6, \ldots$ In that case it has been shown by Sternberg [1], Dragt [2], Nazarathy and Shamir [3], and Bacry and Cadilhac [4], that what we have called *canonical transforms* in [5] is a succint method for describing such systems. Basically, calculations are reduced to 2×2 matrix algebra since the group involved is $Sp(2,\mathcal{R}) \sim SL(2,\mathcal{R})$, a finite-dimensional subgroup of the pseudogroup of all canonical transformations. (So is $W \circledS Sp(2,\mathcal{R})$, its semidirect product with the Heisenberg-Weyl group W [6]. The latter, applicable to systems with prisms or misaligned lenses, does not seem to have been explicitly considered yet).

3. Dragt [2a] has used extensively Lie-algebraic methods to describe magnetic focusing in Tokamaks; he has also shown that the formalism applies to optical systems [2b] allowing computations which involve higher-order terms in (1) with quartic-surface lenses and third-order Seidel aberrations. The problem here involves Baker-Campbell-Hausdorff relations with exponentials of operators in p^4, p^3q, p^2q^2, pq^3, and q^4, which do not close under commutation, but yield still higher-order terms which are replaced by ellipses (...) then then disregarded.

4. Our objective here is to provide a model based on the smidirect product group $I_5 \circledS Sp(2,\mathcal{R})$, where the normal subgroup I_5 is a five-dimensional abelian group with generators

$\mathbb{S}$: *spherical aberration,*	corresponding to	$\mathbb{P}^4$	(2a)
$\mathbb{C}$: *coma,*	"	$\tfrac{1}{2}\{\mathbb{P}^3,\mathbb{Q}\}_+$	(2b)
$\mathbb{A}$: *astigmatism/curvature,*	"	$\tfrac{1}{2}\{\mathbb{P}^2,\mathbb{Q}^2\}_+$	(2c)
$\mathbb{D}$: *distortion*	"	$\tfrac{1}{2}\{\mathbb{P},\mathbb{Q}^3\}_+$	(2d)
$\mathbb{Z}$: *phase,*	"	$\mathbb{Q}^4$	(2e)

On the Lie algebra level we start with the enveloping of the Heisenberg-Weyl albegra $\mathbb{W}$. There, the commutator of mth and n^{th} -order elements is an $(m+n-2)^{th}$-order element. We contract with respect to the elements of orders 1 through 4, so that a commutator lying outside these orders vanishes. These include the vanishing of the 0th order element. We contract with respect to the elements of orders 1 through 4, so that a commutator lying outside these orders vanishes. These include the vanishing of the 0th order element $[\mathbb{Q},\mathbb{P}]$, which allows us to disregard the quantization-scheme problem associated to q^2p^2 (see[6; Sect. IV-B]). Finally, we work only with second- and fourth-order elements, the latter an abelian ideal under the former, and exponentiate these to the group. First- and third-order elements correspond to nonsymmetric optical elements and will appear only in the adjoint action of $\mathbb{I}_5 \supset Sp(2,\mathbb{R})$ on the observables. Although the algebra is no longer $\mathbb{W}$, we may still use the mnemotechnical symbols $\mathbb{P}^2, \frac{1}{2}\{\mathbb{Q},\mathbb{P}\}_+$, and $\mathbb{Q}^2$ for the $Sp(2,\mathbb{R})$ generators —there is no change in their algebra or group properties— and $\mathbb{S}, \mathbb{C}, \mathbb{A}, \mathbb{D}$, and $\mathbb{Z}$ for the fourth-order ones. The transformations in $\mathbb{Q}-\mathbb{P}$ are still canonical up to fourth-order terms, i.e., we may be left with uncancelled commutators of two third-order terms.

5. We may exponentiate the general $\mathbb{I}_5 \supset Sp(2,\mathbb{R})$ Lie algebra element to the $\mathbb{I}_5 \supset Sp(2,\mathbb{R})$ group. The $Sp(2,\mathbb{R})$ part is

$$\exp i[a\mathbb{P}^2 + b\{\mathbb{Q},\mathbb{P}\}_+ + c\mathbb{Q}^2] =: \mathbb{I}\left\{\begin{pmatrix} \cos 2s - bs^{-1}\sin 2s & -as^{-1}\sin 2s \\ cs^{-1}\sin 2s & \cos 2s + bs^{-1}\sin 2s \end{pmatrix}, \mathbf{0}\right\}, \tag{3}$$

where $s := (ac - b^2)^{1/2}$; for the abelian part,

$$\exp i[d\mathbb{Z} + e\mathbb{D} + f\mathbb{A} + g\mathbb{C} + h\mathbb{S}] =: \mathbb{I}\{\mathbf{1}, (d, e, f, g, h)\}. \tag{4}$$

The product law in this semidirect product parametrization appears as

$$\mathbb{I}\{\mathbf{M}_1,\mathbf{v}_1\}\,\mathbb{I}\{\mathbf{M}_2,\mathbf{v}_2\} = \mathbb{I}\{\mathbf{M}_1\mathbf{M}_2,\ \mathbf{v}_1\mathcal{D}_5(\mathbf{M}_2) + \mathbf{v}_2\}, \tag{5a}$$

$$\mathcal{D}_5\begin{pmatrix}\alpha & \beta\\ \gamma & \delta\end{pmatrix} := \begin{pmatrix} \alpha^4 & 4\alpha^3\beta & 6\alpha^2\beta^2 & 4\alpha\beta^3 & \beta^4 \\ \alpha^3\gamma & \alpha^2(\alpha\delta+3\beta\gamma) & 3\alpha\beta(\alpha\delta+\beta\gamma) & \beta^2(3\alpha\delta+\beta\gamma) & \beta^3\delta \\ \alpha^2\gamma^2 & 2\alpha\gamma(\alpha\delta+\beta\gamma) & \alpha^2\delta^2+4\alpha\beta\gamma\delta+\beta^2\gamma^2 & 2\beta\delta(\alpha\delta+\beta\gamma) & \beta^2\delta^2 \\ \alpha\gamma^3 & \gamma^2(3\alpha\delta+\beta\gamma) & 3\gamma\delta(\alpha\delta+\beta\gamma) & \delta^2(\alpha\delta+3\beta\gamma) & \beta\delta^3 \\ \gamma^4 & 4\gamma^3\delta & 6\gamma^2\delta^2 & 4\gamma\delta^3 & \delta^4 \end{pmatrix}. \tag{5b}$$

From here the identity is $\mathbb{I}\{\mathbf{1},\mathbf{0}\}$ and the inverse $\mathbb{I}\{\mathbf{M},\mathbf{v}\}^{-1} = \mathbb{I}\{\mathbf{M}^{-1}, -\mathbf{v}\mathcal{D}_5(\mathbf{M})^{-1}\}$ where $\mathcal{D}(\mathbf{M})^{-1} = \mathcal{D}(\mathbf{M}^{-1})$.

6. Let us remark that the $\mathbb{I}\{\mathbf{M},\mathbf{v}\}$ are operators acting to the right on $\mathcal{L}^2(\mathbb{R})$ complex amplitude functions through (approximate) Huygens integral transforms, and that the composition of optical elements in Latin (left to right) translates into Hebrew (right to left) composition of $\mathbb{I}$ -transforms. The adjoint action of $\mathbb{I}\{\mathbf{M},\mathbf{v}\}$ on the thus-far-unused generators $\mathbb{Q}, \mathbb{P}$, $\mathbb{Q}^3$, $\frac{1}{2}\{\mathbb{Q}^2,\mathbb{P}\}_+$, $\frac{1}{2}\{\mathbb{Q},\mathbb{P}^2\}_+$ and $\mathbb{P}^3$ (we do not use special symbols for the latter four) may be described as follows. In the Gaussian approximation, denoting the column two-vector $\mathbb{X} := (\mathbb{Q},\mathbb{P})^T$, we have $\mathbb{I}\{\mathbf{M},\mathbf{0}\}\,\mathbb{X}\,\mathbb{I}\{\mathbf{M},\mathbf{0}\}^{-1} = \mathbf{M}^{-1}\mathbb{X}$. For the fourth-order enveloping algebra we find it easier to work sometimes with the inverse adjoint action. If we denote the column four-vector $\mathbb{Y} := (\mathbb{Q}^3, \frac{1}{2}\{\mathbb{Q}^2,\mathbb{P}\}_+, \frac{1}{2}\{\mathbb{Q},\mathbb{P}^2\}_+, \mathbb{P}^3)^T$ and $\mathbf{v} := (d,e,f,g,h)$ as before, then

$$\mathbb{I}\{\mathbf{M},\mathbf{v}\}^{-1}\begin{pmatrix}\mathbb{X}\\ \mathbb{Y}\end{pmatrix}\mathbb{I}\{\mathbf{M},\mathbf{v}\} = \begin{pmatrix} \mathbf{M} & \mathbf{M}\begin{pmatrix} -e & -2f & -3g & -4h \\ 4d & 3e & 2f & g \end{pmatrix} \\ \mathbf{0} & \mathcal{D}_4(\mathbf{M}) \end{pmatrix}\begin{pmatrix}\mathbb{X}\\ \mathbb{Y}\end{pmatrix}, \tag{6}$$

where $\mathcal{D}_4(\mathbf{M})$ is the 4x4 representation of $\mathbf{M}$ acting on $\mathbb{Y}$, built along the lines of $\mathcal{D}_5(\mathbf{M})$ in (5).

7. The key relation one would like to obtain, as an analogue to [7; Eqs. (2.3)-(2.4)] is the exponential of the general algebra element $a\mathbb{P}^2 + b\{\mathbb{P},\mathbb{Q}\}_+ + c\mathbb{Q}^2 + d\mathbb{Z} + e\mathbb{D} +$

$+f\mathbb{A}+g\mathbb{C}+h\mathbb{S}$. This has thus far not been possible, but for the purpose of optics it is sufficient to use (6) in order to obtain the 'free' homogeneous medium propagation group element whose adjoint action is $\mathbb{F}_\ell^{(n)}: \mathbb{Q} \to \mathbb{Q}+(\ell/n)\mathbb{P}+(\ell/2n^3)\mathbb{P}^3$, $'' : \mathbb{P} \to \mathbb{P}$. It is

$$\mathbb{F}_\ell^{(n)} := \exp\left[i\ell\left(\tfrac{1}{2n}\mathbb{P}^2+\tfrac{1}{8n^3}\mathbb{S}\right)\right] = \mathbb{I}\left\{\begin{pmatrix}1 & -\ell/n\\ 0 & 1\end{pmatrix}, \left(0,0,0,0,\tfrac{\ell}{8n^3}\right)\right\}. \tag{7}$$

Through up-to-fifth order commutators, we similarly find the general form relevant for Gaussian lens + aberration calculations:

$$\begin{aligned}&\exp i[c\mathbb{Q}^2+d\mathbb{Z}+e\mathbb{D}+f\mathbb{A}+g\mathbb{C}+h\mathbb{S}]\\ &=\mathbb{I}\left\{\begin{pmatrix}1 & 0\\ 2c & 1\end{pmatrix}, \left(d+ce+\tfrac{4}{3}c^2f+2c^3g+\tfrac{16}{5}c^4h,\ e+2cf+4c^2g+8c^3h,\ f+3cg+8c^2h,\ g+4ch,\ h\right)\right\}.\end{aligned} \tag{8}$$

One may directly check that elements on one-parameter subgroups compose properly to verify the suspicious-looking coefficients 4/3 and 16/5. The presence of an interface $z(q)$ between two media with refractive indices n_1 (to the left) and n_2 (to the right) means that the evolution operator is $\exp(i\ell_1\mathbb{H}^{(n_1)})$ between a point z_1 in medium n_1 and $z(q)$, and $\exp(i\ell_2\mathbb{H}^{(n_2)})$ between $z(q)$ and a z_2 in medium n_2. Setting the surface center at $z(0)=0$ and collapsing the points as $z_1=z(0)=z_2$, we have $\ell_1=z(q)=-\ell_2=\ell$. We consider quartic surfaces $z(q)=uq^2+vq^4$. (A sphere of radius ρ and center at $z=\rho$ has thus $u=1/2\rho$, $v=1/8\rho^3$.) The refracting surface operator is obtained as the exponential of $\ell\mathbb{H}^{(n_1)}=(uq^2+vq^4+\cdots)(-n_1+p^2/2n_1+p^4/8n_1^3+\cdots)$ followed (to the left) by the exponential of $-\ell\mathbb{H}^{(n_2)}$. Using (5) and (8), the refracting surface operator/group element is thus obtanied as

$$\begin{aligned}\mathbb{R}^{n_1\to n_2}_{z(q)} &:= \exp i\left(un_2\mathbb{Q}^2+vn_2\mathbb{Z}+\tfrac{u}{2n_2}\mathbb{A}\right)\ \exp i\left(-un_1\mathbb{Q}^2-vn_1\mathbb{Z}+\tfrac{u}{2n_1}\mathbb{A}\right)\\ &=\mathbb{I}\left\{\begin{pmatrix}1 & 0\\ 2un_2 & 1\end{pmatrix}, \left(n_2(v-\tfrac{2}{3}u^3), -u^2, -\tfrac{u}{2n_2}, 0, 0\right)\right\}\mathbb{I}\left\{\begin{pmatrix}1 & 0\\ -2un_1 & 1\end{pmatrix}, \left(-n_1(v-\tfrac{2}{3}u^3), -u^2, \tfrac{u}{2n_1}, 0, 0\right)\right\}\\ &=\mathbb{I}\left\{\begin{pmatrix}1 & 0\\ 2u(n_2-n_1) & 1\end{pmatrix}, \left((n_2-n_1)(v+u^3\{2\tfrac{n_1}{n_2}-\tfrac{2}{3}\}), -2u^2\tfrac{n_2-n_1}{n_2}, \tfrac{1}{2}u\tfrac{n_2-n_1}{n_1n_2}, 0, 0\right)\right\}.\end{aligned} \tag{9}$$

Here the lower-left matrix element gives the Gaussian lens power, while the row-vector elements yield the phase, distortion, astigmatism/curvature, coma, and spherical aberration, respectively. It may be convenient to reverse the Gaussian–aberration order using $\mathbb{I}\{\underset{\sim}{M},\underset{\sim}{v}\}=\mathbb{I}\{\underset{\sim}{1},\underset{\sim}{v}\underset{\sim}{D}_S(\underset{\sim}{M}^{-1})\}\,\mathbb{I}\{\underset{\sim}{M},\underset{\sim}{0}\}$. This changes $n_1/n_2 \mapsto n_2/n_1$ in the d-element above, and $\ldots/n_2 \mapsto \ldots/n_1$ in the e-element.

8. This preliminary account on the use of $I_5 \rtimes Sp(2,R)$ for Gaussian optics and third-order aberrations points to several directions one should explore before claiming a complete, realistic, and economical method for the description of non-Gaussian optics. We give below some of them; for lack of time and expertise in classical optics, the author may have slighted the meaning or importance of others.

- The real optical world is two-dimensional, so $Sp(4,R)$ rather than $Sp(2,R)$ should be considered. The self-adjoint abelian part would then be 35-dimensional or, if axis-rotational symmetry is imposed, six-dimensional [2]: phase $(\underset{\sim}{q}^2)^2$, distortion $\underset{\sim}{q}^2\,\underset{\sim}{q}\cdot\underset{\sim}{p}$, coma $\underset{\sim}{p}^2\,\underset{\sim}{q}\cdot\underset{\sim}{p}$, and spherical aberration $(\underset{\sim}{p}^2)^2$ remain unique; curvature of field $\underset{\sim}{p}^2\underset{\sim}{q}^2$ and astigmatism $(\underset{\sim}{p}\cdot\underset{\sim}{q})^2$ now split. An $Sp(2,R)\otimes SO(2)$ group could be used, but the homogeneous space of first- and third-order Heisenberg-Weyl generators has now more dimensions.

- Our method allows for the refracting index n to depend on $\mathbb{Q}$ for the description of models in fiber optics, without or with z-dependent inhomogeneities. The latter would profit from the results in [7] on $W \rtimes Sp(2,R)$ time-dependent Hamiltonians applicable to Gaussian optics for linear and quadratic $n(q)$, but not yet translated to $I_5 \rtimes Sp(2,R)$.

• Approximate Huygens transform kernels relevant for wave optics follow rather immediately from (6) cast in the form of simultaneous linear but third-order differential equations. The Hilbert space aspects of this metaplectic group and its complex extension -for systems with loss- would be -when implemented- an interesting branch of integral transform theory.

• Gaussian beams remain Gaussian under Gaussian optics, as may be seen applying $\mathbb{I}\{\begin{pmatrix} 1 & -k\omega \\ 0 & 1 \end{pmatrix}, \underset{\sim}{0}\}$ to a Dirac δ (c.f. [5, ¶ 9.3.3]), and coherent states behave in a simple way, as Bargmann transforms of Dirac δ's in the complex plane. What about their behaviour in non-Gaussian systems? Eigenstates of a non-Gaussian system may be defined and found rather easily.

• In our scheme, third-order aberrations transform under the Gaussian part $\underset{\sim}{M}$ of a system through $\underset{\sim}{\mathfrak{D}}_5(\underset{\sim}{M})$ in (5), but add to other aberrations, thus not aberrating further beyond third order. In non-axially-symmetric systems, with second order aberrations, the latter compose with other similar aberrations to yield third-order ones. The group relevant for such systems filling (8) with first- and third-order terms in the generators of $\tilde{W}$ has a more complicated semidirect product structure which should be made explicit in one and two dimensions.

These and other problems for nonlinear optical -and mechanical- systems and technology [8] conform an interesting area for research in the field of canonical transformations.

REFERENCES:

1) S. Sternberg, Lecture Notes on Symplectic Geometry and Optical Systems. Harvard University, unpublished.

2) A. J. Dragt, (a) Lectures on Nonlinear Orbit Dynamics, AIP Conference Proceedings, Vol. 87, 1982. (b) Lie-algebraic theory of geometrical optics and optical aberrations. J. Opt. Soc. Am. 72, 372-379 (1982).

3) M. Nazarathy and J. Shamir, (a) Wavelenght variation in Fourier optics and holography described by operator algebra. Israel J. Tech. 18, 224-231 (1980); (b) Fourier Optics described by operator algebra. J. Opt. Soc. Am. 70, 150-158 (1980); (c) Holography described by operator algebra. J. Opt. Soc. Am. 71, 529-541 (1981); (d) First-order optics -a canonical operator representation: lossless systems. J. Opt. Soc. Am. 72, 356-364 (1982); (e) M. Nazarathy, A. Hardy, and J. Shamir, Generalized mode propagation in first-order optical systems with loss of gain. J. Opt. Soc. Am. 72, 1409-1420 (1982).

4) H. Bacry and M. Cadilhac, Metaplectic group and Fourier optics. Phys. Rev. A 23, 2533-2536 (1981).

5) K. B. Wolf, "Integral Transforms in Science and Engineering". Plenum Publ. Corp. New York, 1979.

6) K. B. Wolf, The Heisenberg-Weyl ring in Quantum Mechanics. In "Group Theory and its Applications", Vol. 3, E. M. Loebl, ed., Academic Press, 1975.

7) K. B. Wolf, On time-dependent quadratic quantum Hamiltonians. SIAM J. Appl. Math. 40, 419-431 (1981).

8) W. Schempp, Radar reception and nilpotent harmonic analysis. I. and II. C. R. Math. Rep. Acad. Sci. Canada 4, 43-48 (1982).

ELEMENTARY PARTICLES AND GAUGE THEORIES

STUDY OF MICHEL'S CONJECTURE*

M. Abud, G. Anastaze
P. Eckert and H. Ruegg
University of Geneva

1211 Geneva 4, Switzerland

* Work partially supported by the Swiss National Science Foundation.

ABSTRACT

We study the little groups of the minima of the Higgs potential built on the representation 75 of SU(5). We find a minimum with a non maximal little subalgebra, but an additional discrete group, so that the little group is maximal. We find a large class of minimas with su(3) + su(2) + u(1) little algebra.

In gauge theories the fundamental interactions of leptons and quarks are invariant under a local compact Lie group or gauge group G. This symmetry may however be spontaneously broken by non vanishing components of the vacuum expectation value of a scalar field ϕ. The remaining invariance is given by the little group H, subgroup of G, which stabilizes the minimum of the scalar, or Higgs potential V. This is a fourth degree polynomial $V(\phi)$, G-invariant, bounded below, and with its lowest values not at the origin. Suppose ϕ is a vector in a representation space E of G. Michel[(1)] has made the following conjecture :

If the representation of the symmetry group G of a Higgs polynomial $V(\phi)$ on E is irreducible (on the real), its minima have little groups maximal in K (the set of conjugation classes of little groups on E-{o}).

For the conjecture to be true it is necessary that G be the largest symmetry group of V (it may be larger than the gauge group one started with).

The conjecture has been explicitly verified for the adjoint representation of SU(n)[(2)]. However, no general proof exists. Indeed, counter-examples have been found when G is a finite group[(3)].

We are studying in detail the representation 75 of SU(5). The latter is the favorite gauge group for grand unified theories. Breaking it with $\phi \in 75$ has certain advantages[(4)] over $\phi \in 24$. ϕ is a tensor ϕ^{ab}_{cd}, where $a,b,c,d = 1 \ldots 5$ is antisymmetric in ab and cd.
The SU(5) invariants are :

$$Q = \phi^{ab}_{cd}\,\phi^{cd}_{ab} \quad ; \quad C = \phi^{ab}_{cd}\,\phi^{ef}_{ab}\,\phi^{cd}_{ef}$$

$$K_2 = \phi^{ab}_{gh}\,\phi^{cd}_{ab}\,\phi^{ef}_{cd}\,\phi^{gh}_{ef} \quad ; \quad K_3 = \phi^{ab}_{cg}\,\phi^{cd}_{ab}\,\phi^{ef}_{dh}\,\phi^{gh}_{ef} \tag{1}$$

$$K_4 = \phi^{ab}_{fg}\,\phi^{cd}_{ab}\,\phi^{ef}_{ch}\,\phi^{gh}_{de} \quad ; \quad K_5 = \phi^{ab}_{de}\,\phi^{cd}_{ag}\,\phi^{ef}_{bh}\,\phi^{gh}_{cf}$$

$$K_6 = \phi^{ab}_{dg}\,\phi^{cd}_{ae}\,\phi^{ef}_{bh}\,\phi^{gh}_{cf}$$

We find the two linear relations

$$K_2 - K_3 + 5K_4 + 2K_5 = o$$
$$K_2 - 2K_3 + 8K_4 + 4K_6 = o \tag{2}$$

Hence the most general Higgs potential is :

$$V(\phi) = -\mu^2 Q + cC + \lambda_1 Q^2 + \sum_{i=1}^{4} \lambda_i K_i \tag{3}$$

Consider now the four inequivalent chains of subalgebras of the Lie algebra su(5) :

$$su(3) + su(2) + u(1) \tag{4}$$

$$su(4) + u(1) \supset sp(4) + u(1) \tag{5}$$

$$su(4) + u(1) \supset so(4) + u(1) \tag{6}$$

$$so(5) \supset so(3) \tag{7}$$

The representation 75 of su(5) will decompose into sums of irreducible representations R_i of the subalgebras. The first subalgebra in a chain, for which one of the R_i's is a singlet of this subalgebra, is called maximal in the sense of Michel. The last subalgebras of the four chains written above are maximal. For example, su(4) + u(1) is not maximal in this sense, but $sp(4) + u(1)$ is.

In order to test Michel's conjecture, we have to look for non maximal subgroups. Hence, we must be sure that we didn't miss some other maximal subgroups. It is helpful to introduce some geometric concepts.

A <u>stratum</u> is the set of all points of the representation space E with conjugated little groups. It can be parametrized by a minimal set of invariants (the integrity basis). The criterion of stratification is completely equivalent to the classification of little groups. Strata in the space of invariants are the image of strata in E by a certain map.[5] In particular, to one dimensional strata defined on the space of invariants (i.e. all invariants of an I.B. being constrained by a set of polynomial equations) correspond strata in E with the property that at every point of it the gradient of an invariant function is parallel to ϕ[5,6].

The little group of a point on a one-dimensional stratum is maximal[5],[6]. Conversely, we have verified that the strata corresponding to the little algebra $su(3) + su(2) + u(1)$, $sp(4) + u(1)$, $so(4) + u(1)$ and $so(3)$ are one-dimensional. However, we have found other one-dimensional strata.

Consider the chain of subalgebras

$$su(3) + su(2) + u(1) \supset su(2) + u(1) + su(2) + u(1) \tag{8}$$

The representation 75 of $su(5)$, when decomposed according to the last subalgebra of (8), contains two singlets S_1 and S_1'. To study the two-dimensional stratum $\Sigma_{\alpha\beta} = \alpha S_1 + \beta S_1'$, we compute the matrix[5]

$$P_{AB} = \sum_{i=1}^{75} \frac{\partial I_A}{\partial \phi_i} \frac{\partial I_B}{\partial \phi_i} \bigg|_{\phi = \phi^0} \tag{9}$$

where I_A, I_B are $su(5)$ invariant functions of ϕ. Here, we consider only the invariants of equation 1. When $\phi^0 \in \Sigma_{\alpha\beta}$, we find that the rank of P_{AB} is at maximum two. This is due to the relations, which hold on $\Sigma_{\alpha\beta}$:

$$K_3 = 4K_4 \quad ; \quad Q^2 + 2K_2 = 6K_3 \tag{10}$$

$$\text{grad}\,[8(\alpha-\beta)^2 Q + 8\alpha C - K_2] = 0 \tag{11}$$

This shows that at most two gradients of invariants are linearly independent on $\Sigma_{\alpha\beta}$.

Furthermore, we find that $\Sigma_{\alpha\beta}$ contains at its boundary two one-dimensional strata. One contains the $su(3) + su(2) + u(1)$ singlet S_1.

$$S_1 : \phi^{45}_{45} = \frac{1}{2} t^{45}_{45} - \frac{1}{6} \sum_{i=1}^{3} (t^{i4}_{i4} + t^{i5}_{i5}) + \frac{1}{6} (t^{12}_{12} + t^{13}_{13} + t^{23}_{23}) \tag{12}$$

Here t has the same symmetry properties as ϕ, but need not be traceless. The gradient of C is now parallel to the gradient of Q.

The second one-dimensional stratum is represented by the $su(2) + u(1) + su(2) + u(1)$ singlet S_1' :

$$S_1' : 2\phi^{45}_{45} + \phi^{34}_{34} + \phi^{35}_{35} = \frac{2}{3} (t^{12}_{12} + t^{45}_{45}) - \frac{1}{3} (t^{14}_{14} + t^{24}_{24} + t^{25}_{25}) \tag{13}$$

The little algebra is not maximal. However, (13) is, in addition, invariant under the discrete transformations $D \in SU(5)$, which transforms the pair of indices (12) into the pair (45). The little group is therefore maximal.

An interesting case arises when the coefficient c of the cubic invariant is zero. In this case, the potential has the additional symmetry under the operation $P : \phi \to -\phi$, where P is not in $SU(5)$. The vector

$$S_1'' : \phi^{34}_{34} + \phi^{35}_{35} = \frac{1}{3} (t^{12}_{12} + t^{34}_{34} + t^{35}_{35} - t^{13}_{13} - t^{23}_{23} - t^{45}_{45}) \tag{14}$$

is invariant under $su(2) + u(1) + su(2) + u(1)$ and under the discrete operation DP. The little group of S_1'' is therefore maximal in $SU(5) \times Z_2$.

We now consider the two conditions for the minimum.

$$\partial_i V = 0 \quad ; \quad \partial_i \partial_j V \geq 0, \quad i,j = 1 \ldots 75 \tag{15}$$

For one-dimensional strata, the first equation (15) determines the norm of ϕ in function of the parameters of the potential μ^2, c, λ_i.

We find for S_1, with little algebra $su(3) + su(2) + u(1)$, an open set of values in parameter space for which (15) is satisfied. Our result is quite general and in disagreement with ref. (7), where our relations (2) are not satisfied. For S_1', with little algebra $su(2) + u(1) + su(2) + u(1)$ and discrete invariance D, we get in general a minimum for $c \neq o$ only.

For the general two-dimensional stratum $\Sigma_{\alpha\beta}$, there is not minimum, except for very special values of the parameters of the potential. In that case, the value of the potential at the minimum point is the same for all strata on the boundary of $\Sigma_{\alpha\beta}$ and corresponding interior points.

REFERENCES

(1) L. Michel, in Regards sur la Physique Contemporaine, p. 157-203, CNRS, Paris (1980).

(2) H. Ruegg, Phys. Rev. D22, 2040 (1980).

(3) M. Jaric, these Proceedings.

(4) H. Georgi, Phys. Lett. 108B, 283 (1982).

(5) M. Abud and G. Sartori, Phys. Lett 104B, 147 (1981).

(6) L. Michel and L. Radicati, Ann. Phys. 66, 758 (1981).

(7) T. Hübsch and S. Pallua, Zagreb preprint (1983).

CONFORMALLY INVARIANT SOLUTIONS OF YANG-MILLS EQUATIONS IN MINKOWSKI SPACE

J-P. Antoine and M. Jacques

Institut de Physique Théorique
Université Catholique de Louvain

B-1348-Louvain-la-Neuve, Belgium

1. Introduction

Following Harnad, Shnider and Vinet [1,2], we study the SU(2) Yang-Mills (YM) equations directly in Minkowski space, restricting our attention to solutions with a large invariance group. Specifically we look for solutions invariant under a maximal subgroup of the conformal group $C(3,1)$ of space-time. With this simplifying assumption, the YM equations reduce to purely algebraic relations and are readily solved. As shown by Beckers et al.[3] (hereafter noted BHPW), there are, up to Poincaré conjugation, nine different types of such maximal subgroups. One of them is compact, namely $O(2) \times O(4)$, and it is the only type considered by Harnad et al.[1,2]. In this paper we shall extend their analysis to the other eight, noncompact, maximal subgroups of $C(3,1)$. The result is that, in all eight cases, the SU(2) YM equations admit no nonzero solutions, except the familiar, abelian, Maxwell solutions.

2. Compactified Minkowski space

A well-known difficulty arises from the fact that the conformal group $C(3,1)$ does not act globally on Minkowski space M, thus preventing any use of global methods, as opposed to infinitesimal ones. As usual [3] we embed M in a compact space $\overline{M}$, with global $C(3,1)$ action, the so-called <u>conformally compactified Minkowski space</u> $\overline{M}$. The image of M, denoted M again, is a dense open submanifold of $\overline{M}$ and $\overline{M} \setminus M$ may be viewed as a light cone at infinity. The space $\overline{M}$ is most easily represented as a projective null cone in $\mathbb{R}^6$, with metric (+ - - - - +). The action on $\overline{M}$ of the conformal group $C(3,1) \sim O(4,2)/Z_2$ is the one induced by the linear action of $O(4,2)$ on $\mathbb{R}^6$, and it is transitive and effective.

For all the nine maximal subgroups of $C(3,1)$, the orbital analysis of their action on $\overline{M}$ has been described in full detail by BHPW. For each such subgroup G, the space $\overline{M}$ has a unique generic dense stratum M', consisting of two orbits at most. Thus it suffices to consider G-invariant solutions of the YM equations on a single orbit G/G_o, where G_o is the isotropy subgroup at some point on the orbit. The singular strata, which make up $\overline{M} \setminus M'$, are lower dimensional submanifolds, contained in M, that coincide with the regions where the YM fields (if any) are singular, and thus they may be identified with the possible locations of sources. For a further discussion of this point, see BHPW.

3. General procedure

Our analysis follows the general, coordinate-free, method developed by Harnad et al.[4], that we briefly summarize here for completeness. Let $\mathfrak{M}$ be a differentiable manifold and H a compact Lie group, with Lie algebra $\mathfrak{h}$. A gauge theory on $\mathfrak{M}$, with gauge group H, will be described in terms of a H-principal bundle $P \to \mathfrak{M}$. Let $\{U_\alpha\}$ be an open covering of $\mathfrak{M}$. Then a gauge potential is a set of $\mathfrak{h}$-valued 1-forms ω_α on U_α, namely the pull-back under some local section $\sigma_\alpha : U_\alpha \to P$ of a connection 1-form ω on P. The corresponding gauge field F_α is the pull-back of the curvature 2-form $F \equiv D\omega \equiv d\omega + \frac{1}{2}[\omega,\omega]$. Assume $\mathfrak{M}$ has a (pseudo) Riemannian metric, which allows to define the (Hodge) duality *. Then the pure YM equations read simply $D * F = 0$, where $D \equiv d + [\omega,.]$ is the covariant derivative.

Let now a Lie group G act smoothly on $\mathfrak{M}$: $(g,x) \to f_g(x)$. We say that the gauge potential ω_α is G-invariant (that is, up to a gauge transformation), if there exists a smooth function $\rho_\alpha : G \times U_\alpha \to H$, such that, locally :

$$(\dagger) \qquad f^*_g\, \omega_\alpha = \mathrm{Ad}\, \rho_\alpha^{-1} \cdot \omega_\alpha + \rho_\alpha^{-1}\, d\rho_\alpha \;.$$

ω_α is called strictly G-invariant if $f^*_g\, \omega_\alpha = \omega_\alpha$. The function ρ_α defines on the principal bundle P a smooth G-action $\tilde{f}_g$ which projects on the given action f_g on $\mathfrak{M}$. In terms of $\tilde{f}_g$, the invariance condition (†) is simply the (strict) invariance of the connection ω :

$$\tilde{f}^*_g\, \omega = \omega \;.$$

Using this language, the general procedure for finding all G-invariant solutions of the YM equations decomposes into three steps :

(i) classify all principal H-bundles $P \to \mathfrak{M}$, with G-action projecting onto the given action f_g on $\mathfrak{M}$;

(ii) for each such bundle P, classify all G-invariant connections ω;

(iii) for each ω, use $F = D\omega$ as Ansatz in the YM equation $D * F = 0$.

For the problem at hand, we have seen above that $\mathfrak{M}$ may be taken as a single orbit G/G_o, and the action f_g is left multiplication. In that case, the situation simplifies considerably, thanks to two remarkable theorems [4].

(i) The homomorphism theorem : there is a one-to-one correspondence (up to equivalence) between principal H-bundles $P \to G/G_o$ and homomorphisms $\lambda : G_o \to H$. The bundle associated to a given λ is $P_\lambda = G \times_{G_o} H$, the quotient of $G \times H$ by the equivalence relation

$$(g,h) \sim (gg_o, \lambda(g_o)^{-1} h) \quad , \quad g_o \in G_o \;.$$

(ii) Wang's theorem : given λ and the associated bundle P_λ, there is a one-to-one correspondence between invariant connections ω on P_λ and linear maps $A : \mathfrak{g} \to \mathfrak{h}$ verifying the following two conditions, written for the case where G_o is connected ($\mathfrak{g}$ denotes the Lie algebra of G) :

(W1) $A(\zeta) = \lambda_*(\zeta)$, $\forall \zeta \in \mathfrak{g}_o$

(W2) $A([\zeta,\eta]) = [\lambda_*(\zeta), A(\eta)]$, $\forall \zeta \in \mathfrak{g}_o, \forall \eta \in \mathfrak{g}$.

Here $\mathfrak{g}_o$ is the Lie algebra of G_o, and $\lambda_* : \mathfrak{g}_o \to \mathfrak{h}$ is the differential of λ at e_G. For a given map A, the corresponding connection on $P_\lambda = G \times_{G_o} H$, pulled back successively to $G \times H$ (with the canonical projection $\pi : G \times H \to P_\lambda$), to G (with the section $g \to (g,e_H)$) and to G/G_o (with a section $\sigma : G/G_o \to G$) is given by $\omega_A = \sigma^* (A \cdot \theta_G)$, where θ_G is the canonical Maurer-Cartan 1-form on G.

Thus, in the case of a single orbit, our program reads :

(i) classify, up to conjugacy, all homomorphisms $\lambda : G_o \to H$;

(ii) given λ, list all possible maps $A : \mathfrak{g} \to \mathfrak{h}$ verifying (W1),(W2);

(iii) given A, compute ω_A and $F_A = D\omega_A$, and solve the YM equation $D * F_A = 0$.

Remark : if G_o is not connected, Wang's theorem still holds, with (W2) replaced by an appropriate, non-infinitesimal, condition [4].

4. Results for the Yang-Mills system

The procedure just described yields all solutions of the pure YM equations, invariant under any noncompact maximal subgroup of C(3,1), with help of two crucial observations.

(1) First we notice that the only nontrivial Lie subgroup of SU(2) is U(1) (up to conjugation), and it is compact and abelian. Then, for any homomorphism $\lambda : G_o \to H$, Ker λ is a closed normal subgroup of G_o, and G_o/Ker λ is isomorphic to Im λ; if it is nondiscrete, Im λ must be a Lie subgroup of SU(2), i.e. either U(1) or SU(2) itself. Therefore, if G_o is noncompact, Ker λ must be noncompact if it is continuous, or infinite if it is discrete, for the isomorphism $G_o/\mathrm{Ker}\,\lambda \simeq \mathrm{Im}\,\lambda$ to be possible. If Im λ is discrete, $\lambda_* \equiv 0$ and the situation is even simpler.

This observation yields easily all possible homomorphisms λ, simply by listing the possible kernels (G_o is indeed noncompact for all eight noncompact maximal subgroups G).

(2) There are two basic homomorphisms :

$$\lambda_n : SO(2) \to U(1) : \begin{bmatrix} \cos\rho & \sin\rho \\ -\sin\rho & \cos\rho \end{bmatrix} \mapsto \begin{bmatrix} e^{in\rho} & 0 \\ 0 & e^{-in\rho} \end{bmatrix} , \quad n \in \mathbb{Z}$$

$$\hat{\lambda}_m : SO_o(1,1) \to U(1) : \begin{bmatrix} \cosh\mu & \sinh\mu \\ \sinh\mu & \cosh\mu \end{bmatrix} \mapsto \begin{bmatrix} e^{im\mu} & 0 \\ 0 & e^{-im\mu} \end{bmatrix} , \quad m \in \mathbb{R}$$

One has Ker $\lambda_n = Z_n$, but Ker $\hat{\lambda}_m \simeq \mathbb{Z}$, i.e. discrete and infinite as it should, since $SO_o(1,1)$ is noncompact. Arbitrary powers $n \in \mathbb{Z}$, resp. $m \in \mathbb{R}$, are allowed since the image is an abelian subgroup of SU(2) in both cases. It turns out that all possible homomorphisms $\lambda : G_o \to SU(2)$ are combinations of λ_n and $\hat{\lambda}_m$, and none of them maps G_o on the whole of SU(2).

Finally we come to the explicit results. The case of the compact maximal subgroup O(2) x O(4) has been discussed by Harnad, Shnider and Vinet [1,2], and all solutions found. The eight noncompact maximal subgroups may be subdivided into three classes, according to the number of solutions.

Class A : $\omega \equiv 0$

. O(4,1) and O(3,2) : G_o = O(3,1) in both cases.

Since O(3,1) is simple and noncompact, Ker λ = O(3,1) and every G-invariant connection must be strictly invariant [4]. According to BHPW, there are none. Indeed, $A \equiv 0$ is the only solution of Wang's equations (W1),(W2) (where $\lambda_* = 0$).

Class B : $\omega \neq 0$, $F = 0$

. O(2) x O(2,2) : $G_o = O(2,1)$

. SIM (3,1) : $G_o = O(3,1) \times O(1,1)$

. OPT (3,1) : $G_o = E(2) \times \tilde{E}(2) = (\mathbb{R}^2 \wedge SO(2)) \times (\mathbb{R}^2 \wedge SO(1,1))$

. S(U(2,1) x U(1)) : G_o complicated (see BHPW)

In each of the four cases, Wang's map A is realized, with an appropriate choice of bases for $\mathfrak{g}$ and $\mathfrak{h}$ = su(2), by a matrix with only one nonvanishing row. This means that the range of A, where ω takes its values, is contained in an abelian subalgebra u(1) of su(2). Furthermore, choosing an adequate parametrization on G, one finds that every possible 1-form ω is exact : $\omega = (\sum_i d\phi_i)\, \sigma_1$, where σ_1 generates u(1) = Im A. Therefore, in each case, $F \equiv D\omega = d\omega = 0$.

Class C : $\omega \neq 0$, $F \neq 0$

. SO(3) x SO(2,1))
. SO(2,1) x SO(2,1)) G_o = SO(2) x SO(1,1) in both cases.

Here Wang's analysis yields a two-parameter family of nonzero connections $\omega_{(n,m)}$ ($n \in \mathbb{Z}$, $m \in \mathbb{R}$) and corresponding nonabelian gauge fields $F_{(n,m)}$. However, the only nonzero fields that verify the YM equation $D * F_{(n,m)} = 0$ have the form $F_{(n,m)} = F^{(j)}_{(n,m)}\, \sigma_1 \in u(1)$, where :

. for SO(3) x SO(2,1) :

$$F^{(1)}_{(n,m)} = \frac{M}{|\underline{x}|^3} (\varepsilon_{ijk}\, x^i\, dx^j \wedge dx^k) + \frac{N}{|\underline{x}|^3}\, dx^o \wedge (\underline{x} \cdot d\underline{x})$$

(M,N real constants, depending on n,m)

. for SO(2,1) x SO(2,1) :

$$F^{(2)}_{(n,m)} = \frac{M}{\ell^3} (\varepsilon_{abc}\, x^a\, dx^b \wedge dx^c) + \frac{N}{\ell^3}\, dx^3 \wedge (x^o dx^o - x^1 dx^1 - x^2 dx^2)$$

where (a,b,c) = (0,1,2) and $\ell^2 = (x^o)^2 - (x^1)^2 - (x^2)^2$.

Clearly, these fields are abelian (they take their values in a subalgebra u(1) $\subset$ su(2)), and are simply the Maxwell fields embedded in the YM theory. Indeed, $F^{(1)}$ is the electromagnetic field of a pointlike static source, both electric and magnetic, whereas $F^{(2)}$ describes a pointlike source moving along the z-axis at constant velocity. These fields are thus singular on the submanifolds $\underline{x} = 0$, resp. $\ell^2(x) = 0$, and these are precisely the singular orbits in the respective cases (see BHPW). Notice finally that the fields $F_{(n,m)}$ are strictly invariant (and as such listed already by BHPW), whereas the connections $\omega_{(n,m)}$ are not.

5. Concluding remarks

We have shown that the pure SU(2) Yang-Mills equations on Minkowski space have no nontrivial solutions invariant under a noncompact *maximal* subgroup of C(3,1), in sharp contrast with the compact case. It is worth noticing that exactly the same result was obtained by Légaré [5] for spinor fields.

It is unclear whether the same situation will prevail in more general situations. If we take a larger gauge group H, still compact but of rank at least two, it will contain nonclosed noncompact subgroups, namely the familiar irrational helices winding around a torus U(1) x U(1) $\subset$ H. Then there might exist homomorphisms λ mapping G_0 onto such a noncompact subgroup of H, and the argument used above fails. In such case, the classification of the homomorphisms $\lambda : G_0 \to H$ will be much more difficult. On the other hand, if we keep H = SU(2), but consider smaller subgroups of C(3,1) as invariance groups, we face other problems : there are many more such subgroups, the orbital structure of their action on $\overline{M}$ is more complicated, and, in general, the YM equations will no longer be purely algebraic, but genuine differential equations. This approach has been taken by Harnad et al.[1,2] in the compact case, and they have found a host of solutions. A similar analysis in the noncompact case remains to be done.

References

[1] J. HARNAD, S. SHNIDER, L. VINET, J. Math. Phys. 20 (1979) 931

[2] J. HARNAD, L. VINET, S. SHNIDER, in "Complex Manifold Techniques in Theoretical Physics", ed. by D. Lerner and P. Sommers (Pitman, New York 1979), pp. 219-230

[3] J. BECKERS, J. HARNAD, M. PERROUD, P. WINTERNITZ, J. Math. Phys. 19 (1978) 2126 (denoted BHPW in the text)

[4] J. HARNAD, S. SHNIDER, L. VINET, J. Math. Phys. 21 (1980) 2719

[5] M. LEGARE, J. Math. Phys. 24 (1983) 1219

TWO BODY RELATIVISTIC SCATTERING WITH AN O(1,1) SYMMETRIC SQAURE WELL POTENTIAL

R. Arshansky and L.P. Horwitz
Tel Aviv University, Ramat Aviv, Israel.

In the framework of a manifestly covariant relativistic quantum theory,[1)] the generalized eigenvalue equation for the scattering wave functions[2)] is exactly soluble in the case in which the direct action potential $V(x^2)$ is piecewise constant. Blaha[3)] has studied the bound state problem when $V(x^2)\propto \theta(x^2)$, where $x^2=\vec{x}^2-t^2$. We study this equation in one space and one time dimension for a potential of the form

$$V(x^2) = \begin{cases} V_S & 0 \leq x^2 \leq a^2 \\ V_T & 0 \leq -x^2 \leq b^2 \end{cases}$$

and zero for x^2 outside of these regions, i.e., an O(1,1) symmetric generalization of the non-relativistic square well.

The evolution equation for the two-body problem,[1)]

$$i\frac{\partial\psi_2}{\partial\tau} = K\psi_2 = [\frac{P^2}{2M} + \frac{p^2}{2m} + V(x^2)]\psi_2 \quad , \tag{1}$$

where

$$P^\mu = p_1{}^\mu + p_2{}^\mu , \quad p^\mu = \frac{M_2p_1{}^\mu - M_1p_2{}^\mu}{M_1 + M_2} \qquad x = x_1 - x_2 \quad , \tag{2}$$

and τ is the invariant evolution parameter ($\psi_2 \in L^2(R^2)$, separate to a direct sum over the conserved values of the total center-of-mass momentum, and one must then solve the generalized eigenvalue equations

$$\pm(\frac{\partial^2}{\partial\rho^2} + \frac{1}{\rho}\frac{\partial}{\partial\rho} - \frac{1}{\rho^2}\frac{\partial^2}{\partial\beta^2})\psi(\rho,\beta) = 2m(k - V(\pm\rho^2))\ \psi \tag{3}$$

where, respectively, in the space-like and time-like regions,

$$\begin{aligned} t &= \pm\rho\cosh\beta \\ x &= \pm\rho\sinh\beta \end{aligned} \qquad (SL) \tag{4}$$

$$\begin{aligned} t &= \pm\rho\sinh\beta \\ x &= \pm\rho\cosh\beta \end{aligned} \qquad (TL) \tag{5}$$

On the light lines, $\partial_\mu\partial^\mu = \frac{\partial^2}{\partial x_+\partial x_-}$, and by integrating across these lines, one obtain the continuity condition

$$\lim_{\varepsilon\to 0} \frac{\partial\psi}{\partial x_+}(x_+,\varepsilon) - \frac{\partial\psi}{\partial x_+}(x_+,-\varepsilon) = 0 \tag{6}$$

$$\lim_{\varepsilon\to 0} \quad \frac{\partial\psi}{\partial x_-}(\varepsilon, x_-) - \frac{\partial\psi}{\partial x_-}(-\varepsilon, x_-) = 0 \tag{7}$$

Further, defining

$$\psi(\rho,\beta) = \int d\lambda\, e^{i\lambda\beta}\; \psi(\rho,\lambda) \quad , \tag{8}$$

where $e^{i\lambda\beta}$ is the generalized eigenfunction of the boost operator

$$A = i(t\frac{\partial}{\partial x} + x\frac{\partial}{\partial t}) = i\frac{\partial}{\partial\beta} \quad , \tag{9}$$

Eq. (3) can be further separated to a "radial" equation for $\psi(\rho,\lambda)$, with $-\frac{\partial^2}{\partial\beta^2} \to \lambda^2$. The asymptotic form of the solutions of (3) near the light lines ($\rho\to 0$) are

$$\psi_\pm(\rho,\beta) = \int e^{i\lambda\beta}\, \rho^{\pm i\lambda}\, f_\pm(\lambda)\, d\lambda + O(\rho^2) \; , \tag{10}$$

and the light line continuity condition become, along the x_+ axis,

$$f_{I,+}(\lambda) = f_{II,+}(\lambda) \quad ; \quad f_{IV,+}(\lambda) = f_{III,+}(\lambda) \tag{11}$$

and along the x - axis,

$$f_{I,-}(\lambda) = f_{III,-}(\lambda) \quad ; \quad f_{IV,-}(\lambda) = f_{II,-}(\lambda) \quad , \tag{12}$$

where we have labelled the solutions in the space-like regions II(x>0), III(x<0), and in the time-like regions, I(t>0) and IV(t<0). (See figure 1).

For $\kappa_R^2 = 2m(k-V_R) \gtrless 0$, R = 0,s,T, the general solutions[4] are

$$\psi_{II,III} = \begin{cases} \hat{a}_\pm^{s'}\, H_{i\lambda}^{(1)}(\rho\kappa_s) + \hat{b}_\pm^{s'}\, H_{i\lambda}^{(2)}(\rho\kappa_s) & \rho \lessgtr a \\ \hat{a}_\pm^{s}\, H_{i\lambda}^{(1)}(\rho\kappa_0) + \hat{b}_\pm^{s}\, H_{i\lambda}^{(2)}(\rho\kappa_0) & \rho \gtrless a \end{cases} \tag{13}$$

where $\hat{a}_\pm = e^{-\pi\lambda/2}\, a_\pm$, $\hat{b}_\pm = -e^{\pi\lambda/2}\, b_\pm$ takes into accounts factors multipying the amplitudes in the asymptotic regions of large ρ, and

$$\psi_{I,IV} = \begin{cases} a_\pm^{T'}\, H_{i\lambda}^{(1)}(i\rho\kappa_T) + \\ \quad + b_\pm^{T'}\, I_{i\lambda}(\rho\kappa_T) & \rho \lessgtr b \\ a_\pm^{T}\, H_{i\lambda}^{(1)}(i\rho\kappa_0) & \rho \gtrless b \end{cases} \tag{14}$$

Since the τ-dependence of these solutions in the space-like regions is $\exp\{-i\frac{\mu^2}{2m}\tau \pm i\mu\rho\}$, $\hat{a}_\pm^{s,s'}$ correspond to outgoing waves, and $\hat{b}_\pm^{s,s'}$ to incoming waves. Requiring continuous differentiability on the hyperbolic boundaries of the square well, and the conditions (11),(12), we find, for the relation be-

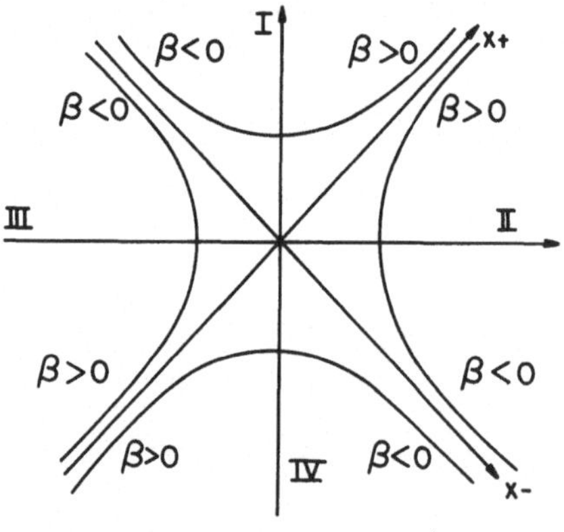

Figure 1. Hyperbolic and light line coordinates.

tween coefficients in the inner space-like regions, an internal S-matrix reflecting the effect of scattering through the time-like regions (tunnelling in the ρ parameter), relating $b_-^{s'}$, $b_+^{s'}$ to $a_+^{s'}$, $a_-^{s'}$:

$$S_{int} = \frac{1}{r^2-1} \begin{pmatrix} r(e^{\pi\lambda} - e^{-\pi\lambda}) , & e^{\pi\lambda} - r^2 e^{-\pi\lambda} \\ e^{\pi\lambda} - r^2 e^{-\pi\lambda} , & r(e^{\pi\lambda} - e^{-\pi\lambda}) \end{pmatrix} , \tag{15}$$

where

$$r = \frac{1}{D} \left(\frac{\kappa_s}{\kappa_T}\right)^{2i\lambda} e^{\pi\lambda} \quad ; \quad D = 1 - e^{-\pi\lambda/2} \sin h\pi\lambda \, \frac{W_{To}}{V_{To}} \tag{16}$$

$$W_{To} = H_{i\lambda}^{(1)}(ib\kappa_T)\, H_{i\lambda}^{(1)'}(ib\kappa_o) - \frac{\kappa_T}{\kappa_o} H_{i\lambda}^{(1)'}(ib\kappa_T)\, H_{i\lambda}^{(1)}(ib\kappa_o)$$

$$V_{To} = I_{i\lambda}(b\kappa_T)\, H_{i\lambda}^{(1)'}(ib\kappa_o) + i\frac{\kappa_T}{\kappa_o} I_{i\lambda}'(b\kappa_T)\, H_{i\lambda}^{(1)}(ib\kappa_o) \quad . \tag{17}$$

It follows from the properties of the modified Bessel functions that $|D|^2 = 1$, and hence is S_{int} is unitary. The full S-matrix, relating b_-^s, b_+^s to a_+^s, a_-^s, is

$$S = \frac{1}{\Delta} \begin{pmatrix} r(e^{\pi\lambda} - e^{-\pi\lambda}) , & u \\ u, & r(e^{\pi\lambda} - e^{-\pi\lambda}) \end{pmatrix} , \tag{18}$$

where

$$\Delta = -\frac{\pi^2 a^2 \kappa_o \kappa_s}{16} \left(\frac{\kappa_s}{\kappa_o}\right)\left\{ r^2\left(W_{os}^{12} - e^{-2\pi\lambda} \frac{\kappa_s}{\kappa_o} W_{so}^{11}\right)^2 - \left(W_{os}^{12} - \frac{\kappa_o}{\kappa_s} W_{so}^{11}\right)^2\right\}$$

$$u = -\frac{\pi^2 a^2 \kappa_o \kappa_s}{16}\left\{ \frac{\kappa_s}{\kappa_o} e^{\pi\lambda} (r^2 - 1) W_{os}^{12} W_{os}^{22} + (e^{\pi\lambda} - r^2 e^{-\pi\lambda})(W_{os}^{12} W_{so}^{12} + W_{so}^{11} W_{os}^{22}) - \left(\frac{\kappa_o}{\kappa_s}\right)(e^{\pi\lambda} - r^2 e^{-3\pi\lambda}) W_{so}^{11} W_{so}^{12} \right\}$$

$$W_{RR'}^{ij} = H_{i\lambda}^{(i)}(a\kappa_R)\, H_{i\lambda}^{(j)'}(a\kappa_{R'}) - \frac{\kappa_R}{\kappa_{R'}} H_i^{(i)'}(a\kappa_R)\, H_{i\lambda}^{(j)}(a\kappa_{R'}) \tag{19}$$

The reflection and transmission coefficients are

$$R = \frac{u}{\Delta} \qquad T = \frac{r}{\Delta}(e^{\pi\lambda} - e^{-\pi\lambda}) \tag{20}$$

Unitarity of the full S-matrix follows easily from the unitarity of S_{int} and conservation of current,(it then follows that $|T|^2 + |R|^2 = 1$, and that there is no Klein paradox[6]). For $a\kappa_o$, $b\kappa_o$, $a\kappa_s$, $b\kappa_T >> 1$, when the potential has a long range compared to the de Broglie wave-length of the reduced motion, one finds

(taking $V_s = V_T \neq 0$)

$$S_{res} = M(M + 2V_s) + \frac{n^2\pi^2}{4} \left(\frac{h}{ac}\right)^2 \left(\frac{M}{m}\right) \tag{21}$$

and

$$\Gamma_{res} \propto \left(\frac{h}{ac}\right)^2 \frac{n}{m_{res}} \left(\frac{M}{m}\right) \xrightarrow[n\ \text{large}]{} \propto \left(\frac{h}{ac}\right) \sqrt{\frac{M}{m}} \quad . \tag{22}$$

A long range sequence,giving an approximately linear mass spectrum for large n, has some resemblance to known low lying meson spectra. Although we are using solutions of O(1,1) potential problem, it is interesting to compare (as in certain classes of non-relativistic problems, one uses the one-dimensional square well as a qualitative model) the sequence (21) with for example, the ρ, g, f, h spectrum (the starred quantities are input data):

	m_{exp}	Γ_{exp}	M	Γ
ρ	769 ± 3	154 ± 5	769*	139
f	1273 ± 5	179 ± 20	1200	185
g	1691 ± 5	200 ± 20	1691*	200
h	2040 ± 20	150 ± 20	2200	204

Assuming that we are dealing with $\pi\pi$ resonances, we find V_s=410 MeV and the size of the potential region $a \sim 30\times10^{-13}$cm.

Acknowledgements:

We are grateful to Avi Soffer for many helpful discussions on this work, and to Yehiahu Lavie (ז"ל)[7] for his participation in the early stages of this work, and for sharing with us his deep physical insights.

References.

1. E.C.G. Stuecklberg, Helv. Phys. Acta 14,372 (1941);585(1941).
   L.P. Horwitz and C. Piron, Helv. Phys. Acta 46, 316 (1973) .
2. L.P. Horwitz and Y. Lavie, Phys. Rev. D26, 819 (1982).
3. S. Blaha, Phys. Rev. D12, 3921 (1975).
4. W. Ya. Vilenkin, "Special Functions and the Theory of Group Representations", Amer. Math. Soc., Providence (1968) .
5. R. Arshansky, L.P. Horwitz and Y. Lavie, Tel Aviv University preprint TAUP 1053-82, and to be published in Found. of Physics.
6. B. Tahller, Lett. Nuovo Cimento 31, 4391 (1981) .
   J.R. Fanchi, Foundations of Physics 11, 493 (1981).
7. Fell in Lebanon, January 2, 1983.

EMERGENCE OF CENTRAL EXTENSION OF KAC-MOODY ALGEBRA IN QUANTUM INTEGRABLE MODELS*

Gautam Bhattacharya
Department of Theoretical Physics
Madras University
Madras 600025, India

and

Ling-Lie Chau
Physics Department
Brookhaven National Laboratory
Upton, NY 11973

ABSTRACT

It is pointed out how quantum effects can alter the Kac-Moody algebra in the spectrum space for integrable nonlinear systems and cause the emergence of the central extension of the algebra.

INTRODUCTION

Recently it has been established that many of nonlinear systems, to which linear systems (or Lax pair) exist, have symmetry transformations forming a Kac-Moody algebra [1,2,3]

$$[Q_a{}^m, Q_b{}^n] = i\, C^c_{ab}\, Q_c{}^{m+n}, \quad -\infty < m,n < \infty , \tag{1}$$

with m,n integers. This algebra comes about, in some sense, in the expansion of the spectrum variable. We can resum it back by multiplying Eq. (1) with $(1/2\pi)^2 e^{im\lambda}\, e^{im\lambda'}$ and summing m and n, and obtain

$$[Q_a(\lambda), Q_b(\lambda')] = i\, C^c_{ab}\, Q_c(\lambda)\, \delta(\lambda-\lambda') , \tag{2}$$

which is like a "current" algebra in the spectrum space. We shall call such algebra Kac-Moody algebra from the spectrum space, in contrast to those from the coordinate space. It has been known in other theoretical examples [4] and to mathematicians [5] that such infinite dimensional algebras can have central extensions, i.e. Eq. (1) becomes,

---

* Invited talk delivered by Ling-Lie Chau.

$$[Q_a{}^m, Q_b{}^n] = i\, C^c_{ab}\, Q_c{}^{m+n} + C\, m\, \delta_{a,b}\, \delta_{m,-n} , \tag{3}$$

and the corresponding "current" algebra in the spectrum space becomes

$$[Q_a(\lambda), Q_b(\lambda')] = i\, C^c_{ab}\, Q_c(\lambda)\, \delta(\lambda'-\lambda) + \frac{iC}{2\pi}\, \delta_{a,b}\, \delta'(\lambda'-\lambda) . \tag{4}$$

However, so far no such central extension has been found in these classical nonlinear systems. Here we would like to show how such central extension can emerge in the quantum version of these nonlinear systems. Here we shall give a generic illustration of how the Kac-Moody algebra comes about, and how the quantum version can provide a mechanism for the emergence of the central extensions. Our discussions for the derivation of Kac-Moody algebra should apply to all those systems to which linear systems (Lax pair) exist, and our discussion on the emergence of the central extension should apply to those systems quantum formulations can be made for their transition matrices. Since so far the quantum formulation has only been done for two dimensional models we shall restrict our discussions to the two dimensional models.

Let's begin with the typical linear systems

$$\left(\frac{d}{du} - U\right)\chi(u,v,\lambda) = 0, \qquad \left(\frac{d}{dv} - V\right)\chi(u,v,\lambda) = 0 , \tag{5}$$

where U, V are functions of the original nonlinear fields, and depend also on the spectral parameter λ. The operators:

$$L = \frac{d}{du} - U, \quad M = \frac{d}{dv} - V , \tag{6}$$

are the Lax pair. The integrability condition for χ is the zero curvature condition on the Lax pair

$$[L, M] = 0 , \tag{7}$$

which in turn gives the original nonlinear equations for the original fields. In the following, three examples are given.

Nonlinear Schrödinger model:

$$U = i \begin{pmatrix} \frac{\lambda}{2} & \sqrt{g}\,\psi^* \\ -\sqrt{g}\,\psi & -\frac{\lambda}{2} \end{pmatrix} , \tag{8}$$

$$V = i \begin{pmatrix} \frac{\lambda^2}{2} + g|\psi|^2 & \sqrt{g}\,(-i\frac{d\psi^*}{du} - \lambda\psi^*) \\ \sqrt{g}\,(-i\frac{d\psi}{dv} + \lambda\psi & -\frac{\lambda^2}{2} - g|\psi|^2 \end{pmatrix} ; \tag{9}$$

and the corresponding nonlinear equation

$$i \frac{\partial}{\partial v} \psi = - \frac{\partial^2}{\partial u^2} \psi + 2g|\psi|^2\psi \quad . \tag{10}$$

Sine-Gordon model:

$$U = - \begin{pmatrix} i\lambda & \frac{1}{2} \partial_u \theta \\ - \frac{1}{2} \partial_u \theta & - i\lambda \end{pmatrix} \quad , \tag{11}$$

$$V = - \frac{1}{i\lambda} \begin{pmatrix} \cos \theta & \sin \theta \\ \sin \theta & - \cos \theta \end{pmatrix} \quad ; \tag{12}$$

and the nonlinear equation

$$\partial_u \partial_v \; \theta = 4 \sin \theta \quad . \tag{13}$$

Principal chiral model:

$$U = g^{-1} \partial_u g \quad , \tag{14}$$

$$V = g^{-1} \partial_v g \quad ; \tag{15}$$

and the nonlinear equation

$$\partial_v(g^{-1} \partial_u g) + \partial_u(g^{-1} \partial_v g) = 0 \quad , \tag{16}$$

It is well known that one can generate a new solution χ' of the linear system from an old one χ through the Riemann-Hilbert transform [7],

$$\chi'(\zeta) - \chi(\zeta) = \left(\frac{1}{2\pi i} \int \frac{d\zeta'}{\zeta' - \zeta} \chi(\zeta')(E(\zeta') - 1)\chi(\zeta')^{-1}\right)\chi(\zeta) \quad , \tag{17a}$$

where $E(\zeta')$ is a group element.

Considering an infinitesimal transformation so that $E(\zeta')-1$ becomes an element in the Lie algebra, e.g. choose

$$E(\zeta') - 1 = \eta^m(\zeta') \; t_a \quad , \tag{17b}$$

where

$$\eta(\zeta') = \frac{\zeta' - \zeta_i^*}{\zeta' - \zeta_i} \quad , \tag{18}$$

$$[t_a, t_b] = iC_{ab}{}^c \; t_c \quad , \tag{19}$$

so that Eq. (17) becomes

$$\delta_b^n \chi(\zeta) = \frac{1}{2\pi i} \left[\int \frac{d\zeta'}{\zeta' - \zeta} \eta^n(\zeta') \; \chi(\zeta') \; t_b \chi(\zeta')^{-1}\right]\chi(\zeta) \quad . \tag{20}$$

To study the algebraic structure of such variations, we make variation again, using the original equation of variation, and obtain

$$\delta_a^m \delta_b^n \chi(\zeta) = \left(\frac{1}{2\pi i}\right)^2 \int_{\Gamma_1} \frac{d\zeta'\ \eta^n(\zeta')}{\zeta' - \zeta} \int_{\Gamma_2} \frac{d\zeta''\ \eta^m(\zeta'')}{\zeta'' - \zeta'}$$

$$\cdot \left\{S_a(\zeta'')\ S_b(\zeta') - S_b(\zeta')\ S_a(\zeta'') + \frac{\zeta''-\zeta'}{\zeta''-\zeta}\ S_b(\zeta')\ S_a(\zeta'')\right\}\chi(\zeta), \tag{21}$$

where

$$S_a(\zeta) \equiv \chi(\zeta)\ t_a\ \chi(\zeta)^{-1} \tag{22}$$

Interchange a ↔ b, and m ↔ n in the previous equation we obtain $\delta_b^n\delta_a^m\chi(\zeta)$. Taking the difference, and carefullly shifting contours, finally we arrive at

$$(\delta_a^m\delta_b^n - \delta_b^n\delta_a^m)\chi(\zeta) = -\frac{1}{(2\pi i)^2} \int_\Gamma d\zeta'' \oint_{\zeta''} \frac{d\zeta'}{(\zeta'-\zeta'')(\zeta'-\zeta)}\ I(m,n;a,b), \tag{23}$$

where

$$I(m,n;a,b) = \eta^m(\zeta'')\ \eta^n(\zeta')\ S_a(\zeta'')\ S_b(\zeta') - \eta^n(\zeta'')\ \eta^m(\zeta')\ S_b(\zeta'')\ S_a(\zeta'),$$

$$\equiv I_- + I_+ , \tag{24a}$$

where

$$I_\mp \equiv \frac{1}{2}\left\{\eta^m(\zeta'')\ \eta^n(\zeta')[S_a(\zeta''),S_b(\zeta')]_\mp - \eta^n(\zeta'')\ \eta^m(\zeta')[S_b(\zeta''),S_a(\zeta')]_\mp\right\} \tag{24b}$$

where $[\ ]_-$, $[\ ]_+$ denote commutator and anticommutator respectively. The contribution from the anticommutator term I_+ vanishes as we take the residue of the ζ' integration around ζ''. The contribution from the commutators, if they do not have any more singularities as $\zeta' \to \zeta''$, becomes

$$I_{-,\zeta'\to\zeta'',\text{regular}} = \eta^{m+n}(\zeta'')[S_a(\zeta''),S_b(\zeta'')]_-$$

$$= \eta^{m+n}(\zeta'')\ \chi(\zeta'')\ [t_a,t_b]_-\ \chi(\zeta'')^{-1}$$

$$= iC_{ab}^{\ \ c}\ \eta^{m+n}(\zeta'')\ S_c(\zeta'') . \tag{25}$$

Substituting this into Eq. (23), we obtain

$$(\delta_a^m\delta_b^n - \delta_b^n\delta_a^m)\ \chi(\zeta) = iC_{ab}^{\ \ c}\ \delta_c^{m+n}\ \chi(\zeta) . \tag{26}$$

This is the derivation of the affine Lie algebra, a specific class of the general ∞-dimensional Kac-Moody algebra in these classical nonlinear systems.

In the quantum case, the $\chi(\zeta)$'s, thus the $S(\zeta)$'s become non-commuting fields. They now become operators in two spaces, one is the original matrix space and the other is the space to which each matrix component of the field $Tr(t_a\chi(\zeta))$ operates on. It is the commutator of these fields as operators may introduce additional singularities. These singular terms need to be evaluated in addition to what has been done for Eq. (25). So we do expect additional terms. The commutation relation actually now has two parts:

$$[S_a(\zeta''), S_b(\zeta')]_- = [S_a(\zeta'') \underset{,}{\tilde{\otimes}} S_b(\zeta')] + [S_a(\zeta'') \underset{,}{\otimes} S_b(\zeta')]_- . \tag{27}$$

The first commutation takes care of the commutation in matrix space, the second takes care of the commutation in the operator space. Here we have adopted the notation used in Ref. [6]. The singularities arise only from the second term

$$[S_a(\zeta''), S_b(\zeta')]_{-,\ \zeta' \to \zeta'',\text{sing.}} = [S_a(\zeta'') \underset{,}{\otimes} S_b(\zeta')]_{-,\text{sing.}} \tag{28}$$

Now let's separate out the $(a \leftrightarrow b)$-symmetric part from the $(a \leftrightarrow b)$-antisymmetric part:

$$[S_a(\zeta'') \underset{,}{\otimes} S_b(\zeta')]_{-,\text{sing.}}$$

$$= \frac{1}{2(\zeta' - \zeta'')} \{(\zeta'-\zeta'')([S_a(\zeta'') \underset{,}{\otimes} S_b(\zeta')]_- + [S_b(\zeta'') \underset{,}{\otimes} S_a(\zeta')]_-)$$

$$+ (\zeta'-\zeta'')([S_a(\zeta'') \underset{,}{\otimes} S_b(\zeta')]_- - [S_b(\zeta'') \underset{,}{\otimes} S_a(\zeta')]_-)\}$$

$$\equiv \frac{1}{\zeta' - \zeta''} \{F_+(a,b;\zeta'',\zeta') + F_-(a,b,;\zeta'',\zeta')\} . \tag{29}$$

Note that F_+ is now both symmetric in $a \leftrightarrow b$, and symmetric in $\zeta'' \to \zeta'$; while F_- is antisymmetric in $a \leftrightarrow b$, and in $\zeta'' \leftrightarrow \zeta'$. Now the additional terms to Eq. (26) are $\Delta_+ + \Delta_-$, so that Eq. (26) becomes

$$(\delta_a{}^m \delta_b{}^n - \delta_b{}^n \delta_a{}^m)\chi(\zeta) = C_{ab}{}^c\ \delta_c{}^{m+n}\chi(\zeta) + \Delta_+ + \Delta_- , \tag{30a}$$

where

$$\Delta_\pm = - \frac{1}{(2\pi i)^2} \int_\Gamma d\zeta'' \int_{\zeta''} \frac{d\zeta'}{(\zeta'-\zeta'')(\zeta'-\zeta)} \cdot \frac{1}{(\zeta'-\zeta'')}$$

$$\cdot \left[\eta^m(\zeta'')\ \eta^n(\zeta') \mp \eta^n(\zeta'')\ \eta^m(\zeta')\right] F_\pm(a,b;\zeta'',\zeta'). \tag{30b}$$

Calculating the double-pole residue in the ζ' integration, we obtain:

$$\Delta_+ = - \frac{1}{2\pi i} \int \frac{d\zeta''}{\zeta''-\zeta} (n-m) \frac{d\eta(\zeta'')}{d\zeta''} \cdot \eta^{m+n-1}(\zeta'')\ F_+(a,b;\zeta'',\zeta')_{\zeta'=\zeta''} , \tag{30c}$$

$$\Delta_- = - \frac{1}{2\pi i} \int \frac{d\zeta''}{\zeta''-\zeta} 2 \cdot \eta^{m+n}(\zeta'') \left(\frac{d}{d\zeta'} F_-(a,b'\zeta'',\zeta')\right)_{\zeta'=\zeta''} . \tag{30d}$$

Note that Δ_+ is symmetric in $a \leftrightarrow b$ but antisymmetric in $m \leftrightarrow n$; while Δ_- is antisymmetric in a $\leftrightarrow$ b but symmetric in m $\leftrightarrow$ n, like the first term on the right hand side of Eq. (30a). We see here Δ_+ is the candidate for the central extension; while Δ_- modifies the original algebra. The commutation relation $[S_a(\zeta'') \underset{,}{\otimes} S_b(\zeta')]_-$ can be related to the commutation relation of $[\chi(\zeta'') \underset{,}{\otimes} \chi(\zeta')]_-$ which in turn can be related to the commutation relation between transition matrix $[T(\zeta'') \underset{,}{\otimes} T(\zeta')]_-$ which has been calculated for the Sine-Gordon and the nonlinear Schrödinger model [6,8,9]. The commutator $[T(\zeta'') \underset{,}{\otimes} T(\zeta')]_-$ can either be Poisson bracket or operator commutation relation. They in general have $1/(\zeta''-\zeta')$ singularity. The calculations are very involved, and the detailed results depend upon specific systems [10,11]. They will be published elsewhere.

Here we have demonstrated that the quantum effects in the integrable nonlinear systems possibly can modify the algebra relation and make appearances of the central extension of the Kac-Moody algebra.

ACKNOWLEDGMENTS

Authored under Contract DE-AC02-76CH00016 with the U.S. Department of Energy.

REFERENCES

[1] For the chiral model see: L. Dolan, Phys. Rev. Lett. 47 (1981) 1371; M.L. Ge and Y.S. Wu, Phys. Lett. 108B (1982) 411; C. Devchand and D.B Fairlie, Nucl. Phys. B194 (1982) 232; K. Ueno and Y. Nukamura, Phys. Lett. 117 (1982) 273; Y.-S. Wu, Nucl. Phys. B211 (1983) 160;
For the self-dual Yang-Mills theory see: L.-L. Chau, M.-L. Ge and Y.-S. Wu, Phys. Rev. D25 (1982) 1086; K. Ueno and Y. Nakamura, Phys. Lett. 107B (1982) 273; L.-L. Chau and Y.-S. Wu, Phys. Rev. D26 (1983) 3581; L.-L. Chau, M.-L. Ge, A. Sinha, Y.-S. Wu, Phys. Lett. 121B (1983) .

[2] For the interplay between two models: L.-L. Chau, "Chiral Fields, Self-dual Yang-Mills Fields as Integrable Systems and the Role of Kac-Moody Algebra", Lecture in the Proceedings of "Nonlinear Phenomena", Oaxtepec, Mexico, Nov. 29-Dec. 17, 1982, Lecture Notes in Physics #189, edited by K.B. Wolf.

[3] For the Sine-Gordon model see: H. Eichenherr, Phys. Lett. 115B (1982) 385; T. Koikawa, R. Sasaki, Phys. Lett. 124B (1983) 85.

[4] S. Fubini and G. Veneziano, Nuovo Cimento 64A (1969) 811.
See review: Dual Theory, editor, M. Jacob, North-Holland/American Elsevier Pub.

[5] I.B. Frenkel and V.G. Kac, Inventiones Math. 62 (1980) 23.

[6] See review, L.D. Fadeev, Les Houches Lectures 1982.

[7] V.E. Zakharov and A.D. Shabat, Func. Anal. and Appl. 13 (1979) 13; V.E. Zakharov and A.V. Mikhailov, Comm. Math. 74 (1980) 21. For a brief review see, e.g., A.V. Mikhailov, CERN preprint TH-3194, 1981, and references therein; see also works by Nakamura and Ueno in Ref. [1], and see discussions in Ref. [2].
For application in general relativity see: I. Hauser and F.J. Ernst, Phys. Rev. D20 (1979) 362 and 978; W. Kinnersley and D.M. Chitre, J. Math. Phys. 18 (1977) 1538 and 19 (1978) 1926.

[8] S.A. Tsyplaev, Theor. Math. Phys. 48 (1981) 24.

[9] P.P. Kulish and E.K. Skylanin, Lecture Notes in Physics, 1982.

[10] The quantum inverse scattering formulation for the principle chiral has yet to be completed, see Ref. [11]. It is obvious that for these systems in which the commutators of the transition matrices have no singularities in the spectrum variable, the mechanism discussed here will not give rise to central extension.

[11] H.J. de Vega, H. Eichenherr, J.M. Maillet, LPTHE preprint 83-9 (1983).

COHOMOLOGICAL INTERPRETATION OF ANOMALIES
THE EXAMPLE OF THE TRACE ANOMALY

L. Bonora: Istituto di Fisica, Università di Padova and INFN Sez. Padova

P. Cotta-Ramusino: Istituto di Fisica, Università di Milano and INFN Sez. Milano

C. Reina: Istituto di Fisica, Università di Milano

To develop a proper interpretation of anomalies, one can resort to the geometric set-up for the calculus of variations as differential calculus on Banach manifolds. Let F be a Banach manifold, whose points ϕ are fields over a 4-manifold M, and let S be a smooth function on F, playing the rôle of the action of a physical therory. We assume that S is local, i.e. it can be written as the integral over M of a polynomial lagrangean density. The space of such functions will be denoted by Ω^o_{loc}. For the sake of simplicity we also assume that M is compact and without boundaries.

Let now $Z \in \chi(F)$ be a vector field over F. The first variation of S along Z is given by $L_Z S = dS(Z)$, where L_Z denotes the Lie derivative and d is the exterior differential. The invariance group of the theory is a subgroup G of the diffeomorphism group of F, which acting on F leaves the action S invariant. We imagine that G can be given the structure of a Lie group and we denote by $\mathfrak{g}$ its Lie algebra. The action of G on F induces a Lie algebra homomorphism $\alpha: \mathfrak{g} \to \chi(F)$, which associates to any $\tau \in \mathfrak{g}$ the fundamental vector field $\alpha(\tau)$. The variation of S induced by τ is then given by $\delta S(\tau) = dS(\alpha(\tau))$. Notice that $\delta S(\ )$ can be seen as a map $\delta S(\ ): \mathfrak{g} \to \Omega^o_{loc}$. The invariance of S under G amounts to requiring that $\delta S(\tau) = 0$, for any $\tau \in \mathfrak{g}$.

We now consider the complex

$$\Omega^o_{loc} \xrightarrow{\delta} \Omega^1_{loc} \xrightarrow{\delta_1} \cdots \xrightarrow{\delta_{p-1}} \Omega^p_{loc} \xrightarrow{\delta_p} \cdots$$

where Ω^p_{loc} is the space of p-linear skew-symmetric and local maps $W: \mathfrak{g} \to \Omega^o_{loc}$ and $\delta_p W(\tau_1, \dots, \tau_{p+1}) = dW(\alpha(\tau_1), \dots, \alpha(\tau_{p+1}))$. It is clear that $\delta_{p+1}\delta_p = 0$, so that we can construct the cohomology sets $H^p(\mathcal{g},\Omega^o_{loc}) = \ker \delta_p/\mathrm{im}\ \delta_{p-1}$. In the following we shall omit the subscripts and we will write simply δ for δ_p. As to the relevance of this cohomology of the Lie algebra $\mathcal{g}$ with values in the space of local action functions, note that, obviuosly enough, G-invariant action functions belong to $H^o(\mathcal{g},\Omega^o_{loc})$. Indeed, the fact that $\delta S = 0$ amounts to state - via Noether's theorem - the existence of conserved currents. Conversely, conserved currents, once integrated over M, are cohomologous to 0 in $H^1(\mathcal{g},\Omega^o_{loc})$.

Let us now consider a quantized version of the condition $\delta S(\tau) = 0$. Although the existence of a quantum theory for a generic action S is not guaranteed, if the quantum action principle holds, the quantum analogue Γ of the action should obey an equation of the form

$$\delta\Gamma = \hbar\,\Delta + O(\hbar^2),$$

where $\Delta:\mathcal{g}\to\Omega^o_{loc}$. It is obvious that Δ should satisfy the consistency condition $\delta\Delta = 0$, so that it represents a cohomology class $|\Delta|\in H^1(\mathcal{g},\Omega^o_{loc})$. If $|\Delta| \neq |0|$, there does not exist any function $\Gamma'\in \Omega^o_{loc}$ such that $\Delta = \delta\Gamma'$. Accordingly, there is no redefinition $\tilde{\Gamma} = \Gamma-\Gamma'$ of the action such that $\tilde{\Gamma}$ is invariant under G. We call any function Δ of this kind an <u>anomaly</u>.

As an example, consider the trace anomaly in the theory of a scalar field ϕ interacting with an external gravitational field $g_{\mu\nu}$. The classical action is

$$S(\phi,g) = \int_M d^4x\ \sqrt{g}\ (g^{\mu\nu}\partial_\mu\phi\ \partial_\nu\phi - \frac{1}{6}\ R\ \phi^2),$$

where R is the scalar curvature of $g_{\mu\nu}$. S is invariant under the conformal transformation $g_{\mu\nu}\to e^{2\sigma}g_{\mu\nu}$, $\phi\to e^{-\sigma}\phi$, where σ is a real function on M. We can easily construct the coboundary operator relevant for this problem by setting $\sigma = \varepsilon\xi$, where ε is a Grassmann number and ξ is a Grassmann field. The coboundary operator δ is then represented by $\Xi = \Xi_g+\Xi_\phi$, where

$$\Xi_g = \int_M d^4x\ 2\xi\ g_{\mu\nu}\ \frac{\delta}{\delta g_{\mu\nu}}$$

$$\Xi_\phi = -\int_M d^4x\ \xi\ \phi\ \frac{\delta}{\delta\phi}\ .$$

At the one-loop level, $\Xi\Gamma = \hbar\Delta + O(\hbar^2)$ and $\Xi\Delta = 0$. Since ϕ is not self-interacting, we can disregard ϕ-dependent terms and the most general function Δ with dimension 4 and "ghost" number 1 turns out to be

$$\Delta = \int_M d^4x \sqrt{g}\, \xi \,(a_1 R_{\mu\nu\lambda\rho}R^{\mu\nu\lambda\rho} + a_2 R_{\mu\nu}R^{\mu\nu} + a_3 R^2 + a_4 \Box R).$$

Now one has $\Xi\Delta = \int_M d^4x \sqrt{g}\,(4a_1+4a_2+12a_3)\, R\, \xi\Box\xi$. Accordingly the condition $\Xi\Delta = 0$ implies that $a_1+a_2+3a_3 = 0$, while a_4 is arbitrary. On the other hand, the first three terms of Δ cannot be obtained as Ξ-transform of anything. As to the last term, one has $\Xi \int_M d^4x \sqrt{g}\,\Box R = -2 \int_M d^4x \sqrt{g}\, \xi \Box R$, and it can be absorbed by a redefinition $\tilde{\Gamma}$ of Γ.

The trace of the energy momentum tensor is thus

$$\theta^{\mu}_{\mu} = \frac{\delta}{\delta\xi}\, \Xi_g\, \tilde{\Gamma} = a\, F + b\, G$$

where

$$F = R_{\mu\nu\lambda\rho}R^{\mu\nu\lambda\rho} - 2\, R_{\mu\nu}R^{\mu\nu} + \frac{1}{3} R^2$$

$$G = R_{\mu\nu\lambda\rho}R^{\mu\nu\lambda\rho} - 4\, R_{\mu\nu}R^{\mu\nu} + R^2$$

and the constants a and b have to be determined by complementary methods.

The analogous construction for the Adler-Bell-Jackiv anomalies has been given in ref. |1| while in ref. |2| the case of the trace anomaly for interacting scalar fields has been considered in some detail.

References

|1| L. Bonora and P. Cotta-Ramusino."Some Remarks on BRS Transformations, Anomalies and the Cohomology of the Lie Algebra of the Group of Gauge Transformations" Commun. Math. Phys. 87, 589-603 (1983)

|2| L. Bonora, P. Cotta-Ramusino and C. Reina " Conformal Anomaly and Cohomology" Phys. Lett. 126B, 305-308 (1983)

ON PURE, CONFORMAL AND EXOTIC SPINORS

P. Budinich

International School for Advanced Studies, Trieste,

and

International Centre for Theoretical Physics, Trieste

1. Introduction

Albert Einstein once said: "The most surprising fact about the Universe is that we understand it". These words express the highest target of theoretical physics: to "understand" natural phenomena; but not only "how" they appear and are correlated but also, and above all, "why" they appear so or, better, from which general principle the necessity of that particular appearance and correlation follows.

Recently extraordinarily successful progress has been made concerning the "how" of elementary physical phenomena: quantum colour dynamics; electroweak interactions; quarks, preons; their confinements; gluons; Higgs mesons, etc. Not so much progress, however, has been achieved concerning the "whys": why internal symmetry in both weak and strong interactions; why its breaking; why leptons steadily appear in pairs of charged and neutral fermions; why the high stability of the proton ; why baryon-lepton families, etc

A common feature of the many attempts to explain the "whys" is to assign to spinors a central role in elementary laws; but, if spinors are fundamental, then it would be highly surprising if one of their basic and wonderful geometrical properties, that of pureness[1], should also not play a fundamental role in the explanation of the mysterious, not yet understood, geometrical structure of those elementary physical laws and allow us to answer, some at least, of those "whys".

To suggest that perhaps this possibility may indeed exist is the aim of this paper.

2. $R_{4+n,2+n}$-Conformal spinors: origin of O(2+2n) internal symmetry

Spinors may be considered as vectors of the representation spaces of given Clifford algebras(or, alternatively, as their appropriately defined left and right ideals). "Pure"or simple spinors build up a subset of spinors equivalent to the maximal,totally null,polarized planes of the corresponding complex euclidean spaces.

The geometry of pure spinors presents remarkable properties of elegance and simplicity: as E. Cartan himself suggested[1], one might construct euclidean geometry starting from the one of"pure spinors" (or of totally null planes) but only of these. This suggestion might be of great relevance in view of the present difficulties of theoretical physics underlined in the Introduction.

That pure spinors might have relevance for physics might indeed be already seen in the elementary case of Majorana spinors which, being always pure, are equivalent to polarized totally null bidimensional (real) planes of (real) space-time apt to represent the electromagnetic field of light as propagating in a vacuum, and containing the geometrical explanation of most of the known properties of anastigmatic optics[2]. However to construct the whole space-time geometry, and consequently the Poincaré group, Majorana or Dirac spinors are not enough[3]; one needs spinors of higher Clifford algebras. Because of reality properties we shall choose real,conformal Clifford algebras.

We shall name $R_{4+n,2+n}$ or $R_{5+n,2n}$-conformal spinor ξ the vectors of the representation space of the corresponding real Clifford algebras. Here 4+n and 2+n refer to space-like and time-like terms of the corresponding quadratic forms.

Because of the important rule of $R_{p,q}$ Clifford algebras:

$$R_{p+1,q+1} = R_{p,q} \otimes R(2), \tag{2.1}$$

where R(2) is the algebra of 2x2 real matrices, we have:

Prop. 2.1: The $2^{(3+n)}$ component $R_{5+n,2+n}$-spinor ξ may be considered*

* We shall call this the pin-splitting, to be compared with the spin-splitting where ξ is considered as a doublet of semi-spinors: one of the I and one of the II kind.

as a doublet of $R_{4+n,1+n}$-spinors and so on up to a multiplet of $2^{(n+2)}$ Dirac spinors. ξ is real iff the $2^{(n+2)}$ Dirac spinors are real or Majorana spinors.

A $R_{4+n,2+n}$-conformal spinor field $\xi(Z)$ may be defined[1] in $R^{4+n,2+n}$ though the O(4+n,2+n) covariant equation:

$$G_A \frac{\partial}{\partial Z_A} \xi(Z) = 0 \qquad A = 0,1,2,3,5 \ldots 6+2n \tag{2.2}$$

where G_A are the generators of $R_{4+n,2+n}$. The spinor ξ is pure iff (2.2) is of rank 3+n. For the moment we shall not adopt this restriction, and ξ need not be considered pure.

Introducing the 6+2n new variables:

$$\eta_a = \frac{Z_a}{K} \; ; \; K = Z_{5+2n} + Z_{6+2n} \; ; \; \alpha^2 = \frac{Z_A Z^A}{K^2} \; ; \; a=0,1,\ldots 4+2n \tag{2.3}$$

(2.2) may be brought to the O(4+n,2+n) covariant form[4]:

$$\begin{aligned} & i\Gamma_a \frac{\partial}{\partial \eta_a} \psi_+(\eta,\alpha^2) + 2 \frac{\partial}{\partial \alpha^2} \psi_-(\eta,\alpha^2) = 0 \\ & i\Gamma_a \frac{\partial}{\partial \eta_a} \psi_-(\eta,\alpha^2) - 2\alpha^2 \frac{\partial}{\partial \alpha^2} \psi_+(\eta,\alpha^2) = 0 \end{aligned} \tag{2.4}$$

where $\psi_\pm$ are $R_{3+n,1+n}$-conformal spinor (homogeneous) fields, and Γ_a the generators of $R_{3+n,1+n}$.

It is known that $R^{3+n,1+n}$ may be densely mapped on the quadric $Z_A Z^A = 0$. In fact its coordinates η_a are obtained from (2.3) taking:

$$\eta_a = Z_a \; ; \; K = 1 \; ; \; \alpha^2 = 0 \tag{2.5}$$

It is then interesting to study if (2.2) or (2.4) imply a corresponding spinor field equation in $R^{3+n,1+n}$. Certainly we cannot simply take the limit $\alpha^2 \to 0$ $(K \neq 0)$ due to the derivatives, however we may adopt the semi-trivial solution:

$$\psi_+(\eta,\alpha^2) = 0 \; ; \; \frac{\partial \psi_+}{\partial \alpha^2}(\eta,\alpha^2) = 0 \; ;$$

by which, defining $\psi_-(\eta, 0) = \psi_-(\eta)$,(2.4) becomes:

$$\Gamma_a \frac{\partial\psi_-(\eta)}{\partial\eta_a} = 0 \qquad a = 0,1 \ldots 4 + 2n \tag{2.6}$$

that is the equation (2.2) in the lower conformal space $R^{3+n,1+n}$; and it is both manifestly O(3+n,1+n)and SO(4+n,1+n) covariant. However it is not surprising that the semi-trivial solution has reduced the faithfulness of the O(4+n,2+n) representations and in fact on $\psi_-(\eta = 0)$ O(4+n,2+n) may not be faithfully represented; furthermore (2.6) is only locally SO(4+n,2+n) covariant on $R^{3+n,1+n}$.

In fact SC(4+n,2+n) transformations, nonlinearly represented on $R^{3+1,1+n}$, present the known singularities which extend $R^{3+n,1+n}$ to the compactified form (on $Z_A Z^A = 0$)

$$\bar{R}^{3+n,1+n} \simeq \frac{S^{1+n} \otimes S^{3+n}}{Z_2} \tag{2.7}$$

obtained from (2.5) adding K = 0,and from $R^{3+n,1+n}$ adding the light cone at infinity.

This extension of $R^{3+n,1+n}$ to $\bar{R}^{3+n,1+n}$ allows global O(4+n,2+n) covariance of (2.6). However on $\bar{R}^{3+n,1+n}$, $R_{3+n,1+n}$ spinors may have two non-equivalent spinor structures[5], consequently we might need two non-equivalent equations like (2.6) to represent the full O(4+n,2+n) covariance. We represent all this by affirming that (2.6) alone is only <u>restricted</u> O(4+n,2+n) covariant[4]*.

An alternative approach to obtain from (2.2) and (2.4) equations in $R^{3+n,1+n}$ is to try to solve (2.4) with respect to the extra α^2 variables.

Iterating (2.4) we obtain:

$$\begin{aligned} \Box_\eta \psi_+(\eta,\alpha^2) &= 4\left(\alpha^2 \frac{\partial^2}{\partial(\alpha^2)^2} + \frac{\partial}{\partial\alpha^2}\right) \psi_+(\eta,\alpha^2) \\ \Box_\eta \psi_-(\eta,\alpha^2) &= 4\alpha^2 \frac{\partial^2}{\partial(\alpha^2)^2} \psi_-(\eta,\alpha^2) \end{aligned} \tag{2.8}$$

---

* For n = 0 all this refers to the conformal covariance of massless Dirac equation [4].

which suggests the separation of variables:

$$\psi_{\pm}(\eta,\alpha^2) = F_{\pm}(\alpha^2)\ \psi_{\pm}(\eta) \tag{2.9}$$

by which

$$\Box_{\eta}\ \psi_{\pm}(\eta)=M_{\pm}^2\ \psi_{\pm}(\eta)$$

$$B_{+}(\alpha^2)F_{+}(\alpha^2) \equiv (\alpha^2 \frac{\partial^2}{\partial(\alpha^2)^2} + \frac{\partial}{\partial\alpha^2})\ F_{+}(\alpha^2) = M_{+}^2F_{+}(\alpha^2) \tag{2.10}$$

$$B_{-}(\alpha^2)F_{-}(\alpha^2) \equiv \alpha^2 \frac{\partial^2}{\partial(\partial\alpha^2)}\ F_{-}(\alpha^2) = M_{-}^2 F_{-}(\alpha^2)$$

The last equations are the well-known ones of Bessel functions; for reasons which will become evident in the next paragraphs we choose, without loss of generality, the form of the exact solutions to be:

$$\begin{aligned} F_{+} &= C\ \sqrt{M^2}\ [I_o(\sqrt{\alpha^2 M^2}) + Y_o(\sqrt{\alpha^2\ M^2})] = C\sqrt{M^2}\ f_o(\sqrt{\alpha^2\ M^2}) \\ F_{-} &= \sqrt{M^2\alpha^2}\ [I_1(\sqrt{\alpha^2\ M^2}) + Y_1(\sqrt{\alpha^2\ M^2}] = f_1(\sqrt{\alpha^2\ M^2}) \end{aligned} \tag{2.11}$$

where $M^2 = M_{+}^2 = M_{-}^2$; C is an arbitrary integration constant, $f_0\, f_1$ known Bessel functions and ψ_{+}, ψ_{-} in (2.9) become both canonical $R^{3+n,1+n}$ spinors

It is easily seen that $F_{\pm}(\alpha^2)$ may be factorized and eliminated and we obtained in $R^{3+n,1+n}$, with coordinates η_a:

$$\begin{aligned} &i\Gamma_a \frac{\partial}{\partial\eta_a}\ \psi_{+}(\eta) + \sqrt{M^2}C^{-1}\psi_{-}(\eta) = 0 \\ &i\Gamma_a \frac{\partial}{\partial\eta_a}\ \psi_{-}(\eta) + \sqrt{M^2}\ C\ \psi_{+}(\eta) = 0 \end{aligned} \tag{2.12}$$

or

$$iG_a \frac{\partial}{\partial\eta_a}\ N(\eta) + (\sigma_{+}C^{-1} + \sigma_{-}C)\ \sqrt{M^2}\ N(\eta) = 0 \tag{2.13}$$

where

$$N(\eta) = \begin{vmatrix} \psi_{+}(\eta) \\ \psi_{-}(\eta) \end{vmatrix}\ ,\quad \sigma_{\pm} = \tfrac{1}{2}(\sigma_1 \pm i\sigma_2) \tag{2.14}$$

which, for

$$C = e^{-i\frac{\beta}{2}} = (C^{-1})^* , \qquad (2.15)$$

is hermitian and presents an O(2) internal symmetry. To make this evident we may express the constants M^2 and β in terms of bilinear polinomia of a given $R_{4+n,2+n}$ spinor θ :

$$s^{5+2n} = \bar{\theta}\, \Gamma^{5+2n}\, \theta ; \; s^{6+2n} = \bar{\theta} i \Gamma^{6+2n}\, \theta \qquad (2.16)$$

and define

$$\frac{s^{6+2n}}{s^{5+2n}} = \mathrm{tg}\beta \quad ; \quad M^2 = (s^{5+2n})^2 + (s^{6+2n})^2 \qquad (2.17)$$

and (2.13) becomes:

$$(iG_a \frac{\partial}{\partial \eta_a} + s^{5+2n}\, \Gamma_{5+2n} + i\, s^{6+2n}\, \Gamma_{6+2n})\, N(\eta) = 0 \qquad (2.18)$$

which is manifestly U(1) covariant with respect to:

$$N \to N' = e^{\frac{i\omega}{2}\Gamma^{6+2n}\Gamma^{5+2n}}\, N \qquad (2.19)$$

and the same for θ.

The transformation (2.19) corresponds to a complex intrinsic dilatation or a rotation in the plane (Γ^{6+2n}, Γ^{5+2n}) as compared to real dilatations which correspond to hyperbolic transformations in that plane: leaving invariant the form: $Z_6^2 - Z_5^2$.

Since the solutions (2.9) of (2.4) fix M^2 in (2.10) as a scalar, $\sqrt{M^2}$ appearing in (2.12), (2.13) and the following may be a matrix:

$$\sqrt{M^2} = \phi\, \Gamma_{7+2n} ; \; \sqrt{M^2} = W^a \Gamma_a \; ; \; \sqrt{M^2} = A^a \Gamma_a \Gamma_{7+2n} \; \cdots$$

where ϕ, W^a, A^a... may represent a $R_{3+n,1+n}$ pseudoscalar, vector, axial-vector etc. respectively. Correspondingly the s^5, s^6 terms in (2.18) may represent O(2)-symmetric scalar, p-scalar, vector, a-vector, etc. fixed external sources.

The method may be iterated and from (2.2) the manifestly O(3,1) a O(2+2n) covariant equation in $R^{3,1}$, to be possibly interpreted as space-time, may be obtained[6]:

$$(iG_\mu \frac{\partial}{\partial x_\mu} + s^5 G_5 + is^6 G_6 + \dots is^{6+2n} G_{6+2n}) N(x) = 0$$

where N is a 2^{n+2} multiplet of Weyl spinors.

The method may be equally well developed in the framework of Lagrangian formalism[6].

3. <u>Pure $R_{4+n,2+n}$-conformal spinors: breaking of internal symmetry</u>

It is conceivable that among conformal spinors,pure (and real) ones, as equivalent to the maximal totally null (polarized) planes of the corresponding (real) spaces, may play a special role as building blocks of space-time geometry and then of physics.

Since a pure $R_{5+n,2+n}$-conformal spinor ξ is a 2^{3+n}-component vector which is equivalent to a 3+n dimensional totally null plane, it is obvious that, for high enough n, the 2^{3+n} spinor components may not be independent and must obey constraint equations, the number of which will increase with n.

In fact[1] the $R_{5+n,2+n}$-spinor ξ is pure iff:

$$V_{i_1 i_2 \dots i_p} = \xi^T C_{(3+n)} G_{[i_1} G_{i_2} \dots G_{i_p]} \xi = 0 \qquad (3.1)$$

for 3+n-p = 3 (mod 4)

where

$$C_{3+n} = (-i)^{(3+n)} G_2 G_0 \dots G_{6+2n} .$$

As anticipated, the number of constraint equations (3.1) increases with n, in fact for n = 0, 1, 2, 3, 4 the number of such equations is 1, 10, 66, 364, 1821 respectively. These constraints may obviously reduce the internal symmetry found in the previous paragraph, insofar as internal symmetry may be expressed in the form of inner products where bilinear spinor polinomia appear; and, due to (3.1) some of these may have to be zero for pure spinors.

Let us consider some simple examples: for $R_{5,2}$ spinors ξ the only pureness equation is

$$\xi^T C_3 \xi = i\xi^T \Gamma_2 \Gamma_0 \Gamma_6 \xi = 0 \tag{3.2}$$

which, for ξ of the form

$$\xi = \begin{vmatrix} \psi_+ \\ \psi_- \end{vmatrix}$$

and real, by which ψ_+, ψ_- are Majorana spinors, reduce to the form:

$$\bar{\psi}_\pm \psi_\mp = \bar{\psi}_\pm \gamma_5 \psi_\mp \tag{3.3}$$

and is valid for both commuting and anti-commuting spinor components[6]. This means that the O(2) internal symmetry in (2.13), derived from the O(4,2) covariant (2.2), is broken for scalar, p-scalar interactions while it is allowed for vector-axialvector ones; for these (2.12) takes the form:

$$\begin{aligned} i\gamma^\mu \partial_\mu \psi_+(x) + (a+b\gamma_5)\gamma^\mu W_\mu^+ \psi_-(x) = 0 \\ i\gamma^\mu \partial_\mu \psi_-(x) + (a+b\gamma_5)\gamma^\mu W_\mu^- \psi_+(x) = 0 \end{aligned} \tag{3.4}$$

where $W_\mu^+ = (W_\mu^-)^*$, which for $a = b = 1$ may represent charged weak (external) interactions*. Had we started from O(5,2) covariant (2.2) we would have found solutions with neutral weak interactions presenting an internal SU(2) symmetry.

For higher n we may exploit the:

Proposition 3.1: A necessary condition for a $R_{5+n,2+n}$ spinor to be pure is that the two $R_{4+n,1+n}$ spinors it contains be pure.

The $R_{6,3}$-conformal spinor Λ is pure iff:

$$\bar{\Lambda}\Lambda = 0 = \bar{\Lambda} G_a \Lambda \qquad a = 0,1,2,3,\ldots 9 \tag{3.5}$$

---

* We are not able to see, in the frame of $R_{5,2}$-conformal spinors, the reason for the particular left-handed form of weak interactions.

two of these ten equations express the condition of pureness for ξ_+, ξ_- $R_{5,2}$-spinors where $\Lambda = \begin{vmatrix} \xi_+ \\ \xi_- \end{vmatrix}$ in the pin splitting.

The remaining 8 give, in terms of $\xi_\pm$, the equations, valid for both commuting and anti-commuting spinors:

$$\begin{aligned} &\bar{\xi}_\pm \Gamma_a \xi_\mp = 0 = \bar{\xi}_\pm \Gamma_a \Gamma_7 \xi_\mp \\ & && a = 0,1,2,3,5,6 \qquad (3.6) \\ &\bar{\xi}_\pm \xi_\mp = 0 = \bar{\xi}_\pm \Gamma_7 \xi_\mp \end{aligned}$$

where $\xi_\pm$ were supposed real.

As a consequence we obtain[6] that for $R_{6,3}$ spinors the O(5) internal symmetry in a space-time is broken for scalar, pseudoscalar, vector, axial-vector, interactions. The only allowed ones are the U(1) (or SU(2)) internal symmetry for vector-axial vector interactions inside each of the two $R_{5,2}$ conformal spinors $\xi_\pm$ contained in Λ. The procedure may be extended[6] and we obtain that for $R_{5+n,2+n}$ real spinors the O(2n+3) internal symmetry in space-time is broken for all interactions but the U(1) or SU(2) vector axial-vector ones inside each of the 2^n $R_{5,2}$-conformal real spinors contained in the $R_{5+n,2+n}$ one.

4. Extended external sources: Exotic or twisted spinors

Up to now we have only used exact solutions of our differential equations (the same solutions may be used in the lagrangian formalism[6]). The price we paid was that the interactions we obtained in space time were apt to represent "fixed external sources". To make these equations apt to represent physical phenomena we need two further steps:

I°) to make the "external" interactions space-time dependent instead of fixed: that is to substitute the integration "constants" $\sqrt{M^2}$ and β with external x-dependent fields and local phases respectively;

II°) to express these fields in terms of our only dynamical variables, that is, spinors.

The II° step will not be dealt with here; we will only give some results of the attempt to introduce "extended external sources".

In order to simplify our considerations let us consider $R_{4,2}$ or $R_{5,2}$-conformal spinors ξ ; in the pin splitting they are doublets of Dirac spinors; and let us substitute the ansatz (2.9) with:

$$\psi_+(x,\alpha^2) = e^{\frac{i\beta(x)}{2}} \sqrt{\phi^2(x)}\; f_0(\sqrt{\phi^2(x)\alpha^2})\; \chi_+(x,\alpha^2)$$
$$\psi_-(x,\alpha^2) = f_1(\sqrt{\phi^2(x)\alpha^2})\; \chi_-(x,\alpha^2) \tag{4.1}$$

where now x_μ represent space-time coordinates and $\phi^2(x)$, substituting M^2 of (2.11), represents a scalar field subject to dilatation, since we want $\phi^2\alpha^2 = \phi^2 Z^2/K^2$ to be invariant and then $\phi^2/K^2 = \phi_0^2(x)$ must be invariant. This also explains the origin of the phase factor $e^{i\beta/2}$ since, in fact, we have seen that O(2) internal covariance may be interpreted as complex dilatations and then the arbitrary, but dilatating, field ϕ^2 must be considered complex (in order for eq. (2.13) to be hermitian). Substituting (4.1) in (2.4) and taking into account the limiting properties of the Bessel functions we obtain[4] for $\alpha^2 \approx 0$:

$$i\gamma_\mu\left[\frac{\partial}{\partial x_\mu} + \frac{\partial \log\phi(x)}{\partial x_\mu} + \frac{i}{2}\frac{\partial\beta(x)}{\partial x_\mu}\right]\psi_+(x) + \sqrt{\phi^2}\; e^{-i\frac{\beta}{2}}\; \psi_-(x) + O(\alpha^2) = 0$$
$$i\gamma_\mu \frac{\partial}{\partial x_\mu}\psi_-(x) + \sqrt{\phi^2}\; e^{i\frac{\beta}{2}}\; \psi_+(x) + O(\alpha^2) = 0 \tag{4.2}$$

where $\psi_\pm(x) = \lim_{\alpha^2\to 0} \chi_+(x,\alpha^2)$ are canonical Dirac spinor fields.

It is seen that the equations for ψ_+ and ψ_- are different (the $O(\alpha^2)$ terms may be eliminated under non-restrictive conditions[4]); and this may be traced back to the different behaviour of χ_+, χ_- under dilatations (they had conformal degree 5/2 and 3/2 respectively) and corresponding asympotic behaviour of f_0, f_1 for $\alpha^2 \to 0$.

Eq. (4.2) (brought to the form (2.18)) is still covariant with respect to the transformation

$$N \to N'(x) = e^{i\frac{\omega}{2}(x)(1+\Gamma_6\Gamma_5)}\; N(x) \tag{4.3}$$

which however implies a local phase transformation of the field $\psi_+(x)$ only:

$$\psi_+(x) \to \psi'_+(x) = e^{\frac{i\omega}{2}(x)}\ \psi_+(x) \qquad (4.4)$$

and consequently we have in (4.2) not only a rotation in "isotopic spin space" $\beta \to \beta + \omega$ but also the local phase transformation (4.4) and consequent change of the gradient term in the first eq. (4.2).

In order to preserve the original covariance we must then have also (4.4) covariance which means that $\psi_+(x)$ must be charged and the local phase transformation of $\psi_+(x)$ must be compensated by the gauge transformation of a vector gauge field A_μ to appear in the first eq. (4.2). We will admit it as an "ad hoc" hypothesis*. In this way, as an example, starting from an O(5,2) covariant equation (2.2) in $R^{5,2}$ one obtains in $M^{3,1}$:

$$\{i\gamma_\mu\ [\frac{\partial}{\partial x_\mu} + \frac{ie}{2}(1+\sigma_3)\ A_\mu\ ] + i\gamma_5\ \vec{\tau}\ \vec{\pi}\}N(x) = 0 \qquad (4.5)$$

apt to properly represent pion-nucleon pseudoscalar interaction (γ_5 derives necessarily from Γ_7) where the charge matrix for N is obtained from (4.3) to be:

$$Q = \tfrac{1}{2}(1 + I_3) \qquad (4.6)$$

being $I_3 = \Gamma_6\Gamma_5$ the third component of isotopic spin.

The similarity of the first eq. (4.2) with the Dirac equation for exotic spinors[7] on non-simply connected manifolds may induce the search for a purely geometrical origin of the system (4.2).

In fact, introducing the variables x_μ, K, α^2 and solving eq.(2.4) with respect to the variable α^2 we have broken the original O(4,2)-covariance which reduced to Poincaré times O(2) internal covariance, and this latter one may be envisaged as transformations which leave invariant the form:

---

* One might hope to obtain the charge of ψ_+ by including our, for the moment purely geometrical structure,on a wider, possibly Riemaniann, frame where dynamics may be included. It would then be conceivable that the field A_μ, necessary to preserve both ψ_+ local phase covariance and momentum conservation, could be obtain from the geometrical structure without any further "ad hoc" hypothesis.

$$\eta_5^2 + \eta_6^2 = \frac{K^2}{2} [1 + (x^2 - \alpha^2)^2] \quad ; \tag{4.7}$$

when we finally will go to ordinary Minkowsky space (not compatified) we will set:

$$\alpha^2 = 0 \text{ and } K = 1$$

and

$$\eta_5^2 + \eta_6^2 = \tfrac{1}{2}(1 + x^4) \tag{4.8}$$

Despite the fact that we take $\alpha^2 = 0$; $K = 1$ our spinor

$$\xi = \begin{vmatrix} \psi_+(x,0,0) \\ \psi_-(x,0,0) \end{vmatrix}$$

is still a doublet of $R_{4,1}$ Dirac spinors: as such a $R_{5,2}$-conformal spinor and we may act on it with the intrinsic part of SO(5,2) Lie algebra. Generators of dilatations (imaginary ones)originate the O(2) internal symmetry group and we may visualize them as rotations on the circle (4.8). The space of definition of our $R_{4,1}$-Dirac spinors is then space-time $R^{3,1}$ times the circle (4.8) whose minimum radius is $\frac{1}{\sqrt{2}}$.

It will be shown separately that on this manifold two non-equivalent spinor structures exist and that eq. (4.2) may be derived from the fact that ψ_+ is exotic while ψ_- is a normal spinor on the given manifold.

We have seen that starting from a O(4,2) covariant eq. (2.4)in $R^{4,2}$ and solving it with respect to the extra-variable α^2 and then going to the limit $\alpha^2 \to 0$ we arrive in ordinary space-time $R^{3,1}$ with(4.2): two Dirac-like equations, one of which may be interpreted as being due to the exotic property of one of the two Dirac spinors. If instead we start from a locally conformally covariant massless Dirac equation in ordinary space-time $R^{3,1}$ and we act on it with the global conformal group, because of the singular conformal transformations,$R^{3,1}$ is brought to $\bar{R}^{3,1}$ where two non-equivalent spinor structures[5] and then two non-equivalent Dirac equations may exist. The difference between the two approaches is that while in the latter one the conformal covariance is not broken and the exotic spinor structure may be only found on the light cone[5] where the only conformal physics may take place, in the former the breaking of conformal symmetry may introduce masses and correspondingly the region of existence of exotic spinors expands to the whole space-time.

These considerations and the corresponding formalism of exotic spinor structures may be extended to $R_{5+n,2+n}$ with $n \neq 0$.

5. The spinor-structure of a possible elementary-particle model

We have seen that the study of $R_{4+n,2+n}$ and $R_{5+n,2+n}$-conformal spinors is suggested firstly by the relevance of space-time $R_{3,1}$ Majorana spinors for both elementary spinor physics and optics and secondly by conformal covariance and by reality properties of those Clifford algebras and spinors appropriate to represent physical quantities (and eventually open the way to the ambitious programme of deriving space-time geometry from a more elementary one). The periodicity theorem[8] on isomorphic real Clifford algebras gives us also an objective criterion to fix the maximal n on which to base a model for elementary-particles. In fact, due to that theorem:

$$R_{p+4,q+4} \simeq R_{p,q+8} \simeq R_{p+8,q} \simeq R_{p,q} \otimes R(16)$$

where R(16) is the algebra of 16x16 real matrices. Taking p = 3,4; q=1 we have that the maximal (beyond which the sequence is repeated) spinors to take are $R_{7,5}$ or $R_{8,5}$-(or $R_{11,1}$) conformal spinors.

Let S be a $R_{7,5}$-spinor; the pin-splitting may be represented by:

$$S = \begin{vmatrix} \theta_+ \\ \theta_- \end{vmatrix} = \begin{vmatrix} \Lambda_{++} \\ \Lambda_{+-} \\ \Lambda_{-+} \\ \Lambda_{--} \end{vmatrix} = \begin{vmatrix} \xi_{+++} \\ \xi_{++-} \\ \xi_{+-+} \\ \xi_{+--} \\ \xi_{-++} \\ \xi_{-+-} \\ \xi_{--+} \\ \xi_{---} \end{vmatrix} = \left.\begin{vmatrix} \psi_{++++} \\ \psi_{+++-} \\ \vdots \\ \psi_{---+} \\ \psi_{----} \end{vmatrix}\right\} 16 = \left.\begin{vmatrix} \psi_{++++L} \\ \psi_{++++R} \\ \vdots \\ \psi_{----L} \\ \psi_{----R} \end{vmatrix}\right\} 32 \tag{5.1}$$

where θ, Λ, ξ, ψ, $\psi_{\substack{L\\R}}$ are $R_{6,4}$; $R_{5,3}$; $R_{4,2}$; $R_{3,1}$-spinors and $R_{3,1}$ Weyl spinors respectively.

A necessary and sufficient condition for S to be pure and Majorana is that θ, Λ, ξ, ψ be pure and Majorana and all scalar, pseudoscalar, vector, axial, tensor $\theta_+\theta_-$ matrix elements to be zero for both commuting and anticommuting spinor components.

θ_+ obeys exotic type of equations in $R^{6,4}$ with respect to θ_-; with the same arguments as in §4 we shall consider it to be canonical, charged* and coupled to a D_A ($A = 0,1,2,3.....10$) gauge (Abelian) field. If we further suppose that the consequent charge interaction of θ_+ generates mass dynamically, we must assume that θ_+ spinors are more massive than θ_- spinors. It is then natural to suppose that:

θ_+ represent baryons

θ_- represent leptons

---

* Here again we admit that since an exotic spinor may be subject to a local phase transformation, induced by the transformations of the internal group, in order to preserve covariance we impose this spinor to be charged. This is, at this stage, an "ad hoc" hypothesis which has interesting consequences. This hypothesis should however become the consequence of the initial postulates in a more comprehensive geometrical-dynamical theory.

If we admit that all interactions are mediated through the elementary ones among <u>pure</u> spinors we have as a consequence that, for purely geometrical reasons, <u>baryons may not decay into leptons</u> and the gauge field D_A implies baryon conservation (stability of the proton).

Since θ_+ and θ_- each contain 8 Majorana spinors, if we identify them with quarks and leptons we have as a first qualitative prediction that 2 more leptons (in ξ_{-++}) and 4 more elementary baryons (in ξ_{+++} and ξ_{++-}) should be discovered.

Let us now consider the leptons: they may be split in two $R_{5,3}$- and four $R_{4,2}$-conformal spinors:

$$\theta_- = \begin{vmatrix} \Lambda_{-+} \\ \Lambda_{--} \end{vmatrix} = \begin{vmatrix} \xi_{-++} \\ \xi_{-+-} \\ \xi_{--+} \\ \xi_{---} \end{vmatrix} \qquad (5.2)$$

whose identification could be:

			CHARGE-GAUGE FIELD			EXOTIC		
			eA_μ	ϵB_a	$\mathcal{E}C_A$	$R_{4,1}$	$R_{5,2}$	$R_{6,3}$
$\theta_- =$	$\xi_{-++} =$	χ	-1	-1	-1	X	X	X
		ν_χ	0	-1	-1		X	X
	$\xi_{-+-} =$	τ	-1	0	-1	X		X
		ν_τ	0	0	-1			X
	$\xi_{--+} =$	μ	-1	-1	0	X	X	
		ν_μ	0	-1	0		X	
	$\xi_{---} =$	e	-1	0	0	X		
		ν_e	0	0	0			

Table 1.

The leptons, their charge and their exotic structure

Inside each of the conformal spinors the standard weak interaction may act, while no scalar, vector, axial vector between ξ_{--+}, ξ_{---} etc. are allowed (geometrical explanation of $\mu \to e + \gamma$ absence).

Furthermore Λ_{-+} obeys exotic type equations (with respect to Λ_{--}) and then is charged, (charge $\mathcal{E}$), coupled to a $C_A(A = 0,1\ldots8)$ gauge vector field in $R^{5,3}$ and for this reason the 4 leptons τ, ν_τ, χ, ν_χ should be of higher mass than e, ν_e, μ, ν_μ. But also ξ_{-++} and ξ_{--+} are exotic with respect to ξ_{-+-} and ξ_{---} in $R^{4,2}$ space, and for this reason charged (charge ε) coupled to a field B_a $(a = 0,1,2,3,5,6)$ gauge vector field in $R^{4,2}$ and of higher mass as the corresponding non exotic ones. Finally each space-time spinor field ψ_+ is exotic with respect to ψ_- and then coupled to the familiar four vector potentials A_μ and electrically charged, (charge e).

It is seen that the picture is not too different from what we observe. In particular, both the increase of masses from ν_e to τ (and χ), the electric charge property, and the universality of weak interactions seem to fit well the geometrical frame implied in the model.

In Table I we have indicated the charges (in the above-mentioned hypothesis) and the Clifford algebras of the exotic spinors.

A further possible prediction of the model would be the existence of a B_a vector interaction of the (ν_μ, μ) $R_{4,2}$-conformal doublet. It would give visible effects in $\nu_\mu - \nu_\mu$ and $\mu - \mu$ scattering (very visible but difficult to perform), B_a bremsstrahlung and creation of ν_μ and $\mu^\pm$ pairs and radiative corrections in ν_μ, μ physics.

The main predicted difference between the θ_- octet of leptons and the θ_+ octet of baryons lies in the θ_+ interaction with D_A gauge field which, being coupled to all baryons (rather strongly, presumably, as the high baryon masses, as compared to the electron one, seems to suggest) should give rise to composite systems among the components of θ_+; in particular, colour and QCD could be introduced at this stage and also electroweak interactions would be obtained reversing the standard procedure: here we introduce, as a result of the exotic $R_{3,1}$ spinor in the $R_{4,2}$ conformal one, the standard A_μ e.m. potential, the neutral weak currents may be obtained both directly from the odd-dimensional Clifford algebra $R_{5,2}$ (the Γ_7 terms) and from the B_a gauge field for the exotic ξ_{--+} and ξ_{+-+} $R_{4,2}$ spinors.

For the baryons contained in θ_+, due to the possibility of composite states which may not decay in leptons if S is pure, the charge structure may be more complex. In fact as we have seen a charged-neutral doublet (proton-neutron) may be inserted in a conformal doublet obeying eq. (4.5) and presenting standard-universal weak interactions, however the existence of proton-neutron + pion transition indicate that proton-neutron may not constitute a pure $R_{5,2}$ conformal spinor*. It can be shown that rotations in the isotopic spin plane may generate fractional charges for the elements of the doublet[6] (and possibly anomalous dimensions). We may then build up a table for baryons where we shall only indicate the ϵ and $\mathcal{E}$ charges:

			CHARGE-GAUGE FIELD			EXOTIC		
			eA_μ	ϵB_a	$\mathcal{E} C_A$	$R_{4,1}$	$R_{5,2}$	$R_{6,3}$
$\theta_+ =$	$\xi_{+++} =$	x	Fractional for pure spinors	+1	+1	X	X	X
		y		+1	+1		X	X
	$\xi_{++-} =$	t		0	+1	X		X
		b		0	+1			X
	$\xi_{+-+} =$	c		+1		X	X	
		s		+1			X	
	$\xi_{+--} =$	u		0		X		
		d		0				

Table 2.

The baryons, charges and exotic structure

For the pure (Majorana) spinor only the vector-axial vector matrix element among the Majorana doublet in each $R_{5,2}$ conformal spinor ξ is allowed: they should be responsible for weak interactions.

In the same way as rotations inside each ξ could be responsible for fractional charges, U(1) rotations in the plane (Z_7, Z_8) (rotation of the weak sources inside ξ_{+-+} and ξ_{+--}) could be responsible for the

* It is amusing to observe that it is possible to reverse the role of $\psi_+ \psi_-$ in $R_{5,2}$-conformal $\xi_{\mp++}$ as exotic spinors. In this case $\xi_{+--} = \left|{P \atop N}\right|$ and $\xi_{-++} = \left|{\nu_e \atop e}\right|$; together they build up an $R_{6,3}$ spinor. If we combine the U(1) deriving from $R_{6,3} \to R_{5,2}$ with the exotic properties inside each ξ we obtain the charge matrix $Q = \frac{1}{2}(\sigma_3 \otimes 1 + 1 \otimes \sigma_3)$ which gives the charges (+e, 0, 0, -e) for proton, neutron, neutrino and electron respectively.

Calibbo angle.

Comparing Table 1 and 2 we see immediately that we can group families of baryons and leptons bearing the same weak (apart from Calibbo angles) and exotic properties. They are:

	FAMILY	WEAK	EXOTIC
I	e, ν_e ; u , d	same	$R_{4,1}$
II	μ, ν_μ ; c , s	"	$R_{5,2}$
III	τ, ν_τ ; t , b	"	$R_{6,3}$
IV	χ, ν_χ , x , y	"	$R_{5,2}$ and $R_{6,3}$

Table 3.
Lepton-baryon families

The III family seems to be partially discovered, the IV is only predicted by the model.

Before trying to develop a dynamical model, further geometrical prescriptions could come from the equivalence of the various pure conformal spinors with totally null, polarized planes, and from the triality properties of $\Lambda_{++}, \Lambda_{+-}, \Lambda_{-+}, \Lambda_{--}$ $-R_{5,3}$-spinors.

REFERENCES

1. E. Cartan, Leçon sur la théorie des spineurs, Hermann, Paris 1938.
2. I. Robinson and A. Trautman, J. Math. Phys. (1983) in print.
3. P. Budinich and P. Furlan, Proceedings of the Conference on Differential Geometric Methods in Theoretical Physics (1981); R. Ablamowicz, J. Mozrzymas, Z. Oziewicz, and J. Rzewoski, Reports on Mathematical Physics, 14, 89 (1978).
4. Y. Muray, Nuovo Cimento 23, 122 (1962); A.O. Barut, Lettere al Nuovo Cimento 7, 625 (1973); P. Budinich and P. Furlan, ISAS Preprint 40/82/E.P. (Trieste).
5. L. Dabrowski, ISAS preprint 29/83/E.P., Trieste.
6. P. Budinich, ISAS preprint 40/83/E.P., Trieste.
7. H. R. Petry, J. Math. Phys. 20(2), 231 (1979).
8. M.F. Atiyah, R. Bott and A. Shapiro, Clifford Modules, Topology, Vol. 3, Sup. 1 (1964); R. Coquereaux, Preprint TH.3279-CERN.

POHLMEYER-TYPE TRANSFORMATIONS IN GENERAL RELATIVITY

F.J. Chinea
Departamento de Métodos Matemáticos de la Física
Facultad de Ciencias Físicas, Universidad de Madrid
Madrid-3, Spain

Bäcklund transformations have been introduced in recent years in General Relativity, in order to provide a method for generating new solutions of the Einstein field equations when symmetries are present.[1,2] Such transformations may be interpreted as gauge transformations, which preserve a zero-curvature condition on an appropriate fibre bundle with an SU(2) or SU(1,1) structure group.[3]

The Einstein equations in vacuum reduce to (essentially) a single partial differential equation for a complex variable, when gravitational fields admitting two commuting isometries are considered; and it is for that equation (the Ernst equation[4]) that Bäcklund transformations have been developed.

On the other hand, the vacuum Einstein equations may also be written in such a case as (essentially) a single partial differential equation for a real, vector-valued field[5]; the components of the vector field are *algebraic* functions of the components of the space-time metric, and so it seems that Bäcklund transformations for the vector variable could be very useful for the purpose of solution generation. Transformations of such a type have been found recently[6]; they resemble the vector Bäcklund transformations introduced in Ref. 7 in connection with the nonlinear σ-model, and in fact reduce to the latter in a special case. They possess an associated permutability property, which permits the construction of new solutions by means of algebraic operations; the solutions of the field equations thus generated are asymptotically Minkowskian.

The geometric meaning of the transformations just referred to will be discussed here. First notice that the integrability condition for the system

$$A_u = 1/2(1+f^{-1})qq_uA \qquad\qquad A_v = 1/2(1+f)qq_vA \tag{1}$$

is just the spinor version of the field equation of Ref. 5 in the case of two spacelike Killing fields (the one spacelike-one timelike case can be treated along the same lines):

$$q_{uv} + 1/2(u+v)^{-1}q_u +1/2(u+v)^{-1}q_v = 1/2(q_uq_v + q_vq_u)q \tag{2}$$

where subscripts denote partial derivatives; q is an $sl(2,R)$ matrix function of u and v satisfying the constraint $q^2 = -1$; A is a function of the coordinates, with values in the SL(2,R) matrix group; and $f = (c+u)^{1/2}(c-v)^{-1/2}$, where c is a fixed arbitrary real constant. The $1/2(u+v)^{-1}$ factor in front of q_u in (2) is a special determination of the expression $1/2\tau^{-1}\tau_u$, where τ is a solution of the d'Alembert equation $\tau_{uv} = 0$ (similarly for $1/2\tau^{-1}\tau_v$). Barring degenerate cases ($\tau_u = 0$ or $\tau_v = 0$ or both), which are relevant for the discussion of the nonlinear σ-model as a subcase, one may take $\tau = u + v$ due to the conformal invariance of the equations under $u \to U(u)$, $v \to V(v)$, so that no generality is lost by this seemingly special choice.

Equations (1) may be written as the exterior equation

$$dA = \Gamma A \tag{3}$$

with an appropriate sl(2,R)-valued 1-form Γ. By exterior differentiation of (3), one obtains Eq. (2) in the form

$$d\Gamma = \Gamma\wedge\Gamma \tag{4}$$

Γ may thus be considered as a connection with vanishing curvature on a fibre bundle. This suggests that gauge transformations leaving invariant Eq. (4),

$$\Gamma \to \Gamma' = S\Gamma S^{-1} + dS\,S^{-1} \tag{5}$$

(where S is an SL(2,R)-valued function) may play an important role in the process of solution generation. This is in fact the case. Let us take

$$S = \alpha I + \beta(pq - qp) \tag{6}$$

where p is a coordinate-dependent matrix in sl(2,R), with $p^2 = -1$, and

$$\alpha = (c-\lambda)^{-1/2}(u+v)^{-1}\{(\lambda+u)(c-v)^{1/2} + (\lambda-v)(c+u)^{1/2}\}$$

$$\beta = 1/4(c-\lambda)^{-1/2}\{(c-v)^{1/2} + (c+u)^{1/2}\}$$

where λ is an arbitrary real constant. In such a case, Eq. (5) reduces to

$$\begin{aligned} pp_u - qq_u + (u+v)^{-1}(pq - I) + p_u q + pq_u &= 0 \\ pp_v - qq_v - (u+v)^{-1}(pq + I) - p_v q - pq_v &= 0 \end{aligned} \tag{7}$$

with

$$pq + qp = 2(u-v+2\lambda)(u+v)^{-1} \tag{8}$$

Equations (7) and (8) are just a spinor version of the Bäcklund transformation found in Ref. 6. Notice that condition (8) assures that $\det S = 1$ as required. A gauge transformation of the special type (6) is then equivalent to a Bäcklund transformation relating q to a new solution p of Eq. (2), corresponding to Γ'.

Finally, if we use the notation $S(\lambda;q\to p)$ for the matrix S defined in (6), it may be shown that the permutability theorem or superposition principle associated with the Bäcklund transformation (7)-(8)[6] is equivalent to

$$S(\mu;p\to w)S(\lambda;q\to p) = S(\lambda;s\to w)S(\mu;q\to s) \tag{9}$$

for appropriate values of the integration constants. From (9), one can solve algebraically for w in terms of q and two of its Bäcklund transforms, p and s. This provides a method for calculating new solutions using algebraic computations only.

References

1. B.K. Harrison, Phys. Rev. Lett. 41, 1197, 1835 (E) (1978)
2. G. Neugebauer, J. Phys. A12, L67 (1979)
3. F.J. Chinea, Phys. Rev. D 24, 1053 (1981), and 26, 2175 (E) (1982); Physica (Utrecht) 114A, 151 (1982)
4. F.J. Ernst, Phys. Rev. 167, 1175 (1968)
5. D. Maison, Phys. Rev. Lett. 41, 521 (1978); J. Math. Phys. (N.Y.) 20, 871 (1979)
6. F.J. Chinea, Phys. Rev. Lett. 50, 221 (1983)
7. K. Pohlmeyer, Commun. Math. Phys. 46, 207 (1976)

ON GROUP COVARIANCE AND THE LAW OF MOTION IN A GENERALIZED METRIC THEORY

Leopold Halpern
Dept. of Physics
Florida State University
Tallahassee, Fla 32306

and

Institutionen för teoretisk fysik
Stockholms universitet
113 46 Stockholm

Riemann's complementarity of geometry and the laws of physics, in the light of the geometrical unification of space-time with the manifold of the invariance group of a Kaluza-Klein gauge theory[1], suggests even a complementarity of the invariance group and the physical laws. The gravitational field in the general theory of relativity is already geometrized but the right hand member of Einstein's field equations, which is an alien to the theory remains adapted to the Poincaré group. Comparison of the economy of the description in terms of other invariance groups, as the De Sitter groups SO (3,2) and SO (4,1) should be instructive; in most cases where such groups were considered however, return to the "reality" of the Poincaré group for all observable situations is imposed. The different mathematical structure of the simple De Sitter groups is then hardly incorporated in the physical laws.

Group covariance is assumed to hold rigorously in idealized limiting situations which are already disturbed when an experiment is made. The culprits for the disturbance, the forces acting on the test system, should however themselves comply in general with this covariance.

An example for this approach is the principle of inertia which says (in a modernized form):

"A body moves along a timelike orbit of the translation group in Minkowski space if no forces act on it".

We compare this with the different version:

"A body moves along a timelike orbit of the De Sitter group in the De Sitter universe if no forces act on it".

The latter version includes all possible geodesic motion and in addition a six parameter family of orbits which fulfill the following equations:

$$\ddot{x}^k + \left\{ {k \atop hf} \right\} \dot{x}^h \dot{x}^f + \frac{1}{2} R^k{}_{hij} \dot{x}^h S^{ij} = 0 \tag{1a}$$

$$\frac{D}{ds} S^{ij} = S^{ij}{}_{;k} \dot{x}^k = 0 \qquad S^{ij} = -S^{ji} \tag{1b}$$

The first version is generalized globally to the geodesic motion in the metric spaces of general relativity. A metric theory with the corresponding generalization of the second version has been suggested at the 1981 Canterbury colloquium.[2 a,b,c] It has the form of a 10-dimensional Kaluza-Klein theory with non-Abelian gauge group H = SO (3,1) which has the peculiar property that the Cartan-Killing metric of the De Sitter group G forms a special solution of the field equations with cosmological member. The natural projection π of this metric on the base manifold: B = G/H is then that of the De Sitter universe in space-time. More general solutions of the field equations in ten dimensions have a right hand member T^{rs}. The solutions must be restricted so that the ten dimensional manifold P still forms a principal fibre bundle with the same typical fibre H and space-time as the base manifold and the projection $\pi' \gamma = g$ with γ the metric on P and g that of space-time. Horizontal vectors are perpendicular to the vertical vectors w. r. t. γ, thus defining a connection form ω.

For any local trivialization we shall denote indices pertaining to the base manifold by letters from a to k, indices pertaining to the fibre by l to q and indices that run over all ten dimensions by letters r to z. This applies in each case also to the summation convention. The principal fibre bundle P can be extended to the frame bundle. A base of horizontal vector fields is chosen which commutes with the left invariant vector fields A_M on each fibre as follows:

$$[A_M, A_E] = C^F_{ME} A_F \tag{2}$$

where C^F_{ME} are structure constants of the De Sitter group. In the special case of the De Sitter manifold the A_E form a left invariant base of the horizontal vector space. They define in the general case a soldering form θ with the torsion form Θ. To Θ and the curvature form Ω there correspond differential two forms T and F on the base. F gives rise to six Yang-Mills fields F^M.

The Lagrangian of the theory in ten dimensions:

$$\mathcal{L}^{(10)} = \mathcal{L}^{(10)}_E + \Lambda + \mathcal{L}^{(10)}_M \tag{3}$$

is the Einstein Lagrangian with a cosmological member and a term giving rise to the r. h. s. T^{rs}. It is equivalent to the Lagrangian in four dimensions:

$$\mathcal{L}^{(4)} = \mathcal{L}_E + \Lambda' + \frac{1}{4}\sqrt{g}\, F^M_{ik} F^{ik}_M + \mathcal{L}_M \qquad (3\,a)$$

with Einstein, cosmological and Yang-Mills term and $\mathcal{L}_M$ depending on matter fields, the metric and the Yang-Mills potentials A^M.

The physical content of the theory becomes more transparent if we consider a restricted form with vanishing torsion: The connection is then Riemannian and the Yang-Mills term equals $\frac{1}{4}\sqrt{g}\, R_{ijhk} R^{ijhk}$, quadratic in the Riemann tensor. Remarkably the general dependance of $\mathcal{L}_M$ on the metric is now only expressible in terms of tetrades - an indication for a spin structure.

The equations of motion of a test particle for a general four dimensional metric are now exactly eq. (1 a, b) (torsion vanishes on the group manifold). These equations fulfill Papapetrou's conditions for the relativistic motion of a spinning test particle. Eq. (1 b) excludes also anomalous motions. The relation of S^{ik} to the spin term in Papapetrou's theory is less transparent for spaces of complicated curvature which results in a mass dipole moment. This is related to the problem of the center of mass in the "rest frame".

There are strictly spoken no point partcles in tendimensions because the conditions

$$[A_M, \gamma] = 0 \qquad [A_M, T] = 0 \qquad (2\,a)$$

have to be fulfilled for metric and matter.

The situation is better described by the ten - dimensional matter conservation law which has the four dimensional form:

$$(\sqrt{g}\, T^{ik})_{;\,k} = F^{i}{}_{k}{}^{M}\, j^{k}_{M} \qquad (4)$$

with the conserved current density j_M related to the vertical part of the momentum tensor. For vanishing torsion the tetrade dependent j_M should be included in T^{ik} as a kind of spin term. The general theory with torsion has an additional (long range) spin-spin interaction apart from the general relativistic one.

The topology of SO (3, 1) which is not simply connected gives rise to a quantization of the Yang-Mills charge.

The theory which is based completely on group covariance seems to yield a rather realistic description of the phenomenon of inner angular momentum.

1. Kaluza TH. Sitzb. Pruss. Acad Berlin p 966 (1921)
   Klein O. Z. Physik 37 p 895 (1926)

2. Halpern L. a. Physica 114 A p. 146 (1982)
b. Found. of Physics Vol. 13 p 237 (1983)
c. FSU preprint (1983), to appear in B. De Witt birthday Volume

3. Papapetrou A. Proc. Roy. Soc. Vol. 209 p 248 (1951)

MINIMALIZATION OF HIGGS POTENTIALS WITH APPLICATION TO THE SU(5) MODEL*

T. Hübsch**, S. Meljanac*** and S. Pallua**
** Zavod za teorijsku fiziku, Prirodoslovno-matematički fakultet, University of Zagreb, Marulićev trg 19, Yugoslavia
*** "Rudjer Bošković" Institute, Zagreb, Bijenička 54, Yugoslavia

Abstract

We study the symmetry-breaking pattern for grand unified theories. We suggest a method suitable for higher-rank tensors. Among applications we develop an alternative minimal SU(5) model where the usual role of the 24-dimensional Higgs field is taken by the 75-dimensional Higgs tensor.

In this paper we study the group-theoretical properties of minima of potentials invariant under a given group. This problem is relevant to grand unified theories of strong, weak and electromagnetic interactions (GUT's). It has been solved for many cases with lower-rank tensors[1]. However, little is known of higher-rank tensors due to a large number of variables.

Suppose a symmetry group is given as well as the irreducible representation D(G) under which the field transforms. We assume that the field is a Lorentz scalar. The most general renormalizable potential is a fourth-order invariant polynomial in the fields:

$$V = V(\phi^a), \quad a = 1 \dots m, \tag{1}$$

where m denotes the dimension of the representation.

Suppose we are dealing with a problem in which we suspect, for physical reasons, that there is a minimum invariant under a particular subgroup H. Then we can simplify the problem significantly. We can decompose a given tensor into irreducible tensors under H. Symbolically,

$$T = \sum_j S_j + \sum_i X_i \ , \tag{2}$$

where S_j are singlets under H and X_i are the other irreducible tensors under H. The decomposition (3) can be used in Eq. (1). For group-theoretical reasons, only quadratic terms present in the potential are of the type S^2 and X^2, only cubic terms present are of the type S^3, SX^2, X^3 and only quartic terms are of the type S^4, S^2X^2, SX^3, X^4. In these variables the H-invariant minimum is parametrized by

* This project was assisted by the U.S. National Science Foundation Grant No. YOR 82/051.

$$X_i = 0. \tag{3}$$

The conditions for the H-invariant point to be the minimum of the potential read

$$\left.\frac{\partial V}{\partial S}\right|_{X=0} = 0, \quad \left.\frac{\partial^2 V}{\partial S^2}\right|_{X=0} > 0, \quad \left.\frac{\partial^2 V}{\partial X^2}\right|_{X=0} > 0 . \tag{4}$$

Once we have performed the change of variables, we have reduced the original problem in many variables to a problem in a small number of variables (number of singlets). In many cases this number is just one.

As an application of this method, we shall try to replace the 24-dimensional representation (adjoint) in the standard SU(5) model[2] by the 75-dimensional tensor and search for an SU(3)xSU(2)xU(1) invariant minimum[3].

The 75-dimensional tensor may be described by a twice covariant and twice contravariant tensor satisfying the following conditions:

$$\Phi^{AB}_{CD} = -\Phi^{AB}_{DC} = -\Phi^{BA}_{CD} ; \quad \Phi^{AB}_{AC} = 0 . \tag{5}$$

The corresponding most general fourth-order potential reads

$$\begin{aligned} V(\Phi) = & -\frac{\mu^2}{2}\Phi^{AB}_{CD}\Phi^{CD}_{AB} + \frac{a^2}{4}\Phi^{AB}_{CD}\Phi^{CD}_{AB}\Phi^{EF}_{GH}\Phi^{GH}_{EF} + \frac{b_1}{2}\Phi^{AB}_{GH}\Phi^{CD}_{AB}\Phi^{EF}_{CD}\Phi^{GH}_{EF} \\ & + \frac{b_2}{2}\Phi^{AB}_{GD}\Phi^{CD}_{AF}\Phi^{EF}_{CH}\Phi^{GH}_{EB} + \frac{b_3}{2}\Phi^{AB}_{GF}\Phi^{CD}_{AB}\Phi^{EF}_{CH}\Phi^{GH}_{ED} + \frac{b_4}{2}\Phi^{AB}_{CF}\Phi^{CD}_{AH}\Phi^{EF}_{GD}\Phi^{GH}_{EB} \\ & + \frac{b_5}{2}\Phi^{AB}_{CH}\Phi^{CD}_{AB}\Phi^{EF}_{GD}\Phi^{GH}_{EF} . \end{aligned} \tag{6}$$

We have disposed of cubic terms by invoking the discrete symmetry of the potential $\Phi^{AB}_{CD} \to -\Phi^{AB}_{CD}$. The decomposition of the 75-dimensional tensor under the SU(3)xSU(2)xU(1) subgroup is

$$\begin{aligned} 75 = & (1,1,0)+(8,1,0)+(8,3,0)+[(3,2,-5/6)+(6,2,-5/6)+(3,1,-5/3) \\ & + \text{compl.conj.}] . \end{aligned} \tag{7}$$

We denote an explicit form of the singlet subtensor:

$$\left.\Phi^{AB}_{CD}\right|_{(1,1,0)} = \frac{s}{12}\{(\delta^{\alpha\beta}_{\gamma\delta}-\delta^{\alpha\beta}_{\delta\gamma})+3(\delta^{ab}_{cd}-\delta^{ab}_{dc})-(\delta^{\alpha b}_{\gamma d}-\delta^{\alpha b}_{\delta c}+\delta^{\beta\alpha}_{\delta c}-\delta^{\beta\alpha}_{\gamma\delta})\}. \tag{8}$$

The result for the value of the field in the required minimum together with the stability conditions is

$$s^2(108a^2+60b_1+37b_2+14b_3+15b_4+56b_5)/216 = \mu^2 , \tag{9}$$

$$\begin{aligned} &(1,1,0): && 108a^2+60b_1+37b_2+14b_3+15b_4+56b_5 \geq 0 , \\ &(8,1,0): && -360b_1-201b_2+69b_3-135b_4+48b_5 \geq 0 , \\ &(8,3,0): && -240b_1-104b_2+104b_3-240b_4-832b_5 \geq 0 , \end{aligned}$$

$(\bar{3},1,-5/3)$: $\quad 60b_1-28b_2-32b_3-27b_4-8b_5 \geq 0$ and

$$(\bar{6},2,-5/6):\quad -48b_1-34b_2-8b_3-15b_4-20b_5 \geq 0\,. \tag{10}$$

These inequalities have a straightforward physical meaning. Multiplying them by the factor $S^2/216$, one obtains the squares of the scalar particle masses.

Another possible application is e.g. to try to obtain the SU(n)x xSU(n)xU(1) invariant minimum for the SU(2n) invariant potential by using the traceless symmetric fourth-rank tensor T^{AB}_{CD} [4]. This tensor is required in the SU(16) model, for example[5].

Finally, let us recall that according to the theorems and conjectures of Michel and Radicatti[6], in order to find subgroups under which minima are invariant, one has to enumerate the maximal little groups. Such an approach can be combined with the approach of this paper to obtain a general and explicit solution .

We would like to thank Professors J.C. Pati, R. Barbieri, G. Senjanović and P. Senjanović for useful conversations. We also wish to thank Professors A. Salam and G. Furlan and the particle physics group at ICTP , Trieste and Istituto di Fisica Teorica di Trieste for their kind hospitality during several short visits when part of this work was done.

References:

1. L.F. Li, Phys. Rev. D9, 1723 (1974);
   F. Bucella, H. Ruegg and C. Savoy, Nucl. Phys. B169, 68 (1980);
   H. Ruegg, Phys. Rev. D22, 2040 (1980).
2. H. Georgi and S.L. Glashow, Phys. Rev. Lett. 32, 438 (1974);
   A.J. Buras, J. Ellis, M.K. Gaillard and D.V. Nanopoulos, Nucl. Phys. B135, 66 (1978).
3. T. Hübsch and S. Pallua, An alternative minimal SU(5) model, submitted for publication.
4. S. Meljanac, M. Milošević and S. Pallua, Phys. Rev. D26, 2936 (1982).
5. J.C. Pati, A. Salam and J. Strathdee, Nucl. Phys. B185, 445 (1981).
6. L. Michel, Report No. CERN-TH-2716 (1979), contribution to the A. Visconti Seminar (unpublished);
   L. Michel, Rev. Mod. Phys. 52, 617 (1980);
   R. Slansky, Phys. Rev. 79, 1 (1981).

SELF-DUAL MONOPOLES AND CALORONS

W. Nahm

Physikalisches Institut
der Universität Bonn
Nussallee 12, D-5300 Bonn
W. Germany

1. The ADHM construction for instantons, self-dual monopoles and calorons

The study of nonlinear partial differential equations remained outside the mainstream of mathematics, because their solution spaces seemed to be rather arbitrary and complicated. But physicists discovered that some of those equations occur naturally, and a closer study by both physicists and mathematicians reveals more and more beautiful structures.

Among them, the Yang-Mills equations in four dimensions are today the most outstanding ones. Results on general solutions are still scanty, but quite a lot is known about the more specialized solutions of the self-duality equation for Yang-Mills fields on euclidean four-manifolds. Indeed, this equation already became a valuable tool in the study of differentiable four-manifolds.

A basic step in the investigation of this equation was the ADHM construction[1)] of all instantons, i.e. of all gauge potentials in R^4 with self-dual and square integrable field strengths. The construction uses the cohomology of certain sheaves over the twistór space, which does not yet belong to the tool kit of many physicists. Thus we shall give an elementary modification of it, which also has the advantage of being easily generalizable to self-dual monopoles and calorons. We only consider the gauge group SU(n), but it is easy to specialize to the other classical Lie groups.

For the space coordinates and the covariant derivatives we use the standard quaternionic notation

$$x = q_\mu x^\mu, \quad (1)$$

$$D = q^\mu \mathcal{D}_\mu, \quad (2)$$

and we represent the quaternions by (2,2) matrices. The self-duality equation for the field strength may be written in the form

$$D^+ D = \mathcal{D}^2 \cdot 1_2 \quad (3)$$

i.e. D^+D commutes with the quaternions.

Let ψ be a matrix whose columns form an orthonormal basis of the solutions of the Weyl equations in the background of the gauge field, such that

$$D^{+}\psi = 0 \tag{4}$$

and

$$\int \psi^{+}\psi \, d^4x = 1_k \tag{5}$$

where k is the number of linearly independent square integrable solutions of eq.(4). Because of eq. (3) the adjoint equation has no solution, such that k is the index of D^{+}.

The projector onto the solution space of eq. (4) may be written in the form

$$\psi\psi^{+} = 1 - DGD^{+} \tag{6}$$

with

$$G = (\mathcal{D}^2)^{-1} \tag{7}$$

Because of the conformal invariance of the Weyl equation it is easy to determine the asymptotic behaviour of ψ . We write it in the form

$$\psi = \pi^{-1} x \varepsilon \sigma \alpha / r^4 + O(r^{-4}) \tag{8}$$

where α is a constant (n,2,k) matrix and

$$\varepsilon = \begin{pmatrix} 0 & 1 \\ -1 & 0 \end{pmatrix} \tag{9}$$

transforms the $q\mu$ to their complex conjugates. Moreover σ is a matrix of orthonormal covariant constants at the sphere at infinity. We may write

$$\mathcal{D}^2\sigma = 0 \tag{10}$$

and

$$\sigma(x_\infty) = 1_n \tag{11}$$

for a point x_∞ at infinite distance.

We shall need the asymptotic behaviour of G, which is given by

$$G\psi = -\frac{1}{4}x^2\psi + O(r^{-2}). \tag{12}$$

Now the hermitian matrices

$$a_\mu = -\int \psi^+ x_\mu \psi \, d^4x \tag{13}$$

can be interpreted as translationally invariant U(k) potentials which satisfy the self-duality equation up to a known source term. Indeed, eq. (6) yields

$$[a_\mu a_\nu] = \int \psi^+ (q_\mu q_\nu^+ - q_\nu q_\mu^+) G \psi \, d^4x - \frac{1}{8}(\alpha\varepsilon)^+ (q_\mu^+ q_\nu - q_\nu^+ q_\mu)(\varepsilon\alpha). \tag{14}$$

The first term on the r.h.s. is self-dual.

Besides eq. (4) and the fact that the quaternions commute with D^+D, the main ingredient in the calculation is the simple commutator

$$[x_\mu D] = -q_\mu . \tag{15}$$

The surface term of the partial integration has been evaluated using eq. (12)

Interpreting the (2k + n, 2k) matrix

$$\Delta = \begin{pmatrix} q_\mu (a^\mu + x^\mu) \\ \alpha \end{pmatrix} \tag{16}$$

as the analogon of D for the new connection, we find that

$$\Delta^+\Delta = \left[(a_\mu + x_\mu)(a^\mu + x^\mu) + \frac{1}{2} tr_2 (\alpha^+\alpha) \right] \cdot 1_2 \tag{17}$$

commutes with the quaternions, in analogy to eq. (3).

As Δ satisfies the hypotheses of the ADHM construction, we may use it to construct a new self-dual connection. The construction completely follows the one given above, which in fact was modelled on it. Put

$$\Delta^+ v = 0, \tag{18}$$

$$v^+ v = 1_n , \tag{19}$$

$$v v^+ = 1 - \Delta F \Delta^+ , \tag{20}$$

$$F \cdot 1_2 = (\Delta^+ \Delta)^{-1} . \tag{21}$$

With

$$\bar{\mathcal{D}}_\mu = v^+ \partial_\mu v \tag{22}$$

we obtain

$$[\bar{\mathcal{D}}_\mu \bar{\mathcal{D}}_\nu] = v^+ b (q_\mu q_\nu^+ - q_\nu q_\mu^+) F b^+ v , \tag{23}$$

which is self-dual. Here we used the commutator

$$[\partial_\mu , \Delta] = b q_\mu \tag{24}$$

with

$$b = \begin{pmatrix} 1_k \\ 0 \end{pmatrix} \tag{25}$$

The eqs. (18-24)are completely analogous to eqs. (4-7) and (13-15).

Now we would like to prove that the new connection (22) is the same as the old one. Using eqs. (6), (15) and

$$(G D^+ G)(x,y) = \frac{1}{2} (x-y)^+ G(x,y) \tag{26}$$

one finds that

$$v^+ = (4\pi \varepsilon G \psi , \sigma) \tag{27}$$

satisfies eq. (18). Eq. (26) is easily proved by acting with D^2 on both sides. Now we want to prove

$$v^+(x) v(y) = -4\pi^2 (x-y)^2 G , \tag{28}$$

which in particular implies that v is normalized, by taking the limit y to x. Both sides of eq. (28) have the same asymptotic behaviour for large x, and due to eqs. (6) and 26) one obtains the same function, when one applies $\mathcal{D}^2$ to them. Thus eq. (28) is proved.

Now the equality of the new connection with the old one is obvious. In fact, for a normalized v the inverse of the covariant Laplacian of the new connection is given by[2)]

$$\bar{G}(x,y) = -\left(4\pi^2(x-y)^2\right)^{-1} v^+(x)\, v(y). \tag{29}$$

Actually we even do not need this result, as we only have to evaluate eq. (28) with y=x + dx.

We have shown that each instanton connection can be obtained by eqs. (18-19) and (22), i.e. by the ADHM construction.
Again we may dualize this procedure. Using

$$F \Delta^+ b F = -\frac{1}{2} q_\mu^+ \partial^\mu F \tag{30}$$

it is easy to see that

$$\psi = \pi^{-1} v^+ b F \varepsilon \tag{31}$$

fulfills the Weyl equation for the connection (22). Then one may prove

$$tr_2\, \psi^+ \psi = -(4\pi^2)^{-1} \partial^2 F \tag{32}$$

by application of $\Delta^+\Delta$ to both sides and use of eqs. (20) and (30). Eqs. (30-32) are anlogous to eqs. (26-28). Integrating eq. (32) immediately shows that ψ is normalized and that eq. (13) is fulfilled, such that we indeed went full circle.

What have we achieved? On one hand we have proved the ADHM construction for instantons. On the other hand we have an involutory construction which basically associates two solutions of different versions of the self-duality equation to each other. Each solution is given by a simple bilinear expression in terms of the solutions of the Weyl equation associated to the other self-dual connection.

This procedure also works for self-dual monopoles[3)] and calorons. The monopoles can be described as solutions of the self-duality equation which are translationally invariant in one direction and have field strengths which are square integrable over the R^3 orthogonal to this direction. The Weyl equation (4) is formally unchanged, but we have to put

$$\mathcal{D}_0 = \varphi + iz \tag{33}$$

where ϕ is the Higgs field, and z is a new real constant. Of course we now have to normalize by an R^3 integration.

Eq. (13) is replaced by

$$T^m(z) = -i\int \psi^+ x^m \psi \, d^3x \, , \quad m = 1,2,3, \tag{34}$$

and

$$\frac{d}{dz} + T^0(z) = \int \psi^+ \frac{\partial}{\partial z} \psi \, d^3x \tag{35}$$

In analogy to eq. (14) one finds

$$\left[\frac{d}{dz} + T^0 , T^k\right] = \frac{1}{2} \varepsilon_{klm} \left[T^l T^m\right] \tag{36}$$

where we have used the analogon of eq. (15), including

$$\left[i\frac{\partial}{\partial z} , D\right] = -1 . \tag{37}$$

As before let v be the normalized kernel of Δ^+, with Δ given by

$$\Delta = \frac{d}{dz} + q_\mu T^\mu(z) + i q_m x^m , \tag{38}$$

in analogy to eq. (16), and replace eq. (22) by

$$A_m = \int v^+ \partial_m v \, dz \, , \tag{39}$$

$$\varphi = -i\int v^+ z \, v \, dz \, . \tag{40}$$

As before one may show that the connection determined by the latter two equations yields a self-dual field strength, using the analogon of eq. (24), which includes

$$[-z , \Delta] = 1 . \tag{41}$$

Otherwise there are no major changes. For more details see ref. 3-7.

Now we shall see that self-dual calorons, which may be described as instantons on $S^1 x R^3$, can be treated in the same way. Self-dual monopoles may be considered as special calorons which do not depend at all on x°, whereas in general we only require periodicity. Let ξ be the period.

For calorons we put

$$\mathcal{D}_0 = \partial_0 + A_0 + iz \, . \tag{42}$$

For the Weyl equation (4) we require solutions which are onevalued on $S^1 x R^3$. Thus different values of z yields quite different solution spaces. However, a shift in z of $2\pi/\xi$ just yields

$$\psi(x, z + 2\pi/\xi) = \exp(-2\pi i x^0/\xi) \, \psi(x,z) . \tag{43}$$

Thus z is a periodic variable with period $2\pi/\xi$.

We require

$$\int_{S^1 x R^3} \psi^+ \psi \, d^4x = 1_k \tag{44}$$

and put

$$i\,T^m(z) = s^{-1} \times R^3 \int \psi^+ x^m \psi \, d^4x , \tag{45}$$

$$T^0(z) = s^{-1} \times R^3 \int \psi^+ \frac{\partial \psi}{\partial z} \, d^4x . \tag{46}$$

As for monopoles, the $T^\mu(z)$ fulfil eq. (36), because eq. (37) is still true. Eq. (38) is replaced by

$$\Delta = \frac{d}{dz} + q_\mu \left(T^\mu(z) + i x^\mu\right). \tag{47}$$

The non-trivial x^0 dependence of the solutions of

$$\Delta^+ v = 0 \tag{48}$$

comes from the requirement that v is a periodic function of z. One finds for normalized v

$$A_\mu = s^{-1} \int v^+ \frac{\partial v}{\partial x^\mu} \, dz , \tag{49}$$

where the z integration goes over one period. Exactly as for the monopoles, eq. (48) in certain cases has to be supplemented by delta function terms on the r.h.s.[4)] For simplicity we omit this modification.

For monopoles, $T^0(z)$ may be gauged to zero, such that eq. (36) simplifies. For calorons this is still possible, but one has to take into account the possibility that the group element

$$g = P \exp \int_{z_0}^{z_0 + 2\pi/\xi} T^0(z) \, dz \tag{50}$$

is nontrivial. Thus for vanishing $T^0(z)$ one has to use the modified periodicity conditions

$$T^m(z_0 + 2\pi/\xi) = g\, T^m(z_0)\, g^{-1}, \tag{51}$$

$$v(z_0 + 2\pi/\xi) = g\, v(z_0). \tag{52}$$

This completes the ADHM construction of calorons.

2. The spectral curve and its envelope

Obviously, calorons share many properties with self-dual monopoles. In particular, they also have spectral curves. For each z interval for which eq. (36) is valid, a spectral curve is defined by the equation

$$p(y,\eta) = \det\,(y_m T^m(z) + \eta) = 0. \tag{53}$$

Here η , y are homogeneous complex coordinates satisfying

$$y^2 = 0. \tag{54}$$

Because of eq. (36) the spectral curve does not depend on z.

The points η, y parametrize the oriented lines in R^3. To the usual description by a direction u and a point x mod u on the line they are related by

$$\eta = i y_m x^m \tag{55}$$

$$u^m = i\varepsilon_{mnr}\, y^n y_r^+ / (yy^+). \tag{56}$$

For (k,k) matrices $T^i(z)$ one obtains 2k distinct lines through each generic point $x\varepsilon R^3$, or k lines, if one neglects the orientation.

The spectral curves are basic for the solution of the monopole equations and reappear in different treatments of the monopoles[3,5]. As they also occur for calorons it becomes even more desirable to obtain a more physical interpretation. We shall see that the curves are closely related to the asymptotic behaviour of the monopoles and calorons. A spectral curve in the space of lines of R^3 has an envelope in R^3 itself, given by the set of points through which go less than k distinct lines. This envelope is given by the conditions (53-55) plus the equation

$$\delta\eta = -\frac{\partial p}{\partial y_m}\,\delta y_m \Big/ \frac{\partial p}{\partial \eta} = i x^m \delta y_m\,, \tag{57}$$

which must be satisfied for some δy orthogonal but not proportional to y:

$$y_m \delta y^m = 0, \tag{58}$$

$$u^m \delta y_m \neq 0. \tag{59}$$

For δ y proportional to y, eq. (57) is fulfilled trivially, as η , y are homogeneous coordinates.

For a point x of the envelope the components perpendicular to u are given by eq. (55), the remaining component by

$$u^m x_m = i\,\frac{\partial p}{\partial y_m}\,u^m \Big/ \frac{\partial p}{\partial \eta}\,. \tag{60}$$

An arbitrary line of the spectral curve meets the envelope, if the r.h.s. of eq.(60) is real. When η is given by eq. (55), the eqs. (53) and (60) together are equivalent to

$$\varepsilon_{mnr}\, y^n \left(\frac{\partial p}{\partial y^r} + i x^r \frac{\partial p}{\partial \eta} \right) = 0 \tag{61}$$

as one sees by taking scalar products with u and y^+. Eq. (61) shows that the envelope is given by a complex polynomial equation for x, i.e. by the real points of a complex surface in the complexification of R^3. These real points form a curve in R^3, which may contain disconnected pieces and isolated points.

Geometrically it is clear that the tangents of the envelope belong to the spectral curve. Algebraically this is easy to see, too. Let η , y and $\eta + d\eta$, y + dy be two neighbouring lines of the spectral curve for which the r.h.s. of eq. (60) is real. Let x and x + dx be the corresponding points of the envelope. One has

$$d\eta = i\,(x^m dy_m + y_m dx^m). \tag{62}$$

On the other hand eq.(57) is fulfilled by some nontrivial δy satisfying eq. (58), and by linear combination it is satisfied for all such δ y including dy. Thus

$$y_m dx^m = 0, \tag{63}$$

which shows that the slope of the tangent of the envelope is u, such that this tangent indeed belongs to the spectral curve, as it coincides with the line given by η ,y.

As the spectral curve is given by a polynomial equation, the subset given by the tangents of the envelope is sufficient to determine it completely, or at least, for composite spectral curves, the corresponding irreducible component.

For a spherically symmetric monopole the envelope obviously consists only of the point at the monopole center, with some multiplicity. The corresponding spectral curve consists of all the lines passing through this point. For axially symmetric monopoles, the envelope may consist of isolated points on the symmetry axis, plus circles around this axis. The component of the spectral curve corresponding to such a circle consists of the straight lines contained in a family of hyperboloids filling the R^3.

For SU(2) monopoles of charge 2 the envelope turns out to be an ellipse, the major axis of which contains the zeros of the Higgs field. In each case the continuous and discrete symmetries of the solutions can be read off from the symmetries of the envelope.

3. The asymptotic behaviour of self-dual monopoles and calorons

Asymptotically, the structure of monopoles and calorons simplifies considerably. The x° dependent terms of calorons decay exponentially with the distance. Similarly, the full gauge group G of a monopole also is realized only for exponentially decaying terms. If one neglects these terms, the gauge group reduces to a subgroup H which commutes with the value of the Higgs field at infinity.

The reduced monopoles and calorons are obtained, if one expands the fields into power series in $1/r$ and subsequently sums up the series. Apparently, these asymptotic expansions are summable in all known cases, but no general proof is known. For the reduced calorons the gauge group also may reduce to a subgroup, but this is not always the case. The new fields still are self-dual, but they are singular.

How can the reduction be explained in the formalism which we described above? For all self-dual monopoles and calorons, the range of z contains a finite number of jumping points z_i across which the spectral curves change. In general, even the dimension of the $T^1(z)$ and correspondingly the number of components of $v(z)$ jumps at these points. These are also the points where the solutions of the differential eq. (48) for v may develop singularities which are not square integrable, which restricts the solution space.

Now for large r a basis for the solutions of eq. (48) is given by functions which have support in the neighbourhood of a single z_i, up to exponentially decaying terms. In particular, the product terms for solutions belonging to different z_i which occur in eqs. (39-40) and (49) vanish for the reduced configurations. If the reduced gauge group H is abelian, one only has one solution for each z_i, such that the gauge fields become diagonal for this choice of base. In any case the reduction of the gauge group into as many factors as there are different z_i becomes obvious. For calorons there may be a single jumping point, in which case G is not reduced asymptotically. But as the support is basically restricted to the neighbourhood of jumping points, the periodicity condition for the $v(z)$ becomes trivial. Consequently the x° dependence of the solutions of eq. (48) is just a phase and drops out in the expression (49) for the potential, again up to exponentially suppressed terms.

Let us scetch the proof for these assertions. Eq.(48) may be solved with the help of the adjoint equation

$$\left(\frac{d}{dz} + q_\mu (T^\mu + i x^\mu)\right) w(x,z) = 0 \qquad (64)$$

as for solutions v and w

$$\frac{\partial}{\partial z} v^+ w = 0. \tag{65}$$

Let u_ℓ and $-u_\ell$, $\ell=1, \ldots, k$ be the orientations of the 2k lines of the spectral curve passing through x. Then 2k linearly independent solutions of eq. (64) may be characterized by[3)]

$$(1 \pm i q_m u_\ell^m) \, w_{\pm \ell}(z,x) = 0 , \quad \ell = 1, .., k. \tag{66}$$

At large distance one has

$$u_\ell^m = x^m / r + O(r^{-1}) \tag{67}$$

and

$$w_{\pm\ell}(z,x) \sim \exp(\pm r z). \tag{68}$$

Thus the support is indeed restricted to the neighbourhood of one jumping point, up to exponentially suppressed terms. Because of eq. (65) this carries over to v(z), for which we may put

$$v^+ w_{-\ell} = 0, \quad \ell = 1, .., k. \tag{69}$$

To obtain the reduced potentials we must put for those v(z)

$$\int_{z_i}^{\infty} v^+ v \, dz = 1, \tag{70}$$

$$A_\mu = \int_{z_i}^{\infty} v^+ \frac{\partial v}{\partial x^\mu} \, dz , \tag{71}$$

with a suitably chosen integration path. Complex integration is possible, as the dependence on z is analytic.

Where are the singularities of the reduced monopoles and calorons? The procedure outlined above should yield regular potentials as long as the distinction between the directions u_ℓ and $-u_\ell$ works, i.e. as long as one stays away from the envelope of the spectral curve.

The various factors of H yield different reduced configurations, such that the different spectral curves come into play. For an abelian H the envelopes of the spectral curves may be considered to carry magnetic charge acting as source of the reduced abelian magnetic field. Each envelope carries an integer charge, but the integers depend on the representation of H under consideration.

One may hope that the surprising and beautiful relationship between the spectral curves and the asymptotic behaviour of the self-dual monopoles and calorons will open up new possibilities for their investigation.

References

1) M. Atiyah, N. Hitchin, V. Drinfeld, and Yu. Manin, Phys. Lett. 65A (1978) 185.

2) E. Corrigan, D. Fairlie, P. Goddard,and S. Templeton, Nucl. Phys. B140 (1978)31.

3) W. Nahm, in: Group Theoretical Methods in Physics, M. Serdaroglu et al. eds., Istanbul 1982, p.456.

4) W. Nahm, in: Monopoles in Quantum Field Theory, N. Craigie et al. eds., Trieste 1981, p.87.

5) N. Hitchin, Comm. Math. Phys. 89 (1983) 145.

6) W. Nahm, in: Structural Elements in Particle Physics and Statistical Mechanics, J. Honerkamp et al. eds., NATO ASI B82 (1981) 301.

7) W. Nahm, in:Symposium on Particle Physics, Z. Horváth et al. eds., Visegrád 1981, p. 397.

U(1) INVARIANT HIERARCHY THEORIES IN d-DIMENSION ANTISYMMETRIC GAUGE TENSOR FIELDS

H. Nencka-Ficek

Institute of Molecular Physics, Polish Academy of Sciences, Poznań, Smoluchowskiego 17/19, Poland

Recently, a lot of experiments have been carried out to investigate the nature of the topological excitations in the ferromagnetic and amorphous superconductors [1,2]. Therefore, we deal with the generalised Savit model /U(1) invariant theory/ in d-dimensional hypercubic lattice without and with frustrations.

THE GENERALISED SAVIT MODEL. Our model is devoted to U(1) invariant theory and it is described by the action

$$A[\varphi] = A_o[\varphi] + A_{int}[\varphi] , \qquad (1)$$

$$A_o[\varphi] = \sum_{i=1}^{d-1} \sum_{\{b_i\}} K_i \cos\left\{ \frac{1}{(d-i)!} \epsilon_{\mu_1 \dots \mu_1, \alpha_1 \dots \alpha_{d-i}} \epsilon_{\alpha_1 \dots \alpha_{d-i} \beta \gamma_1 \dots \gamma_{i-1}} \Delta_\beta \varphi_{\gamma_1 \dots \gamma_{i-1}} \right\}$$

$$A_{int}[\varphi] = \sum_{j=1}^{d-2} \sum_{\{b_j\}} \beta_j \cos\{ \epsilon_{\mu_1 \dots \mu_j, \alpha_1 \dots \alpha_{d-j}} \epsilon_{\alpha_1 \dots \alpha_{d-j}, \beta, \gamma_1 \dots \gamma_{j-1}} (\Delta_\beta \varphi_{\gamma_1 \dots \gamma_{j-1}} - \varphi_{\beta \gamma_1 \dots \gamma_{j-1}})\}$$

where the sum $\sum_{\{b_i\}}$ runs over all i-dimensional nearest neighbours simplices, K and β are the coupling constants, Δ is a finite difference operator, $\varphi_{\gamma_1 \dots \gamma_r}$ is a generalised angle $(-\pi \leq \varphi \leq \pi)$ which is the usual angle when r = 0. Using the standard method for obtaining the topological excitations of a model [3], we obtain the following partition function

$$Z = Z_o \sum_{\{J,Q\}} \exp\Big\{\pi\Big[\sum_{i=1}^{d-2} \beta_i \sum_{\tilde{i} \neq \tilde{j}} Q_{\mu_1 \dots \mu_i}(\tilde{i}) D|\tilde{i}-\tilde{j}| Q_{\mu_1 \dots \mu_i}(\tilde{j})$$
$$+ \sum_{i=1}^{d-2} \bar{K}_{i+1} \sum_{\tilde{i} \neq \tilde{j}} Q_{\mu_1 \dots \mu_{i+1}}(\tilde{i}) D|\tilde{i}-\tilde{j}| Q_{\mu_1 \dots \mu_{i+1}}(\tilde{j})$$
$$+ \sum_{j=1}^{d-1} K'_j \sum_{\tilde{i} \neq \tilde{j}} J_{\mu_1 \dots \mu_j}(\tilde{i}) D|\tilde{i}-\tilde{j}| J_{\mu_1 \dots \mu_j}(\tilde{j})\Big]\Big\} \qquad (2)$$

where $\bar{K}$, K' are the coupling constants $/\bar{K}_i \neq K'_j$, $i \neq j/$. There are two types of the multi-dimensional topological excitations: J and Q

/the integer-valued topological excitations/. The J' s are associated with the j-dimensional surfaces and $\Delta_\mu J_{\mu_1 \ldots \mu_j} = 0$/ the so-called free topological excitations/.

However, the $Q_{\mu_1 \ldots \mu_i}$ s and the $Q_{\mu_1 \ldots \mu_{i+1}}$'s associated with the i- and (i+1) -dimensional surfaces respectively are such that: $\Delta_\mu Q_{\mu_1 \ldots \mu_{i+1}} = Q_{\mu_1 \ldots \mu_i}$. These topological excitations are pair bounded in such a sense that the (i+1)-dimensional surfaces with the $Q_{\mu_1 \ldots \mu_{i+1}}$ are bounded by the i-dimensional with the $Q_{\mu_1 \ldots \mu_i}$'s.For instance: the surface like topological excitations $Q_{\mu\beta}$ are bounded by the line-like topological excitations Q_μ ; or the strings Q_μ are bounded by the vortices Q.

Let us consider the various cases of the model (1).

/i/ $A[\varphi] = A_0[\varphi]$, it means that the coupling constants $\overline{K}, \beta = 0$ but $K' \neq 0$. Here, there are free the topological excitations J, only.

/ii/ $A[\varphi] = A_0^-[\varphi] + A_{int}[\varphi]$, but the coupling constants $\beta_i \neq 0$ if $\overline{K}_{i+1} \neq 0$ and $\beta_i = 0$ if $\overline{K}_{i+1} = 0$; $K' = 0$. Now, there are no the topological excitations of the type J. There are the bounded /in the above sense/ topological excitations, only.

In general, if there are no the restrictions on parameters, there are both the bounded and the free topological excitations.

THE MODEL WITH GENERALISED FRUSTRATIONS. We define the generalised frustration in the following manner

$$2\pi\, \Phi_{\mu_1 \ldots \mu_{k+1}} = \Delta_\nu \psi_{\mu_1 \ldots \mu_k} - \Delta_\mu \psi_{\nu_1 \ldots \nu_k} \qquad (3)$$

where ψ is the so-called disorder parameter. The frustration function $\Phi_{\mu_1 \ldots \mu_{k+1}}$ is considered on a k-dimensional, closed hypercontour in the lattice. If k=1, then we have the usual frustration for the links.However, introduced by us the generalised frustrations measure the disorder among the antisymmetric gauge fields on the multi-dimensional simplices.

We consider the model (1) with the generalised frustration. The model is the following now

$$\begin{aligned} A = & \sum_{i=1}^{d-2} \sum_{\{b_i\}} \beta_i \cos\{\epsilon_{\mu_1 \ldots \mu_i \alpha_1 \ldots \alpha_{d-i}} \epsilon_{\alpha_1 \ldots \alpha_{d-i} \beta \gamma_1 \ldots \gamma_{i-1}} (\Delta_\beta \varphi_{\gamma_1 \ldots \gamma_{i-1}} - \varphi_{\beta \gamma_1 \ldots \gamma_{i-1}})\} \\ & + \sum_{i=1}^{d-2} \sum_{\{b_{i+1}\}} \overline{K}_{i+1} \cos\{\tfrac{1}{(d-i-2)!} \epsilon_{\mu_1 \ldots \mu_{i+1} \alpha_1 \ldots \alpha_{d-i-1}} \epsilon_{\alpha_1 \ldots \alpha_{d-i-1} \beta \gamma_1 \ldots \gamma_{i-2}} (\Delta_\beta \varphi_{\gamma_1 \ldots \gamma_{i-2}} - \psi_{\beta \gamma_1 \ldots \gamma_{i-2}})\} \\ & \sum_{\substack{j=1 \\ j \neq i}}^{d-1} \sum_{\{b_j\}} K'_j \cos\{\tfrac{1}{(d-j)!} \epsilon_{\mu_1 \ldots \mu_j \alpha_1 \ldots \alpha_{d-j}} \epsilon_{\alpha_1 \ldots \alpha_{d-j} \beta \gamma_1 \ldots \gamma_{j-1}} (\Delta_\beta \varphi_{\gamma_1 \ldots \gamma_{j-1}} - \psi_{\beta \gamma_1 \ldots \gamma_{j-1}})\} \end{aligned} \qquad (4)$$

The partition function is now

$$
\begin{aligned}
Z = Z_0 \sum_{\{Q J\}} \exp\{\pi [& \sum_{i=1}^{d-2} \beta_i \sum_{\tilde{i}=\tilde{j}} (Q_{\mu_1 \cdots \mu_i}(\tilde{i}) - \frac{1}{\pi} \xi_{\mu_1 \cdots \mu_i}(\tilde{i})) D|\tilde{i}-\tilde{j}| (Q_{\mu_1 \cdots \mu_i}(\tilde{j}) - \frac{1}{\pi} \xi_{\mu_1 \cdots \mu_i}(\tilde{j})) \\
& + \sum_{i=1}^{d-2} \bar{K}_{i+1} \sum_{\tilde{i} \neq \tilde{j}} (Q_{\mu_1 \cdots \mu_{i+1}}(\tilde{i}) + \Phi_{\mu_1 \cdots \mu_{i+1}}(\tilde{i})) D|\tilde{i}-\tilde{j}| (Q_{\mu_1 \cdots \mu_{i+1}}(\tilde{j}) + \Phi_{\mu_1 \cdots \mu_{i+1}}(\tilde{j})) \\
& \sum_{\substack{j=1 \\ i \neq j}}^{d-1} K'_j \sum_{\tilde{i} \neq \tilde{j}} (J_{\mu_1 \cdots \mu_j}(\tilde{i}) + \Phi_{\mu_1 \cdots \mu_j}(\tilde{i})) D|\tilde{i}-\tilde{j}| (J_{\mu_1 \cdots \mu_j}(\tilde{j}) + \Phi_{\mu_1 \cdots \mu_j}(\tilde{j}))]\}
\end{aligned}
$$

There are two types of the topological excitations connected with the appearance of the disorder in the original model (4). There are the multi-dimensional, half-integer-valued frustrational topological excitations Φ and the multi-dimensional called by us disorder topological excitations ξ. The relation between them and the disorder parameter is: $\xi_{\mu_1 \cdots \mu_\mu} = \frac{1}{2} \epsilon_{\mu_1 \cdots \mu_\mu \cdots \mu_d} \psi_{\mu_{\mu+1} \cdots \mu_d}$. The frustrational topological excitations are caused by the frustrations in the original model, only. However, the coupling between i- and (i+1)-dimensional, frustrated fields in the original lattice is the necessary condition for the disorder topological excitations to exist. We can easily see that the 1-dimensional, disorder topological excitations can exist in the model (4) if d=3. The dimension of the frustrational topological excitations is then 0.

Finally let us present some possible applications of our model with generalised frustration. Let us refer to Aubry [4]. He has shown that the frustration is determined in the defectible model by such parameters as pressure and an amplitude of the periodic potential. Therefore, one can consider frustration for k=2 as a disorder caused by the pressure /for d=3,4,../. On the other hand, one can easily verify that there is the correspondence between 3D model with frustrations for k=2 and the 2D XY model with the frustrations for the links [5]. Hence, one can think that the tensions /due to the surface effects/ will cause a creation of the disorder topological excitations in the 2D superconductors and these disorder topological excitations will have an influence on the behaviour of the resistance in this system.

REFERENCES

1 C.Kuper,M.Revzen and A.Ron, Phys.Rev.lett.44, 1545 (1980).
2 K.Epstein,A.Goldman and A.Kadin, Phys.Rev.Lett.47, 534 (1981).
3 R.Savit, Rev.Mod.Phys.52, 453 (1980).
4 S.Aubry, J.Phys./Paris/ 44, 147 (1983).
5 H.Nencka-Ficek, Proc of the XI th Int.Colloq.on Group Theor. Methods in Phys.,Istanbul,Aug.1982, to be published.

GENERALIZED CONNECTION FORMS WITH LINEARIZED CURVATURE

F.B. Pasemann
Institut für Theoretische Physik
Technische Universität Clausthal
D-3392 Clausthal, F.R.Germany

A quantization program for non-abelian gauge theories was outlined in [1], and the resulting linearized Gauge Quantum Field Theories have been discussed in [2]. Linearization is due to the fact that a pure quantum gauge field interacts with its classical counterpart. The generic objects for these theories are generalized connection forms with linearized curvature defined on the Kaluza-Klein space corresponding to a classical solution of the Yang-Mills equation under consideration. The properties of generalized connection forms then determine the explicit structure of test form algebras [2] and state spaces [3] in the algebraic and the indefinite-metric formulation of linearized Gauge Quantum Field Theories, respectively. In the following we summarize the main results on generalized connection forms with linearized curvature. The general notation and definitions are due to [2],[4].

By (M,g) we denote Minkowski space, and by $P(M,G)$ a principal G-bundle over M, with G a compact connected, semi-simple Lie group with Lie algebra $\mathfrak{G}$. The Kaluza-Klein space associated to a given connection form ω on $P(M,G)$ is denoted by $(P,\hat{g})$.

In the following we fix a triple (ω,Ω,j), where ω is a connection form with curvature Ω and j is a source for ω, i.e.

$$\Omega = \nabla\omega \quad \text{and} \quad \not\nabla\,\Omega = j\ .$$

(∇ denotes the covariant derivative, and $\not\nabla$ the covariant coderivative on $(P,\hat{g})$. Recall that a $\mathfrak{G}$-valued p-current T (in the sense of deRham [5]) on $(P,\hat{g})$ can be viewed as a linear functional on $D^p(P,\mathfrak{G})$, the space of $\mathfrak{G}$-valued p-forms on P with compact support, i.e.

$$T(\alpha) = \int_P T^a \wedge *\alpha^b \otimes \tilde{g}(e_a,e_b)\ , \qquad \alpha \in D^p(P,\mathfrak{G})\ ,$$

where $\tilde{g}$ is the canonical metric on G, and $\{e_a\}$ denotes a basis in $\mathfrak{G}$, which is chosen to be orthonormal in what follows. We then give the

<u>Definition 1:</u> A <u>generalized connection form</u> T^{ω} on $P(M,G)$ is a 1-current on $(P,\hat{g})$ satisfying

$$L(Z_h)T^{\omega} = -\,\mathrm{ad}(h)\,T^{\omega} \quad , \quad h \in \mathbb{G},$$

$$di(Z_h)T^{\omega} = 0 \quad , \quad h \in \mathbb{G},$$

$$T^{\omega}(\alpha) = 0 \quad , \quad \alpha \in D^1_M(P,\mathbb{G}),$$

with $D^p_M(P,\mathbb{G})$ denoting the space of basic p-forms on P, and Z_h the fundamental vector field on P generated by $h \in \mathbb{G}$.

<u>Definition 2:</u> The <u>curvature</u> T^{Ω} of T^{ω} <u>linearized</u> with respect to the classical connection form ω is a 2-current on $(P,\hat{g})$ defined by

$$(1) \qquad T^{\Omega} := \nabla T^{\omega} \quad .$$

<u>Lemma 1:</u> The curvature T^{Ω} is basic, and it satisfies

$$(2) \qquad T^{\Omega} = dT^{\omega} + \tfrac{1}{2}[\![\omega, T^{\omega}]\!] \quad ,$$

$$(3) \qquad \nabla T^{\Omega} = 0 \quad .$$

A 1-current T^{j} on $(P,\hat{g})$ is called a <u>source</u> for T^{ω} iff it satisfies

$$(4) \qquad T^{j} = \nabla^{*} T^{\Omega} \quad .$$

For a given triple $(T^{\omega},T^{\Omega},T^{j})$ satisfying equation (1) and (4) we have :

$$(5) \qquad T^{\omega} = T^{\omega}_a \wedge \omega^a \ , \text{ i.e. } \ T^{\omega}(\alpha) = T^{\omega}(e(\omega^a)i(\omega^a)\alpha) \ , \ \alpha \in D^1(P,\mathbb{G}),$$

$$(6) \qquad T^{\Omega} = T^{\omega}_a \wedge \Omega^a \ , \text{ i.e. } \ T^{\Omega}(\alpha) = T^{\omega}(e(\omega^a)i(\Omega^a)\alpha) \ , \ \alpha \in D^2(P,\mathbb{G}),$$

$$(7) \qquad T^{j} = T^{\omega}_a \wedge j^a \ , \text{ i.e. } \ T^{j}(\alpha) = T^{\omega}(e(\omega^a)i(j^a)\alpha) \ , \ \alpha \in D^1(P,\mathbb{G}) \ ,$$

where $e(\alpha)$ and $i(\alpha)$ denote the left exterior product and the substitution with $\alpha \in D^p(P)$, respectively, and summation over repeated indices is understood. From these equations we get

<u>Lemma 2:</u> If ω is flat, then T^{ω} is flat, i.e. $\Omega = 0 \Longrightarrow T^{\Omega} = 0$.
If ω is pure, then T^{ω} is pure, i.e. $\nabla^{*}\Omega = 0 \Longrightarrow \nabla^{*} T^{\Omega} = 0$.

<u>Lemma 3:</u> A source T^{j} for T^{ω} is covariantly conserved, i.e. $\nabla^{*} T^{j} = 0$.

Let σ denote a section in $P(M,G)$, and (A,F,J) the corresponding local representative of the triple (ω,Ω,j). A push forward σ_{*} from $A^p(M)$ to $A^p_G(P)$ can be defined by using σ, the metrics g and $\hat{g}$, and the principal action of G on P. It induces a pull-back of p-currents on $(P,\hat{g})$ to p-currents on (M,g). Let (T^A,T^F,T^J) denote the local representative of the triple $(T^{\omega},T^{\Omega},T^{j})$. Then the following equations are satisfied:

(8) $$T^F = dT^A + \frac{1}{2} [\![A, T^A]\!] \quad ,$$

(9) $$dT^F + [\![A, T^F]\!] = 0 \quad ,$$

(10) $$\delta T^F - \bar{\kappa}^1 [\![A, * T^F]\!] = T^J \quad .$$

The following properties hold:

(11) $$T^A(\alpha) = T^\omega(e(\omega^a)\pi^*(i(A^a)\alpha)) \quad , \qquad \alpha \in D^1(M,\mathbb{G}) \quad ,$$

(12) $$T^F(\alpha) = T^\omega(e(\omega^a)\pi^*(i(F^a)\alpha)) \quad , \qquad \alpha \in D^2(M,\mathbb{G}) \quad ,$$

(13) $$T^J(\alpha) = T^\omega(e(\omega^a)\pi^*(i(J^a)\alpha)) \quad , \qquad \alpha \in D^1(M,\mathbb{G}) \quad ,$$

where $\pi : P \longrightarrow M$ is the bundle projection.

Given a classical solution of a Yang-Mills equation, a linearized Gauge Quantum Field Theory can be constructed as described in [2] . Properties (5) - (7) and (11) - (13) will explicitely specify the structure of the test form algebras and the properties of the n-point functions of the theory. Using a generalized reconstruction theorem the corresponding indefinite-metric operator theory can be given. The effect of the non trivial topological and geometrical structures of classical Yang-Mills theories can be traced through the whole construction from the classical to the quantum description of the theory. In the case of pure non-abelian gauge fields the theory will describe in a rigorous mathematical way quantum fluctuations in a classical background field.

References:

[1] Doebner, H.D., Pasemann, F.B.: Czech.J.Phys., B32, 430 (1982)

[2] Pasemann, F.B.: "Linearized non-Abelian Gauge Quantum Field Theories" , Proceedings of the "Int. Conf. in Math.Phys.", Clausthal 1981, Lect.Notes in Math., Springer, Berlin 1983

[3] Pasemann, F.B.: "State spaces for linearized gauge quantum field theories", (to appear)

[4] Greub, W., Halperin, S., Vanstone, R.: "Connections, Curvature, and Cohomology", Vol.I,II, Academic Press, New York, 1973/74

[5] deRham, G.: "Variêté Differentiables, Herman, Paris 1955

see [2] for further references.

DYNAMICAL SYMMETRY BREAKING IN S^4 DE SITTER SPACE

E. SPALLUCCI - Istituto di Fisica Teorica dell'Università - Trieste

Several inconsistencies of the Standard Cosmological Model can be solved if the Universe is assumed to have undergone one or more phase transitions just after the Big-Bang. The initial Grand (or Super) Unified Symmetry breaks as the Universe expands in a vacuum energy dominated De Sitter phase. This is the basic idea of the " Inflationary Cosmology" [1]. In order to have a self-consistent theory it is compelling to study symmetry breaking in a background De Sitter geometry [2].

We consider here a massless, self-interacting, scalar field $\Phi(x)$, non minimally coupled to the De Sitter metric

$$L[\Phi(x); g_{\mu\nu}] = \frac{1}{2} g^{\mu\nu}\partial_\mu\Phi\,\partial_\nu\Phi - \frac{1}{2}\xi R\Phi^2 - \frac{\lambda}{24}\Phi^4 \tag{1}$$

$R = 12/a^2$ is the Ricci scalar; ξ and λ are positive coupling constants, and $g_{\mu\nu}$ is the metric tensor.

The one-loop effective potential for this model can be formally written as [3]

$$V(\varphi;R) = \frac{1}{2}(\xi+\delta\xi)R\varphi^2 + \frac{\delta m^2}{2}\varphi^2 + \frac{\lambda+\delta\lambda}{24}\varphi^4 - \frac{3}{16\pi^2 a^4}\left[\zeta'(0;\Delta) + \zeta(0;\Delta)\ln\mu^2 a^2\right] \tag{2}$$

μ is an arbitrary regularization mass, $\delta\xi$, δm^2, $\delta\lambda$ are finite counterterms.

$\zeta(s,\Delta)$ is the generalized Riemann zeta function for the scalar D'Alembertian operator on the S^4 De Sitter space [2]

$$\zeta(s,\Delta) = \frac{1}{6}\sum_0^\infty {}_m \frac{(m+1)(m+2)(2m+3)}{\left[(m+3/2+\sqrt{\Delta})(m+3/2-\sqrt{\Delta})\right]^s} \tag{3}$$

$$\zeta'(s,\Delta) \equiv d\zeta(s;\Delta)/ds \quad \text{and} \quad \Delta \equiv 9/4 - a^2(\xi R + \lambda\varphi^2/2).$$

The first order gravitational quantum corrections (linear terms in the Ricci scalar) are obtained by expanding (2) for large a

$$V(\varphi;R) = \frac{1}{2}(\xi+\delta\xi)\varphi^2 + \frac{\delta m^2}{2}\varphi^2 + \frac{\lambda+\delta\lambda}{24}\varphi^4 - \frac{\lambda^2\varphi^4}{256\pi^2}\left(\frac{3}{2} - \ln\frac{\lambda\varphi^2}{2\mu^2}\right) +$$
$$- \frac{\lambda\varphi^2}{32\pi^2}\xi R\left(\frac{3}{2} - \ln\frac{\lambda\varphi^2}{2\mu^2}\right) + \frac{\lambda\varphi^2}{32\pi^2}\frac{R}{12}\left(\frac{25}{16} - \ln\frac{\lambda\varphi^2}{2\mu^2}\right) \tag{4}$$

In order to fix the arbitrary mass μ in (4) we have to impose the renormalization conditions

$$\lambda = \left.\frac{\partial^4 V(\varphi;R)}{\partial \varphi^4}\right|_{\substack{\varphi=\langle\phi\rangle \\ R=0}} \qquad (5)$$

$$\xi = \left.\frac{\partial^3 V(\varphi;R)}{\partial \varphi^2 \partial R}\right|_{\substack{\varphi=\langle\phi\rangle \\ R=0}} \qquad (6)$$

$$\left.\frac{\partial^2 V(\varphi;R)}{\partial \varphi^2}\right|_{\substack{\varphi=0 \\ R=0}} = f(\hbar)\, b \qquad (7)$$

(5) and (6) are off-shell definitions of λ and ξ because of the infrared divergences in $\varphi=0$; (7) is the generalized Coleman-Weinberg condition [4]: $f(\hbar)$ is a function which vanishes at the tree-level, $\hbar=0$, and equals one at the one-loop level $\hbar=1$; b is an arbitrary parameter fixed by the requirement that a physically sensible dynamical minimum of $V(\varphi,R)$ is located in $\varphi=\langle\phi\rangle \neq 0$ for $R=0$. Then $f(\hbar)b = -\frac{\lambda}{6}\langle\phi\rangle^2 + O(\lambda^2)$. (5), (6), (7) replace μ by $\langle\phi\rangle$ in (4):

$$V(\varphi;R) = -\frac{\lambda\langle\phi\rangle^2}{12}\varphi^2 + \frac{\lambda}{24}\varphi^4 + \frac{\lambda^2\varphi^4}{256\pi^2}\left(\ln\frac{\varphi^2}{\langle\phi\rangle^2} - \frac{25}{6}\right) + \frac{\xi}{2}R\varphi^2 + \frac{\lambda\varphi^2}{32\pi^2}\left(\xi - \frac{1}{6}\right)R\left(\ln\frac{\varphi^2}{\langle\phi\rangle^2} - 3\right) \qquad (8)$$

The one loop approximation in order to be reliable requires not only $\lambda \ll 1$ but also $\lambda \ln \varphi^2/\langle\phi\rangle^2 \ll 1$, therefore only some terms in (8) are physically meaningful

$$V(\varphi;R) \simeq -\frac{\lambda}{12}\langle\phi\rangle^2\varphi^2 + \frac{1}{2}\xi R\varphi^2 + \frac{\lambda}{24}\varphi^4 + O\left(\lambda^2;\ \lambda\ln\frac{\varphi^2}{\langle\phi\rangle^2}\right) \qquad (9)$$

From (9) we recover a new curvature dependente minimum :

$$\bar{\varphi}^2 = \langle\phi\rangle^2 - \frac{6\xi}{\lambda}R \qquad (10)$$

This new minimum continuously moves towards the origin $\varphi=0$ as the spacetime curvature increases. When the Ricci scalar attains the critical value

$$R_c = \frac{\lambda\langle\phi\rangle^2}{6\xi} \qquad (11)$$

$\bar{\varphi}$ vanishes, and a 2nd order phase transition occurs.

The model we have discussed is a sort of Higgs model with a dynamically generated " negative mass square " term $-\lambda\langle\phi\rangle^2/6$ and a classical, positive,

" effective mass " ξR provided by the non minimal gravitational coupling. At high curvature this second quantity dominates and the dynamically broken symmetry in flat spacetime becomes restored.

REFERENCES

1) A.H. Guth, Phys. Rev. D23 (1981) 347; A.D. Linde, Phys.Lett. 108B(1982) 389.
2) G.M. Shore, Ann.Phys. 128 (1980) 376; B. Allen, "Phase Transition in De Sitter Space " Cambridge preprint (1983).
3) S.W. Hawking, Comm.Math.Phys. 55 (1977) 133.
4) P. Ghose, J. Phys. G8 (1982) 193, and "Scalar Loops and the Higgs Mass in the Salam - Weinberg- Glashow Model" IC/82/118 Trieste preprint.

APPLICATIONS OF CONFORMAL INVARIANCE TO GAUGE QUANTUM FIELD THEORY

I.T. Todorov
International Centre for Theoretical Physics and
International School for Advanced Studies, Trieste
and
Institute of Nuclear Research and Nuclear Energy,
Bulgarian Academy of Sciences, Sofia 1184*

INTRODUCTION

The ups and downs in the applications of group theory methods to particle physics during the last twenty years have taught me one lesson: such applications stand a better chance of being fruitful when intertwined with dynamical considerations.

The applications of conformal invariance to Quantum Field Theory (QFT) provide, in my opinion, a good example of a mutually beneficial interaction between dynamics and symmetry principles.

Let me cite two reasons why the conformal group is expected to have a part in field theory. (1) It is the maximal group of local causal automrphisms of a pseudo-Riemannian space-time.[1]. (2) It is the symmetry group of Maxwell's electrodynamics [2], and more generally, of a large class of renormalizable massless field theories (that involve no dimensional parameters) [3].

Unlike Poincaré invariance and conventional internal symmetries, however, scale (and <u>a fortiori</u> conformal) invariance is broken by QFT renormalization. It follows that QFT Green functions can only be conformal invariant in a point of phase transition in which the coupling constant assumes a renormalization group stable value that annihilates the Callan-Symanzik β-function [4]. Since conformal invariance determines 2- and 3-point functions [5] and leads to a skeleton diagram expansion free of ultraviolet divergences, in the presence of anomalous dimensions [6] , the hope arose that it may help construct a critical type QFT (provided that such a nontrivial theory does exist). Although <u>this most ambitious goal</u> has not been achieved, some interesting partial

* Permanent address.

results were obtained [7-12], notably a dynamical derivation of a conformal invariant operator product expansion (OPE) acting on the vacuum vector. Short distance and light-cone OPE have been earlier applied, in particular, to the deep inelastic proton electron scattering [13,14] and are still an effective tool in studying both theoretical and phenomenological problems of particle dynamics.

I shall review in this lecture some basic facts about a class of "local field representations" of the conformal group [15,16], which are needed in QFT as well as two recent developments: one, concerning the canonical formulation of conformal quantum electrodynamics [17] (QED) and another, on the application of conformal techniques for the construction of composite operators in quantum chromodynamics [18] (QCD).

1. LOCAL FIELD REPRESENTATIONS OF THE CONFORMAL GROUP

1A. The quantum mechanical conformal group

The connected component of the conformal group of Minkowski space M is the simple group

$$SO_0(4,2)/\mathbb{Z}_2 \simeq G/\mathbb{Z}_4 \simeq U(2,2)/U(1) \qquad \text{where } G = SU(2,2). \tag{1.1}$$

We shall also consider the extension G_{ex} of G by space reflections I_s (regarded as an outer automorphism of G).

The group U(2,2) consists of all linear transformations of $\mathbb{C}^4$ which preserve a non-degenerate hermitian form with two positive and two negative eigenvalues. For a suitable choice of basis in $\mathbb{C}^4$ we can characterize a $g \in U(2,2)$ by

$$g^*\beta g = \beta \quad \text{for} \quad \beta = \begin{pmatrix} 0 & -1 \\ -1 & 0 \end{pmatrix} \equiv -\begin{pmatrix} 0 & 0 & 1 & 0 \\ 0 & 0 & 0 & 1 \\ 1 & 0 & 0 & 0 \\ 0 & 1 & 0 & 0 \end{pmatrix}. \tag{1.2}$$

If we write the 4x4 matrix $g \in U(2,2)$ in a 2x2 block matrix form

$$(U(2,2)\ni)\, g = \begin{pmatrix} a & b \\ c & d \end{pmatrix} \text{ where } a^*c + c^*a = 0 = b^*d + d^*b, \quad a^*d + c^*b = 1, \tag{1.3}$$

then its action on a point x in M, represented by a "pure imaginary quaternion"

$$i\underset{\sim}{x} = i\begin{pmatrix} x^0+x^3 & x^1-ix^2 \\ x^1+ix^2 & x^0-x^3 \end{pmatrix} \qquad \left(\underset{\sim}{x} = x^\mu \sigma_\mu \right) \tag{1.4}$$

is given by a fractional linear transformation

$$g:\ i\underset{\sim}{x} \rightarrow (a\, i\underset{\sim}{x} + b)(c\, i\underset{\sim}{x} + d)^{-1} \qquad (\text{for } \det(c\, i\underset{\sim}{x}+d) \neq 0). \tag{1.5}$$

Clearly, the centre U(1) of U(2,2) leaves $i\underset{\sim}{x}$ invariant. G acts without singularities on compactified Minkowski space $\overline{M}$ which is isomorphic to the group space of U(2) (the imbedding of M into U(2) being realized through the Cayley transform [19] $i\underset{\sim}{x} \rightarrow u = (1-ix)(1+ix)^{-1}$).

The quantum mechanical(ray) representations of G can be lifted to single valued representations of its (infinite sheeted) universal covering $\tilde{G}$ which is not a matrix group. Roughly speaking $\tilde{G}$ is obtained from G by replacing the centre U(1) of the maximal compact subgroup K = S(U(2)xU(2)) of G by its universal covering, which is isomorphic to the (non-compact!) additive group of reals. In the basis characterized by (1.2)

$$U(1) = \left\{ \begin{pmatrix} \cos\frac{\tau}{2} & i\sin\frac{\tau}{2} \\ i\sin\frac{\tau}{2} & \cos\frac{\tau}{2} \end{pmatrix},\ -2\pi < \tau \leq 2\pi \right\},\ \widetilde{U(1)} = \{\tau \in \mathbb{R}\}. \tag{1.6}$$

The (infinite) centre of $\tilde{G}$ can be identified with $Z \times Z_2$ where Z is built out of all integer powers of a central element ζ_1 that is a superposition of the τ-translation $\tau \rightarrow \tau - \pi$ with the Weyl inversion

$$w \text{ defined by } w^{-1} g w = g^{*-1} \text{ or } w = \beta \ ; \tag{1.7}$$

Z_2 is the centre of the quantum mechanical Lorentz group SL(2,$\mathbb{C}$) ($\subset$G).

1B. Local elementary representations of $\tilde{G}$ with a lowest weight subrepresentation

The classification of unitary irreducible representations of SU(2,2) is rather complicated and has been completed only recently [20] using high brow mathematical techniques (an earlier study [21] that exploits more conventional tools is not complete). The restriction to the class of representations encounted in QFT on one side simplifies the problem, since the set of relevant irreducible representations is indeed easier to handle, but it involves another complication: some non-decomposable representations of G appear in the field-theoretic framework and have a significance of their own which is not exhausted by the knowledge of their irreducible components.

An important step in the study of physically interesting representations was made by G. Mack [22] who classified all irreducible unitary ray representations of SU(2,2) with positive energy. We shall briefly review here the next step in this direction: the study of elementary induced representations containing a lowest weight subrepresentation [15,16].

The representation of G associated with local fields should act in a space of spin-tensor valued functions on Minkowski space. We are thus led to consider induced representations of $\tilde{G}$ with inducing subgroup $\tilde{H}$, the stability subgroup of a point, say $x = 0$, of Minkowski space. According to (1.5) the stability subgroup $H \subset G$ of the point $x = 0$ is the (11-parameter) subgroup of lower block triangular matrices. It is compounded of Lorentz (SL(2,C)) transformations, dilatations (A_1), and (4-parameter, nilpotent) special conformal transformations (N_4); $\tilde{H}$ is the direct product of its connected component of the identity H_o with the central subgroup Z:

$$\tilde{H} = Z \times H_o , \qquad H_o = N_4 \, A_1 \, SL(2,\mathbb{C}). \tag{1.8}$$

On the basis of (1.2)(1.3) the various subgroups of H_o are identified as follows:

$$N_4 = \left\{ h_c = \begin{pmatrix} 1 & 0 \\ i\tilde{c} & 1 \end{pmatrix}, \quad \tilde{c} \equiv c^0 - \underline{c}\,\underline{\sigma} = \tilde{c}^* \quad (c^\mu \in M) \right\} \tag{1.9a}$$

$$A_1 = \left\{ h_\alpha = \begin{pmatrix} e^{\alpha/2} & 0 \\ 0 & e^{-\alpha/2} \end{pmatrix}, \quad \alpha \in \mathbb{R} \right\}, \quad SL(2,\mathbb{C}) = \left\{ h_\Lambda = \begin{pmatrix} \Lambda & 0 \\ 0 & \Lambda^{*-1} \end{pmatrix}, \det\Lambda = 1 \right\}. \tag{1.9b}$$

We define the <u>local elementary representations</u> of $\tilde{G}$ as the representations of $\tilde{G}$ induced by irreducible finite dimensional representations of $\tilde{H}$. The latter are trivial on N_4 and are labelled by four numbers:

$$\chi = (d; j_1, j_2 ; \delta) \qquad 2j_{1,2} = 0, 1, \dots, \quad \delta \in \mathbb{R}. \tag{1.10}$$

Here δ and d give the characters of Z and A_1; (j_1, j_2) label the $(2j_1+1)(2j_2+1)$-dimensional representation of $SL(2,\mathbb{C})$ which can be realized in the space $\mathcal{H}_{j_1 j_2}$ of homogeneous polynomials $f(\kappa,\bar{\kappa})$ of degree $2j_1$ in $\kappa = (\kappa_A, A = 1,2)$ and $2j_2$ in $\bar{\kappa}$:

$$\left[D_\chi(\zeta_1^\nu h_c h_\alpha h_\Lambda) f\right](\kappa,\bar{\kappa}) = e^{-\alpha d - i\pi\nu\delta} f(\kappa\Lambda, \Lambda^* \bar{\kappa}). \tag{1.11}$$

There is a transparent explicit formula for the action of the corresponding induced representation T_χ of $\tilde{G}$, reminiscent of the familiar realization [23] of the elementary representations of SL(2,C). Using the block matrix notation (1.3) this time for g^{-1} (rather than for g) we can write $(\text{for } \delta = d + 2j_2 \ (\mathrm{mod}\, 2))$

$$\left[T_\chi(g) f\right](i\underset{\sim}{x}; \kappa, \bar{\kappa}) = \left[\det(b_3 i\underset{\sim}{x} + b_4)\right]^{-d-j_1-j_2} f(i\underset{\sim}{x}_g; \kappa(b_4^* - i\underset{\sim}{x} b_3^*), (b_4 + i b_3 \underset{\sim}{x})\bar{\kappa}),$$

with

$$i\underset{\sim}{x}_g = (b_1 i\underset{\sim}{x} + b_2)(b_3 i\underset{\sim}{x} + b_4)^{-1} \quad \text{for} \quad g^{-1} \in \begin{pmatrix} b_1 & b_2 \\ b_3 & b_4 \end{pmatrix} \in G_0$$
$$(\det(b_3 i\underset{\sim}{x} + b_4) \neq 0) \tag{1.12a}$$

$$\left[T_\chi(\zeta_1^\nu) f\right](i\underset{\sim}{x}; \kappa, \bar{\kappa}) = D_\chi(\zeta_1^\nu) f(i\underset{\sim}{x}; \kappa, \bar{\kappa}) = e^{-i\pi\nu\delta} f(i\underset{\sim}{x}; \kappa, \bar{\kappa}). \tag{1.12b}$$

A basis of infinitesimal operators of T_χ is given by the Poincaré generators

$$P_\mu (\equiv J_{\mu 6} - J_{\mu 5}) = -i\nabla_\mu \equiv -i \frac{\partial}{\partial x^\mu},$$

$$J_{\mu\nu} = L_{\mu\nu} + s_{\mu\nu}, \quad L_{\mu\nu} = i(x_\nu \nabla_\mu - x_\mu \nabla_\nu) \tag{1.13a}$$

$$s_{\mu\nu} = \frac{1}{2}\left(\kappa\, \sigma_{\mu\nu} \frac{\partial}{\partial \kappa} - \frac{\partial}{\partial \bar{\kappa}} \sigma_{\mu\nu}^* \bar{\kappa}\right), \quad \sigma_{\mu\nu} = \frac{i}{2}(\sigma_\mu \tilde{\sigma}_\nu - \sigma_\nu \tilde{\sigma}_\mu)$$

$$s_{\mu\nu} = \frac{1}{2}\left(\kappa\,\sigma_{\mu\nu}\frac{\partial}{\partial\kappa} - \frac{\partial}{\partial\bar{\kappa}}\,\sigma^*_{\mu\nu}\bar{\kappa}\right), \qquad \sigma_{\mu\nu} = \frac{i}{2}(\sigma_\mu\tilde{\sigma}_\nu - \sigma_\nu\tilde{\sigma}_\mu)$$

$$\sigma_{12} = i\sigma_{03} = \sigma_3 (= -i\sigma^*_{03}), \tag{1.13b}$$

and the generators of dilatation (J_{65}) and special conformal transformations

$$J_{65} = -i(d + x\nabla)$$

$$K_\mu (\equiv J_{\mu 6} + J_{\mu 5}) = i\{2x_\mu(d + x\nabla) - x^2\nabla_\mu\} + 2x^\lambda s_{\lambda\mu}. \tag{1.13c}$$

Since $P^0 > 0$ yields $K^0 = T_\chi(w)P^0T_\chi(w)^{-1} > 0$ (for w given by (1.7)) energy positivity implies positivity of the compact generator

$$J^0{}_6 = J_{60} = \frac{1}{2}(P^0 + K^0) =$$

$$= i\left\{x^0(d + x\nabla) + \frac{1+x^2}{2}\nabla_0 - \frac{1}{2}\left(\kappa\,\underline{x}\,\underline{\sigma}\frac{\partial}{\partial\kappa} + \frac{\partial}{\partial\bar{\kappa}}\underline{x}\,\underline{\sigma}\,\bar{\kappa}\right)\right\} \tag{1.14}$$

(Segal's "conformal Hamiltonian" [1]). According to the analysis of ref. [16] (which extends the result of Mack [22]) a necessary and sufficient condition for the existence of a subrepresentation of χ for which J_{60} is bounded below is

$$\delta = (d + 2j_2) (\mathrm{mod}\, 2) \qquad (d \text{ real}) \tag{1.15}$$

(then $\min J_{60} = d$ for the subrepresentation). Similarly, for a highest weight (negative energy) representation

$$\delta = (-d + 2j_1)(\mathrm{mod}\, 2) \qquad (\max J_{60} = -d). \tag{1.16}$$

1C. Sextets of nondecomposable elementary representations involving a finite-dimensional invariant subspace

Dual representations of $\tilde{G}$

$$\chi = (2 + c (= d); j_1, j_2; \delta) \quad \text{and} \quad \tilde{\chi} = (2 - c; j_2, j_1; \delta) \tag{1.17}$$

have the same Casimir invariants. For generic χ they are equivalent and topologically irreducible (with respect to an appropriate Fréchet space topology - see Sec. 2A of Ref. [16]. For δ given by (1.15) or (1.16) T_χ admits - as stated - an invariant subspace. If in addition $d+j_1+j_2$ is integer, then the structure of invariant subspaces and intertwining maps is more complex. We shall be particularly concerned with integer points for which

$$\max(2j_2+c,\ 2j_1-c) \geq 2 + \min\left(|j_1-j_2|,\ ||j_1-j_2|-1|\right),\quad c=d-2. \tag{1.18}$$

For each such point there are exactly six partially equivalent elementary representations with the same values of the (three) Casimir invariants which can be labelled in the way shown on Fig. 1

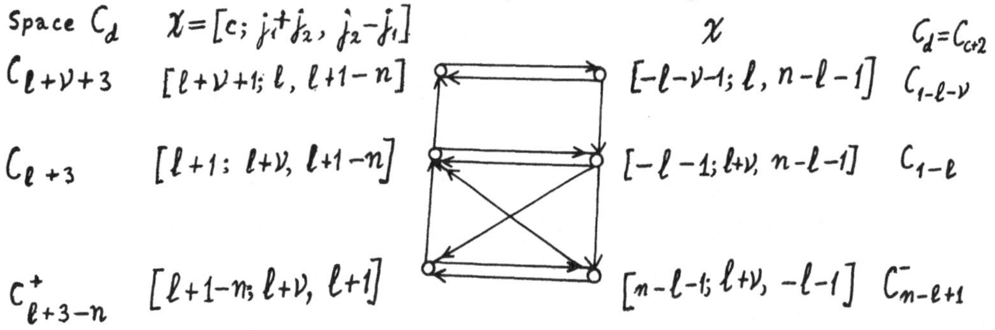

Fig. 1. Sextet of exceptional integer points. Arrows indicate intertwining maps.

Range of (ℓ,ν,n):

$$\ell=0,\tfrac{1}{2},1,\dots;\quad \nu=1,2,\dots;\quad n=1,2,\dots,2\ell+1. \tag{1.19}$$

Each such sextet involves precisely one finite dimensional representation (with given values of the Casimir invariants). It is the space $C_{1-\ell-\nu}$ that contains a $\frac{\nu n(\nu+n)}{12}(2\ell+2+\nu)(2\ell+2+\nu-n)(2\ell+2-n)$ - dimensional invariant subspace $E_{\ell\nu n}$ of polynomials $f(i\underset{\sim}{x};\kappa,\bar\kappa)$ satisfying the differential equation

$$(\kappa\underset{\sim}{\nabla}\bar\kappa)^{\nu} f(i\underset{\sim}{x};\kappa,\bar\kappa)=0. \tag{1.20}$$

Special cases of (1.20) are the Penrose twistor equation [24] obtained for $\nu = 1 = n, \ell = \frac{1}{2}$, and the conformal Killing equation that appears for $\nu = 1 =$, $n = 2$ (see for more detail Chapter 3 of ref. [16]).

The simplest of the sextets, the one with $\ell = 0$, $n = \nu = 1$, gives room for the electromagnetic potential, the Maxwell tensor and the current along with dimensionless scalar fields that can be used to define gauge transformations and scalar densities (of dimension 4).

Historically, elementary representations have been studied (by Gel'fand and Naimark and by Harish Chandra and his school - see for recent reviews [25, 26] as means to classify unitary irreducible representations of semisimple Lie groups. The appearance of nondecomposable elementary representations at some exceptional "integer points" came out as a nuisance from the point of view of this programme. It is remarkable that these exceptional elementary representations (as well as more general non-decomposable representations) appear in the physical applications with all their additional structure. Each point and every arrow on the electromagnetic sextet diagram has a physical meaning: horizontal arrows correspond to the Knapp-Stein intertwining maps [27] whose kernels are identified with appropriate 2-point Wightman functions (or with the Schwinger functions, if we replace G with the Euclidean conformal group Spin (5,1)); downward pointing vertical arrows give the expression $F_{\mu\nu} = \nabla_\mu A_\nu - \nabla_\nu A_\mu$ of the Maxwell field $F_{\mu\nu}$ in terms of the vector potential A_μ and a map of dimensionless scalar fields onto the invariant subspace of C_1 of longitudinal potentials; vertical arrows pointing upwards give rise to the Maxwell equation $\nabla_\nu F^{\mu\nu} = j^\mu$ and to the current conservation law.

We shall discuss more general non-decomposable representations of G in the context of a specific example of physical interest: conformal (massless) QED.

2. NONSINGULAR CONFORMAL INVARIANT QUANTUM ELECTRODYNAMICS

2A. Introductory remarks. The price for a conformal gauge fixing

The gauge invariant part of the Lagrangian of massless spinor electrodynamics

$$L_{inv} = \frac{1}{2} F^{\mu\nu}\left(\frac{1}{2}F_{\mu\nu} - \nabla_\mu A_\nu + \nabla_\nu A_\mu\right) - \bar{\Psi} \not{D} \Psi \tag{2.1}$$

where $\not{D}$ is the covariant derivative*

$$\not{D} = \gamma^\mu D_\mu, \quad [\gamma^\mu, \gamma^\nu]_+ = 2\eta^{\mu\nu}, \quad D_\mu = \nabla_\mu - ieA_\mu \tag{2.2}$$

is known to be conformal invariant (at least classically) since prehistoric times [2,28]. It is, however, singular, since $\frac{\partial L_{inv}}{\partial(\nabla_0 A_0)} = 0$, and the standard Lorentz invariant gauge fixing term ($\sim(\nabla A)^2$) violates conformal invariance. Standard canonical quantization does require a gauge fixing. The problem of finding a conformal invariant gauge condition has been solved for the free electromagnetic field in the mid 70's [29]. The case of interacting QED has only been attacked recently from this point of view [30, 31, 17].

The key to the solution advanced in [17] (which will be previewed in the following sections) lies in a systematic use of the manifestly covariant formalism [28, 11, 32]. Limiting our discussion to a canonical (local) Lagrangian picture we end up with a (nonsingular) theory in which the 4-potential A_μ is combined in a non-decomposable multiplet with a dimensionless scalar field $A_-(x)$. The 5-potential $(A_N(x)) = (A_\nu, A_-)$ is coupled to a conserved 5-current $(J^N(x)) = (J^\nu, J_+)$.

As noted in the introduction conformal invariance is, in general, destroyed by renormalization. The question arises: what is canonical conformal invariant QED good for? We mention two kinds of possible applications. One is perturbative and renormalization group calculations that do not involve dimensional parameters. Such an application - one loop renormalization and evaluation of anomalous dimensions of gauge invariant composite operators [18] - is reviewed in §4.

Another, would be the search for a "finite QED" corresponding to a renormalization group fix point [33,34,35]. A programme of this type (using a decomposable representation of G) is being pursued in [36,37]

---

* We are using the spacelike metric: $(\eta_{\mu\nu}) = \mathrm{diag}(-,+++)$.

(see also [38]). We note that an interacting theory with conformal invariant Green functions would require a change in the representation of the basic fields (including a non-linear dilatation law for the charged field) and the introduction of the gradient of a new scalar field (which transforms under a non-decomposable representation of the subgroup of dilatations - see Sec. 3).

2B. Manifestly covariant connection form. Conformal Lorentz condition

We use the projecture light-cone realization [28] of compactified Minkowski space:

$$\bar{M} = Q/\mathbb{R}^* \qquad (\mathbb{R}^* = \mathbb{R}\setminus\{0\}),$$

$$Q = C_{4,2} \equiv \{\xi = (\xi^0, \underline{\xi}, \xi^5, \xi^6) \in \mathbb{R}^6;\ \xi^a \eta_{ab} \xi^b \equiv \xi^\mu \eta_{\mu\nu} \xi^\nu + \xi_5^2 - \xi_6^2 = 0, \xi \neq 0\}. \tag{2.3}$$

Minkowski space M with coordinates $x = (x^\mu,\ \mu = 0,1,2,3)$ is imbedded in a dense open set of $\bar{M}$ whose complement is the "light cone at infinity" $\xi^5 + \xi^6 = 0 = \xi_\mu \xi^\mu$:

$$x^\mu = \frac{1}{\kappa}\xi^\mu \quad \text{for } \kappa = \xi^5 + \xi^6 = \xi_5 - \xi_6 \neq 0 \quad (\xi^6 - \xi^5 = \kappa x^2). \tag{2.4}$$

The conformal electromagnetic potential can be defined as a homogeneous 1-form $\mathcal{A}_a(\xi)d\xi^a$ on the 5-dimensional quadric Q; for points of $M(\subset \bar{M})$

$$\mathcal{A}_a(\xi)\, d\xi^a = A_\mu(x)\, dx^\mu + A_-(x)\frac{d\kappa}{\kappa} \qquad \text{for} \qquad \xi_a\, d\xi^a = 0. \tag{2.5}$$

The Minkowski space potentials $(A_M(x)) = (A_\mu(x), A_-(x))$ are expressed from (2.5) in terms of $\mathcal{A}_a(\xi)$ using $d\xi^\mu = \kappa dx^\mu + x^\mu d\kappa$, $d\xi^5 + d\xi^6 = d\kappa$, $d\xi^6 - d\xi^5 = d(\kappa x^2)$:

$$A_-(x) = \xi^a \mathcal{A}_a(\xi), \qquad A_\mu(x) = \kappa[\mathcal{A}_\mu(\xi) + (\mathcal{A}_6(\xi) - \mathcal{A}_5(\xi))x_\mu]. \tag{2.6}$$

The fields (2.6) as well as

$$A_+(x) = \kappa\left(\mathcal{A}_5(\xi) - \mathcal{A}_6(\xi)\right) \tag{2.7}$$

have the standard transformation law $U(a)A_M(x)U(a)^{-1} = A_M(x+a)$ under translations (unlike the $\mathcal{A}_a(\xi)$ which require an accompanying index transformation); moreover they reduce the representation of the 11-parameter Weyl subgroup Aut$\mathcal{P}$ of G (that is, the Poincaré subgroup, extended by dilatations). We have, in particular, the dilatation law

$$U(h_\alpha)\begin{pmatrix} A_+(x) \\ A_\mu(x) \\ A_-(x) \end{pmatrix} U(h_\alpha)^{-1} = \begin{pmatrix} e^{2\alpha} A_+(e^\alpha x) \\ e^\alpha A_\mu(e^\alpha x) \\ A_-(e^\alpha x) \end{pmatrix}. \tag{2.8}$$

Note that the curvature form corresponding to (2.5),

$$d(\mathcal{A}_a(\xi)\, d\xi^a) = \tfrac{1}{2} F_{\mu\nu}(x)\, dx^\mu \wedge dx^\nu + F_{\mu-}(x)\, dx^\mu \wedge \frac{d\kappa}{\kappa} \tag{2.9a}$$

where

$$F_{\mu\nu}(x) = \nabla_\mu A_\nu(x) - \nabla_\nu A_\mu(x), \quad F_{\mu-}(x) = \nabla_\mu A_-(x), \tag{2.9b}$$

also transforms under a nondecomposable representation of G (with $F_{\mu-}=0$ singling out an invariant subspace); however, it is not generated naturally from the manifestly covariant formalism.

Since the field $A_+(x)$ (2.7) (of dimension 2 in mass units) enters neither the connection form (2.5) nor the nondecomposable conformal Maxwell field (2.9) the question arises whether it cannot be expressed in terms of A_μ and A_-. There is a unique conformal invariant way to do so which consists in imposing the 6-dimensional Lorentz condition on A_N:

$$\kappa^2 \delta_a \mathcal{A}^a(\xi) = \delta_N A^N(x) \equiv 2A_+(x) + \nabla_\mu A^\mu(x) - \tfrac{1}{2}\Box A_-(x) = 0. \tag{2.10}$$

Here

$$\delta_a = \left(2 + \xi\frac{\partial}{\partial\xi}\right)\frac{\partial}{\partial\xi^a} - \frac{1}{2}\xi_a \Box_6 \tag{2.11}$$

is the interior derivative [39] on the quadric Q (2.3); it satisfies $\delta_a\xi^2 = \xi^2(\delta_a + 2\frac{\partial}{\partial\xi^a})$ so that $\delta_a[f(\xi) + \xi^2 f_1(\xi)]|_{\xi^2=0} = \delta_a f(\xi)|_{\xi^2=0}$ and $[\delta_a, \delta_b] = 0 = \delta_a\delta^a$.

For a more general manifestly covariant vector field $V^a(\xi)$ of degree of homogeneity -d, if we set

$$V_N(x) = \kappa^d\, \mathcal{J}_{Na}^{(x)}\mathcal{V}^a(\xi) \quad \text{where} \quad (\mathcal{J}_{Na}) = \begin{pmatrix} 0 & 1 & 1 \\ \eta & -x & -x \\ x & \frac{1-x^2}{2} & -\frac{1+x^2}{2} \end{pmatrix} \tag{2.12}$$

where $\eta = (\eta_{\mu\nu})$ is the Minkowski space metric tensor, we find the following extension of (2.10):

$$\kappa^{d+1}\delta_a \mathcal{V}^a(\xi) = \delta_N V^N(x) = (3-d)(2-d)V_+ + (2-d)\nabla V - \tfrac{1}{2}\Box V_- . \tag{2.13}$$

In particular, for d = 3 we have the following conformal extension of the current conservation law:

$$\kappa^4\delta_a \mathcal{J}^a(\xi) = \delta_N J^N(x) \equiv -\nabla J - \tfrac{1}{2}\Box J_- . \tag{2.14}$$

If $\mathcal{J}$ satisfies the conformal invariant subsidiary condition

$$\kappa^2\xi^a\mathcal{J}_a(\xi) \equiv J_-(x) = 0, \tag{2.15}$$

then (2.14) reduces to the standard current conservation law $\nabla J (= \nabla_\mu J^\mu) = 0$. Conditions (2.10) and (2.15) provide (as we shall see in the next section) a most economical option for writing down a non-singular conformal invariant action for an electromagnetic field interacting with a conserved (external) current.

2C. Photon propagator matrix and a conformal action integral

The expression for a conformal invariant 2-point function

depends crucially on the corresponding field transformation law. If A_μ transforms under the elementary representation $\chi_1 = (1;\frac{1}{2},\frac{1}{2};0)$ then its (Euclidean*) 2-point function is proportional to the longitudinal expression

$$\frac{1}{8\pi^2}\,\frac{r^\mu{}_\nu(x-y)}{(x-y)^2} = \int \frac{p^\mu p_\nu}{p^4}\, e^{ip(x-y)}\, d_4p = \frac{1}{(4\pi)^2}\nabla^\mu \nabla_\nu \log\left(\frac{x-y}{l}\right)^2 \tag{2.16a}$$

$$\left(d_4p = \frac{d^3\underline{p}\, dp_4}{(2\pi)^4}\,,\quad p_4 = ip_0 \text{ is real}\right)$$

where

$$r^\mu{}_\nu(x) = \delta^\mu{}_\nu - 2\,\frac{x^\mu x_\nu}{x^2} \qquad \left(r^\mu{}_\sigma r^\sigma{}_\nu = \delta^\mu{}_\nu\right). \tag{2.16b}$$

By contrast the 5-potential $(A_\mu(x), A_-(x))$ which transforms under the non-decomposable representation of G that leaves the 1-form (2.5) invariant has a O(5,1)-invariant Schwinger function

$$\left\langle \begin{pmatrix} A^\mu(x) \\ A_-(x)\end{pmatrix} (A_\nu(0), A_-(0))\right\rangle_E =$$

$$= \int d_4p\, e^{ipx} \begin{pmatrix} \frac{1}{p^2}\delta^\mu{}_\nu - (1-\beta)\frac{1}{p^4}p^\mu p_\nu & 2i\frac{p^\mu}{p^4} \\ -2i\frac{p_\nu}{p^4} & -2\pi^2\delta(p)\end{pmatrix} \tag{2.17}$$

with a non-trivial transverse part (which we have normalized to its canonical value in (2.17)).

The inverse p-space propagator

$$\mathcal{M}(p) = \begin{pmatrix} p^2\delta^\mu{}_\nu - p^\mu p_\nu & \frac{i}{2}p^\mu p^2 \\ -\frac{i}{2}p_\nu p^2 & -\frac{1}{4}\beta p^4\end{pmatrix} \tag{2.18}$$

---

* The energy positivity condition (which, as we saw, restricts the class of physically admissible representations of $\tilde{G}$) allows to continue analytically quantum field theoretic Green functions to the Euclidean "Schwinger functions" (with pure imaginary times). In a $\tilde{G}$-invariant QFT the Schwinger functions are invariant with respect to the Euclidean conformal group Spin (5,1).

gives rise to the conformal invariant action integral

$$I(A,J) = \int \left(L_A(x) + A_\mu(x) J^\mu(x) + A_-(x) J_+(x) \right) d^4x \tag{2.19}$$

where L_A can be reduced to the following canonical form

$$\begin{aligned} L_A(x) = \tfrac{1}{2} F^{\mu\nu} \left(\tfrac{1}{2} F_{\mu\nu} - \nabla_\mu A_\nu + \nabla_\nu A_\mu \right) - \\ - \tfrac{1}{2} \nabla_\mu A^\mu \Box A_- + \tfrac{1}{8} \beta (\Box A_-)^2 \end{aligned} \tag{2.20a}$$

that differs by a 4-divergence from

$$L_A^{\mathcal{M}}(x) = -\tfrac{1}{2} \left(A_\mu(x), A_-(x) \right) \mathcal{M}(-i\nabla) \begin{pmatrix} A^\nu(x) \\ A_-(x) \end{pmatrix} . \tag{2.20b}$$

(We omit the normal product sign which is assumed to be there in the quantum case.) A necessary and sufficient condition for the invariance of the action (2.19) under the restricted gauge transformations

$$\begin{aligned} A_\mu(x) \to A_\mu(x) + \nabla_\mu S(x), \quad A_-(x) \to A_-(x) + q \\ \text{where} \qquad \Box^2 S(x) = 0 = \nabla_\mu q \end{aligned} \tag{2.21}$$

is the current conservation law

$$\nabla_\mu J^\mu(x) = 0 \tag{2.22a}$$

supplemented by the conformal invariant integral property

$$\int J_+(x) d^4x = 0. \tag{2.22b}$$

The equations of motion

$$\mathcal{M}(-i\nabla) \begin{pmatrix} A^\nu \\ A_- \end{pmatrix} = \begin{pmatrix} J^\mu \\ J_+ \end{pmatrix} \tag{2.23}$$

are actually independent of the gauge parameter β. Indeed, the first line, the conformal Maxwell equation

$$\begin{aligned} \nabla^\mu \nabla A - \Box A^\mu - \tfrac{1}{2} \nabla^\mu \Box A_- \equiv \\ \equiv \nabla_\nu F^{\mu\nu} - \tfrac{1}{2} \Box F^\mu{}_- = J^\mu \end{aligned} \tag{2.24}$$

implies (because of current conservation) the free field equation

$$\Box^2 A_- = 0 \tag{2.25}$$

for A_-; then the second line of (2.23) can be written in the form

$$\frac{1}{2}\Box \nabla A = J_+ . \tag{2.26}$$

Clearly, Eqs. (2.23) are invariant under the gauge transformations (2.21).

3. CONFORMAL QED OF AN ELECTRON-POSITRON FIELD INTERACTING WITH A 5-POTENTIAL

3A. Implications of the Ward Identity

Up to this point $J^N(x)$ has been a given (external) conserved current. We shall now consider the possibility that it is built out of a charged spinor field.

The peculiarities of an interacting conformal charged field are indicated by the following statement.

Proposition 1. The conventional transformation law

$$U(h_\alpha)\, \psi(x)\, U(h_\alpha)^{-1} = e^{\alpha d}\, \psi(e^\alpha x) \qquad \left(d \geq d_{can} = \tfrac{3}{2}\right) \tag{3.1}$$

for a Dirac field $\psi(x)$ under dilatation and the conformal Maxwell equation (2.24) contradict the standard Ward identity*

---

* The Minkowski space picture used in this section appears more straightforward to us when spinor fields (and/or "operator valued dimensions" as in (3.4)) are involved. The Ward identity in this picture is a direct consequence of the commutation relations between a charged density and a pair of oppositely charged fields:

$$\delta(x^0)\left[\psi(x), J^0(0)\right] = e\, \delta(x)\, \psi(x), \qquad \delta(x^0)\left[\bar\psi(x), J^0(0)\right] = -e\, \delta(x)\, \bar\psi(x)$$

The rules for going from Minkowski (M-) space to Euclidean (E-) space (if needed) include: $-(i d_4 p)_M = (d_4 p)_E$; $\gamma^0 p_0 = \gamma_4 p_4$ ($p_4 = i p_0$ real, $\gamma_4 = i\gamma_0 = -i\gamma^0 = \gamma_4^*$).

$$\nabla_{3\mu} G^{\mu}(x_1, x_2; x_3) \equiv \frac{\partial}{\partial x_3^{\mu}} \langle T\, \psi(x_1) \overline{\psi}(x_2)\, J^{\mu}(x_3) \rangle_0 =$$

$$= e \langle T\, \psi(x_1) \overline{\psi}(x_2) \rangle_0 \left[\delta(x_1 - x_3) - \delta(x_2 - x_3) \right] \tag{3.2}$$

for e $\neq$ 0. Eq. (2.25) is, however, consistent with the 3-point function

$$\langle T\, \psi(x_1) \overline{\psi}(x_2) A_-(x_3) \rangle_0 =$$

$$= \frac{ie}{8\pi^2} \langle T\, \psi(x_1) \overline{\psi}(x_2) \rangle_0 \left[\log \frac{(x_2 - x_3)^2 + i0}{(x_1 - x_3)^2 + i0} + C \right] \tag{3.3}$$

which is invariant (for any value of C) under the more general "operator law of dilatation" (cf. [30])

$$U(h_\alpha)\, \psi(x)\, U(h_\alpha)^{-1} = e^{\frac{3}{2}\alpha}\, e^{i\alpha d^{(+)}}\, \psi(e^{\alpha} x)\, e^{i\alpha d^{(-)}} \tag{3.4a}$$

where $d^{(\pm)}$ are constant creation and annihilation operators,

$$d^{(-)}|0\rangle = 0 = \langle 0| d^{(+)}, \qquad d^{(+)} = d^{(-)*}, \tag{3.4b}$$

satisfying

$$[d^{(-)}, d^{(+)}] = 0,\ [d^{(\pm)}, A_-(x)] = ie\, q^{(\mp)},\ [d^{(+)} + d^{(-)}, A_N(x)] = ie\,\delta_N S, \tag{3.5}$$

$$[d^{(\pm)}, \psi(x)] = \mp i b\, \psi(x), \qquad [d^{(\pm)}, \overline{\psi}(x)] = \pm i b\, \overline{\psi}(x). \tag{3.6}$$

Here S is an associate homogeneous function on the quadric Q (2.3)

$$(\mathcal{S}(\xi) =)\ S(x, \kappa) = S(x) + q \log \kappa, \quad q = q^{(+)} + q^{(-)},$$
$$\nabla_\mu q = 0 \tag{3.7}$$

$$(\delta_N S(x, \kappa)) = \begin{pmatrix} (1 + \kappa \frac{\partial}{\partial \kappa})\, \partial_\nu \\ (1 + \kappa \frac{\partial}{\partial \kappa}) \frac{\partial}{\partial \kappa} \end{pmatrix} S(x, \kappa) = \begin{pmatrix} \partial_\nu S(x) \\ q \end{pmatrix}. \tag{3.8}$$

Finally, it is required that $[d^{(\pm)}, \delta_N S] = ie\,\delta_N A_-^{(\mp)}$ while

$$\langle \psi(x_1)\bar\psi(x_2)\, q^{(+)}\rangle_0 = \langle q^{(-)}\, \psi(x_1)\bar\psi(x_2)\rangle_0 = \frac{1}{8\pi^2}\langle \psi(x_1)\bar\psi(x_2)\rangle_0 \; ; \tag{3.9a}$$

on the other hand, the 3-point function of $\nabla_\mu S$ with ψ and $\bar\psi$ is

$$\langle \psi(x_1)\bar\psi(x_2)\nabla_\mu S^{(+)}(x^3)\rangle_0 =$$

$$= \frac{1}{(4\pi)^2}\langle \psi(x_1)\bar\psi(x_2)\rangle_0 \nabla_{3\mu}\Bigg[\log \frac{(x_{13}^2 + i0x_{13}^0)(x_{23}^2 + i0x_{23}^0)}{\ell^4} +$$

$$+ e^2 \log^2 \frac{x_{23}^2 + i0x_{23}^0}{x_{13}^2 + i0x_{13}^0} + C \log \frac{x_{23}^2 + i0x_{23}^0}{x_{13}^2 + i0x_{13}^0}\Bigg] .$$

$$(x_{ik} = x_i - x_k) \tag{3.9b}$$

Remark. A field S(x) with a non-homogeneous dilatation (and special conformal transformation) law that follows from the representation (3.7) has been introduced by Salam and Strathdee [40] in the context of non-linear realizations of the conformal group. Non-decomposable representations of dilatations have been studied systematically by Dell'Antonio [41]; they have been incorporated into non-decomposable representations of the conformal group (associated with the manifestly covariant formalism) by Ferrara et al. [42].

Sketch of proof. (1) The term $\nabla_\nu F^{\mu\nu}$ in the Maxwell equation (2.24) gives rise to a purely transverse part of G^μ (since we are using the covariant /Wick/ time-ordered product in which the derivatives can be taken outside the vacuum expectation value); therefore, it does not contribute to the Ward identity (3.2).

(2) A straightforward application of the analysis of Sec. IV2-B of ref. [11] shows that the general conformal invariant 3-point function $\langle T\psi(x_1)\bar\psi(x_2)\, A_-(x_3)\rangle_0$ is a constant multiple of the electron 2-point function and hence, $-\frac{1}{2}\Box\nabla^\mu A_-$ also gives a vanishing contribution to the Ward identity.

(3) It is readily verified that applying $-\frac{1}{2}\Box_3^2$ to (3.3) reproduces the right-hand side of the Ward identity (3.2).

(4) To verify that (3.3) is invariant under the dilatation law (3.4) and the corresponding infinitesimal conformal transformation

law

$$[\psi(x), K_\mu] = i\{2x_\mu(x\nabla + \tfrac{3}{2}) - x^2\nabla_\mu + \tfrac{1}{2}[\gamma_\mu, \not{x}]\}\psi(x) - \\ - 2x_\mu\left(d^{(+)}\psi(x) + \psi(x)d^{(-)}\right) \tag{3.10}$$

(cf. (1.13c)), and moreover, is determined up to a constant factor from this invariance property, we first consider the corresponding Wightman function and then use the rule $\log(x^2+iox^0) \to \log(x^2+io)$ for passing from a Wightman to a time ordered Green function.

Remark. We note by passing that the 2-point function of the Dirac field is found in this way to be

$$w_{\psi\bar{\psi}}(x) = \langle \psi(x)\bar{\psi}(0)\rangle_0 = -Z_\psi^{(b)} \frac{\not{x}}{(x^2+i0x^0)^2}\left(\frac{\ell^2}{x^2+i0x^0}\right)^b \tag{3.11}$$

in other words, the operator dilatation law (3.4) produces an effective anomalous dimension.

3B. Construction of the conformal Lagrangian in the local gauge b = 0

The (real) constant b in (3.6) and (3.11) plays the role of a gauge parameter (unlike the coefficient in the right-hand side of (3.5) for S normalized by (3.9), b is not determined from the Ward identity). There is a distinguished choice of gauge, b = 0, for which (3.11) (with $Z_\psi(b=0) = 1$) reduces to the free 2-point Wightman function for a massless spin ½ field. One can speculate that such a choice corresponds to the generalized Landau gauge studied by several authors [43]:

$$b = 0 \quad \leftrightarrow \quad \beta = -\frac{3e^2}{2(4\pi)^2} + \frac{15e^4}{16(4\pi)^4} + \cdots \; . \tag{3.12}$$

It turns out that the field S introduced in (3.5) in order to save the Ward identity also appears in the correct conformal generalization of the Dirac equation for a field ψ obeying the operator homogeneity relation (3.4). We shall first indicate why the naive approach to the interaction problem goes wrong and then will write the correct spinor part of the Lagrangian refering for a derivation to the original paper [17].

The manifestly covariant formalism gives rise to a conventional free Dirac equation and to a conserved 5-current if we start with an 8-component conformal spinor field $\psi(\xi)$ that is a homogeneous function of degree -2 in ξ (cf. [28,32]) and set

$$\mathcal{L}_{\psi A}(\xi) = \mathcal{L}_{\psi}(\xi) + \mathcal{L}_I(\xi), \tag{3.13}$$

where $\mathcal{L}_\psi$ is the manifestly covariant counterpart of the free Dirac Lagrangian

$$L_\psi(x) = \kappa^4 \mathcal{L}_\psi(\xi) = -\frac{\kappa^4}{2}\overline{\Psi}(\xi)\left(\Gamma\xi\,\Gamma\frac{\partial}{\partial\xi} + \Gamma\frac{\overleftarrow{\partial}}{\partial\xi}\Gamma\xi\right)\Psi(\xi) \tag{3.14a}$$

$$= -\frac{1}{2}\,\overline{\Psi}_-(x)\,\gamma^\mu \overleftrightarrow{\nabla}_\mu\,\Psi_-(x) \tag{3.14b}$$

while $\mathcal{L}_I$ is related to the interaction Lagrangian

$$L_I(x) = \kappa^4 \mathcal{L}_I(\xi) = \kappa^4 \frac{ie}{2}\,\overline{\Psi}(\xi)\left[\Gamma\xi, \Gamma A(\xi)\right]\Psi(\xi) \tag{3.15a}$$

$$= A_N(x)\,J^N(x), \quad \text{where } J^\nu(x) = ie\,\overline{\Psi}_-(x)\gamma^\nu\Psi_-(x), \quad J_+(x) = e\left(\overline{\Psi}_-(x)\,\Psi_+(x) + \overline{\Psi}_+(x)\,\Psi_-(x)\right). \tag{3.15b}$$

Here Γ_a are the O(4,2)-Clifford units which have the following tensor product realization

$$\Gamma_\mu = \tau_3 \otimes \gamma_\mu, \quad \Gamma_5 = \tau_2 \otimes \mathbb{1}, \quad \Gamma_6 = -i\tau_1 \otimes \mathbb{1} \quad (\Gamma_7 = \tau_3 \otimes \gamma_5). \tag{3.16}$$

The Dirac conjugate $\overline{\Psi}$ is defined by

$$\overline{\Psi}(\xi) = \Psi^* \mathcal{B} \quad \text{where } \Gamma_a^* \mathcal{B} = \mathcal{B}\Gamma_a, \ \mathcal{B} = \mathcal{B}^*. \tag{3.17a}$$

In the basis (3.16) we can set

$$\mathcal{B} = i\Gamma_0\Gamma_6 = -\tau_2 \otimes \beta, \quad \beta = i\gamma^0 \quad (\overline{\psi}(x) = \psi^*(x)\beta, \ \gamma_\mu^*\beta = -\beta\gamma_\mu). \tag{3.17b}$$

The relation between $\Psi(\xi)$ and $\psi_\pm(x)$ is

$$\begin{pmatrix} \psi_+(x) \\ \psi_-(x) \end{pmatrix} = \kappa^2 \mathcal{V}(-x)\, \Psi(\xi), \quad \mathcal{V}(-x) = \begin{pmatrix} 1 & 0 \\ i\not{x} & 1 \end{pmatrix} \tag{3.18a}$$

$$\kappa^2 \overline{\Psi}(\xi)\, \mathcal{V}(x) = -i\left(\overline{\psi}_-(x), -\overline{\psi}_+(x)\right). \tag{3.18b}$$

(Here $\mathcal{V}(a) = \exp\{a^\mu(\Gamma_{\mu 6} - \Gamma_{\mu 5})\} = 1 + \frac{\Gamma_6 - \Gamma_5}{2}\Gamma_\mu a^\mu$ is the 8-dimensional representation of the subgroup of translations, $\Gamma_{ab} = \frac{1}{4}[\Gamma_b, \Gamma_a]$.)

The Lagrangian (3.15) is not satisfactory for at least two reasons. First, the field ψ_+ has no kinetic part; if we regard it as an independent field and vary with respect to it, we would find $e\underline{A}(x)\psi_-=0$ which is, clearly, undesirable. Secondly, the resulting Dirac equation (obtained by varying in $\bar{\psi}_-$) is not conformal invariant if $\Psi(\xi)$ satisfies the operator homogeneity property

$$\rho^2 \Psi(\rho\xi) = \rho^{id^{(+)}}\, \Psi(\xi)\, \rho^{id^{(-)}}, \qquad [d^{(\pm)}, \Psi(\xi)] = 0 \tag{3.19}$$

that is the extension of (3.4) to the manifestly covariant picture.

It is demonstrated in [17] that both defects are cured if we impose the constraint

$$q\psi_+(x) + \tfrac{i}{2}\not{\nabla} S(x)\psi_-(x) = 0\,. \tag{3.20}$$

To summarize: conformal invariance of interacting QED with a non-singular photon propagator requires a considerable extension of the conventional formalism. Not only do we use a non-decomposable representation for the electromagnetic potential (which thus acquires a dimensionless fifth component A_-); the reconciliation of conformal invariance with a non-trivial Ward identity leads (according to our Proposition 1) to a non-linear (operator) dilatation law (3.4) for the charged field and to the introduction (through (3.5)) of yet another dimensionless scalar field S which then also appears in the constraint (3.20). I believe that one should not be afraid of these complications. The point is that dilatations (and special conformal transformations) are inter-related with renormalization and their proper account requires a deeper understanding of the interacting theory.

4. CONFORMAL COMPOSITE OPERATORS

Gauge dependent local fields do not correspond to physical particles. This is particularly clear for non-abelian gauge theories like QCD. A natural way to realize the idea that hadrons are made out of quarks and gluons is to associate to each observable (particle) state a local field operator that is a product of the underlying (quark and gluon) fields. More generally, a set of gauge invariant ("colourless") composite local tensor field operators appears in Wilson (short distance or light cone) OPE used in studying asymptotic properties of hadron form factors (for large transverse momenta).

Two problems immediately come to mind when writing down composite local field operators.

The first question arises even at the level of free constituent fields; it is concerned with the appropriate choice of such operators. Is there a distinguished basis of composite fields, especially when arbitrary derivatives of the constituent fields are involved?

The second problem is to give a precise meaning of local products of interacting fields; in other words, to find the renormalized expressions for such products and to evaluate their anomalous dimensions.

It turns out that conformal invariance provides a clue to the solution of both problems.

In order to address the first question we consider for a moment the theory of a free massless Dirac field ψ . It is completely integrable and therefore possesses an infinite set of conservation laws. They are generated by a series of conserved tensor currents built as products of derivatives of ψ and $\bar{\psi}$. Such currents indeed form a distinguished set of composite (neutral) tensor fields.

One can find the infinite set of conserved currents explicitly by replacing symmetric traceless tensors with homogeneous polynomials of a light-like vector z and using the techniques of ref. [39]. We set

$$O_\ell(x,z) = :\bar{\psi}(x)\, i\gamma z\, D_{\ell-1}(iz\overleftarrow{\nabla}, iz\overrightarrow{\nabla})\, \psi(x):, \quad \nabla = \frac{\partial}{\partial x}, \quad z^2 = 0 \tag{4.1}$$

where $D_n(\alpha,\beta)$ is a homogeneous polynomial of α and β of degree n. (The correspondence between such polynomials and the symmetric traceless tensors $O_\ell^{\mu_1\ldots\mu_\ell}(x)$ given by $O_\ell(x,z) = O_\ell(x)^{\mu_1\ldots\mu_\ell} z_{\mu_1}\ldots z_{\mu_\ell}$ is one-to-one). The (local) conservation law

$$\nabla_\mu O_\ell^{\mu\mu_2\ldots\mu_\ell}(x) = 0 \tag{4.2}$$

is equivalent to the differential equation

$$\nabla\delta\, O_\ell(x,z) \equiv \left[(1+z\partial)\nabla\partial - \tfrac{1}{2} z\nabla\, \partial^2\right] O_\ell(x,z) = 0 \tag{4.3}$$

where $\partial_\mu \equiv \frac{\partial}{\partial z^\mu}$, $\partial^2 = \partial_\mu \partial^\mu$ (δ_μ is the interior derivative [39] on the light cone $C_{3,1}$ in 4 dimensions - cf. (2.11)). Eq. (4.3) leads to an ordinary differential equation for the homogeneous polynomial D_n. Its solution is expressed in terms of a Gegenhauer polynomial

$$D_n(\alpha,\beta) = N_n(\alpha+\beta)^n\, C_n^{3/2}\left(\frac{\beta-\alpha}{\alpha+\beta}\right). \tag{4.4}$$

The first two fields O_1 and O_2 reproduce the electromagnetic current and the free stress energy tensor.

It is remarkable that the fields $O_\ell(x,z)$ are conformally covariant and are in fact determined (up to normalization) from the infinitesimal transformation law

$$\left[O_\ell(x,z), K_\mu\right]_{x=0} = 0 \tag{4.5}$$

where K_μ is the sum of the special conformal generators acting on ψ and $\bar{\psi}$. (The law (4.5) is characteristic for elementary representations of $\tilde{G}$.)

Going to the case of interacting fields and to evaluation of anomalous dimensions of the composite fields we shall just sketch the results; the interested reader can find more details and further references) in [18].

Let the field ψ interact with a gauge field. We shall use for the sake of simplicity the language of QED although the physically interesting applications appear in QCD (where short distance OPE, are believed to be relevant for deep inelastic scattering experiments).

The change of a composite operator O_ℓ due to the interaction is given (in terms of the scattering operator S) by

$$\delta O_\ell(x,z) = T(O_\ell(x,z)S)S^{-1} - O_\ell(x,z). \tag{4.6}$$

In 1-loop approximation δO_ℓ is divergent and can be assigned a regularized finite value by means of dimensional regularization (seeting the space-time dimension which appears as a parameter in the Schwinger-Feynman α-integrals equal to $4-2\varepsilon$). The $\frac{1}{\varepsilon}$ term in the expansion of δO_ℓ can, in principle, involve total derivatives of O_n with $n < \ell$, as well as fields with the same transformation properties built out of a photon field:

$$\delta O_\ell^e(x,z) = \frac{1}{\varepsilon}\sum_{n\leq\ell}(z\nabla)^{\ell-n}\left[Z_{\ell n}^{ee}\, O_n^e(x,z) + Z_{\ell n}^{e\gamma} O_n^\gamma(x,z)\right]. \tag{4.7}$$

(the superscripts e and γ refer to electron and photon constituents). It was discovered by Efremov and Raduyshkin [44] that no total derivative appears in (4.7): the choice of conformal covariant operators O_ℓ reduces a (potentially) 2ℓ dimensional mixing problem for the anomalous dimension matrix Z to a 2-dimensional one, which is exactly soluble.

The above constructions apply, in principle, to every conformal QFT. They are, however, particularly useful for asymptotically free theories like QCD where renormalization group improved perturbative calculations are reliable in the short distance limit.

Acknowledgements

It is a pleasure to thank Professor P. Budinich for his hospitality at the International School for Advanced Studies in Trieste where the text of this lecture was prepared. I would like to thank Professor G.C. Ghirardi and the Local Organizing Committee of the XII International Colloquium on Group Theoretical Methods in Physics for their kind hospitality at the Colloquium.

The present talk reviews recent work with Drs. V. Petkova, N. Craigie and V. Dobrev and current work with P. Furlan, V. Petkova and G. Sotkov. They all have, in various degrees, contributed to my understanding of the subject matter of the talk. I would also like to thank G. Sotkov for his part in the preparation of these notes.

REFERENCES

[1] M. Flato, D. Sternheimer, Comptes Rendus Acad. Sci., Paris 263A, 935 (1966);
I.E. Segal, Bull. Am. Math. Soc. 77, 958 (1971).

[2] E. Cunningham, Proc. London Math. Soc. 8, 77 (1909);
H. Bateman, Proc. London Math. Soc. 8, 223 (1909).

[3] G. Mack, Abdus Salam, Ann. Phys. (N.Y.) 53, 174 (1969);
M. Flato, J. Simon, D. Sternheimer, Ann. Phys. (N.Y.) 61, 78(1970);
D.J. Gross, J. Wess, Phys. Rev. D2, 753 (1970).

[4] C. Callan, Jr. Phys Rev. D2, 1541 51970);
K. Symanzik, Commun. Math. Phys. 18, 227 (1970);
For a modern review see D.J. Gross, Applications of the renormalization group to high-energy physics, in: Methods in Field Theory Les Houches, 1975, Eds. R. Balian and J. Zinn-Justin (North Holland, Amsterdam 1976) pp. 141-250.

[5] A.M. Polyakov, Zh. Eksp. Teor. Fiz., Pis. Red. 12, 538 (1970) (Engl. transl.: JETP Lett. 12, 381 (1970));
A.A. Migdal, Phys. Letters 37B, 386 (1971);
G. Parisi, L. Peliti, Lett. Nuovo Cim. 2, 627 (1971).

[6] G. Mack, I.T. Todorov, Phys. Rev. D8, 1764 (1973).

[7] G. Mack, Group theoretical approach in conformal invariant quantum field theory, in: Renormalization and Invariance in Quantum Field Theory, Ed. E.R. Caianiello (Plenum Press, N.Y. 1974) pp. 123-157;
S. Ferrara, R. Gatto, A. Grillo, G. Parisi, General consequences of conformal algebra (see also earlier work of Ferrara et al. cited there) and G. Mack, Conformal invariant quantum field theory, in: Scale and Conformal Symmetry in Hadron Physics, ed. R. Gatto (Wiley, N.Y. 1973) pp. 59-108 and 109-130.

[8] A.M. Polyakov, Zh. Eksp. Teor. Fiz. 66, 23 (1974) (Engl. Transl. JETP 39, 10 (1974)).

[9] V. Dobrev, V. Petkova, S. Petrova, I. Todorov, Phys. Rev. D13, 887 (1976).

[10] V. Dobrev, G. Mack, V. Petkova, S. Petrova, I. Todorov, Harmonic Analysis on the n-Dimensional Lorentz Group and its Application to Conformal Quantum Field Theory, Lecture Notes in Physics, 63 (Springer, Berlin 1977).

[11] I.T. Todorov, M.C. Mintchev, V. B. Petkova, Conformal Invariance in Quantum Field Theory (Scuola Normale Superiore, Pisa 1978) (this book contains a comprehensive bibliography).

[12] E.S. Fradkin, M. Ya. Palchik, Phys. Reports 44C, 249 (1978).

[13] K. Wilson, Phys. Rev. 179, 1499 (1969);
W. Zimmermann, Local operator products and renormalization in quantum field theory, Lectures on Elementary Particles and Quantum Field Theory. Vol. 1 (MIT Press, Cambridge 1970) pp. 359-589.

[14] R.A. Brandt, G. Preparata, Nucl. Phys. B27, 541 (1971);
S.A. Anikin, O.I. Zavialov, Ann. Phys. (N.Y.) 116, 135 (1978).

[15] V.B. Petkova, G.M. Sotkov, Bulg. J. Phys. 10, 144 (1983); The six-point families of exceptional representations of the conformal group, ISAS preprint 15/83/E.P., Trieste (1983).

[16] V.B. Petkova, I.T. Todorov, Local field representations of the conformal group and their physical interpretation, ISAS preprint 14/83/E.P., Trieste (1983).

[17] P. Furlan, V.B. Petkova, G.M. Sotkov, I.T. Todorov, Conformal quantum electrodynamics with a 5-potential, ISAS preprint 52/83/E.P., Trieste (1983).

[18] N.S. Craigie, V.K. Dobrev, I.T. Todorov, Conformally covariant composite operators in quantum chromodynamics, ICTP preprint IC/83/35, Trieste (1983).

[19] A. Uhlmann, Acta Phys. Pol. 24, 295 (1963);
W. Rühl, B.C. Yunn, Fortschr.d. Physik, 25, 83 (1977).

[20] A.W. Knapp, B. Speh, J. Func. Anal. 45, 41 (1982);
E. Angelopoulos, Lett. Math. Phys. 7, 121 (1983) and Commun. Math. Phys. 89, 41 (1983)

[21] T. Yao, J. Math. Phys. 8, 1931 (1967) and 9, 1615 (1968).

[22] G. Mack, Commun. Math. Phys. 55, 1 (1977).

[23] I.M. Gel'fand, M.I. Graev, N. Ya. Vilenkin, Generalized Functions, Vol 5 (Academic Press, N.Y. 1966).

[24] E. Cartan, Leçons sur la théories des spineurs (Hermann, Paris 1937);
R. Penrose, J. Math. Phys. 8, 345 (1967).

[25] A.W. Knapp, G. Zuckerman, Classification theorems for representations of semi-simple Lie groups, in: Non-Commutative Harmonic Analysis, Lecture Notes in Mathematics 587 (Springer, Berlin 1977) pp. 138-159;
A.W. Knapp, B. Speh, States of classification of irreducible unitary representations, in: Harmonic Analysis, Lecture Notes in Mathematics 808 (Springer, Berlin 1982) pp. 1-38.

[26] J. Rawnsley, W. Schmid, J.A. Wolf, Singular unitary representations and indefinite harmonic theory, preprint (November 1981).

[27] A.W. Knapp, E.M. Stein, Ann. Math. 93, 489 (1971).

[28] P.A.M. Dirac, Ann. Math. 37, 429 (1936).

[29] D.H. Mayer, J. Math. Phys. 16, 884 (1975);
F. Bayen, M. Flato, J. Math. Phys. 17, 1112 (1976).

[30] G. Sotkov, D. Stoyanov, J. Phys. A13, 2807 (1980); -,-, A conformally invariant gauge fixing condition in quantum electrodynamics, JINR Report P2-81-665, Dubna (1981); J. Phys. A16 (1983).

[31] B. Binegar, C. Fronsdal, W. Heidenreich, Conformal QED, J. Math. Phys. 24, 2828 (1983).
R.P. Zaikov, On conformal invariance in gauge theories: quantum electrodynamics, JINR Report E2-83-28, Dubna (1983).

[32] P. Budinich, P. Furlan ., Nuovo Cim. 70A, 243 (1982) and ISAS preprint 49/82/E.P., Trieste 1982;
P. Furlan , Nuovo Cim. 71A, 43 1982).

[33] S. Adler, Phys. Rev. D5, 3021 (1972) and D7, 1948 (1973); ibid. D6 3445 (1972) and D7, 3821 (1973).

[34] M. Baker, K. Johnson, Physica 96A, 120 (1979).

[35] M.C. Mintchev, V.B. Petkova, I.T. Todorov, Reports on Math. Phys. 9, 355 (1976).

[36] E.S. Fradkin, A.A. Kozhevnikov, M. Ya.Palchik, A.A. Pomeransky, Maxwell equations in conformal invariant electrodynamics, Commun. Math. Phys. 91, 529 (1983).

[37] A.A. Kozhevnikov, M. Ya. Palchik, A.A.Pomeransky, Jad. Fiz. 37, 481 (1983); M. Ya. Palchik, J. Phys. A16, 1523 (1983).

[38] I.T. Todorov, Conformal invariance in (gauge) quantum field theory, in: Mathematical Problems in Theoretical Physics, Lecture Notes in Physics 153 (Springer, Berlin 1982)pp. 319-323.

[39] V. Bargmann, I.T. Todorov, J. Math. Phys. 18, 1141 (1977).

[40] Abdus Salam, J. Strathdee, Phys. Rev. 184, 1760 (1969);
C.J. Isham, Abdus Salam, J. Strathdee, Ann. Phys. (N.Y.) 62, 98 (1971).

[41] G.F. Dell'Antonio, Nuovo Cim. 12A, 756 (1972).

[42] S. Ferrara, R. Gatto, A. Grillo, Phys. Letters 42B, 264 (1972).

[43] S. Adler, W.A. Bardeen, Phys. Rev. D4, 3045 (1971) and D6, 734 (1972);
F. Englert, Nuovo Cim. 16A, 557 (1973);
M. Fry, Nuovo Cim. 31A, 129 (1976).

[44] A.V. Efremov, A.V. Radyushkin, Theor. Math. Phys. (transl.) 42, 97 (1980).

ON THE NECESSITY OF BREAKING COLOUR $SU_C(3)$ SYMMETRY

J. Werle

International School for Advanced Studies

and

International Centre for Theoretical Physics, Trieste, Italy.

ABSTRACT

A class of quark field models of hadrons is discussed. It is assumed that the quarks are the only carriers of colours and that the forces between the quarks are described by local or nonlocal interaction terms. It is shown that in order to have colour confinement - and in particular inseparability of colours- one must break $SU_c(3)$ symmetry. The proposed way of this breaking maintains rigorous conservation laws only for observable white currents and implies manifest inseparability of colours.

The problem of quark confinement is still far from being completely solved and well understood. It certainly involves two basic and in general quite independent conditions of:spatial localization and inseparability of the constituent fields. Inseparability means that none of the constituent fields can appear isolated from all the others but only certain combinations of several fields are realized in nature . Localization means that these combinations have the form of clusters of the typical hadronic sizes $\sim 10^{-13}$ cm (measured from the respective centre of mass of the hadron).

Although QCD is a very elegant theory of strong interactions which has scored several partial successes, it faces still too many mathematical and experimental difficulties to be regarded already as the correct theory of all the strong interactions. In particular the field equations of QCD do not permit any direct and simple proof of confinement.

Thus it seems to be well justified to look for other quark field models with more transparent structure. Of particular interest are the models in which hadrons are described as soliton solutions of suitable non-linear, relativistic field equations in the 3 + 1 dimensional Minkowski space. There are two types of soliton solutions called respectively non-topological [1] and topological [2]. It seems that these two approaches are not mutually exclusive but rather complementary.

In this paper we shall follow the non-topological line of approach and outline a relatively simple class of quark field models that seem to be much closer than QDC to the - so far very succesfull - naive quark model with colours. Thus it is assumed that the quark fields $\psi_{cf}(x)$ - with the colour index $c = 1,2,3$ and the flavour index $f = 1,\dots,6$ - are the only carriers of colours. Next we assume that all the hadrons and their strong interactions can be described by a suitable set of non-linear equations for these basic Dirac quark fields.

We shall start from a $SU_C(3)$ symmetric Lorentz invariant Lagrangian that conserves also all 6 flavours.

$$\mathcal{L}_0 = -i \sum_f \sum_c \bar{\psi}_{cf} \gamma^\mu \partial_\mu \psi_{cf} - \mathcal{U}(\varkappa) . \tag{1}$$

For the argument of the non-linear function $U(\varkappa)$ we take tentatively the expression

$$\varkappa = \sum_f A_f \sum_c \bar{\psi}_{cf} \psi_{cf} \tag{2}$$

that is invariant with respect to $SU_C(3)$. Flavour symmetries (exact or approximate) higher than the Abelian groups $U_f(1)$ can be easily introduced by a suitable choice of the real coefficients A_f. However, in this paper we shall not discuss flavour symmetries restricting our attention to the problem of colour confinement. It is worth while adding the remark that the same general conclusions can be reached with an other choice of $\varkappa(x)$ that may be even non-local but it must be a $SU_C(3)$ scalar. E.g. we may take $\varkappa(x)$ of the form

$$\varkappa = \sum_f \sum_c \bar{\psi}_{cf}(x) \gamma^\mu \psi_{cf}(x) \sum_{f'} \sum_{c'} \int d^4x' \, S(x-x') \, \bar{\psi}_{c'f'}(x') \gamma_\mu \psi_{c'f'}(x') \tag{3}$$

without any essential change of the general conclusions. For the particular form

(2) one gets the following field equations

$$D_f \psi_{cf} = 0 \quad \text{with} \quad D_f = i\gamma_\mu \partial^\mu + A_f\, U'(x)\,, \quad U' = \frac{dU}{dx}\,, \tag{4}$$

which imply 9x6 conservation laws

$$\partial_\mu\, j^\mu_{cc'f} = 0 \tag{5}$$

for the currents

$$j^\mu_{cc'f} = \bar{\psi}_{cf}\, \gamma^\mu\, \psi_{c'f}\,. \tag{6}$$

Obviously the conservation laws for the flavours as well as the electric and baryonic currents

$$\partial_\mu\, j^\mu_f = 0 \quad , \quad \partial_\mu\, j^\mu_e = 0 \quad , \quad \partial_\mu\, j^\mu_B = 0 \;,$$
$$j^\mu_f = \sum_c j^\mu_{ccf} \quad , \quad j^\mu_e = \sum_f e_f\, j^\mu_f \;, \quad j^\mu_B = \sum_f j^\mu_f \;, \tag{7}$$

are a direct consequence of (5). However , apart from the 6 independent conservation laws for white (i.e. observable) flavour currents, the set (5) contains 8 x 6 redundant colour octet currents which have not been observed in nature and by the confinement hypotesis cannot describe any isolated hadronic system. Thus we infer that due to the too high amount of colour symmetry the requirement of inseparability cannot be satisfied. E.g. equations (4) have solutions with only one quark field different from zero and all others fields vanishing identically.

However, the requirement of localization can be met by a suitable choice of the non-linear function $U(x)$ as shown in [3].

The enforcement of inseparability of colours by restricting the initial conditions to suitable white states is, infortunately, not sufficient. In fact, at a collision between two white hadrons several non-overlapping (isolated) coloured objects may emerge in agreement with the redundant conservation laws. These objects should be of course intercorrelated so that the final state is also white but only in the global sense. Thus we see that rigorous $SU_C(3)$ symmetry does not exclude creation of isolated coloured hadronic objects, even if we restrict the initial states to proper white ones. It seems that the only way out of this difficulty consists in breaking the $SU_C(3)$ symmetry in order to get rid of the troublesome redundant conservation laws.

Consider the following two Lorentz scalars that are also invariant under $U_f(1)$ but violate $SU_C(3)$ symmetry

$$W = \sum_f \sum_{c' \neq c} \sum_c a_{cc'f} \bar{\psi}_{cf} \psi_{c'f} \, , \quad Z = \sum_f \sum_c b_{cf} \bar{\psi}_{cf} \psi_{cf} \, . \tag{8}$$

The complete Lagrangian is taken in the following tentative form

$$\mathcal{L} = \mathcal{L}_0 + \mathcal{L}_1 \quad , \quad \mathcal{L}_1 = W + \tfrac{1}{2} Z^2 . \tag{9}$$

Without any essential loss of generality we have taken such a colour representation in which W contains only off-diagonal elements and Z only diagonal elements in the colour indices. The field equations implied by (8) and (9) are

$$D_f \psi_{cf} = Z b_{cf} \psi_{cf} + \sum_{c' \neq c} a_{cc'f} \psi_{c'f} \, . \tag{10}$$

To simplify notation we shall drop in the following the (fixed) flavour indices and for the coefficients a and b we shall make the following ansatz:

$$\begin{aligned} &a_{12} = a_{23} = a_{31} = i a \, , \quad a_{cc'} = a^*_{c'c} \\ &b_c \neq b_{c'} \quad for \quad c \neq c' \quad , \quad b^*_c = b_c \end{aligned} \tag{11}$$

The modified field equations (10) acquire now a much more transparent form

$$\begin{aligned} D\psi_1 &= Z b_1 \psi_1 + i a (\psi_2 - \psi_3) \, , \\ D\psi_2 &= Z b_2 \psi_2 + i a (\psi_3 - \psi_1) \, , \\ D\psi_3 &= Z b_3 \psi_3 + i a (\psi_1 - \psi_2) \, . \end{aligned} \tag{12}$$

The currents $j^\mu_{cc'}$ are now no more separately conserved. In fact, instead of (5) we now have e.g.:

$$\begin{aligned} \partial_\mu j^\mu_{11} &= a (s_{12} + s_{21} - s_{13} - s_{31}) \\ \partial_\mu j^\mu_{12} &= -i (b_1 - b_2) Z s_{12} + a (s_{13} - s_{11} + s_{22} - s_{32}) \end{aligned} \tag{13}$$

with

$$s_{cc'} = \bar{\psi}_c \psi_{c'} \,. \tag{14}$$

The divergencies of the remaining 7 currents can be easily obtained from (13) by the cyclic permuatations of the colour indices or by complex conjugation. However, the sum of diagonal currents, i.e. all the white currents, are rigorously conserved. Thus all the observable currents $j^\mu_f, j^\mu_e, j^\mu_B$ satisfy the rigorous conservation laws (7) and hence can describe isolated hadronic systems. On the other hand coloured (octet) currents are - in the discussed class of models- not conserved and, therefore, they cannot describe any isolated hadronic objects.

We shall non outline a proof that our equations of motion (12) imply directly inseparability of colours. Consider a finite region Γ of space time

$$\Gamma = \{(x) : \vec{x} \in S_R \,,\, t_o < t < t_o + \Delta t\} \tag{15}$$

where Δt is a small but non-vanishing time interval and S_R is, e.g., the interior of a sphere of radius R. The three colour fields (for any fixed flavour) are regarded as being separable if only one or two of them can be different from zero in Γ, while the remaining ones vanish in Γ, i.e. if either

$$I: \ \psi_1(x) \neq 0 \,, \text{ or } II: \psi_1(x) \neq 0, \psi_2(x) \neq 0 \tag{16}$$

at least for some $x \in \Gamma$, but respectively

$$I: \ \psi_2(x) = \psi_2(x) = 0 \,, \text{ or } II: \psi_3(x) = 0 \tag{17}$$

for all $x \in \Gamma$. In case I our equations (12) reduce to

$$D\psi_1 = Z b_1 \psi_1 \,, \quad 0 = -ia\psi_1 \,, \quad 0 = ia\psi_1 \,. \tag{12 I}$$

In case II we obtain from (12)

$$D\psi_1 = Z b_1 \psi_1 + ia\psi_2 \,; \quad D\psi_2 = Z b_2 \psi_2 - ia\psi_1 \,, \quad \psi_1 = \psi_2 \,. \tag{12 II}$$

It can easily be seen that both assumptions I and II are incompatible with our field equations (12). This means that if one colour field is different from zero in S_R for t from the interval $t_o < t < t_o + \Delta t$ and is vanishing outside S_R, then the same applies to the remaining two colour fields. Thus we conclude that our

$SU_C(3)$ breaking field equations (12) imply inseparability of colours in the sense that for each value of t the three colour fields are always different from zero in the same regions of space time and, moreover, all the conserved currents that describe isolated systems must be white .

Obviously, the $SU_C(3)$ violating past $\mathcal{L}_1$ of the Lagrangian that implies this general property of solutions is not unique. For example one could take some other polynomial in Z or introduce a coupling to a neutral scalar boson field ϕ like $Z\phi$. Instead of a scalar boson one can also use pseudoscalar or vector fields as well. The problem of the best choice will be discussed in another paper.

REFERENCES

1) R. Finkelstein et al. Phys.Rev. 83, 3261 (1951), Phys.Rev. 103, 1571 (1956).
M. Soler; Phys.Rev. D1, 2766 (1970); D8, 3424 (1973).
J. Werle ,Phys.Lett. 71B, 357 and 367 (1977), Ac.Phys.Pol. B12,601 (1981).
T.F. Morris, Phys.Lett. 76B, 337 (1978).
A.F. Rañada and M.F. Rañada, Confinement in Nonlinear Classical Theory (to appear in Physica D, 1983).

2) T.H.R. Skyrme, Proc.Roy.Soc. A260, 127 (1961)
D. Finkelstein and J. Rubinstein, J.Math.Phys. 9, 1762 (1968).
E. Witten, Nucl.Phys. B160, 57 (1979).
E. Witten, Current Algebra, Baryons and Quark Confinement; G.S. Adkins et al. Static Properties of Nucleons in the Skyrme Model.(Princeton University Preprints 1983).

3) see J. Werle, l.c .

SUPERSYMMETRY AND SUPERGRAVITY

Massive Vector Superfields with SU(2) Internal Symmetry*

Changkeun Jue
Department of Physics
Kyungpook National University
Taegu 635, Korea

Byung-Ha Cho
Department of Physics
Korea Advanced Institute of Science and Technology
Seoul 131, Korea

Abstract:

A Lagrangian for massive vector superfields is constructed in such a way that it can be related to the ordinary field equations for the constituent fields, and eliminate the ghost terms.

In a previous paper [1], we have formulated the Lagrangian for massive vector superfields without internal symmetry. This Lagrangian formalism should be extendible to include internal symmetry. However, for superfields with internal symmetry a direct generalization from the case without internal symmetry evokes a severe problem, namely the presence of higher order derivative field equations requires a Hilbert space with indefinite metric [2]. For the chiral superfields with SU(2) internal symmetry, CAPPER and LEIBBRANDT [3], DONDI [4], and JUE et al. [5] found a way to eliminate the ghost terms. Just similarly to the case of chiral superfields, we have found a way to eliminate the ghost terms for the vector superfields with SU(2) internal symmetry, and constructed a Lagrangian such that the equations of motion for vector superfields reduce to the ordinary equations for the constituent fields.

In this note, the notation is introduced as follows. For Grassman parameters $\bar{\theta}^A = \bar{\theta}^{i,\alpha}$ and $\theta_{\dot{B}} = \theta_{i,\dot{\beta}}$ the spinor index α is either 1 or 2, while the dotted index $\dot{\beta}$ is either 3 or 4, and the internal symmetry index i is either 1 or 2; otherwise they are zero. The covariant derivatives D and $\bar{D}$ are given in terms of chiral representation of γ-matrices [6] as

$$D_A = \frac{\partial}{\partial\bar{\theta}^A} + \frac{i}{2}(\gamma_\mu)_A{}^{\dot{B}}\theta_{\dot{B}}\,\partial^\mu \tag{1}$$

$$\bar{D}^{\dot{A}} = -\frac{\partial}{\partial\theta_{\dot{A}}} - \frac{i}{2}\bar{\theta}^B(\gamma_\mu)_B{}^{\dot{A}}\partial^\mu \tag{2}$$

*Work supported in part by the Korea Science and Engineering Foundations where

$$(\gamma_\mu)_A{}^{\dot{B}} = (\gamma_\mu)_{\alpha\dot{\beta}}\,\delta_i^j \quad . \tag{3}$$

For chiral superfields, a suitable action may be given by

$$A = \int d^4x\, d^4\bar{\theta}\, d^4\theta\, (\Lambda + \bar{\Lambda})(D^4 + \bar{D}^4 + \Box + 2m^2)(\Lambda + \bar{\Lambda}) \tag{4}$$

where

$$D\bar{\Lambda} = \bar{D}\Lambda = 0 \quad . \tag{5}$$

The equation of motion can be obtained directly by considering the variation with respect to the superfield itself.

$$(D^4 + \bar{D}^4 + \Box + 2m^2)(\Lambda + \bar{\Lambda}) = 0 \quad . \tag{6}$$

Since we have the following identity equations,

$$D^4\bar{D}^4\bar{\Lambda} = \Box^2\bar{\Lambda} \tag{7-a}$$

$$\bar{D}^4 D^4\Lambda = \Box^2\Lambda \tag{7-b}$$

$$(D^4 + \bar{D}^4)^2(\Lambda + \bar{\Lambda}) = \Box^2(\Lambda + \bar{\Lambda}) \quad . \tag{7-c}$$

One may easily find out that the fields should satisfy the Klein-Gordon equation by operating $D^4 + \bar{D}^4$ to Eq.(6).

$$(\Box + m^2)(\Lambda + \bar{\Lambda}) = 0 \quad . \tag{8}$$

By the help of Eqs.(5), (7), and (8), the equation of motion (6) can be simplified as follows:

$$D^4\Lambda + m^2\bar{\Lambda} = 0 \tag{9-a}$$

$$\bar{D}^4\bar{\Lambda} + m^2\Lambda = 0 \quad . \tag{9-b}$$

For the vector superfield Φ, where

$$\Phi = V + \Lambda + \bar{\Lambda} \tag{10}$$

$$V = \bar{\theta}\theta F(x, \bar{\theta}, \theta)$$

$$= \bar{\theta}^A \theta_{\dot{\beta}}\, V_A{}^{\dot{B}} + \ldots \tag{11}$$

one may put a suitable action as

$$A = \int d^4x\, d^4\bar{\theta}\, d^4\theta\, \Phi(\mathscr{D}^2 + A\mathscr{D} + B)\Phi \tag{12}$$

where

$$\mathscr{D} = D^4 + \bar{D}^4 \tag{13}$$

$$A = A(m,\Box) \tag{14}$$

$$B = B(m,\Box) \tag{15}$$

such that

$$[A,\mathscr{D}] = [B,\mathscr{D}] = [A,B] = 0 \quad , \tag{16}$$

since one does not need any term higher than $\mathscr{D}^2$, because

$$\mathscr{D}^3\Phi = \Box^2\mathscr{D}\Phi \quad . \tag{17}$$

One may easily obtain the equation of motion as

$$(\mathscr{D}^2 + A\mathscr{D} + B)\Phi = 0 \quad . \tag{18}$$

Operating $\mathscr{D}$ to the above equation one obtains the following by the help of Eq.(17).

$$[\Box^2 + B - A^2]\mathscr{D}\Phi - AB\Phi = 0 \quad . \tag{19}$$

We put the operators A and B such that

$$\Box^2 + B - A^2 = 0 \quad . \tag{20}$$

Then, it reduces to

$$AB\Phi = 0 \quad . \tag{21}$$

If we choose

$$A\Phi = 0 \tag{22}$$

the equation of motion reduces to

$$(\mathscr{D}^2 - \Box^2)V = 0 \tag{23}$$

which cannot be valid for the equation of massive fields. Hence, one has to choose

$$B\Phi = 0 \tag{24}$$

while

$$A\Phi \neq 0 \quad . \tag{25}$$

It should be reasonable to put

$$B = C(\Box + m^2) \tag{26}$$

such that Eq.(24) should be identified to the Klein-Gordon equation. From the identity Eq.(20), one may finally obtain A and B as

$$A = \Box + 2m^2 \tag{27}$$

$$B = 4m^2(\Box + m^2) \tag{28}$$

and rewrite Eq.(18) as

$$[\mathscr{D}^2 + (\Box + 2m^2)\mathscr{D} + 4m^2(\Box + m^2)]\Phi = 0 \quad . \tag{29}$$

Operating D^4 or $\bar{D}^4$ to Eq.(29) with the Klein-Gordon equation, they can again be reduced to the chiral superfield equations given by Eq.(9).

$$D^4\Lambda' + m^2\bar{\Lambda}' = 0 \tag{30-a}$$

$$\bar{D}^4\bar{\Lambda}' + m^2\Lambda' = 0 \tag{30-b}$$

where

$$\Lambda' = \bar{D}^4V + m^2\Lambda \tag{31-a}$$

$$\Lambda' = D^4V + m^2\bar{\Lambda} \quad . \tag{31-b}$$

Though it be tedious, it is straightforward to calculate it for the constituent fields, which should be identified to the ordinary equations.

References

1. C.-K. Jue, B.-H., Cho, C.-C. Chiang, E.C.G. Sudarshan: Physica 114A (1982) 134
2. A. Pais, Uhlenbeck: Phys. Rev. 79 (1950) 145; E.C.G. Sudarshan: Phys. Rev. 123 (1961) 2183
3. D.M. Capper, G. Leibbrandt: Nucl. Phys. B85 (1975) 503
4. P.H. Dondi: J. Phys. 8A (1975) 1298
5. C.-K. Jue, C.-C. Chiang, E.C.G. Sudarshan: University of Texas Report No. DOE-394 (1980)

Ref.TH.3711-CERN

SUPERGRAVITY IN ELEVEN-DIMENSIONAL SPACE-TIME +) ")

F. Englert *) and H. Nicolai

CERN - Geneva

A B S T R A C T

A comprehensive review of supergravity in eleven dimensions is presented with particular emphasis on the seven-sphere compactification. The complete mass spectrum and its Osp(8,4) content are exhibited and discussed. We conclude with some speculative remarks about the effects of quantum gravity.

---

*) On leave from the Université Libre de Bruxelles, Campus Plaine, CP225, Boulevard du Triomphe, B-1050 Bruxelles, Belgique.

+) Supported in part by the Belgian State under the contract A.R.C. 79/83-12.

") Invited Contribution to the XII International Colloquium on Group Theoretical Methods in Physics, Trieste (1983).

Ref.TH.3711-CERN

September 1983

SUPERGRAVITY IN ELEVEN-DIMENSIONAL SPACE-TIME

F. Englert and H. Nicolai
CERN
1211 Geneva 23
Switzerland

1. - FROM SUPERSYMMETRY TO SUPERGRAVITY IN ELEVEN DIMENSIONS

Quantum field theory in Minkowski space-time is marred by ultra-violet divergences which apparently preclude a perturbative analysis of non-renormalizable theories such as gravity. One of the most basic quantities which diverge in quantum field theory is the vacuum energy $\langle 0|H|0\rangle$. Bosonic and fermionic degrees contribute to it with opposite sign, and one may therefore search for theories which possess a symmetry between Bose and Fermi fields such that

$$\langle 0|H|0\rangle = 0 \tag{1.1}$$

This possibility is indeed realized in supersymmetric field theories[1]. These are characterized by a fermionic symmetry which relates bosons and fermions and transforms them into each other. In the simplest case, the generators of supersymmetry transformations are given by a Majorana spinor supercharge Q_α which obeys

$$\{Q_\alpha, \bar{Q}_\beta\} = 2\gamma^\mu_{\alpha\beta} P_\mu \qquad (\mu = 0,1,2,3) \tag{1.2}$$

If supersymmetry is not spontaneously broken, the supercharge Q annihilates the vacuum, and this leads directly to (1.1). Supersymmetric field theories have further divergence cancellations, and fully convergent supersymmetric models have been constructed. In general, the weakening of ultra-violet divergences implied by (1.2) reduces the radiative corrections of the scalar fields to logarithmic terms, a feature which persists if supersymmetry is softly broken[2]. In this way one can stabilize large mass scale ratios of the type encountered in grand unified theories. Thus both at the prospective theoretical level and in more direct phenomenological applications, supersymmetry appears promising for further development in the field theoretical description.

It is then natural to investigate invariance under local supersymmetry. By Eq. (1.2), this implies invariance under "local translations" and hence general relativity. In this way a thread leading towards a unified theory of matter and gravitation is brought to light: the graviton helicity state 2(-2) must now be accompanied by a spin-$\frac{3}{2}$ ($-\frac{3}{2}$) degree of freedom, the gravitino, connected to it by the supersymmetric generator Q. This locally supersymmetric graviton-gravitino system constitutes $N = 1$ supergravity[3]. Full unification is achieved if one requires in addition that the two helicity states of the graviton are connected by supersymmetry transformations. If there are N independent supercharges Q^i_α, the irreducible multiplets cover a range of spin of at least $N/4$. To connect the two helicity states of the graviton, one therefore needs eight helicity-$\frac{1}{2}$ flips; this case corresponds to $N = 8$ supersymmetry and the minimal multiplet obtained in this way is CPT self-conjugate[4]. In the presence of eight supersymmetry charges Q^i_α $(i = 1,\dots,8)$, (1.2) generalizes to

$$\{Q^i_\alpha, \bar{Q}_{\beta j}\} = 2\delta^i_j \gamma^\mu_{\alpha\beta} P_\mu \tag{1.3}$$

The field content of such an "$N = 8$ extended supergravity theory" is determined by the states obtained from the helicity 2 state by the anticommuting Q^i; hence it comprises 1 spin-2, $\binom{8}{1} = 8$ spin-$\frac{3}{2}$, $\binom{8}{2} = 28$ spin-1, $\binom{8}{3} = 56$ spin-$\frac{1}{2}$ and $\binom{8}{4} = 70$ spin-0 fields. Clearly, no other field system of spin <2 can be consistently coupled to such an $N = 8$ theory which therefore fully unifies matter and gravitation. Moreover, any supersymmetric theory with $N > 8$ would have to contain particles of spin-$\frac{5}{2}$ or higher and such theories do not appear to be consistent.

We will now rewrite Eq. (1.3) in a more compact and "unified" form. For this purpose, we combine the eight four-component Majorana spinors of (1.3) into one single 32-component Majorana spinor $\tilde{Q}$. Simple counting reveals that the number of space-time dimensions appropriate to accommodate such a spinor is eleven [for a review, see Ref. 5)]. We are thus led to consider an eleven-dimensional space-time of signature $(+-\dots-)$ instead of the usual four-dimensional space-time. The $\tilde{\Gamma}$ matrices in eleven dimensions can be expressed in terms of four-dimensional γ matrices and eight-dimensional Γ matrices, which generate the Clifford algebra in seven dimensions, according to

$$\tilde{\Gamma}^\mu = \gamma^\mu \otimes \mathbb{1}, \quad \tilde{\Gamma}^m = \gamma^5 \otimes \Gamma^m \tag{1.4}$$

where the indices $\mu, \nu, \dots$ take the values $0,1,2,3$ and the indices $m, n\dots$ are $4,\dots,10$. Indices in eleven dimensions will be denoted by $M, N, \dots$ With this, (1.3) becomes (spinor indices α now take the values $1,\dots,32$)

$$\begin{aligned}\{\tilde{Q}_\alpha, \bar{\tilde{Q}}_\beta\} &= 2\tilde{\Gamma}^M_{\alpha\beta} P_M \\ \{\tilde{\Gamma}_M, \tilde{\Gamma}_N\} &= 2 g_{MN}\end{aligned} \tag{1.5}$$

with $g^{MN} = \mathrm{diag}(1,-1,\dots,-1)$ if, in addition, we stipulate that $P_m = 0$, or

$$P_M = 0 \quad \text{for} \quad M = 4, 5, \dots, 10 \tag{1.6}$$

in order to make (1.5) agree with (1.3). From this equivalence, we conclude that $N = 1$ supergravity in eleven dimensions should lead to an $N = 8$ supergravity theory in four dimensions when the fields do not depend on the extra seven co-ordinates. The action of $N = 1$ supergravity in eleven dimensions has been constructed[6)], and an $N = 8$ supergravity in four dimensions has indeed been obtained in this way[7)]. The action in eleven dimensions is unique, and, on the basis of the observation that supergravity in more than eleven dimensions would inevitably contain states of spin higher than two[8)], it has been argued that no supergravity theory exists beyond $d = 11$.

At this point, one may call into question the physical relevance of the constraint (1.6) by which the four-dimensionality of space-time is a priori imposed. There are reasons to expect that, at large scales, the apparent four-dimensionality of space-time is, from quantum theory, the required framework to describe long-range forces. Four is the critical number of dimensions for gauge field theories: in more than four dimensions, these theories are non-renormalizable, so the effective coupling constants should decrease like inverse powers in the low energy limit (or at large distances). It is only in four dimensions that gauge theories are strictly renormalizable and that the effective coupling strengths vary logarithmically. This could explain the existence of long-range forces at present day energies and provide a reason why not all coupling constants are of the same order as the gravitational coupling (whose smallness would be a consequence of the non-renormalizability of gravity). This picture should then naturally emerge from a truly unified theory of all interactions. However, there is no compelling reason why space-time should be four-dimensional at very small scales. Given the uniqueness of the $N = 1$ supergravity in eleven dimensions, this line of argument suggests that this theory is more than just a technical device to construct $N = 8$ supergravity in four dimensions (as it was originally conceived) and should be considered in its own right. We shall therefore drop the constraint $P_m = 0$ and tentatively assume that it is in fact the $d = 11$ theory which describes the physical world.

The assumption that space-time has more than four dimensions is not new and also underlies Kaluza-Klein theories [for reviews and further references, see 9)-12)]. There, the transition from the original higher-dimensional theory at small scales to the required four-dimensional background at large scales can occur naturally already at the classical level through the spontaneous compactification mechanism [13)] which we shall now investigate.

2. - SPONTANEOUS COMPACTIFICATIONS OF ELEVEN-DIMENSIONAL SUPERGRAVITY

2.1 Spontaneous Compactification

The eleven-dimensional $N = 1$ supergravity action is based on the following fields[6]: an elfbein $e_M{}^A$ (flat indices are labelled by the first letters of the alphabet) which describes ordinary gravity in eleven dimensions, a 32-component Majorana spinor ψ_M which is the analogue of the gravitino in four dimensions and an antisymmetric three-index tensor A_{MNP} defined up to a gauge transformation*)

$$\delta A_{MNP} = D_{[M}\Lambda_{NP]} \quad , \quad D^M \Lambda_{MN} = 0 \tag{2.1}$$

This field is required in order to balance bosonic and fermionic degrees of freedom in eleven dimensions. Defining $F_{MNPQ} \equiv 24A_{[NPQ,M]}$ and $\Gamma^{MN\ldots R} \equiv \Gamma^{[M}\Gamma^{N}\ldots\Gamma^{R]}$, we may write the action as follows (up to four-Fermi terms which can be included through supercovariantized elfbein affinities and field strengths F_{MNPQ})

$$S = \int d^{11}x \; e\Big\{-\tfrac{1}{2}R - \tfrac{1}{48}F_{MNPQ}F^{MNPQ} - \tfrac{i}{2}\bar{\psi}_M\tilde{\Gamma}^{MNP}\psi_{P;N} + \frac{4\sqrt{2}}{(4!)^3}\eta^{M_1\ldots M_{11}}F_{M_1\ldots M_4}F_{M_5\ldots M_8}A_{M_9M_{10}M_{11}} + \frac{3\sqrt{2}}{(4!)^2}\left(\bar{\psi}_M\tilde{\Gamma}^{MNPQRS}\psi_N + 12\bar{\psi}^P\tilde{\Gamma}^{QR}\psi^S\right)F_{PQRS}\Big\} \tag{2.2}$$

Here $\eta^{M_1\ldots M_{11}}$ is the eleven-dimensional Levi-Civita tensor. The action (2.2) is invariant under the local supersymmetry transformations

$$\delta e_M{}^A = -\tfrac{i}{2}\bar{\varepsilon}\tilde{\Gamma}^A\psi_M \tag{2.3}$$

$$\delta A_{MNP} = \frac{\sqrt{2}}{8}\bar{\varepsilon}\tilde{\Gamma}_{[MN}\psi_{P]} \tag{2.4}$$

$$\delta\psi_M = \varepsilon_{;M} + \frac{2\sqrt{2}i}{(4!)^2}\left(\tilde{\Gamma}^{NPQR}{}_M - 8\delta^N_M\tilde{\Gamma}^{PQR}\right)F_{NPQR}\,\varepsilon \tag{2.5}$$

Again we have omitted Fermi bilinears which result from supercovariantization of $\delta\psi_M$ in Eq. (2.5).

We have argued that, at large scales, the space-time background should be four-dimensional and it is therefore of interest to inquire whether four-dimensional space-time has a natural classical limit. To this effect, we search for solutions of the classical equations of motion which describe a direct product space

---

*) Covariant derivatives are denoted by $D_M\phi = \phi_{;M}$.

$\mathcal{M}_7\{y^m\} \times \mathcal{M}_4\{x^\mu\}$ where $\mathcal{M}_7$ is compact and $\mathcal{M}_4$ admits local Lorentz invariance. Any such classical solution provides a "vacuum state" configuration $|\Omega\rangle$ for the Fock-space of normal mode excitations $\phi^{(n)}(x)$. Explicitly, one expands all fields $\phi(x,y)$ of the $d = 11$ theory in terms of a complete set of functions on the internal manifold $\mathcal{M}_7$ according to

$$\phi(x,y) = \sum_n \phi^{(n)}(x)\, Y^{(n)}(y) \tag{2.6}$$

where we have suppressed indices which might be attached to the various fields. Four-dimensional fields are associated with functions $\phi^{(n)}(x)$ while $Y^{(n)}(y)$ are eigenfunctions of suitable operators on $\mathcal{M}_7$ (explicit examples will be presented later). Inserting the expansion (2.6) into the kinetic term in the higher dimensional space-time, one sees that, quite generally, the part which operates in the internal space becomes a mass term in four dimensions. It is therefore the mass operator which determines the spectrum and the set of normal modes $\{Y^{(n)}(y)\}$ in (2.6). In general, the relevant operators are elliptic since the signature in the internal space is uniform $(-\ldots-)$, and the compactness of the internal manifold $\mathcal{M}_7$ ensures that the spectrum is discrete and that there exists a gap between massless and massive excitations. The mechanism by which a classical solution leads to a spontaneous decomposition of eleven-dimensional space-time into a product of $\mathcal{M}_4$ and $\mathcal{M}_7$, is referred to as spontaneous compactification[13].

In the absence of fermion condensates, the classical equations which arise from the action (2.2) are

$$R_{MN} - \frac{1}{2} g_{MN} R = -\frac{1}{48}\left\{ 8 F_{MPQR} F_N{}^{PQR} - g_{MN} F_{PQRS} F^{PQRS} \right\} \tag{2.7}$$

$$F^{MNPQ}{}_{;M} = -\frac{\sqrt{2}}{2\cdot(4!)^2}\, \eta^{M_1\ldots M_8 NPQ} F_{M_1\ldots M_4} F_{M_5\ldots M_8} \tag{2.8}$$

and, up to supersymmetry gauge transformations, we must have

$$\psi_M = 0 \tag{2.9}$$

The solutions describing spontaneous compactification on $\mathcal{M}_7 \times \mathcal{M}_4$ are characterized by $g_{\mu m} = F_{\mu mnp} = F_{\mu\nu np} = F_{\mu\nu\sigma p} = 0$. If we impose in addition

$$F_{mnpq} = 0 \tag{2.10}$$

we get from Eqs. (2.8) and (2.9)

$$F_{\mu\nu\rho\sigma} = f\,\varepsilon_{\mu\nu\rho\sigma}\ ,\quad f = \text{constant} \tag{2.11}$$

$$R_{mn} = -\,6\,m_7^2\,g_{mn} \tag{2.12}$$

$$R_{\mu\nu} = 12\,m_7^2\,g_{\mu\nu} \tag{2.13}$$

$$m_7^2 = f^2/18 \tag{2.14}$$

These solutions were obtained by Freund and Rubin[14] who pointed out that the natural splitting of eleven dimensions in $7 + 4$ which follows from Eq. (2.11) is a consequence of the fact that the field strength F_{MNPQ} has four indices. Note that Eq. (2.8) implies the validity of Eq. (2.11) even when $F_{mnpq} \neq 0$.

2.2 Symmetries of the compactified vacuum

We now examine the possible gauge symmetries and four-dimensional supersymmetries of the vacuum obtained after spontaneous compactification. As in the original Kaluza-Klein theory and its non-Abelian generalizations[9),10)], the invariance gauge group on $\mathcal{M}_4$ will contain the isometry group of the internal space $\mathcal{M}_7$ for Freund-Rubin solutions. Indeed, under the transformation

$$\begin{aligned} x'^{\mu} &= x^{\mu} \\ y'^{m} &= y^{m} + \overset{(\alpha)}{\varepsilon}(x)\,\overset{(\alpha)}{\xi}{}^{m}(y) \end{aligned} \tag{2.15}$$

the seven-dimensional metric is left invariant if the vector field $\overset{(\alpha)}{\xi}{}^m(y)\partial_m$ generates an isometry. This is the case if $\overset{(\alpha)}{\xi}{}^m(y)$ is a Killing vector, i.e.,

$$2\,\overset{(\alpha)}{\xi}_{(m;n)} = \overset{(\alpha)}{\xi}_{m;n} + \overset{(\alpha)}{\xi}_{n;m} = 0 \tag{2.16}$$

The Killing vectors generate a group - the isometry group of $\mathcal{M}_7$ - , and (α) is therefore a group index. Thus, we have

$$\left[\overset{(\alpha)}{\xi}{}^m(y)\partial_m\, ,\ \overset{(\beta)}{\xi}{}^n(y)\partial_n\right] = f^{(\alpha)(\beta)}{}_{(\gamma)}\,\overset{(\gamma)}{\xi}{}^p(y)\partial_p \tag{2.17}$$

where $f^{(\alpha)(\beta)}{}_{(\gamma)}$ are the structure constants of the group. Since the parameter $\overset{(\alpha)}{\xi}(x)$ in (2.15) is space-time dependent, the resulting theory in four dimensions will inherit the isometry group of $\mathcal{M}_7$ as a local gauge group. In the presence of matter fields, the actual symmetry may be smaller than the isometry group of $\mathcal{M}_7$.

For $F_{mnpq} \neq 0$, it is reduced to that subgroup of the isometry group which, in addition to (2.16), leaves the tensor F_{mnpq} invariant, viz.,

$$\delta_\xi F_{mnpq} = \xi^r D_r F_{mnpq} + 4 D_{[m} \xi^r F_{npq]r} = 0 \tag{2.18}$$

Global supersymmetries, on the other hand, may survive spontaneous compactification only if there exists at least one 32-component Majorana spinor $\varepsilon(x,y)$ for which[11)]

$$\delta \psi_M = 0 \tag{2.19}$$

This condition is the fermionic analogue of (2.16). We now show that only Freund-Rubin solutions may have supersymmetric ground states if the spinor $\varepsilon(x,y)$ factorizes into a product of a four-component and an eight-component spinor according to $\varepsilon(x,y) = \varepsilon(x)\eta(y)$ [15)] [we will assume that $\varepsilon(x,y)$ factorizes in this way although it is far from clear whether such a factorization is really necessary to isolate a local supersymmetry in four dimensions. If space-time is not described by a global product $\mathcal{M}_4 \times \mathcal{M}_7$ but rather by a non-trivial fibre-bundle, this assumption is presumably no longer valid].

From Eq. (2.5) and local Lorentz invariance which implies Eq. (2.11), we may write Eq. (2.16) as

$$\varepsilon(x) \left\{ \eta(y)_{;m} - \frac{1}{2} \frac{f}{\sqrt{18}} \Gamma_m \eta(y) \right\} + \\ + i\gamma^5 \varepsilon(x) \cdot \frac{\sqrt{2}}{(4!)^2} \left\{ \Gamma_m{}^{npqr} - 8\delta_m^n \Gamma^{pqr} \right\} F_{npqr}(y)\eta(y) = 0 \tag{2.20}$$

$$\left\{ \varepsilon(x)_{;\mu} - \frac{f}{\sqrt{18}} \gamma^5 \gamma_\mu \varepsilon(x) \right\} \eta(y) + \frac{2\sqrt{2}i}{(4!)^2} \gamma_\mu \Gamma^{mnpq} F_{mnpq} \eta(y) \varepsilon(x) = 0 \tag{2.21}$$

Projecting (2.20) onto chiral components, we see that both the coefficients of ε and $i\gamma_5\varepsilon$ in (2.20) must vanish and hence η must be a "Killing spinor" satisfying

$$\eta(y)_{;m} = \frac{1}{2} m_7 \Gamma_m \eta(y) \tag{2.22}$$

where m_7 is related to f by (2.14). The integrability condition of (2.22) reads

$$\left(R_{mnpq} \Gamma^{pq} + 2 m_7^2 \Gamma_{mn} \right) \eta(y) = 0 \tag{2.23}$$

which yields Eq. (2.12) upon contraction with Γ^m.. However, Eqs. (2.11), (2.12) and (2.14) are consistent with the eleven-dimensional Einstein equation (2.7) if and only if $F^{mnpq}F_{mnpq} = 0$. Hence the only admissible solutions of Eq. (2.17) are given by

$$F^{mnpq} = 0 \tag{2.24}$$

and we recover the Freund-Rubin solutions. One then verifies that the integrability condition of (2.21), namely

$$\left(R_{\mu\nu\rho\sigma} \gamma^{\rho\sigma} - 8 m_7^2 \gamma_{\mu\nu} \right) \varepsilon(x) = 0, \tag{2.25}$$

is indeed consistent with Eq. (2.13). Hence, only for Freund-Rubin solutions can supersymmetry survive compactification if $\varepsilon(x,y) = \varepsilon(x)\cdot\eta(y)$. That the residual supersymmetry is in fact local follows from (2.20) which implies $\delta\psi_m = 0$ for arbitrary $\varepsilon(x)$ while such transformations induce only a four-dimensional supersymmetry gauge term in $\delta\psi_\mu$.

The largest number of supersymmetries available in four dimensions is equal to the number of linearly independent solutions of (2.22), namely eight. This can only happen if the integrability conditions (2.23) and (2.25) are identically satisfied. Hence

$$R_{mnpq} = - m_7^2 (g_{mp} g_{nq} - g_{mq} g_{np}) \tag{2.26}$$

$$R_{\mu\nu\rho\sigma} = 4 m_7^2 (g_{\mu\rho} g_{\nu\sigma} - g_{\mu\sigma} g_{\nu\rho}) \tag{2.27}$$

and there are only two fully supersymmetric compactifications[16)]

$$(A) : \quad m_7^2 = 0 \qquad \mathcal{M}_4 \times \mathcal{M}_7 = \text{Minkowski} \times T^7 \tag{2.28}$$

$$(B) : \quad m_7^2 \neq 0 \qquad \mathcal{M}_4 \times \mathcal{M}_7 = (AdS)_4 \times S^7 \tag{2.29}$$

Case (A) corresponds to the original $N = 8$ theory of Cremmer and Julia[7)] when only the zero modes are retained. Case (B) corresponds to the compactification on the seven sphere S^7 of (arbitrary) radius $|m_7|^{-1}$ to an anti-de Sitter $(AdS)_4$ space-time. This solution was first exhibited by Duff and Pope who were led to it by requiring the existence of eight unbroken supersymmetries[12),18)]. Since the seven sphere S^7 is the coset space SO(8)/SO(7) which admits an SO(8) isometry group, this compactification describes a theory with local SO(8) invariance. If only the zero-mass supermultiplet of states is retained, this theory is presumably equivalent to gauged $N = 8$ supergravity[17)]. This expectation has been partially confirmed by explicit calculations at the linearized level[16),18)] where one may identify fields and states.

Solutions with $N < 8$ supersymmetries have also been obtained with $N = 1,2$ and 4, and it has recently been shown that these are the only permitted values[19)]. One example of a solution with $N = 1$ supersymmetry is the "squashed sphere" of

Ref. 20) that corresponds to a compactification on the coset space Sp(2)xSp(4)//Sp(2)'xSp(2)"[21] [the primes indicate that Sp(2)' ≠ Sp(2) ≠ Sp(2)"] with an isometry group Sp(2)xSp(4). An example with N = 2 supersymmetry is obtained with $\mathcal{M}_7$ = (SU(3)xSU(2)xU(1))/(SU(2)xU(1)'xU(1)") for particular embeddings of the U(1) subgroups into SU(3)xSU(2)xU(1)[22]. Both spaces have $m_7 \neq 0$. Finally, an example with N = 4 supersymmetry and $m_7 = 0$ is also known[23].

2.3 Supersymmetry breaking

We shall now show that for each solution having $N \geq 1$ supersymmetry one can associate at least one other solution where supersymmetry is fully broken by a mechanism having a geometrical significance in the unseen dimensions[15]. The new solution which has $F_{mnpq} \neq 0$, can be interpreted as a spontaneously broken version of the old one.

Consider Eq. (2.22) and eventually the equation obtained from it by reversing the sign in front of m_7. We write

$$\eta^{\pm}{}_{;m} \mp \frac{1}{2} m_7 \Gamma_m \eta^{\pm} = 0 \tag{2.30}$$

The η^+ is a covariant constant spinor characterizing a residual supersymmetry. The η^- satisfies a different differential equation in the same co-ordinate system*) and its existence is not guaranteed by the necessary condition (2.23) and in some cases it indeed does not exist. We look for a new solution of (2.8) with $F_{mnpq} \neq 0$ in the old Einstein space $\mathcal{M}_7$ but we allow for a constant rescaling of the $\mathcal{M}_4$ space to cope with an eventual shift of the cosmological constant. We have

$$F_{\mu\nu\rho\sigma} = f' \eta_{\mu\nu\rho\sigma} \, , \quad f' = \text{constant} \tag{2.31}$$

and F^{mnpq} must be a solution of

$$F^{mnpq}{}_{;m} = \frac{\sqrt{2}}{4!} f' \varepsilon^{npqrstu} F_{rstu} \tag{2.32}$$

It is readily checked that

$$A^{\pm}_{mnp} = \xi \bar{\eta}^{\pm} \Gamma_m \Gamma_n \Gamma_p \eta^{\pm} = \\ = \xi \bar{\eta}^{\pm} \Gamma_{mnp} \eta^{\pm} \qquad (\bar{\eta} \equiv \eta^{+}) \tag{2.33}$$

---

*) In order to avoid confusion we distinguish the two equations explicitly despite the fact that the sign of m_7 is not fixed.

solves (2.32) provided that

$$\pm m_7 = - f'/\sqrt{8} \tag{2.34}$$

and is dual or antidual to its field strength F_{pqrs}. The arbitrary constant ξ is determined, up to a sign, by Eq. (2.7) and one finds $\xi^2 = 2/(4!)^2$. The new solution(s) corresponding to η^+ (and η^- if the latter exists) has (have) $F_{mnpq} \neq 0$ and hence all supersymmetries are broken. The metric is unchanged except for a shift of the four-dimensional cosmological constant which amounts to a constant rescaling $\overset{0}{g}_{\mu\nu} \to \frac{6}{5}\overset{0}{g}_{\mu\nu}$ on $\mathcal{M}_4$. The field F_{mnpq} has a geometrical interpretation. Consider the quantities $S^{\pm}_{mnp}$, given by

$$S^{\pm}_{mnp} = \pm m_7 \bar{\eta}^{\pm} \Gamma_{mnp} \eta^{\pm} \tag{2.35}$$

Using

$$1 = \eta\bar{\eta} + \Gamma_p \eta \bar{\eta} \Gamma^p \tag{2.36}$$

one gets

$$\begin{aligned} S_{mnp;q} &= S_{[mnp,q]} = \\ &= S_{mnt} S_{pq}{}^{t} + m_7^2 (g_{mp} g_{nq} - g_{mq} g_{np}) \end{aligned} \tag{2.37}$$

and the curvature tensor associated with the connection $\{^m_{np}\} + S^m_{np}$, where $\{^m_{np}\}$ is the Levi-Civita connection, takes the form

$$\begin{aligned} R_{mnpq}(\{^r_{st}\} + S^r{}_{st}) &= \\ &= R_{mnpq}(\{^r_{st}\}) + m_7^2 (g_{mp} g_{nq} - g_{mq} g_{np}) = \\ &= C_{mnpq}(\{^r_{st}\}) \end{aligned} \tag{2.38}$$

where C_{mnpq} is the Weyl tensor of the Einstein space $\mathcal{M}_7$. Hence

$$R_{mn}(\{^r_{st}\} + S^r{}_{st}) = 0 \tag{2.39}$$

Thus up to a constant, the dual of F^{mnpq} is interpretable as a totally antisymmetric torsion which renders $\mathcal{M}_7$ Ricci-flat. The total antisymmetry of S_{mnp} guarantees that metric properties are preserved, namely $g^{mn}{}_{;p} = 0$ both for the Levi-Civita and for the total connexion.

The new solution(s) differs from the old one by a change in the four-dimensional cosmological constant due to $F^{mnpq} \neq 0$. In this way all the supersymmetries of the old solution are broken by a spontaneously induced torsion which Ricci-"flattens" the compact manifold. These solutions with supersymmetry breaking have been explicitly constructed for S^7 [16], for the squashed seven sphere[15),21),*)] and for the coset space $(SU(3)\times SU(2)\times U(1))/(SU(2)\times U(1)'\times U(1)'')$ [22].

---

*) We find ourselves in disagreement with the authors of Ref. 34) who distinguish between left and right squashing and claim that a solution with squashing and torsion can only be obtained for the right squashed sphere. The difference between left and right squashing corresponds to $m_7 < 0$ and $m_7 > 0$, respectively (P. Spindel, private communication). However, the existence of a covariantly constant spinor is determined by the relative sign between the parameters m_7 and f' in (2.34). If this sign is a priori fixed as in Ref. 34), it is indeed true that a solution exists only for one handedness; if, however, this sign is reversed, a solution with opposite handedness is obtained. The statement that the existence of a solution with squashing and torsion is linked to a definite handedness of the squashed sphere is therefore incorrect.

3. - THE SEVEN SPHERE

3.1 The geometry of S^7

We have seen that $S^7 \times (AdS)_4$ is the only non-"trivial" compactification which preserves full $N = 8$ local supersymmetry in four dimensions. It turns out that S^7 has remarkable geometric properties which fit in a surprising way into the structure of the eleven-dimensional theory. We shall therefore first review these exceptional properties which are easily revealed by the octonion parametrization of the sphere.

The real octonion algebra is an eight-dimensional vector space over the real number field endowed with a non-associative multiplication law[24)]. The latter can be defined by choosing a basis formed by the identity and seven "imaginary" units e_a whose multiplication table is

$$e_a e_b = -\delta_{ab} + \sum_{c=1}^{7} a_{abc} e_c \tag{3.1}$$

Here a_{abc} is totally antisymmetric and equal to $+1$, when the indices a, b, c are equal to 126, 315, 234, 465, 714, 736 and 725; otherwise, $a_{abc} = 0$. To each octonion $X = x_o + \sum_a x_a e_a$ one associates a complex conjugate octonion $\bar{X} = x_o - \sum_a x_a e_a$; the scalar product of X and Y is then defined by

$$\langle XY \rangle = \mathrm{Tr}\, X\bar{Y} = \mathrm{Tr}\, \bar{X}Y \;; \quad \begin{matrix} \mathrm{Tr}\, 1 = 1 \\ \mathrm{Tr}\, e_a = 0 \end{matrix} \tag{3.2}$$

The non-degenerate norm $N(x)$ defined by

$$N(x)^2 = \langle XX \rangle = x_o^2 + \sum_{a=1}^{7} x_a^2 \tag{3.3}$$

satisfies the composition law

$$N(X)\, N(Y) = N(XY) \tag{3.4}$$

This property, Eq. (3.4), is shared with the algebra of real numbers, complex numbers and octonions. In contradistinction to those, however, it is not associative and one verifies from Eq. (3.1) that the associator defined by

$$[X, Y, Z] = (XY)Z - X(YZ) \tag{3.5}$$

is totally antisymmetric in X, Y and Z.

We restrict ourselves to octonions of unit norm which describe Euclidean vectors of unit length in E_8; they may thus be used to parametrize points on the unit seven sphere, $|m_7| = 1$. At each point X of this S^7 we can construct an orthonormal tangent 7-bein $e_a X$ as is easily verified from Eqs. (3.2) and (3.5). Such a 7-bein field which is globally defined because Eq. (3.4) defines an absolute parallellism on S^7: tangent vectors are called parallel if they have the same components in the two frames defined at their respective origin. In fact, one easily constructs two seven-parameter families of parallelism from the 7-bein fields[25)]

$$u_a^+(X) = (e_a A)(\bar{A} X), \quad u_a^-(X) = (X\bar{A})(A e_a) \tag{3.6}$$

where A is an arbitrary constant unit octonion.

The above construction of absolute parallelisms can be viewed as a generalization of the parallelism induced on S^1 and S^3 in an identical manner using complex numbers and quaternions, respectively. However, as S^1 and S^3 are group spaces, their parallelizability also follows from group theory while the S^7 parallelism constitutes a unique feature rooted in the existence of the octonion composition algebra[26)]. Clearly, the occurrence of distinct sets of (+) and (-) parallelisms defined by (3.6) is a consequence of the non-commutative character of (+) while the dependence of each set on seven arbitrary parameters follows from its non-associative character. All the (+) parallelisms [and all the (-) parallelisms] are conjugate under an SO(8) rotation on the seven sphere so that to study the parallelized geometries it suffices to put $A = 1$ in (3.6).

The integral curves of Eq. (3.6) (with $A = 1$) are the unit circles parametrized by

$$\exp(\lambda e_a) X = \cos\lambda \, X + \sin\lambda \, e_a X \tag{3.7}$$

where λ varies from 0 to 2π. Hence, metric geodesies are preserved by the parallelisms and hence the corresponding linear connections are characterized by totally antisymmetric torsion tensors $S^{\pm}_{rst}$. Thus, the two parallelized geometries satisfy

$$R^m{}_{npq}\left(\left\{ {r \atop st} \right\} + S^r{}_{st}\right) = 0 \tag{3.8}$$

Contracting (3.8) to R_{nq} and comparing with (2.39), we see that these geometries are precisely those constructed from the η^+ and η^- spinors defined on the seven sphere. Hence in addition to the Riemann seven sphere S_7, eleven-dimensional supergravity admits compactification on the two exceptional "flat" seven sphere geometries $S_7^{\pm}$[27)]. These may be viewed as spontaneously broken phases of the S_7

compactifications. All the supersymmetries are broken[16),28)] and the SO(8) gauge group is reduced to the isometry invariance group which preserves a parallelism, this is readily shown to be the covering group of SO(7), namely Spin(7) [29)-31),*)]. Moreover it has been shown that, when supersymmetry is defined by η^+ Killing spinors, the breaking induced by the (-) parallelized solution occurs (at the linearized level) within the Fock space of zero-mass excitations defined from the S_7 compactification[16),33)]. This is a peculiar feature which, in general, is not a property of spontaneous compactification. For example, the squashed solution of Ref. 20) cannot be interpreted in this way because the covariantly constant spinor is not a linear combination of the eight covariantly constant spinors of the compactification on the round S^7 [21),34)]. In the case of the parallelized solution of Ref. 27), the identification is made possible by the existence of covariantly constant spinors for both signs in Eq. (2.30). The fact that this solution can be interpreted as a spontaneously broken version of gauged N = 8 supergravity is also strongly suggested by the existence of stationary points of the N = 8 potential with Spin(7) symmetry[35),29)].

3.2 The mass spectrum

The bare mass spectrum of the four-dimensional theory can be obtained by varying the fields g_{MN}, A_{MNP}, ψ_M around their background values. The bosonic field equations (2.7) and (2.8) are linear in δg_{MN}, δA_{MNP} and the corresponding linearized equation for the fermionic field is, from Eq. (2.2),

$$- i \tilde{\Gamma}^{MNP} \psi_{P;N} + \\ + \frac{\sqrt{2}}{288} m_7 \left\{ \tilde{\Gamma}^{MNPQRS} + 12 g^{MP} g^{NQ} \tilde{\Gamma}^{RS} \right\} \overset{\circ}{F}_{PQRS} \psi_N = 0 \tag{3.9}$$

where the superscript o labels a background value. One expands the components of ψ_M, δg_{MN} and δA_{MNP} in complete sets of eigenfunctions of suitable differential operators on S_7 according to Eq. (2.6). The masses are then obtained in terms of the corresponding eigenvalues. However, the actual calculation reveals a mixing between the various fields of the theory and one has in fact to diagonalize a mass matrix.

The diagonalization can be performed by field redefinitions, and such a method was used to obtain fields describing the zero-mass supermultiplet, namely one graviton, eight gravitinos, 28 SO(8) gauge fields, 56 spin-$\frac{1}{2}$ fields, 35 scalars and 35 pseudo-

---

*) The Spin(7) invariance may also be deduced from the results of Ref. 32).

scalars[16),18)]. By retaining only these fields in the full non-linear theory, one would expect to recover the gauged $N = 8$ supergravity theory of Ref. 17). But such an expectation can only be borne out if this truncation is consistent, namely if no massive mode can be varied into a massless mode by a supersymmetry transformation in four dimensions. As such a mixing does indeed occur, a consistent truncation cannot be defined from the ansätze as given in Refs. 16) and 18), and can therefore be obtained only if non-linear modifications are taken into account[29)] (in particular, the supersymmetry transformation parameter $\varepsilon(x,y)$ must be redefined in such a way that it is no longer a Killing spinor for non-vanishing fluctuations, i.e., outside the background).

There is another, and perhaps related problem, which motivates an analysis of the complete four-dimensional field content, including the massive modes, which arise from the seven-sphere compactification. It was conjectured that the supersymmetry breaking induced by different compactifications, which are topologically equivalent to the seven-sphere, may lead to a "level crossing" in the sense that massive modes may become massless in the new compactified geometry[16)]; recently, it was shown that level crossing indeed occurs in the squashed sphere compactification[34),21)] which reduces the supersymmetries from $N = 8$ to $N = 1$ [20)]. Finally, we wish to point out that the emphasis on the zero-mass sector based on its possible physical relevance for a low energy phenomenology does not appear to be justified a priori. Massive states have masses of the order of the inverse seven-sphere radius $|m_7|$ which is also a measure of the cosmological constant. Thus, compactifications topologically equivalent to the seven sphere can acquire physical significance only if quantum effects could cancel the cosmological term and this will be touched upon in the last section. All these arguments indicate that a selection of the zero-mass sector on physical grounds in the seven-sphere compactification appears premature at least.

We now turn to the evaluation of the full bare mass spectrum, which, as we shall see, contains in fact additional zero mass states not included in the basic $N = 8$ supermultiplet[36)]. We first note that the mass matrix contains spurious modes describing the gauge degrees of freedom of the theory. These can be eliminated at the outset by fixing a gauge. Such a procedure has the advantage that a convenient gauge choice can be made to undo mixings between remaining physical fields. In this way, field redefinitions become superfluous; moreover, while these redefinitions were useful at the level of the zero-mass supermultiplet, they would have to be extended in a mass dependent, and hence a non-local way, in order to cope with the full spectrum.

The gauge invariances of the linearized equations for field fluctuations are

$$\delta_G h_{MN} = 2\,\xi_{(M;N)} \tag{3.10}$$

$$\delta_G a_{MNP} = \Lambda_{[MN;P]} \quad , \quad \Lambda^{MN}{}_{;N} = 0 \tag{3.11}$$

$$\delta_G \psi_M = \overset{0}{\mathcal{D}}_M \chi \tag{3.12}$$

Here $h_{MN} \equiv -\delta g_{MN}$ $(h^{MN} \equiv \delta g^{MN})$, $a_{MNP} \equiv \delta A_{MNP}$, and the symbol δ_G labels the eleven co-ordinate, the 45 Maxwell, and the 32 supersymmetry gauge transformations. From hereon, covariant derivatives are taken with respect to the background and indices are raised and lowered by the background metric tensor. Correspondingly, $\overset{0}{\mathcal{D}}_M$ is a background supercovariant derivative and, from (2.5), we have

$$\overset{0}{\mathcal{D}}_\mu \chi = \chi_{;\mu} - m_7 \gamma^5 \gamma_\mu \chi \tag{3.13}$$

$$\overset{0}{\mathcal{D}}_m \chi = \chi_{;m} - \frac{1}{2} m_7 \Gamma_m \chi \tag{3.14}$$

Convenient gauge choices are[36),37)]

$$h^{m\mu}{}_{;m} = h^{\nu\mu}{}_{;\nu} = 0 \qquad \text{(10 conditions)} \tag{3.15}$$

$$h^M{}_M = 0 \qquad \text{(1 condition)} \tag{3.16}$$

$$a^{mnp}{}_{;p} = a^{m\mu p}{}_{;p} = a^{\mu\nu p}{}_{;p} = 0 \tag{3.17}$$

(15 + 24 + 6 = 45 conditions)

$$\tilde{\Gamma}^M \psi_M = 0 \qquad \text{(32 conditions)} \tag{3.18}$$

Now we consider the fermionic sector[37)]. By substituting the Freund-Rubin expectation value for $\overset{0}{F}_{MNPQ}$ in (3.9) and by writing out four- and seven-dimensional indices explicitly, we obtain

$$\gamma^{\mu\nu\rho} \psi_{\rho;\nu} + \gamma^5 \gamma^{\mu\nu} \Gamma^m \psi_{m;\nu} - \gamma^5 \gamma^{\mu\nu} \Gamma^m \psi_{\nu;m} +$$
$$+ \gamma^\mu \Gamma^{mn} \psi_{n;m} + \frac{3}{2} m_7 \gamma^5 \gamma^{\mu\nu} \psi_\nu = 0 \tag{3.19}$$

and

$$\gamma^5 \gamma^{\mu\nu} \Gamma^m \psi_{\nu;\mu} + \gamma^\mu \Gamma^{mn} \psi_{\mu;n} - \gamma^\mu \Gamma^{mn} \psi_{n;\mu} +$$
$$+ \gamma^5 \cdot \Gamma^{mnp} \psi_{p;n} - \frac{3}{2} m_7 \gamma^5 \Gamma^{mn} \psi_n = 0 \tag{3.20}$$

Clearly, (3.19) and (3.20) mix the four-dimensional spin-$\frac{3}{2}$ and spin-$\frac{1}{2}$ fields. In Refs. 16) and 18), Eqs. (3.19) and (3.20) were diagonalized by a linear field redefinition

$$\psi_\mu = \psi'_\mu + \tfrac{1}{2}\gamma^5\gamma_\mu\Gamma^m\psi'_m$$
$$\psi_m = \psi'_m \tag{3.21}$$

However, it is not difficult to see that this procedure only works for the zero modes and fails to decouple the massive modes. Instead of using (3.21), we therefore simply substitute the gauge condition (3.18) which reads

$$\tilde{\Gamma}^M\psi_M = \gamma^\mu\psi_\mu + \gamma^5\Gamma^m\psi_m = 0 \tag{3.22}$$

into (3.19) and (3.20). In this way, we get

$$\begin{aligned}-\gamma^\nu\psi^\mu{}_{;\nu} - m_7\gamma^5\gamma^\mu\gamma^\nu\psi_\nu &=\\ = -\tfrac{3}{2}m_7\gamma^5\psi^\mu &+ \gamma^5\Gamma^m\psi^\mu{}_{;m}\end{aligned} \tag{3.23}$$

and

$$\begin{aligned}-\gamma^5\gamma^\mu\psi^m{}_{;\mu} = \Gamma^n\psi^m{}_{;n} &-\\ - m_7\left(\Gamma^m\Gamma^n\psi_n - \tfrac{3}{2}\psi^m\right)&\end{aligned} \tag{3.24}$$

Consequently, the equations for the four-dimensional spin-$\frac{1}{2}$ and gravitino fields have been completely decoupled by the gauge choice (3.18). Inspection now shows that, up to an additive constant - the canonical mass term for a massless gravitino in $(AdS)_4$ space - the spin-$\frac{3}{2}$ mass matrix is given by the eigenvalues of the Dirac operator on S^7. Similarly, the spin-$\frac{1}{2}$ mass matrix is given by the eigenvalues of the operator on the right-hand side of (3.24). The eigenmodes of the Dirac operator on S^7 can be easily constructed as follows. The spherical harmonics $Y(y)$ on S^7 obey the equations[38)]

$$\Delta Y(y) = k(k+6)\,Y(y), \quad k = 0,1,2,\ldots \tag{3.25}$$

where Δ is the Laplace-Beltrami operator on S^7. To solve the equation $\not{D}\phi = \lambda\phi$ and to find the eigenvalues λ, we make the ansatz

$$\varphi(y) = \alpha\, Y(y)\,\eta(y) + \beta\, Y_{,m}(y)\,\Gamma^m\,\eta(y) \tag{3.26}$$

where $\eta(y)$ is the covariantly constant spinor on S^7. After a little calculation, one finds that, for each $k \geq 1$, there are two eigenvalues whereas there is only one for $k = 0$, namely

$$\lambda = \frac{7}{2} \quad \text{for } k = 0$$

$$\lambda = k + \frac{7}{2} \quad \text{and} \quad -k - \frac{5}{2} \quad \text{for } k \geq 1 \tag{3.27}$$

with

$$\frac{\alpha}{\beta} = \lambda + \frac{5}{2} \qquad \left(\lambda \neq -\frac{5}{2}\right) \tag{3.28}$$

The spin-$\frac{1}{2}$ mass matrix is obtained in an analogous fashion, although the calculation is considerably more tedious [details will be given in Ref. 37)]. To determine the mass spectrum, it follows from (3.24) that one has to solve the eigenvalue equation

$$\Gamma^n \psi^m{}_{;n} - \Gamma^m \Gamma^n \psi_n = \lambda \psi^m \tag{3.29}$$

It can be shown that Eq. (3.29) gives rise to six towers of massive states. Two of these are spurious gauge modes of the form $\psi_m = \overset{0}{\mathfrak{D}}_m \phi$; their eigenvalues are given by

$$\lambda = k + \frac{3}{2} \quad \text{and} \quad -k - \frac{9}{2} \quad , k \geq 1 \tag{3.30}$$

The four other towers are physical and characterized by $\overset{0}{\mathfrak{D}}_m \psi^m = 0$; their mass eigenvalues are

$$\lambda = k + \frac{7}{2} \ (k \geq 1) \quad \text{and} \quad -k - \frac{5}{2} \ (k \geq 2) \tag{3.31}$$

$$\lambda = k - \frac{5}{2} \quad \text{and} \quad -k - \frac{17}{2} \ (k \geq 1) \tag{3.32}$$

It should be understood that the eigenmodes corresponding to (3.27), (3.30), (3.31) and (3.32) belong to irreducible representations of SO(8); we have suppressed their representation labels for simplicity. The symmetry assignments of all the modes will be the subject of Section 3.3.

The existence of gauge mode solutions to Eq. (3.29) follows from the gauge condition Eq. (3.18) which mixes seven- and four-dimensional subspaces. These modes determine four-dimensional constraints which must be imposed on the massive spin-$\frac{3}{2}$ fields to eliminate spurious degrees of freedom. Indeed, combining Eq. (3.18) with the equation of motion (3.9), we get

$$\psi^M{}_{;M} = \psi^\mu{}_{;\mu} + \psi^m{}_{;m} = \tag{3.33}$$
$$= \frac{1}{2} \gamma^5 \gamma^\mu \psi_\mu = -\frac{1}{2} \Gamma^m \psi_m$$

Thus, using (3.22) and (3.33), we may rewrite the gauge condition as

$$\mathring{\mathcal{D}}^M \psi_M = \mathring{\mathcal{D}}^\mu \psi_\mu + \mathring{\mathcal{D}}^m \psi_m = 0 \tag{3.34}$$

For spurious modes $\mathring{\mathcal{D}}^m \psi_m \neq 0$ and hence Eq. (3.34) determines the required constraint on $\mathring{\mathcal{D}}^\mu \psi_\mu$ for a spin-$\frac{3}{2}$ field having the same SO(8) content as the gauge mode. To understand why the solution of Eq. (3.29) naturally splits into a space of gauge modes $\psi_m = \mathring{\mathcal{D}}_m \chi$ and the orthogonal space of physical solutions $\mathring{\mathcal{D}}_m \psi^m = 0$, we apply the operator $\mathring{\mathcal{D}}_m$ to the eigenvalue equation (3.29); we obtain

$$\Gamma^m (\mathring{\mathcal{D}}^n \psi_n)_{;m} = (\lambda + 1)\, \mathring{\mathcal{D}}^n \psi_n \tag{3.35}$$

If one now decomposes ψ_m into pieces which are transversal and longitudinal with respect to the operator $\mathring{\mathcal{D}}_m$, viz.

$$\psi_m = \chi_m + \mathring{\mathcal{D}}_m \chi \,, \quad \mathring{\mathcal{D}}^m \chi_m = 0 \tag{3.36}$$

one infers from (3.34) that this decomposition is maintained by the spin-$\frac{1}{2}$ mass operator. This means that the eigenspace of this operator decomposes into the space of spurious states with $\mathring{\mathcal{D}}^m \psi_m \neq 0$ and the space of physical states associated with the eigenvalues (3.31) which obey

$$\mathring{\mathcal{D}}^m \psi_m = 0 \tag{3.37}$$

Equation (3.37) is a genuine seven-dimensional gauge condition. Note that the decomposition

$$\psi_m = \chi'_m + \Gamma_m \chi' \,, \quad \Gamma^m \chi'_m = 0 \tag{3.38}$$

does not leave these two subspaces invariant. In fact the two towers (3.31) are traceless while the two others, Eq. (3.32), are not. According to the gauge condition (3.18), the modes given by Eq. (3.32) can thus be classified according to the same SO(8) representation as the spin-$\frac{1}{2}$ fields; this correspondence, however, occurs between different supermultiplets, as will become evident in Section 3.3.

The masses of the spin-$\frac{1}{2}$ and spin-$\frac{3}{2}$ particles are defined by the eigenvalues $m_{\frac{1}{2}}$ and $m_{\frac{3}{2}}$ of the four-dimensional differential operators appearing in the left-hand side of Eqs. (3.23) and (3.24). They are given, in units of m_7, by Eqs. (3.27), (3.31) and (3.32), up to an additive constant. We thus have

$$m^{(1)}_{3/2} = k + 2 \,, \quad k \geq 0 \tag{3.39}$$

$$m_{3/2}^{(2)} = -k-4 \quad , \quad k \geq 1 \tag{3.40}$$

$$m_{1/2}^{(1)} = k-1 \quad , \quad k \geq 1 \tag{3.41}$$

$$m_{1/2}^{(2)} = -k-7 \quad , \quad k \geq 1 \tag{3.42}$$

$$m_{1/2}^{(3)} = k+5 \quad , \quad k \geq 1 \tag{3.43}$$

$$m_{1/2}^{(4)} = -k-1 \quad , \quad k \geq 2 \tag{3.44}$$

The superscripts label the towers. The members of the "massless" N = 8 supermultiplet are at the bottom of the towers $m_{\frac{3}{2}}^{(1)}$ and $m_{\frac{1}{2}}^{(1)}$. Note that for convenience the "massless" gravitino which is anyway not conformal[39] has been given the value +2. We do not list the SO(8) content of the modes here as this will be discussed in the next section.

The fermionic sector illustrates all the essential features of the procedure. Thus, for the bosonic sector we shall simply list the results[36]. The $(\text{mass})^2$ operators are defined by the following differential operators

$$\text{spin-0:} \quad \eta_{;\mu}{}^{;\mu} - 8 m_7^2 \eta = -m_0^2 \eta \tag{3.45}$$

$$\text{spin-1:} \quad (\eta_{\mu;\nu} - \eta_{\nu;\mu})^{;\nu} = -m_1^2 \eta_\mu \tag{3.46}$$

$$\text{spin-2:} \quad \eta_{(\mu\nu);\rho}{}^{;\rho} - 16 m_7^2 \eta_{(\mu\nu)} = -m_2^2 \eta_{(\mu\nu)} \quad (\eta^\mu{}_\mu = \eta^{\mu\nu}{}_{;\nu} = 0) \tag{3.47}$$

In this way, the spin-0 and spin-1 members of the "massless" supermultiplet have indeed a zero mass while the (non-conformal) graviton has been assigned a mass +8 in units of m_7^2. In these units, one obtains

$$m_2^2 = (k+3)^2 - 1 \quad , \quad k \geq 0 \tag{3.48}$$

$$m_{1+}^2 = (k+3)^2 - 1 \quad , \quad k \geq 1 \tag{3.49}$$

$$(m_{1-}^{(1)})^2 = k^2 - 1 \quad , \quad k \geq 1 \tag{3.50}$$

$$(m_{1-}^{(2)})^2 = (k+6)^2 - 1 \quad , \quad k \geq 1 \tag{3.51}$$

$$(m_{0-}^{(1)})^2 = k^2 - 1 \quad , \quad k \geq 1 \tag{3.52}$$

$$(m_{0-}^{(2)})^2 = (k+6)^2 - 1 \quad , \quad k \geq 1 \tag{3.53}$$

$$(m_{0+}^{(1)})^2 = (k-3)^2 - 1 \quad , \quad k \geq 2 \tag{3.54}$$

$$(m_{0+}^{(2)})^2 = (k+9)^2 - 1 \quad , \quad k \geq 0 \tag{3.55}$$

$$(m_{0+}^{(3)})^2 = (k+3)^2 - 1 \quad , \quad k \geq 2 \tag{3.56}$$

The massless supermultiplet is given by the lowest value of k in the towers m_2, $m_{1-}^{(1)}$, $m_{0-}^{(1)}$ and $m_{0+}^{(1)}$. Note the appearance of an additional zero-mass supermultiplet for $k = 4$ and a multiplet with $m^2 = -1$ for $k = 3$ in the scalar tower $m_{0+}^{(1)}$.

To end this section, we wish to stress that the expression of the modes in terms of fields depends on the gauge choice while the spectrum is, of course, gauge invariant. This is best illustrated in the case of the massless spin-$\frac{1}{2}$ mode. In Refs. 16) and 18), it was proved that the ansatz

$$\psi_m(x,y) = \chi^{IJK}(x) \left\{ \eta^{[I}(y) \bar{\eta}^{J}(y) \Gamma_m \eta^{K]}(y) - \right.$$
$$\left. - \frac{1}{9} \Gamma_m \Gamma^n \eta^{[I}(y) \bar{\eta}^{J}(y) \Gamma_n \eta^{K]}(y) \right\} \tag{3.57}$$

describes the 56 massless spin-$\frac{1}{2}$ fields of $N = 8$ supergravity [I, J, K are SO(8) indices]; in the proof, use was made of the field redefinition (3.21). In the gauge (3.18), however, the result is different, and one easily checks that the right-hand side of (3.24) vanishes for

$$\psi_m(x,y) = \chi^{IJK}(x) \eta^{[I}(y) \bar{\eta}^{J}(y) \Gamma_m \eta^{K]}(y) \tag{3.58}$$

i.e., the trace term has disappeared. One immediately verifies that (3.58) also satisfies the gauge condition (3.37) whereas (3.57) does not. The notion of ansatz is therefore a gauge-dependent concept in general.

3.3 Group theory: Osp(8,4) classification

Up to this point, the symmetry assignments of the various modes have not been discussed in any detail. The mass spectrum of $N = 8$ supergravity on S^7 in the bosonic[36] and fermionic case[37] has been determined by solving the appropriate eigenvalue equations, and no explicit reference to the SO(8) and supersymmetry content of the modes was necessary. The SO(8) assignments can be deduced from those of the spherical (scalar, vector and tensor) harmonics on S^7 which are known[38], but this is not sufficient to group the various states into supermultiplets. For a complete classification, one has to make use of the full invariance of the S^7 ground state. This group contains not only the 28 rotations of SO(8) corresponding to the 28 Killing vectors on S^7 but also eight spinorial translations which correspond to the eight Killing spinors on S^7. Together, the generators associated with these bosonic and fermionic transformations constitute the graded Lie algebra Osp(8,4), and a rigorous proof of the Osp(8,4) invariance of the S^7 ground state has been given in Ref. 40). The excitations corresponding to the fluctuations about the ground state should therefore form irreducible representations of Osp(8,4). From the general Kaluza-Klein theory[9),10)] and the absence of higher spin fields in eleven-dimensional supergravity, it follows that the relevant representations are those with maximum spin 2. The latter have been classified in Ref. 41); and we will restrict our attention to these representations here. For a comprehensive review of the general construction and properties of unitary irreducible representations of Osp(N,4), we refer the reader to Ref. 42).

We have already mentioned that the masses of the excited states are proportional to the inverse radius $|m_7|$ of the seven sphere. Thus, in the limit $m_7 \to 0$ where the space becomes flat, all masses tend to zero. In this limit, the relevant superalgebra is the Poincaré superalgebra, and we conclude that in this contraction limit, the massive representations of Osp(8,4) become massless representations of $N = 8$ Poincaré supersymmetry[*)]. This has the very important consequence that all massive representations of Osp(8,4) with maximum spin 2 must be obtainable from massless representations of $N = 8$ supersymmetry with the same spin limit. There is only one such multiplet with maximum spin 2, namely the massless $N = 8$ multiplet already mentioned in the introduction. It contains one graviton [$\underset{\sim}{1}$ of SO(8)], eight gravitinos[**)] (= 8_s), 28 spin-1 fields (= 28), 56 spin-$\frac{1}{2}$ fields (= 56_s), 35 scalars

---

*) In $N = 8$ Poincaré supersymmetry, massive (Poincaré) multiplets go up to spin 4, if SO(8) is preserved, and can be restricted to stop at spin 2 only if central charges are introduced which break SO(8); Our argument shows, however, that no inconsistency with this result arises[41)].

**) The assignment of the spin-$\frac{3}{2}$ fields to the 8_s representation (instead of 8_v or 8_c) is a matter of conventions.

(= 35_v) and 35 pseudoscalars (= 35_c) [for the group theoretic conventions, see Ref. 43)]. Hence, one should be able to derive all massive Osp(8,4) multiplets from products of the form

$$R \otimes \{1, 8_s, 28, 56_s, 35_v, 35_c\} \tag{3.59}$$

where R is an as yet unspecified representation of SO(8).

To facilitate the discussion, we next introduce Dynkin labels to classify the representations of SO(8) [43)]. Each irreducible representation of SO(8) can be uniquely labelled by a set $(a_1a_2a_3a_4)$ of four non-negative integers a_1, a_2, a_3, a_4. Since the massless graviton which belongs to the massless N = 8 multiplet is an SO(8) singlet, the charged massive gravitons will carry the same label as the relevant irreducible representation. One now realizes that the representation R which occurs in (3.59) is no longer arbitrary, since we know from the explicit calculations[9),10)] that the massive gravitons are in one-to-one correspondence with the eigenfunctions of the Laplacian on S^7, (3.25), i.e., the spherical harmonics on S^7. These are characterized by the Dynkin labels (n000), $n \in \mathbb{N}$, which correspond to the symmetric and traceless SO(8) tensors with n indices. To obtain the full Osp(8,4) multiplet, we replace R in (5.2) by (n000), perform the multiplication and identify the irreducible components in this product. The Dynkin labels of the massless representation are given by

$$\begin{aligned} 8_s &= (0001) \\ 28 &= (0100) \\ 56_s &= (1010) \\ 35_v &= (2000) \\ 35_c &= (0020) \end{aligned} \tag{3.60}$$

The result of this multiplication, which is given in Ref. 41), is, however, not yet the final answer. One still has to add lower helicity states to the spin-2, spin-$\frac{3}{2}$ and spin 1 fields to make them massive. The lower helicity states which are absorbed must belong to the same representation as the gauge field into which they are absorbed. The final result which is obtained after absorbing these states reads[*)]:

$$\text{spin-}2: \quad (n\,0\,0\,0)$$

$$\text{spin-}\tfrac{3}{2}: \quad (n\,0\,0\,1) \oplus (n{-}1\,0\,1\,0)$$

---

*) For the special case n = 1, this result was first obtained in Ref. 34).

$$
\begin{aligned}
\text{spin}-1^{+} &: \ (n-1\ 0\ 1\ 1) \\
\text{spin}-1^{-} &: \ (n\ 1\ 0\ 0) \oplus (n-2\ 1\ 0\ 0) \\
\text{spin}-\tfrac{1}{2} &: \ (n+1\ 0\ 1\ 0) \oplus (n-1\ 1\ 1\ 0) \oplus \\
&\quad \oplus (n-2\ 1\ 0\ 1) \oplus (n-2\ 0\ 0\ 1) \\
\text{spin}-0^{+} &: \ (n+2\ 0\ 0\ 0) \oplus (n-2\ 2\ 0\ 0) \oplus \\
&\quad \oplus (n-2\ 0\ 0\ 0) \\
\text{spin}-0^{-} &: \ (n\ 0\ 2\ 0) \oplus (n-2\ 0\ 0\ 2)
\end{aligned}
\tag{3.61}
$$

where, whenever an integer is negative, the associated representation does not exist; for example, the second spin-$\frac{3}{2}$ tower starts only at $n = 1$. For each n, (3.61) is an irreducible representation of Osp(8,4), and the integer n therefore labels the "floors" of the massive tower*).

The SO(8) content of the spectrum as calculated directly by solving the relevant eigenvalue equations[36),37)] is in complete agreement with (3.61) which was obtained in a completely different manner. For example, the two eigenmodes of the Dirac operator found before [see (3.27)] exactly correspond to the two representations in (3.61), and the absence of the "ground-floor" for the second tower was also obtained here. Similarly, the four spin-$\frac{1}{2}$ towers of (3.31) and (3.32) have their counterparts (3.61), if one properly adjusts the relation between n and k. Analogous considerations apply to the bosonic sector.

We can express the mass formulas of Section 3.2 in terms of the supersymmetry index n and hence the spectrum is completely characterized by Eqs. (3.39)-(3.44) and Eqs. (3.48)-(3.56). The results are summarized in the Table, where we have marked the Dynkin labels of the towers containing massless multiplets by an asterisk, and relabelled the various towers.

Besides spin and SO(8) content, the complete characterization of the Osp(8,4) states requires the knowledge of the lowest eigenvalues E_0 of the "energy operator" M_{04} of the SO(2,3) subalgebra of Osp(8,4). The energy labels are most easily found by using the Osp(1,4)xSO(7) subalgebra of Osp(8,4) and the known energy labels of Osp(1,4) representations[44)]. The results are given in Ref. 37) and can be summarized by the following universal relations between mass and energy:

---

*) It is perhaps instructive to note that the massive modes on the hypertorus T^7 [whose massless sector corresponds to the $N = 8$ theory of Ref. 7)] have a much simpler structure. The relevant group is $[U(1)]^7$ in this case, and after recombining helicities, one easily sees that the massive multiplets all have the same multiplicities (1,8,27,48,42).

$$E_0 = \frac{3}{2} \pm \frac{1}{2}\sqrt{m^2+1} \quad \text{for bosons} \tag{3.62}$$

$$E_0 = \frac{3}{2} + \frac{1}{2}|m| \quad \text{for fermions} \tag{3.63}$$

where all the bosonic states obey relation (3.62) with the + sign except the 35_v massless scalars for which the - sign is required.

The universality of Eqs. (3.62) and (3.63) implies that E_0 has a dynamical significance. In fact, we know from Ref. 45) that the relation (3.62) for spin-0 fields characterizes modes which die fast enough at infinity to ensure energy conservation in AdS. The reality of E_0 is guaranteed by the fact that $m^2 \geq -1$ [45)], the lower limit being reached in the $0^{+(1)}$ tower for $n = 2$ $(k = 3)$ with a multiplet of 112 scalars. For $n = 3$ $(k = 4)$ the same tower contains again conformal massless modes (294 scalars) with vanishing energy flow at spatial infinity. However, they must satisfy different boundary conditions, characterized by the + sign in Eq. (3.62)[45)], than the 35_v in order not to break supersymmetry. In this way they fit indeed as massless members in the "massive" supermultiplet $n = 2$. Note from the Table that, quite generally, in contrast to Poincaré supersymmetry, states belonging to the same supermultiplet characterized by n may have different masses because of the non-commutativity of the energy operator (M_{04}) with supersymmetry generators. Thus we see that for scalar modes, the supersymmetric spectrum is consistent with a Hilbert space of functions with boundary condition preventing energy flow in and out of AdS and hence admitting well-defined Cauchy data in this otherwise unviable space. We infer that such a property holds for all the modes because of the universality of Eqs. (3.62) and (3.63), a conjecture that could be checked explicitly following the method of Ref. 45). It follows from the positivity of E_0 that the seven sphere is stable against small fluctuations belonging to this Hilbert space, a fact which also follows from supersymmetry[46)].

The emergence of such a Hilbert space in the context of the seven-sphere compactification in an AdS background may have an important consequence. At least we cannot dismiss a priori that inclusion of quantum effects and strong localized disturbances could lead to a smooth Minkowskian limit when $|m_7| \to 0$. This possibility would not only cure the physical problem posed by the AdS cosmological constant but would put into new terms the long-standing chirality issue. Indeed, the arguments preventing complex fermions to appear in this Kaluza-Klein approach[11),47),48)] have to be confronted with the infinite dimensional space of modes which would become relevant in this limit. This will be discussed in Section 4, whose speculative content, needless to say, cannot as yet be based on any firm mathematical basis.

4. - TOWARDS QUANTUM SUPERGRAVITY

Spontaneous compactifications provide classical solutions which play the rôle of a "mean field" average over quantum metric fluctuations. Hence at least a qualitative understanding of the nature of these fluctuations seems necessary in order to appreciate the relevance of such "mean field" configurations. The difficulty which then appears at the non-perturbative level clearly required here is the same as in ordinary gravity. Indeed, even if perturbative ultra-violet behaviour would turn out to be controllable, we cannot avoid the indefinite sign of R (in the Euclidean sector) due to non-perturbative scale fluctuations, which make the gravitational action unbounded from below[49]. How could then the path integral over metrics be defined at all? Moreover, in compactified supergravity, a new question arises which cannot be eliminated even by a fine tuning procedure: what are the fluctuations which can produce the vacuum energy required to make the cosmological constant vanish? This leads naturally to a third question: Is it conceivable that such fluctuations could provide a mechanism to stabilize the path integral? We suggest that the answer to these questions lies in the structure of space-time revealed by the "cosmological puzzle"[50], and we speculate about the possibly related chirality issue raised at the end of the last section.

It is well known that the conventional hot big bang theory is hardly acceptable because of the tuning of initial conditions required. In this theory, the evolution of the early Universe is determined mainly by the set of asymptotically free massless fields in thermal equilibrium with the photon and the expansion is adiabatic. The cosmological state is then defined by the temperature T, the number of degrees of freedom $\nu(T)$ $[\equiv \nu_{bose}(T) + \frac{7}{8}\nu_{fermi}(T)]$ and the scale factor a. These parameters are related by Einstein's equations and by the black-body formulae [see, e.g., Weinberg's book[51]] $(h = c = G = k = 1)$

$$aT = \frac{3}{2\pi}\left[\frac{5\mathcal{S}}{\nu(T)}\right]^{1/3} \tag{4.1}$$

$$\left(\frac{\dot a}{a}\right)^2 + \frac{k}{a^2} = \frac{8}{90}\pi^3\nu(T)T^4 \tag{4.2}$$

where $\mathcal{S}$ is the total conserved entropy within a volume $4\pi a^3/3$ and $k = \pm 1$ (the limiting case $k/a^2 = 0$ need not be considered separately) according to the closed or open character of the Universe. The presently observed background radiation in the volume $(4\pi/3)H^{-3}$ $[H \equiv \dot a/a]$ yields $\mathcal{S} \gtrsim 10^{87}$ [we used $a \gtrsim H^{-1}$ because the Universe is either open $(a > H^{-1})$ or anyway nearly flat]. As $\mathcal{S}$ is conserved, we have $a \gtrsim 10^{29}/\nu^{\frac{1}{3}}(T)$ from Eq. (4.1).when T approaches 1 $(\sim 10^{19}$ GeV). Thus the classical "mean field" values $(\dot a/a)^2$ and $\frac{8}{90}\pi^3\nu(T)T^4$ in Eq. (4.2) have to be tuned to an accuracy of the order of k/a^2, $(a \gtrsim 10^{29}/\nu^{\frac{1}{3}}(1))$, when quantum fluctuations are of the order of 1. If $\nu(1)$ remains bounded, such a tuning appears absurd in the absence of a mechanism producing an initial correlated classical state.

A solution to the tuning problems posed by the "flatness" of the Universe at proper times $t \simeq O(1)$ requires that prior to the adiabatic expansion, the Universe experienced a different régime in which the scale factor could grow rapidly in a natural way from $O(1)$ to, say, $O(10^{30})$. Such a primordial "inflation" was initially proposed in the context of the creation of the Universe itself[52]. Indeed, in some cases one finds a semi-classical solution of the coupled gravity-matter system such that $a(t)$ grows exponentially:

$$a(t) = e^{t/\tau} \qquad \left[H = \frac{1}{\tau}\right] \qquad t \gtrsim 1 \tag{4.3}$$

The initial value $a(1)$ was then attributed to a quantum effect. Such a solution can in fact occur because of the "wrong" sign of R for the scale mode constituted by the cosmological expansion; in more intuitive terms, matter can be created at the expense of a negative expansion energy where the energy is measured in the conformal Minkowskian background. However, matter quanta must be higher than the Planck mass[53] in order to sustain the semi-classical solution and a reinterpretation of these primordial quanta in terms of black holes[50] was proposed. Other inflationary scenarios leading to Eq. (4.3), based on the cosmological constant induced by a phase transition at some grand unified scale, have been proposed[54],[55]. We shall not restrict ourselves here to a specific model leading to Eq. (4.3), but we shall simply assume that this equation correctly describes the primordial phase of the Universe.

Equation (4.3), however, does not fully solve the initial tuning problem because there seems to be no way to stabilize an initial spatial curvature $(k/a^2 \simeq \pm 1)$. The latter is indeed inconsistent with Eq. (4.3) and could only arise therefore from quantum fluctuations at $t \lesssim 1$. As, however, any curvature appears flat for short distances, we may conclude that Eq. (4.3) solves the tuning problem if the initial classical state $a \simeq O(1)$ extends only over a finite proper length $L \simeq a \simeq O(1)$. An observer within the Universe would hardly feel the finiteness of it as a consequence of Birkhoff's theorem[51] and thus the presently seen Universe can emerge from an initial classical state characterized by

$$L = O(1) \quad \text{when} \quad t = O(1) \tag{4.4}$$

Here comes the puzzle; the naturalness of the classical initial condition, Eq. (4.4), means that this initial state may well arise from a quantum metric fluctuation. But a fluctuation on which background? A natural answer appears to be on a "mean field" which should be Minkowskian at scales large compared to the Planck size. This is indeed suggested both by the experimental fact that space is indeed flat for such scales, and by the theoretical fact mentionned earlier that in such a background the expansion energy is negative and can ensure energy conservation for the quantum transition giving birth to the Universe. If this is the case, however, a "Universe" can be born anywhere and at any time within our own "Universe" and quantum theory tells us that "we" must include such Universe-like configurations in the path integral over metrics.

A possible solution to the puzzle is that such configurations do exist, namely that within any "Universe" one can construct other "Universes" of comparable proper space-time extensions but which may be viewed as Planckian fluctuations with respect to the first one. To see that this does not lead to contradictions, we rewrite the (de Sitter) metric describing the exponential expansion of such a fluctuation

$$ds^2 = dt^2 - e^{2t/\tau}\, d\vec{x}^2 \quad . \quad |\vec{x}| \lesssim L \tag{4.5}$$

in the conformally Minkowskian form

$$ds^2 = \left(1 - \bar{t}/\tau\right)^{-2} \left\{ d\bar{t}^2 - d\vec{x}^2 \right\} , \quad |\vec{x}| \lesssim L \tag{4.6}$$

We see that around $\bar{t} = 0$, the co-ordinate system $(\bar{t},\vec{x})$ may be used to parametrize the fluctuation in our Minkowskian background. The whole exponential expansion reduces to a quasi instantaneous event $(\bar{t} < \tau)$ for the external Minkowskian observer and in fact the subsequent adiabatic period will also last for a time $\bar{t}$ of order τ. Indeed, the interior of the Planckian Universe is only protected from the outside in the space-time region bounded by the light cones originating from the "edge" $|x| \simeq L$. From the conformally flat metric, Eq. (4.6), this time $\bar{t}_f = O(L) = O(1)$. After the time $\bar{t}_f$ the Planckian Universe will then be destroyed by the vacuum fluctuations from which it originated. More generally, one can show that no paradox arises from TCP reversed configuration[50)]. Thus, a "Universe" may be viewed as a Planckian fluctuation, hence as an unstable excitation of quantum gravity, defined on the Minkowskian background of some other Universe. Stability for "us" is then simply a consequence of the enormous contraction of the proper time due to the exponential expansion.

The number of distinct metric configurations describing ... Universes within Universes within ... has the power of the continuum and may well dominate the path integral of quantum gravity in such a way as to stabilize it in the vicinity of this "foam of Universes". One may thus adopt the point of view that the lower unboundedness of the Euclidean gravitational action is not just a technical nuisance which may be got rid of by rotating the contour of integration for the scale fluctuations into the complex plane[49] but that it has a physical significance. The unboundedness of the action should therefore be compensated for by a decrease of the functional measure of the associated configurations, and we conjecture that it is precisely the foamy structure of space-time that provides the needed mechanism (the relation of this scenario with Hawking's space-time foam[56] is not clear). This line of thought also suggests that the unperturbed four-dimensional space-time is transmuted to a fractal space-time. The Hausdorff dimension of the fractal[57] set available for quantum mechanical propagations may well be less than four: propagation within any Universe would be damped by a factor involving the ratio of the Planck size to the proper size of the Universe and would therefore be practically unavailable. To test this picture of quantum gravity, one should thus try to construct a fractal space endowed with an intrinsic metric.

Interestingly enough, the above picture of ... Universes within Universes within ... fits quite naturally into the Kaluza-Klein picture as is most easily illustrated by a two-dimensional analogy. Consider the space $\mathbb{R}_1 \times \mathbb{R}_1$ compactified on $\mathbb{R}_1 \times S_1$; this is just a dylinder where $\mathbb{R}_1$ is the analogue of $\mathcal{M}_4$ and S_1 is the analogue of the internal space. Clearly, one can construct new two-dimensional manifolds by locally opening the original cylinder, inserting a new one, and ... repeating the procedure. This construction is illustrated in the Figure. Clearly, as viewed from one cylinder (one "Universe"), other cylinders are either small (Planckian) disturbances or define a background from which it itself originates. In this way, the description of ... Universes within Universes within ... is recovered in the special gauge in which all Universes belong to the same (four-dimensional) space-time but in the general gauge illustrated in the Figure, one gets instead a "space-time proliferation" in which all Universes are manifestly of the same type. Note that matrix elements between different Universes must exist and would already be induced in the tube picture of the Figure by tube loops connecting different tubes.

To conclude this section, we wish to point out that it is conceivable that space-time proliferation in eleven-dimensional supergravity is in fact the mechanism restoring a zero-cosmological constant*). Indeed, when, in a given AdS background, the mass of massive modes exceeds the Planck mass, their vacuum fluctuations may give rise through gravitational interactions to new Universes hence to space-time proliferation. These will thus affect the vacuum energy in a way which is dependent on the radius of the compact manifold and hence on the cosmological constant itself. Hopefully, stability will be achieved for cosmological constants which are small compared to the scale of one Universe, a conjecture strengthened by the fact that the full eleven-dimensional theory does not admit a cosmological constant[8),58)].

---

*) An entirely different mechanism to cancel the cosmological constant through fermion condensates was proposed in Ref. 59).

If the (approximately) zero cosmological constant is stabilized by space-time proliferation, then the whole spectrum of excitations may become relevant. Following this line of thought, we conclude that there is no reason to give a preferred status to the original zero-mass supermultiplet and we suggest instead a different approach to make contact with low energy phenomena. It has been shown in the last paper of Ref. 10) that the massive spectrum of ordinary five-dimensional Kaluza-Klein theory exhibits a non-compact O(2,1) symmetry. In a similar vein, one may conjecture that the full spectrum on the seven-sphere possesses a non-compact symmetry group of which Osp(8,4) is only a subgroup. The full spectrum would then transform as one single irreducible representation of this larger (super)group but would be reducible under Osp(8,4). In fact, the supersymmetry in eleven dimensions provides a hint that such a larger group must exist. For the fermionic generators of Osp(8,4) just correspond to the lowest eigenmode of the Dirac operator on S^7 in the expansion of the eleven-dimensional supersymmetry transformation parameter $\varepsilon(x,y)$. Higher modes in this expansion correspond to more complicated fermionic charges which will transform states at different levels of the massive tower into each other. This infinite-dimensional superalgebra, which may possibly be extended to include dynamical symmetries, may contain non-compact Lie algebras which could be used to classify the physical states. At this point, one may invoke the curious fact that real unitary representations of non-compact groups may contain representations which are complex under a compact subgroup, a property which has no counterpart in the theory of compact groups*). If such subgroups are relevant, chirality may be spontaneously generated by this mechanism, and one may hope that the relevant representations are dynamically isolated. It is remarkable that the concept of "spontaneous breaking of reality" in a vector-like theory seems to require the existence of an infinite number of states and provides one more reason why one should expect the number of space-time dimensions to be greater than four.

Some solutions describing local compactification in eleven-dimensional supergravity have been found and discussed in [60]; these solutions may in fact be of relevance for the question of space-time proliferation. The possible origin of hidden symmetries in compactified supergravities has been discussed in [61]. Finally, recent reviews of the subject which contain further references, have been listed in [62].

---

*) We are grateful to M. Günaydin for bringing this result to our attention.

Spin	SO(8) content		$(\text{Mass})^2$ in units of m_7^2
2^+	$(n\ 0\ 0\ 0)^*$	$n \geq 0$	$(n+3)^2 - 1$
$\frac{3}{2}^{(1)}$	$(n\ 0\ 0\ 1)^*$	$n \geq 0$	$(n+2)^2$
$\frac{3}{2}^{(2)}$	$(n-1\ 0\ 1\ 0)$	$n \geq 1$	$(n+4)^2$
$1^{-(1)}$	$(n\ 1\ 0\ 0)^*$	$n \geq 0$	$(n+1)^2 - 1$
1^+	$(n-1\ 0\ 1\ 1)$	$n \geq 1$	$(n+3)^2 - 1$
$1^{-(2)}$	$(n-2\ 1\ 0\ 0)$	$n \geq 2$	$(n+5)^2 - 1$
$\frac{1}{2}^{(1)}$	$(n+1\ 0\ 1\ 0)^*$	$n \geq 0$	n^2
$\frac{1}{2}^{(2)}$	$(n-1\ 1\ 1\ 0)$	$n \geq 1$	$(n+2)^2$
$\frac{1}{2}^{(3)}$	$(n-2\ 1\ 0\ 1)$	$n \geq 2$	$(n+4)^2$
$\frac{1}{2}^{(4)}$	$(n-2\ 0\ 0\ 1)$	$n \geq 2$	$(n+6)^2$
$0^{+(1)}$	$(n+2\ 0\ 0\ 0)^*$	$n \geq 0$	$(n-1)^2 - 1$
$0^{-(1)}$	$(n\ 0\ 2\ 0)^*$	$n \geq 0$	$(n+1)^2 - 1$
$0^{+(2)}$	$(n-2\ 2\ 0\ 0)$	$n \geq 2$	$(n+3)^2 - 1$
$0^{-(2)}$	$(n-2\ 0\ 0\ 2)$	$n \geq 2$	$(n+5)^2 - 1$
$0^{+(3)}$	$(n-2\ 0\ 0\ 0)$	$n \geq 2$	$(n+7)^2 - 1$

TABLE: The spectrum of supergravity on the seven-sphere. The states marked by an asterisk contain the zero-mass supermultiplet.

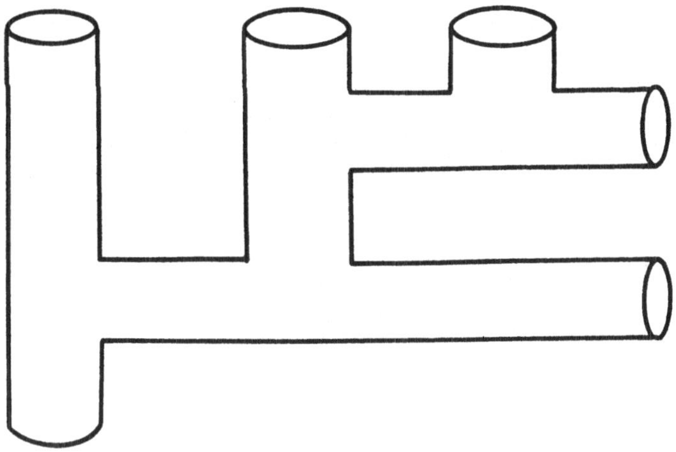

- Figure -

... Universes within Universes within ...
in a two-dimensional Kaluza-Klein analogy.

REFERENCES

1) D. Volkov and V.P. Akulov - Phys.Letters 46B (1973) 109;
J. Wess and B. Zumino - Nuclear Phys. B70 (1974) 39.
2) J. Wess and B. Zumino - Phys.Letters 49B (1974) 52;
J. Iliopoulos and B. Zumino - Nuclear Phys. B76 (1974) 310.
3) S. Ferrara, D.Z. Freedman and P. van Nieuwenhuizen - Phys.Rev. D13 (1976) 3214;
S. Deser and B. Zumino - Phys.Letters 62B (1976) 335.
4) A. Salam and J. Strathdee - Nuclear Phys. B80 (1974) 499;
D.Z. Freedman - in "Recent Developments in Gravitation", Cargèse 1978, eds. M. Levy and S. Deser (Plenum Press, 1979).
5) J. Scherk - same as last reference in 4).
6) E. Cremmer, B. Julia and J. Scherk - Phys.Letters 76B (1978) 409.
7) E. Cremmer and B. Julia - Phys.Letters 80B (1978) 48; Nuclear Phys. B159 (1979) 141.
8) W. Nahm - Nuclear Phys. B135 (1978) 149.
9) Th. Kaluza - Sitzungsber.Preuss.Akad.Wiss. Kl (1921) 966;
O. Klein - Z.Phys. 37 (1926) 895.
10) B. De Witt - Dynamical Theory of Groups and Fields (Gordon and Breach, New York, 1965);
R. Kerner - Ann.Inst. H. Poincaré 9 (1968) 143;
A. Trautmann - Rep.Math.Phys. 1 (1970) 29;
Y.M. Cho and P.G.O. Freund - Phys.Rev. D12 (1975) 1711;
C. Orzalesi - Fortschr.Phys. 29 (1981) 413).
A. Salam and J. Strathdee - Ann.Phys. 141 (1982) 316.
11) E. Witten - Nuclear Phys. B186 (1981) 412.
12) M.J. Duff - in "Supergravity '81", eds. S. Ferrara and J.G. Taylor (Cambridge University Press, 1982);
M.J. Duff and D.J. Toms - in "Unification of the Fundamental Interactions II", eds. S. Ferrara and J. Ellis (Plenum Press, 1982).
13) E. Cremmer and J. Scherk - Nuclear Phys. B103 (1976) 399.
14) P.G.O. Freund and M.A. Rubin - Phys.Letters 97B (1980) 233.
15) F. Englert, M. Rooman and P. Spindel - Phys.Letters 127B (1983) 47.
16) B. Biran, B. de Wit, F. Englert and H. Nicolai - Phys.Letters 124B (1983) 45.
17) B. de Wit and H. Nicolai - Phys.Letters 108B (1981) 285; Nuclear Phys. B208 (1982) 323.
18) M.J. Duff and C.N. Pope - in "Supersymmetry and Supergravity '82", eds. S. Ferrara, J.G. Taylor and P. van Nieuwenhuizen (World Scient.Pub.Comp., 1983).
19) P.G.O. Freund, CERN Preprint TH.3655 (1983).
20) M.A. Awada, M.J. Duff and C.N. Pope - Phys.Rev.Letters 50 (1983) 294.
21) F.A. Bais, H. Nicolai and P. van Nieuwenhuizen - Nuclear Phys. B228 (1983) 333.
22) L. Castellani, R. D'Auria and P. Fré - Torino Preprint IFTT427 (1983).
23) M.J. Duff, B.E. Nilsson and C.N. Pope - "Compactification of d=11 Supergravity on $K_3 \times T^3$", Texas University Preprint (1983).
24) M. Günaydin and F. Gürsey - J.Math.Phys. 14 (1973) 1651.
25) M. Rooman - "Eleven-Dimensional Supergravity, and Octonions", Preprint Université Libre de Bruxelles (1983).
26) E. Cartan and J. Schouten - Proc.Kon.Akad.Wet. Amsterdam 29 (1926) 933.
27) F. Englert - Phys.Letters 119B (1982) 339.
28) R. D'Auria, P. Fré and P. van Nieuwenhuizen - Phys.Letters 122B (1983) 225.
29) B. de Wit and H. Nicolai - Preprint NIKHEF-H/83-8 (1983).
30) F. Englert, M. Rooman and P. Spindel - "Symmetries in Eleven-Dimensional Supergravity Compactified on a Parallelized Seven Sphere", Preprint Université Libre de Bruxelles (1983), to appear in Phys.Letters B.
31) L. Castellani and N.P. Warner - Preprint CALT-68-1033 (1983).
32) J. Lukierski and P. Minaert - Preprint Université de Bordeaux I (1983).
33) M.J. Duff - in the Proceedings of the Marcel Grossmann Meeting, Shanghai (1982), to appear.
34) M.J. Duff, B.E.W. Nilsson and C.N. Pope - Phys.Rev.Letters 50 (1983) 2043.
35) N.P. Warner - Preprint CALT-68-1008 (1983).

36) B. Biran, A. Casher, F. Englert, M. Rooman and P. Spindel - "The Fluctuating Seven Sphere in Eleven Dimensional Supergravity", Preprint Université Libre de Bruxelles (1983).
37) A. Casher, F. Englert, H. Nicolai and M. Rooman - in preparation.
38) S. Gallot and D. Meyer - J.Math. Pures et Appliquées 54 (1975) 259; G.W. Gibbons and M.J. Perry - Nuclear Phys. B146 (1978) 90.
39) S. Deser and R. Nepomechie - "Gauge Invariance Versus Masslessness in de Sitter Space", Preprint Brandeis University (1983).
40) R. D'Auria and P. Fré - Phys.Letters 121B (1983) 141.
41) D.Z. Freedman and H. Nicolai - "Multiplet Shortening in Osp(N,4)", MIT Preprint (1983).
42) M. Günaydin - in Proceedings of the XI International Colloquium on Group Theoretical Methods in Physics, Istanbul (1982). Lecture Notes in Physics 180 (Springer Verlag);
I. Bars and M. Günaydin - CERN Preprint TH. 3350 (1983), to appear in Commun. Math.Phys.
43) R. Slansky - Physics Reports C79 (1981) 1.
44) W. Heidenreich - Phys.Letters 110B (1982) 461.
45) P. Breitenlohner and D.Z. Freedman - Phys.Letters 115B (1982) 197; Ann.Phys. 144 (1982) 249.
46) G.W. Gibbons, C.M. Hull and N.P. Warner - Nuclear Phys. B218 (1983) 173.
47) E. Witten - in Proceedings of the Shelter Island II Conference, to appear.
48) C. Wetterich - Nuclear Phys. B223 (1983) 109.
49) S.W. Hawking - in General Relativity, eds. S.W. Hawking and W. Israel (Cambridge University Press, 1979) and references therein.
50) A. Casher and F. Englert - Phys.Letters 104B (1981) 117.
51) S. Weinberg - "Gravitation and Cosmology", J. Wiley (1972).
52) R. Brout, F. Englert and E. Günzig - Gen.Rel.Grav. 10 (1979) 1.
53) R. Brout, F. Englert and P. Spindel - Phys.Rev.Letters 43 (1979) 417.
54) A. Guth - Phys.Rev. D23 (1981) 347.
55) A. Linde - Phys.Letters 108B (1982) 389.
56) S.W. Hawking - Nuclear Phys. B144 (1978) 349.
57) P. Mandelbrot - "Fractals, Form, Chance and Dimension", Freeman (San Francisco, 1977).
58) H. Nicolai, P. Townsend and P. van Nieuwenhuizen - Lettere al Nuovo Cimento 30 (1981) 315.
59) C. Destri, C.A. Orzalesi and P. Rossi - Ann.Phys. 147 (1983) 147;
M.J. Duff and C.A. Orzalesi - Phys.Letters 122B (1983) 37;
C.A. Orzalesi - CERN Preprint TH. 3647 (1983);
C.A. Orzalesi and G. Venturi - CERN Preprint TH. 3648 (1983).
60) P. van Baal, F.A. Bais and P. van Nieuwenhuizen, Utrecht preprint
P. van Baal and F.A. Bais, Utrecht preprint
61) L. Castellani, R.D'Auria, P. Frè and P. van Nieuwenhuizen, Utrecht preprint
62) L. Castellani, R. D'Auria and P. Frè, in "Supersymmetry and Supergravity '83", proc. 19th Winter School of Theoretical Physics, Karpacz, ed. B. Milewski (World Scientific)
M.J. Duff, B.E.W. Nilssan and C.N. Pope, Imperial College preprint ICTP/82-83/29
B. de Wit, preprint NIKHEF - H/83-18
P. van Nieuwenhuizen, Les Houches Lecture Notes, to appear.

DIMENSIONAL REDUCTION OF EXCEPTIONAL GAUGE GROUPS AND FLAVOR CHIRALITY

Mehmet KOCA

Çukurova University,
Physics Department, P.O.Box 171
Adana, TURKEY.

Abstract

Phenomenologically realistic flavor-chiral Yang -Mills-Higgs theories of 10-dimensional vectorlike gauge theories of F_4, E_6, E_7, and E_8, where the extra 6 dimensions form the caset spaces SO(7)/SO(6) or G_2/SU(3). A three family structure of two 16's and one 144 of SO(10) in 4 dimensions arises from the dimensional reduction of E_8 in 10 dimensions.

The dimensional reduction technique based on the theory of symmetric gauge fields[1] has gained some interest in obtaining a flavor-chiral theory in 4-dimensions from a vectorlike theory in 10 dimensions[2]. Among a number of 6-dimensional compact coset spaces only SO(7)/SO(6) and G_2/SU(3) lead to the compactification of the 10-dimensional supergravity[3]. In this work we give examples of Yang-Mills-Higgs theories in 4-dimensions derived by dimensional reduction of 10-dimensional vectorlike gauge theories of F_4, E_6, E_7, and E_8 using only the compact coset spaces SO(7)/SO(6) and G_2/SU(3).

A. The dimensional reduction of the supersymmetric F_4, E_6, and E_7 gauge groups with the coset space SO(7)/SO(6). Fermions and Gauge bosons are assigned to the adjaint representations of the respective gauge groups. Embedding of SO(6) and SO(7) can be made via orthogonal subgroups.

A.1

$$F_4 \supset SO(9) \supset SU(2) \times SO(6)$$

$$52 = 16 + 36 = (2,4) + (2,\bar{4}) + (1,15) + (3,1) + (3,6)$$

The matching rule of the dimensional reduction technique leads to the following effective theory in 4- dimensions,

The residual gauge group in 4-dimensions: SU(2)

Higgs : 3

Fermions : 2

SU(2) is broken to U(1), the centralizer of SO(7) in F_4 by the geometrical Higgs 3.

A.2 $E_6 \supset SO(10) \times U(1) \supset SO(6) \times SU(2) \times SU(2) \times U(1)$

Both Fermions and the gauge bosons are in the adjoint 78. Considering the SO(6)xSU(2)xSU(2)xU(1) branching rule of 78 and using the matching rule we obtain,

the residual symmetry in 4-dimensions: SU(2)xSU(2)xU(1)

Fermions : (2,1)+(1,2), Higgs: (2,2)

This is the left-right symmetric lepton theory[4].

The geometrical Higgs breaks the symmetry to SU(2)xU(1), the centralizer of SO(7) in E_6.

A.3 $E_7 \supset SU(2) \times SO(12)$ SU(2) x SO(6)xSO(6). Fermions and the gauge bosons are in 133. In this case we obtain an effective theory in 4-dimensions,

the residual gauge group : SU(2)xSU(4)

Fermions : (2,4), Higgs : (1,6)

Here the symmetry is broken to SU(2)xSU(2)xU(1).

B. The dimensional reduction of the supersymmetric F_4, E_6, and E_7 gauge groups with the coset space G_2/SU(3). Here we follow the same procedure above and obtain the following results.

B.1 $F_4 \supset SU(3)xSU(3)$ and $F_4 \supset SU(2)xG_2$

The residual symmetry: supersymmetric SU(3)

Fermions: 8+6, Higgs : 6+6

SU(3) is broken to SU(2) by geometrical Higgs.

B.2 $E_6 \supset SU(3)xSU(3)xSU(3)$, $E_6 \supset SU(3)xG_2$

Here the effective theory in 4-dimensions is a supersymmetric SU(3)xSU(3) which breaks to SU(3) by geometrical Higgs.

B.3 F_7 leads to a supersymmetric SU(6) theory in 4-dimensions with the particle contents, Fermions: 35+15 and Higgs: $15+\overline{15}$. SU(6) breaks to SP(6).

C. Vectorlike E_8 in 10-dimensions

To obtain a realistic theory in 4-dimensions we make the following assignments in 10-dimensions

Gauge bosons : 248

Fermions : 248+3875

Higgs Scalars : 3875

(This theory is not supersymmetric)

The dimensional reduction using the coset space SO(7)/SO(6) leads to an effective SO(10) theory with the particle contents,

Fermions: 16 + 16 + 144

Higgs : 10 + 1 + 54 + 210

Fermions couple to only geometrical Higgs 10.210 and 54 breaks SO(10) to SU(3)xSU(2)xU(1). Finally 10 breaks the symmetry to SU(3)xU(1). This is a three generation model where τ - family is hosted in 144.

The dimensional reduction with the coset space G_2/SU(3) leads to an effective E_6 theory[5)] in 4-dimensions. The symmetry breaking yields a three generation model at low energies.

REFERENCES

1) G.Chapline and N.Manton, Nucl. Phys. B184 (1981) 391; A.S.Schwarz and Y.S.Tyupkin, Nucl. Phys. B187 (1981) 321: J.Harnard, S.Shnider and J.Tafel, Lett. Math. Phys. 4(1980) 107.

2. G.Chapline and R.Slansky, Nucl.Phys.B209(1982) 461; D.Olive and P.West, Nucl.Phys.B217 (1983) 248.

3. S.Ranjbar-Daemi, Abdus Salam and J.Strathdee, Phys.Lett.124B (1983)349.

4. S.Weinberg, Phys.Rev.Lett.29(1972)388;J.C.Pati and Abdus Salam,Phys.Rev.D10(1974)275.

5. M.Koca, ICTP preprint, IC/83/163.

SEVEN - SPHERES FROM OCTONIONS

J. Lukierski
International Centre for Theoretical Physics, 34100 Trieste, Italy.
P. Minnaert
Laboratoire de Physique Théorique, Bordeaux (France)

1. The aim of this lecture is to consider the seven-spheres obtained as cosets of the octonionic realizations of O(7) and O(8) algebras. We get in particular the model of S^7 with geometric torsion and G_2 holonomy group. If we modify the connection by adding the geometric torsion term with suitably chosen coefficient, we shall show that one can obtain the torsionless $S^7 = SO(8)/SO(7)$ with holonomy group SO(7) as well as the curvatureless parallelizable S^7.

The torsion term is described by the octonion multiplication table

$$e_\alpha e_\beta = -\delta_{\alpha\beta} + f_{\alpha\beta}{}^\gamma e_\gamma \tag{1}$$

where $f_{\alpha\beta}{}^\gamma$ is totally antisymmetric [1)] and defines the vector cross product $\hat{x} \times \hat{y} = 1/2 \left[\hat{x},\hat{y}\right]$ in seven-dimensional space of imaginary octonions $\hat{x} = x_\alpha e_\alpha$. If we introduce D=7 Euclidean product $\hat{x} \cdot \hat{y} = 1/2\left\{\hat{x},\hat{y}\right\}$ the torsion term is defined by the mixed scalar product $(\hat{x},\hat{y},\hat{z}) = 1/4\left\{\left[\hat{x},\hat{y}\right],\hat{z}\right\}$. Such a trilinear form exists only in D=3 and D=7 dimensions [1]. In D=3 the sphere S^3 can be identified with the group manifold SU(2) [2,3], and the parallelizable torsion is described by SU(2) structure constants ϵ_{rst}(r,s,t, = 1,2,3) which do define also the three-dimensional mixed scalar product of imaginary quaternions [2)]; We define octonion S^7 as described by unit length octonions (See e.g. [5,6])

$$X = X_0 + X_\alpha e_\alpha = X_A e_A \qquad X_A X_A = 1 \tag{2}$$

Three nonequivalent representation of O(8) algebra - one vectorial denoted by SO(8) and two spinorial denoted by $\mathrm{Spin}^L 8$ and $\mathrm{Spin}^R 8$, related by triality [7,8] - ean be realised by suitable multiplication of x from left and right by octonionic operators. The reduction of the algebra O(8) to O(7) induces the reduction of the representations $SO(8) \to SO(7) \oplus 1$, $\mathrm{Spin}^L 8 \to \mathrm{Spin}7$, $\mathrm{Spin}^R 8 \to \mathrm{Spin}7$. Borel [9] first listed the four manifolds homeomorphic to S^7, which can be described as homogeneous reductive spaces of compact, connected, simple Lie groups [3)]

$$a)\ \frac{SO(8)}{SO(7)} \qquad b)\ \frac{SU(4)}{SU(3)} \qquad c)\ \frac{Sp(2)}{Sp(1)} \qquad d)\ \frac{\mathrm{Spin}7}{G_2} \tag{3}$$

We would like to add that

(i) Only on cosets a) and b) the canonical metric (obtained by identification of the Cartan one-forms on the cosets with the one-forms on S^7) is Einstein [4),]

(ii) The structure group of the tangent bundle over S^7 can be either SO(7) or G_2 [15].

We shall describe the possible connections on S^7, using the decomposition of the Cartan-Maurer equations for O(7) algebra which provide the Cartan structure equations on S^7.

2. The eight-dimensional real algebra O of octonions is alternative, nonassociative, and has an identity $e_o=1$. An arbitrary octonion belongs to $R^8 = R\oplus R^7$, where R denotes the subspace spanned by the identity. Octonions with unit length (see (2)) define the octonionic sphere S^7. The isometries of octonionic S^7 are described by O(8) algebra, preserving Euclidean scalar product in $R^8: x.y=x_A y_A = 1/2(xy+yx)$. One can introduce the following description of 28 generators of O(8), realized on octonionic S^7,

$$O(8): \qquad T = H \oplus L \oplus R \tag{4}$$

where $H=\{H_{\hat a,\hat b}\}$ is the 14-parameter G_2 algebra of the automorphism group of O. It is parametrized by two imaginary octonions $\hat a,\hat b$ as follows (see [6-8,16]).

$$H: \quad \begin{aligned} H_{\hat a,\hat b}\,\hat x &= -\frac{3}{2}(\hat a,\hat b,\hat x) + \frac{1}{2}[[\hat a,\hat b],\hat x] \\ H_{\hat a,\hat b}\,x_o &= 0 \end{aligned} \tag{5}$$

with $(\hat a,\hat b,\hat x)=(\hat a\hat b)x - \hat a(\hat b\hat x)$ the (completely antisymmetric) associator of three octonions. The generators L and R in eq.(4) represent the left and right multiplications by imaginary octonions

$$L,R: \qquad L_{\hat c}\,x = \hat c\,x \qquad\qquad R_{\hat d}\,x = x\,\hat d \tag{6}$$

The G_2 algebra can be written in two equivalent forms. Introducing $H_{\alpha\beta}= H_{e_\alpha,e_\beta} = -H_{\beta\alpha}$ one can write the G_2-algebra in the following way

$$[H_{\alpha\beta}, H_{\gamma\delta}] = \Gamma_{\alpha\beta\gamma}{}^{\varepsilon} H_{\varepsilon\delta} + \Gamma_{\alpha\beta\delta}{}^{\varepsilon} H_{\gamma\varepsilon} \tag{7}$$

where

$$\Gamma_{\alpha\beta\gamma}{}^{\varepsilon} = f_{\alpha\beta}{}^{\delta} f_{\delta\gamma}{}^{\varepsilon} - f_{\beta\gamma}{}^{\delta} f_{\delta\alpha}{}^{\varepsilon} - f_{\gamma\alpha}{}^{\delta} f_{\delta\beta}{}^{\varepsilon} \tag{8}$$

describes the action of $H_{\alpha\beta}$ on the imaginary octonionic basis

$$H_{\alpha\beta} e_\gamma = \Gamma_{\alpha\beta\gamma}{}^{\varepsilon} e_\varepsilon \tag{9}$$

and $H_{\alpha\beta}$ satisfies seven linearly independent relations

$$f_{\alpha\beta}{}^{\gamma} H_{\alpha\beta} = 0 \tag{10}$$

Using the notation $L_{e_\gamma} = L_\gamma$, $R_{e_\gamma} = R_\gamma$ the covariance relations for the generators L_γ, R_γ look as follows:

$$\begin{aligned} [H_{\alpha\beta}, L_\gamma] &= \Gamma_{\alpha\beta\gamma}{}^{\varepsilon} L_\varepsilon \\ [H_{\alpha\beta}, R_\gamma] &= \Gamma_{\alpha\beta\gamma}{}^{\varepsilon} R_\varepsilon \end{aligned} \tag{11}$$

The remaining commutation relations of the generators (4) are,

$$\begin{aligned} &- [L_\alpha, L_\beta] = \tfrac{1}{3}(4H_{\alpha\beta} + 2 f_{\alpha\beta}{}^{\gamma} L_\gamma + 4 f_{\alpha\beta}{}^{\gamma} R_\gamma) \\ &- [R_\alpha, R_\beta] = \tfrac{1}{3}(4H_{\alpha\beta} - 4 f_{\alpha\beta}{}^{\gamma} L_\gamma - 2 f_{\alpha\beta}{}^{\gamma} R_\gamma) \\ &- [L_\alpha, R_\beta] = \tfrac{1}{3}(-2H_{\alpha\beta} + 2 f_{\alpha\beta}{}^{\gamma} L_\gamma - 2 f_{\alpha\beta}{}^{\gamma} R_\gamma) \end{aligned} \tag{12}$$

It is interesting to check that the following three pairs of linear combinations of the generators L_α, R_α close to O(7) algebra,

$$\frac{SO(7)}{G_2}: \quad K^{\pm}_{V\alpha} = \pm\tfrac{1}{2}(L_\alpha - R_\alpha) \tag{13a}$$

$$\frac{Spin7}{G_2}: \quad K^{\pm}_{S\alpha} = \pm(\tfrac{1}{2}L_\alpha + R_\alpha) \tag{13b}$$

$$\frac{\overline{Spin7}}{G_2}: \quad \overline{K}^{\pm}_{S\alpha} = \mp(L_\alpha + \tfrac{1}{2}R_\alpha) \tag{13c}$$

where the equivalent octonionic-conjugated representation is defined by the relation $\bar{S}x = \overline{(S\bar{x})}$. Because $\bar{L}_\alpha = -R_\alpha$, $\bar{R}_\alpha = -L_\alpha$, (13b) is octonion-conjugate to (13c) and (13a) is self-conjugate .

The vector representation SO(7) of the algebra O(7) generated by $H_{\alpha\beta} \oplus K^{\pm}_{V\alpha}$ is seven-demensional, because $K^{\pm}_{V\alpha} e_o = 0$. The spin representation Spin 7 generated by $H_{\alpha\beta} \oplus K^{\pm}_{S\alpha}$ is eight-dimensional. The representation (13) satisfy the duality (or reduced O(7)-triality) relations for SO(7) and Spin 7 representations of O(7) algebra [6-8],

$$K^{\pm}_{V\alpha}(xy) = (K^{\pm}_{S\alpha} x) y + x(\bar{K}^{\pm}_{S\alpha} y) \tag{14}$$

which express in octonionic language the property that the vectorial coordinates on octonionic S^7 can be introduced by considering bilinears of spinorial ones. The generators (13) extend G_2 algebra to O(7) in the same way for all three realizations SO(7), Spin 7 and $\overline{\text{Spin 7}}$. Denoting $K^{\pm}_{\alpha}$ for $K^{\pm}_{V\alpha}$ or $K^{\pm}_{S\alpha}$ or $\bar{K}^{\pm}_{S\alpha}$, the commutation relations are,

$$[H_{\alpha\beta}, K^{\pm}_{\gamma}] = \Gamma_{\alpha\beta\gamma}{}^{\varepsilon} K^{\pm}_{\varepsilon} \tag{15a}$$

$$[K^{\pm}_{\alpha}, K^{\pm}_{\beta}] = H_{\alpha\beta} \mp f_{\alpha\beta}{}^{\gamma} K^{\pm}_{\gamma} \tag{15b}$$

i.e. we obtain torsion terms with positive and negative signs.

3. The nonsymmetric, reductive, homogeneous coset space K=G/H is characterized by the following decomposition of the Lie algebra of the group G (see e.g. [11])

$$
\begin{aligned}
&[H_a, H_b] = C_{ab}{}^{c} H_c \\
&[H_a, K_\mu] = C_{a\mu}{}^{\nu} K_\nu \\
&[K_\mu, K_\nu] = C_{\mu\nu}{}^{a} H_a + C_{\mu\nu}{}^{\rho} K_\rho
\end{aligned}
\tag{16}
$$

The Cartan-Maurer equation for the Cartan one-forms (θ^a, θ^μ) on G decomposes into the structure equations on K=G/H, defining the torsion and curvature two-forms

$$
\begin{aligned}
T^\mu &= d\theta^\mu + \omega^\mu{}_\nu \wedge \theta^\nu \\
R^\mu{}_\nu &= d\omega^\mu{}_\nu + \omega^\mu{}_\rho \wedge \omega^\rho{}_\nu
\end{aligned}
\tag{17}
$$

where the canonical choice of connection is given by $\omega^{\mu}{}_{\nu} = C_{a\nu}{}^{\mu}\theta^{a}$.
Using (16) and Jacobi identities, one obtains [11,16],

$$R^{\mu}{}_{\nu} = \frac{1}{2}R^{\mu}{}_{\nu\varrho\tau}\,\theta^{\varrho}\wedge\theta^{\tau} = -\frac{1}{2}C_{a\nu}{}^{\mu}C_{\varrho\tau}{}^{a}\,\theta^{\varrho}\wedge\theta^{\tau} \tag{18a}$$

$$T^{\mu} = \frac{1}{2}T^{\mu}{}_{\varrho\tau}\,\theta^{\varrho}\wedge\theta^{\tau} = -\frac{1}{2}C_{\varrho\tau}{}^{\mu}\,\theta^{\varrho}\wedge\theta^{\tau} \tag{18b}$$

Let us now consider the case of $S^7 = \mathrm{Spin}\ 7/G_2 \simeq \overline{\mathrm{Spin}\ 7}/G_2$. From eq.(15a,b) for both signs of coset generators, the structure constants are,

$$\begin{aligned} C_{a\nu}{}^{\mu} &= \Gamma_{\alpha\beta\nu}{}^{\mu} \\ C_{\varrho\tau}{}^{a} &= \frac{1}{2}(\delta_{\varrho}{}^{\alpha}\delta_{\tau}{}^{\beta} - \delta_{\varrho}{}^{\beta}\delta_{\tau}{}^{\alpha}) \\ C_{\varrho\tau}{}^{\mu} &= \mp f_{\varrho\tau}{}^{\mu} \end{aligned} \tag{19}$$

and for the canonical choice of the connection we get,

$$R^{\mu}{}_{\nu\varrho\tau} = -\Gamma_{\varrho\tau\nu}{}^{\mu} \tag{20a}$$

$$T^{\mu}{}_{\varrho\tau} = \pm f_{\varrho\tau}{}^{\mu} \tag{20b}$$

The components of the curvature tensor (20) are the matrix elements of the 7-dimensional representation of G_2 (see eq.(9), therefore the holonomy group of the canonical connection is G_2.

On a homogeneous nonsymmetric, reductive coset space with nonvanishing canonical torsion (18b), one can introduce a one-parameter family of connections,

$$\omega(k)^{\mu}{}_{\nu} = C_{a\nu}{}^{\mu}\theta^{a} + \frac{k}{2}C_{\nu\varrho}{}^{\mu}\theta^{\varrho} \tag{21}$$

which, due to Jacobi identities, provides via formula (17) the horizontal torsion and curvature two-forms. The expressions (18) are generalized as follows,

$$\begin{aligned} R(k)^{\mu}{}_{\nu\varrho\tau} = {} & C_{a\nu}{}^{\mu}C_{\varrho\tau}{}^{a} + \frac{k}{2}C_{\nu\sigma}{}^{\mu}C_{\varrho\tau}{}^{\sigma} + \\ & + \frac{k^2}{4}(C_{\varrho\sigma}{}^{\mu}C_{\tau\nu}{}^{\sigma} - C_{\tau\sigma}{}^{\mu}C_{\varrho\nu}{}^{\sigma}) \end{aligned} \tag{22a}$$

$$T(k)^{\mu}{}_{\rho\tau} = (k-1)\, C^{\mu}{}_{\rho\tau} \tag{22b}$$

Substituting the values (19) for the structure constants one obtains,

$$R(k)^{\mu}{}_{\nu\rho\tau} = \frac{1}{2}(k+2)\Big[(k-1)\Gamma_{\rho\tau\nu}{}^{\mu} - \frac{3}{2}k\,(\delta_{\rho\nu}\delta_{\tau}{}^{\mu} - \delta_{\tau\nu}\delta_{\rho}{}^{\mu}) \tag{23a}$$

$$T(k)^{\mu}{}_{\rho\tau} = \mp (k-1)\, f_{\rho\tau}{}^{\mu} \tag{23b}$$

We see from this expression that, for arbitrary k, we obtain a linear combination of the curvatures with holonomy groups G_2 and SO(7) . The canonical choice k=0 provides the annihilation of the part with the holonomy group SO(7). There are however two noncanonical choices, with $k \neq 0$, providing respectively the annihilation of the part with holonomy group G_2 and even the annihilation of the whole curvature tensor. We obtain,

a) k = 1 :

$$R^{\mu}{}_{\nu\rho\tau} = -\frac{9}{4}(\delta_{\rho\nu}\delta_{\tau}{}^{\mu} - \delta_{\tau\nu}\delta_{\rho}{}^{\mu}) \tag{24a}$$

$$T^{\mu}{}_{\nu\rho} = 0 \tag{24b}$$

For the choice of the torsionless connection, the coset Spin $7/G_2$ can be identified with the coset SO(8)/SO(7) ,(see also [12]).

b) k = -2 :

$$R^{\mu}{}_{\nu\rho\tau} = 0 \tag{25a}$$

$$T^{\mu}{}_{\nu\rho} = \pm 3 f_{\nu\rho}{}^{\mu} \tag{25b}$$

Such a choice of the connection on Spin $7/G_2$ identifies it with the curvatureless parallelizable seven-sphere [17,18] and we see that the holonomy group is reduced to the identity. It should be mentioned that this result has been obtained independently using different formalism by the present authors [19] and de-Wit, Nicolai [20] .

c) The Ricci tensor $(\mathrm{Ric})_{\nu\tau} = R^{\mu}{}_{\nu\mu\tau}$ is given by the formula

$$(\mathrm{Ric})_{\nu\tau} = -\frac{3}{2}(k+2)(k-4)\,\delta_{\nu\tau} \tag{26}$$

We see that for k = 4 one obtains the Ricci-flat nonparallelizable sphere. It is possible to have such a solution in the internal sector of D =11 supergravity if we assume that the "fermionic condensate" $\langle \bar{\Psi}_M \Gamma_A \Psi_N \rangle \neq 0$. Such a mechanism was proposed by Duff and Orzalesi [21] for parallelizable S^7, and it implies vanishing of cosmological constant in space-time sector. Its extension to non-paralellizable Ricci-flat seven dimensional manifolds is now under investigation.

FOOTNOTES

I. $f_{\alpha\beta}{}^{\gamma}$ are determined by the choice $f_{\alpha\beta}{}^{\gamma}$ =I for $(\alpha\beta\gamma)$=(123),(516),(624), (435),(471),(673),(572).

2. The existence of parallelizable torsion on S^n (n =3,7) was firstly related with the existence of antisymmetric bilinear product operation for division algebras in a letter of J. Milnor to R. Bott on December 23, 1957 (see [4]).

3. The homogeneous reductive nonsymmetric coset are discussed extensively in [10 - 12].

4. For the existence of Einstein metric on d) see [12]. The canonical metrics on b) and c) are not Einstein, however in case c), as has been shown by Jensen [13] one can find the class of continuous deformations of the canonical metric,which for particular values of the deformation parameter, describes two different Einstein metrics (see also [14]). One of these metrics describes so-called squashed sphere S^7. It was shown in [13] that the analogous deformation in case b) does not produce any new Einstein metrics.

REFERENCES

1. R.P. Brown and A. Gray, Comm.Math.Helv. 42, 222 (1967).
2. M.J. Duff, P. van Nieuwenhuizen and P.K. Townsend, Phys.Lett.122B, 232 (1983).
3. C.H. Tze; Phys.Lett. 128B, 160 (1983).
4. R. Bott and J. Milnor, Bull.Amer.Math.Soc. 64, 87-89;(1958).
5. T. Dereli, M. Panahimoghaddam, A. Sudbery and R.W. Tucker, Phys.Lett. 126B, 33 (1983).
6. F. Gürsey and C?H. Tze, Phys.Lett. 127B, 191 (1983).
7. R.D. Shafer, An Introduction to Nonassociative Algebras, Academic Press (1966).
8. F. Gürsey and M. Günaydin, J.Math.Phys. 14, 1661 (1973).
9. A. Borel, C.R. Acad.Sci. Paris, 230, 1378, (1950).
10. M.Berger, Bull.Soc.Math.France 83, 279 (1955).
11. S. Kobayashi and K. Nomizu, Fundations of Differential Geometry, Vol. II; Chapter X, Interscience Publisher (1969).
12. J. Wolf, Acta Mathematica , 120, 59 (1968).
13. G.R. Jensen, Duke Math.Journ.,42? 397, (1975).
14. R. Coquereaux and A. Jadczyk, CERN preprint, Th. 3483,(1982).
15 R B Brown and A. Gray, Pacific Journ. of Math. 34, 83 (1970).
16. A. Salam and J. Strathdee, Annals of Physics, 141, 316, (1982).
17. E. Cartan and J.A. Schouten, Proc.K. Akad.Wet. Amsterdam, 29, 933 (1926).
18. J.A. Wolf, Jour.Diff.Geom. 6, 317 (1972), ibid. 7, 19 (1972).
19. J. Lukierski and P. Minnaert, Bordeaux preprint PTB-128, April 1983, Phys. Lett. B, in press
20. B. de Wit and H. Nicolai, Amsterdam preprint NIKHEF -H 183-8, June 1983.
21. M. Duff and C. Orzalesi, Phys.Lett. 122B, 37 (1983).

A SOLUTION OF BIANCHI IDENTITIES FOR EXTENDED SUPERGRAVITIES

Sorin Marculescu
Institut für Theoretische Physik, Universität Karlsruhe, Kaiserstraße 12
D-7500 Karlsruhe, Germany

Starting from the Wess-Zumino constraints/1/ for U(N) extended supergravities, a solution of the Bianchi identities (BI) is given in terms of constrained spinor superfields of dimension 1/2. In the limit when the internal symmetry is degauged, the solution reduces to the system of torsions of dimension 1/2.

We use the notations and conventions of ref./2/ and assume the following constraints

$$T_{\alpha}^{A}{}_{B\dot\beta}{}^{c} - 2i\,\delta^{A}_{B}(\sigma_c)_{\alpha\dot\beta} = T_{\underline{\alpha}\,b}{}^{c} = T_{ab}{}^{c} = 0 \quad . \tag{1}$$

We look first for the linear BI of various dimensions:

dim 1/2

$$T^{\dot\alpha\dot\beta}_{ABC}{}^{\gamma} = \varepsilon^{\dot\alpha\dot\beta}\, T^{\gamma}_{[ABC]} \quad ; \qquad T_{\alpha}^{A}{}_{B\dot\beta\dot\gamma}{}^{C} = \varepsilon_{\dot\beta\dot\gamma}\, T_{\alpha}^{A}{}_{B}{}^{C} \, ;$$
$$T_{\alpha\beta}^{AB}{}_{\gamma C} = -\varepsilon_{\alpha\gamma}\, T_{\beta}^{B}{}_{C}{}^{A} - \varepsilon_{\beta\gamma}\, T_{\alpha}^{A}{}_{C}{}^{B} \quad ; \tag{2}$$

dim 1 torsions

$$T_{\alpha\dot\alpha}{}^{BC}{}_{\beta\dot\gamma} = \varepsilon_{\alpha\beta}\, T^{[BC]}_{\underbrace{\dot\alpha\dot\gamma}} - \varepsilon_{\dot\alpha\dot\gamma}\, T^{[BC]}_{\underbrace{\alpha\beta}} + \varepsilon_{\alpha\beta}\,\varepsilon_{\dot\alpha\dot\gamma}\, T^{\underbrace{BC}} \quad ;$$
$$T_{\alpha\dot\alpha}{}^{B}{}_{\beta\gamma C} = -\varepsilon_{\beta\gamma}\, T_{\alpha\dot\alpha C}{}^{B} - \sum_{\beta\gamma} \varepsilon_{\alpha\beta}\,(U_C{}^{B})_{\gamma\dot\alpha} \quad ; \tag{3}$$

dim 1 Lorentz curvatures

$$R_{\alpha\beta}^{AB}{}_{\underbrace{\gamma\delta}} = -2i\,\varepsilon_{\alpha\beta}\, T^{[AB]}_{\underbrace{\gamma\delta}} + 2i\Big(\sum_{\gamma\delta} \varepsilon_{\alpha\gamma}\,\varepsilon_{\beta\delta}\Big)\, T^{\underbrace{AB}} \quad ;$$
$$R_{\alpha\beta}^{AB}{}_{\underbrace{\dot\gamma\dot\delta}} = -2i\,\varepsilon_{\alpha\beta}\, T^{[AB]}_{\underbrace{\dot\gamma\dot\delta}} \quad ; \qquad R_{\alpha}^{A}{}_{B\dot\beta\,\underbrace{\gamma\delta}} = 2i \sum_{\gamma\delta} \varepsilon_{\alpha\gamma}\,(U_B{}^{A})_{\delta\dot\beta} \; ; \tag{4}$$

dim 3/2: Linear BI allow to express Lorentz curvatures through torsions of dimension 3/2.

dim 2 /3/

$$R_{\underbrace{\alpha\beta}\,\underbrace{\gamma\delta}} = R_{\underbrace{\gamma\delta}\,\underbrace{\alpha\beta}} \quad ; \qquad R_{\underbrace{\dot\alpha\dot\beta}\,\underbrace{\gamma\delta}} = R_{\underbrace{\gamma\delta}\,\underbrace{\dot\alpha\dot\beta}} \quad ; \tag{5}$$

$$R^{\underbrace{\varepsilon\varphi}}{}_{\underbrace{\varepsilon\varphi}} = R_{\underbrace{\dot\varepsilon\dot\varphi}}{}^{\underbrace{\dot\varepsilon\dot\varphi}} \tag{6}$$

All the other BI are non-linear. A certain simplification occurs by modifying the internal symmetry connection $\phi_A{}^B$ according to

$$\tilde{\phi}_A{}^B \equiv \phi_A{}^B - E_E^{\varepsilon} T_{\varepsilon A}^{E\ B} + E_{\dot{\varepsilon}}^{E} T_{E\ \ A}^{\dot{\varepsilon}\ B} + E^e T_{eA}{}^B \quad . \tag{7}$$

The corresponding covariant derivative, torsion and curvature are denoted by $\tilde{\mathcal{D}}$, $\tilde{T}$ and $\tilde{R}$, respectively. The non-linear BI of dimension 1 give the conditions

$$\tilde{\mathcal{D}}_{\alpha}^{A} T_{\dot{\beta}}^{[BCD]} \equiv P_{\alpha\dot{\beta}}^{[ABCD]} \quad ;$$

$$\sum_{\alpha\beta} \tilde{\mathcal{D}}_{\alpha}^{A} T_{\beta\,[BCD]} = -4i \oint_{(BCD)} \delta_B^A \; T_{\underset{\smile}{\alpha\beta}\,[CD]} \quad ; \tag{8}$$

as well as the internal symmetry curvatures (ISC)

$$\tilde{R}_{\alpha\beta C}^{AB\quad D} = \varepsilon_{\alpha\beta}\left\{\tfrac{1}{2}\, \tilde{\mathcal{D}}_C^{\dot{\varepsilon}}\, T_{\dot{\varepsilon}}^{[ABD]} + 2i\left(\delta_C^A\, T^{\underset{\smile}{D}B} - \delta_C^B\, T^{\underset{\smile}{D}A}\right)\right\} + $$

$$+ 2i\left(\delta_C^A\, T_{\underset{\smile}{\alpha\beta}}^{[DB]} + \delta_C^B\, T_{\underset{\smile}{\alpha\beta}}^{[DA]}\right) \quad ;$$

$$\tilde{R}_{\alpha\dot{\beta}B\,C}^{A\qquad D} = -T_{\alpha\,[BCE]}\, T_{\dot{\beta}}^{[ADE]} + 2i\,\big(\delta_B^A\, U_C{}^D + $$

$$+ \delta_C^D\, U_B{}^A - 2\,\delta_C^A\, U_B{}^D - 2\,\delta_D^B\, U_C{}^A\big)_{\alpha\dot{\beta}} \quad . \tag{9}$$

The non-linear identities of dimension $\geqslant$ 3/2 are not all independent/4/. The independent ones have the following property: There is a set of BI for ISC completely equivalent with them. Hence, the torsions and the differential constraints of dimension $\geqslant$ 3/2 are consequences of BI for ISC, Ricci identity and BI of dimension 1.

Moreover, eqs.(5) are identically satisfied, while (6) yields

$$\left(\tilde{\mathcal{D}}^{\varepsilon E}\, \tilde{\mathcal{D}}_{\varepsilon}^{F} + T_{\dot{\varepsilon}}^{[EFH]}\, \tilde{\mathcal{D}}_H^{\dot{\varepsilon}}\right)\left(\tilde{\mathcal{D}}^{\varphi G}\, T_{\varphi\,[EFG]}\right) =$$

$$= \left(\tilde{\mathcal{D}}_{\dot{\varepsilon}E}\, \tilde{\mathcal{D}}_F^{\dot{\varepsilon}} + T_{[EFH]}^{\varepsilon}\right)\left(\tilde{\mathcal{D}}_{\dot{\varphi}G}\, T^{\dot{\varphi}[EFG]}\right) \quad . \tag{10}$$

The BI for ISC are fulfilled if we use the representation

$$\tilde{R}_{AB\,C}{}^{D} = \tilde{T}_{AB}{}^{\mathcal{F}}\, \Sigma_{\mathcal{F}C}{}^{D} + \left(\tilde{\mathcal{D}}_A + \Sigma_A ,\, \tilde{\mathcal{D}}_B + \Sigma_B\right)_C{}^D \quad . \tag{11}$$

Here, (,) is the graded commutator and $\Sigma_{\underset{=}{\alpha}B}{}^{C}$ are dimension 1/2 spinor superfields. They reduce to $T_{\underline{\alpha} B}{}^{C}$ (see (2)). $\Sigma_{\alpha B}{}^{C}$ and the other quantities of dimension 1 are computed from (9) and (11) in terms of $\Sigma_{\underline{\alpha} B}{}^{C}$ and $T_{\alpha [ABC]}$. The constraints upon $\Sigma_{\underline{\alpha} B}{}^{C}$ follow then algebraically, reinserting these results into (9).

Acknowledgements. I would like to thank Richard Grimm, Wolf Lang and Julius Wess for fruitful discussions and suggestions.

References

/1/. J.Wess and J.Bagger, Supersymmetry and Supergravity, Princeton University Press 1983.
/2/. To save space as much as possible, we do not write the complex conjugate quantities or equations. Also we limit our discussion to $N \geqslant 3$. Complete results for N=2 are given in:
R.Grimm, Ettore Majorana International Science Series, Physical Sciences vol. 7, 1981, Ed. A.Zichichi, p.509.
/3/. R.Penrose, Ann.Phys. 10 (1960) 171.
/4/. P.Howe, Nucl. Phys. B 199 (1982) 309.

N=2 UNCONSTRAINED SUPERFIELD SUPERGRAVITY

FROM HYPERMULTIPLET

B. Milewski and K. Pilch
Institute of Theoretical Physics
University of Wroclaw, Wroclaw, Poland*)

Abstract: Generalizing the N=1 superfield Lagrangian for N=2 hypermultiplet in the presence of N=2 local supersymmetry we derive the multiplet of N=2 supergravity in terms of N=1 superfields and the transformation rules.

1. Following the idea of ref.[1] we start from the simplest multiplet of N=2 supersymmetry-the hypermultiplet[2] and demanding that it transform under local N=2 supersymmetry we find the minimal multiplet of compensating superfields composed of N=1 supergravity superfields H^M and the spin (3/2,1) superfield Ψ_α. This method is a generalization of the program of Siegel and Gates[3] utilized in the case of N=1 supergravity.
2. The hypermultiplet is most easily described as a doublet of N=1 chiral superfields[4] S and T. The action

$$I = \int d^4x\, d^4\theta\, (S\bar{S} + \bar{T}T) \tag{1}$$

is invariant under the following transformations

$$\delta S = i\bar{D}^2(\varepsilon\, \bar{T}) \qquad\qquad \delta T = -i\bar{D}^2(\varepsilon\, S) \tag{2}$$

where $\varepsilon(\theta,\bar{\theta})$ is the x- independent superfield parameter

$$\varepsilon = z - \bar{z} + \theta\zeta - \bar{\theta}\bar{\zeta} + \theta^2 g - \bar{\theta}^2\bar{g} \tag{3}$$

comprising the central charge (z-z), second supersymmetry $\zeta_\alpha, \bar{\zeta}_{\dot\alpha}$ and internal symmetry $g,\bar{g}$ parameters.
3. The next step is to couple the hypermultiplet to the N=1 supergravity and covariantize the second supersymmetry transformation laws (2). It turns out that although the coupling to the minimal ($n=-\frac{1}{3}$) supergravity is possible it does not lead to N=2 supergravity. Thus we choose the simplest formulation of nonminimal supergravity (n=-1 in ref.[3]; see also [5]) with the set of superfields H^M appearing through the combination

$$\exp\{iH\}\ , \qquad H=H^M\partial_M = H^m\partial_m + H^\mu\partial_\mu + \bar{H}_{\dot\mu}\bar{\partial}^{\dot\mu}\ . \tag{4}$$

The covariantization of (1) and (2) reads

$$I = \int d^4x d^4\theta\, (E\, S\, e^{-iH}\bar{S} + \bar{E}\,\bar{T}\, e^{iH} T), \tag{5}$$

$$\delta S = \bar{\partial}^2(\Omega\, e^{-iH}\bar{T}) \qquad \delta T = -\bar{\partial}^2(\Omega e^{-iH}\bar{S}), \tag{6}$$

where $E = \mathrm{sdet}\, E_M{}^A$ (denoted by E^{-1} in ref.[3]), $\bar{E}$ its complex conjugate, $\bar{\partial}^2 = \bar{\partial}_{\dot\alpha}\bar{\partial}^{\dot\alpha}$ and Ω is the generalization of the superparameter ε (cf.(3)), with the following reality condition $\bar{\Omega} = -e^{iH}\Omega e^{-iH}$. However, the action (5) is no longer invariant under (6). We need one more compensating superfield, the transformation of which would cancel the variation of (5). This superfield Ψ_α enters the lagrangian in the following way

$$I = d^4xd^4\ ES(2\,\Psi^\alpha\nabla_\alpha - (\nabla_\alpha\Psi^\alpha) + \Psi^\alpha T_\alpha)T + \text{h.c.} \tag{7}$$

---

*) Presented by B. Milewski

where ∇_α is the covariant derivative of N=1 supergravity and $T_\alpha = T_{a\alpha}{}^a = \nabla_\alpha \ln (e^{-iH} \bar{E} e^{iH})$ is the remnant of the dilatation gauge field. The following transformation of Ψ_α compensates the action of (6) upon (5):

$$\delta \Psi^\alpha = \nabla^\alpha \Omega - T^\alpha \Omega \tag{8}$$

The action (7) is itself invariant, in the linearized limit, under additional gauge transformation

$$\delta \Psi^\alpha = \lambda^\alpha \quad , \quad \bar{D}_{\dot\alpha} \lambda^\alpha = 0 \tag{8a}$$

which is necessary to obtain the representation of pure spin (3/2, 1) from the superfield Ψ^α.

4. The most difficult problem is to find the transformation rules of the first supergravity superfields H^M, as they appear also through the superdeterminat of the achtbein E. The variation (6) when substituted into (7) must cancel the variation of supergravity fields entering (5). The following trick is crucial in performing this calculation. We observe that the variation of H^M under second (local) supersymmetry appears only through the combination

$$(\delta_{II}\, e^{-iH})\, e^{iH} = \Delta^A E_A = \Delta^a E_a + \Delta^\alpha E_\alpha + \bar{\Delta}_{\dot\alpha} \bar{\partial}^{\dot\alpha} \tag{9}$$

which is the first order differential operator. We expand it in the basis of semicovariant derivatives $E_A = E_A{}^M \partial_M$ ($E_{\dot\alpha} = \bar{\partial}_{\dot\alpha}$, $E_\alpha = e^{-iH} \partial_\alpha e^{iH}$ and $\frac{i}{2} E_{\alpha\dot\alpha} = \frac{i}{2} \sigma^a_{\alpha\dot\alpha} E_a = \nabla_\alpha \bar{\partial}_{\dot\alpha} + \bar{\nabla}_{\dot\alpha} E_\alpha$). The variation of E may be calculated in terms of Δ

$$E^{-1} \delta E = E_\alpha \Delta^\alpha - 2 \bar{\partial}^{\dot\alpha} \bar{\Delta}_{\dot\alpha} + \bar{T}^{\dot\alpha} \bar{\Delta}_{\dot\alpha} - i \bar{\partial}_{\dot\alpha} \nabla_\alpha \Delta^{\dot\alpha\alpha}$$
$$-\frac{i}{2} (\bar{T}_{\dot\alpha} \nabla_\alpha \Delta^{\dot\alpha\alpha} + \bar{\partial}_{\dot\alpha} (T_\alpha \Delta^{\dot\alpha\alpha}) + \frac{1}{2} \bar{T}_{\dot\alpha} T_\alpha \Delta^{\dot\alpha\alpha}) - \frac{1}{2} \Delta^{\dot\alpha\alpha} T_{\alpha\dot\alpha,\beta,}{}^\beta \tag{10}$$

where $\bar{T}_{\dot\alpha} = T_{a\dot\alpha}{}^a = \bar{\partial}_{\dot\alpha} E$ and $T_{a,\beta,}{}^\beta$ are the contracted torsions.

Variation of (7) under (6) is cancelled by the following transformations of e^{iH} (cf.(9))

$$\Delta^{\dot\alpha\alpha} = 4i\, \bar{\nabla}^{\dot\alpha} \Psi^\alpha \Omega + 2i \bar{T}^{\dot\alpha} \Psi^\alpha \Omega + \text{h.c.}$$
$$\Delta^\alpha = 2 \bar{\nabla}^2 (\Psi^\alpha \Omega) - \Psi^\alpha \bar{\partial}^2 \Omega + \frac{1}{2} \bar{T}^2 \Psi^\alpha \Omega + (\bar{\Delta}^{\dot\alpha})^+ ,$$
$$\bar{\Delta}_{\dot\alpha} = \bar{\partial}_{\dot\alpha} E_\alpha \Psi^\alpha \Omega - E_\alpha \Psi^\alpha \bar{\partial}_{\dot\alpha} \Omega - 2 \nabla_\alpha \bar{T}_{\dot\alpha} \Psi^\alpha \Omega - T_\alpha \bar{\nabla}_{\dot\alpha} \Psi^\alpha \Omega$$
$$- \bar{\nabla}_{\dot\alpha} T_\alpha \Psi^\alpha \Omega + 2\bar{T}^{\dot\alpha} E_\alpha \Psi^\alpha \Omega + T_\alpha \Psi^\alpha \bar{\partial}_{\dot\alpha} \Omega - 2i T_{\alpha\dot\alpha,\beta,}{}^\beta \Psi^\alpha \Omega + (\Delta_\alpha)^+, \tag{11}$$

provided that the following constraint on Ω holds:

$$E_\alpha \bar{\partial}^2 \Omega + T_\alpha \bar{\partial}^2 \Omega + 2\bar{T}_{\dot\alpha} \nabla_\alpha \bar{\partial}^{\dot\alpha} \Omega + (-\bar{\partial}\bar{T} + \bar{T}^2) E_\alpha \Omega - T_\alpha \bar{T}_{\dot\alpha} \bar{\partial}^{\dot\alpha} \Omega$$
$$+ \bar{T}_{\dot\alpha} \bar{\nabla}^{\dot\alpha} T_\alpha \Omega + 2i T_{\alpha\dot\alpha,\beta,}{}^\beta \bar{\partial}^{\dot\alpha} \Omega + \frac{1}{2} T_\alpha \bar{T}^2 \Omega$$
$$-i \bar{T}^{\dot\alpha} (2 T_{\alpha\dot\alpha,}{}^{\dot\beta,}{}_{\dot\beta} + T_{\alpha\dot\alpha,\beta,}{}^\beta) \Omega + 2i \bar{\nabla}^{\dot\alpha} (T_{\alpha\dot\alpha,\beta}{}^\beta + T_{\alpha\dot\alpha,}{}^{\dot\beta}{}_{\dot\beta}) \Omega = 0 \tag{12}$$

In the linearized limit (12) reads simply $D_\alpha \bar{D}^2 \Omega = 0$, which may be also derived from ref.[6]. This completes the coupling of the hypermultiplet to N=2 supergravity.

5. Having established the multiplet (H^M, Ψ_α) and the transformation rules (8), (11) of N=2 supergravity we are now in a position to write down the kinetic terms for these. The first part will be the usual action for N=1 supergravity

$$I = \int d^4x\, d^4\theta\, E \tag{13}$$

Its variation under (10),(11) must cancel the variation of the action for the superfield Ψ_α (taking into account the constraints (12)) which is of the generic form

$$I = \int d^4x d^4\theta\, E(2\nabla_\alpha \bar{\Psi}^{\dot\alpha} \bar{\nabla}_{\dot\alpha} \Psi^\alpha - \frac{1}{2}(\nabla_\alpha \bar{\Psi}_{\dot\alpha})^2 - \frac{1}{2}(\bar{\nabla}_{\dot\alpha} \Psi_\alpha)^2 + \text{other terms}). \qquad (14)$$

The covariant derivatives in (14) must be altered by the inclusion of dilatation gauge fields $T_\alpha, \bar{T}_{\dot\alpha}$, and also some four-linear terms are needed. The full result will be given in the forthcoming publication.

References:

(1) B. Milewski "Towards an Unconstrained Superfield Formulation of N=2 Supergravity" University of Wroclaw, preprint No. 584 (May 1983) (unpublished).

(2) P. Fayet, Nucl. Phys. B 113 (1973) 135;
M.F. Sohnins, Nucl. Phys. B 138 (1978) 109.

(3) W. Siegel and S.J. Gates, Jr., Nucl. Phys. B 147(1979)77.

(4) B. Milewski, "Representations of Extended Supersymmetry on Simple Superfields", University of Wroclaw, preprint No. 581 (April 1983), to appear in "Supersymmetry and Supergravity 1983", Proceedings of the Winter School and Workshop on Theoretical Physics, Karpacz 1983, ed. B. Milewski, World Scientific (1983).

(5) M. Brown and S.J. Gates, Jr., Nucl. Phys. B 165 (1980) 445.

(6) S.J. Gates, Jr. and W. Siegel, Nucl. Phys. B 164 (1980) 484.

EUCLIDEAN SUPERSYMMETRIES IN THREE AND FOUR DIMENSIONS

A. Nowicki
Institute of Teachers Training - ODN
50-527 Wrocław, Poland

It is known [1], that one can get D=4 Poincaré supersymmetry from the Minkowski conformal supersymmetry SU(2,2;N) . On the other hand, D=4 Euclidean supersymmetry can be obtained in similar way, as a contraction of the Euclidean conformal supersymmetry [2] given by the quatrnionic supergroup SL(2,N;H) with the bosonic sector $\bar{O}(5,1)\times U^*(2N)$. D=3 Euclidean supersymmetry can be obtained as a contraction of D=3 Euclidean anti de Sitter supersymmetry represented by OSp(N,2;C) or SL(2;N;C) , or as a contraction of D=3 Euclidean de Sitter supersymmetry given by $UU_\alpha(1,2;H)$ ($\bar{O}(4)\times\bar{O}(2,1)$) or SL(1,1;H) ($\bar{O}(4)\times\bar{O}(2,1)$). In the first case we get the Weyl supercharges but in the second one obtains three-vectorial supercharges.

I. Further we shall use the quaternionic superalgebras. Let us recall therefore some basic facts related with quaternionic superalgebras and their supergroups. We shall use the following notations:

(1) - quaternions
$$q_i = q_i^o + q_i^r e_r \qquad ; q_i^A \in \mathbb{R}$$
$$e_r e_s = -\delta_{rs} + \epsilon_{rst} e_t \quad ; \ r,s,t = 1,2,3$$

the quaternionic units e_r can be represented by the Pauli matrices σ_r as $e_r = -i\sigma_r$ or $e_r = i\sigma_r^T$. The quaternionic Grassmann variables as $\xi_\alpha = \xi_\alpha^o + \xi_\alpha^r e_r$ where ξ_α^A are the real Grassmann variables i.e. $\{\xi_\alpha^A, \xi_\beta^B\} = 0$.

(2) - two involutions
$$q_i \longrightarrow \bar{q}_i = q_i^o - q_i^r e_r$$
$$q_i \longrightarrow \tilde{q}_i = q_i^o + q_i^1 e_1 - q_i^2 e_2 + q_i^3 e_3$$

Quaternionic supergroups arise as an extension (quaternionic extension) of real or complex supergroups. As in the complex case, one can introduce two families of <u>the metric preserving quaternionic supergroups</u>:

(3) OSp(n+m;2k;H) : (O(n+m;H)×Sp(2k;H)) which leave invariant the form
$$\tilde{q}_i g_{ij} q_j + \tilde{\xi}_\alpha C_{\alpha\beta} \xi_\beta = \text{inv.}$$

(4) $UU_\alpha(n,m;k;H)$: ($U(n,m;H)\times U_\alpha(k;H)$) which leave invariant the form
$$\bar{q}_i g_{ij} q_j + \bar{\xi}_\alpha C_{\alpha\beta} \xi_\beta = \text{inv.}$$

where $g_{ij} = \mathrm{diag}(\underbrace{1,\ldots,1}_{n},\underbrace{-1,\ldots,-1}_{m})$, $C_{\alpha\beta} = (I_k \otimes i\sigma_2)$ and

(5) $O(m;H) = U_\alpha(m;H) = O^*(2m;\mathbb{R})$; $Sp(2k;H) = U(k,k;H) = USp(2k,2k;C)$

From (3)-(5) follows that in the quaternionic case one has the unique family of metric preserving supergroups, on the contrary to the complex case, because

(6) $OSp(m;2k;H) = UU_\alpha(k,k;m;H)$

The family of <u>the volume preserving quaternionic supergroups</u>

(7) $SL(n;m;H) \quad : \quad (SL(n;H) \times GL(m;H))$

is unique as in the complex case. The following relations hold

(8) $SL(n;H) = SU^*(2n) \quad ; \quad GL(n;H) = U^*(2n)$

The sequence of extensions of the metric preserving supergroups

(9) $OSp(1;2;R) \subset OSp(1,2;C) \subset OSp(1,2;H)$

has been considered in [3] and the extensions of the volume preserving supergroups in [4]

(10) $SL(2;N;R) \subset SL(2;N;C) \subset SL(2;N;H)$

II. From [1] and [2] one can see that the following diagram is valid:

Minkowski SUSY		Euclidean SUSY
$SU(2,2;2N)$	⟨∿∿⟩	$SL(2;N;H)$
⇓		⇙ ⇘
(11) $OSp(2N;4)$	⟨∿∿⟩	$OSp(N;2;H)$ ⟨∿∿⟩ $UU_\alpha(2;N;H)$
↓		↓ ↓
i) 2N - extended D=4 super-Poincare´ with O(2N) internal symmetry	⟨∿∿⟩	i) N - extended D=4 super-Euclidean with O*(2N) internal symmetry
ii) D=4 super-Poincaré with N(2N-1) central charges	⟨∿∿⟩	ii) D=4 super-Euclidean with N(2N-1) central charges

where we denote: restriction ⟹ ; contraction ⟶ ; analytic continuation ⟨∿∿⟩
The choice of (i) or (ii) depends on the way the contraction is performed. Let us notice that in considered Minkowski supersymmetries the internal symmetry is described by the compact group on the contrary to the Euclidean case. Further, from this diagram follows that the analytic continuation from Minkowski to Euclidean supersymmetries exists only for even n=2N. In particular Wess-Zumino superalgebra $su(2,2;1)$ does not have an Euclidean counterpart.

III. D=3 Euclidean supersymmetry we can get as the contraction of

a) D=3 anti de Sitter supersymmetry:

(12a) $\dfrac{OSp(N;2;C)}{(\bar{O}(4,1) \times O(N;C))} \longrightarrow$
i) N-extended D=3 super-Euclidean with O(N;C) internal symmetry
ii) D=3 super-Euclidean with N(N-1) central charges

(12b) $\dfrac{SL(2;N;C)}{(\bar{O}(4,1) \times GL(N;C))} \longrightarrow$
i) N-extended D=3 super-Euclidean with GL(N;C) internal symmetry
ii) D=3 super-Euclidean with $2N^2$ central charges

Like in the previous case, the choce of (i) or (ii) depends on the way the contraction is performed. In both cases, we get the Weyl spinorial supercharge.

b) D=3 de Sitter supersymmetry:

(13a) $UU_\alpha(1;2;H)$ ($\bar{O}(4)\times O(2,1)$) $\longrightarrow$ i) D=3 super-Euclidean with O(2,1) internal symmetry
ii) D=3 super-Euclidean with 3 central charges

(13b) SL(1,1;H) ($\bar{O}(4)\times O(1,1)$) $\longrightarrow$ i) D=3 super-Euclidean with O(1,1) internal symmetry
ii) D=3 super-Euclidean with 1 central charge

It is interesting to notice that only in <u>the case (13b) the supercharges form O(3) vector</u>, while in the other ones they behave like <u>the Weyl spinors</u>.

Let us discuss in detail the cases (12a) and (13b) assuming for the first possibility N=1 .

i) The superalgebra osp(1,2;C) can be written down in the form [3] : (R denotes O(3,1) radius)

$$[X_{ij},X_{kl}] = i(\delta_{ik}X_{jl}+\delta_{jl}X_{ik}-\delta_{il}X_{jk}-\delta_{jk}X_{il}) \qquad i,j,k,l = 1,2,3$$
$$[X_{ij},P_k] = i(\delta_{ik}P_j-\delta_{jk}P_i) \quad ; \quad [P_i,P_j] = -iR^{-2}X_{ij}$$
$$(14) \quad [X_{ij},Q_A] = -\tfrac{1}{2}(\sigma_{ij})_{AB}Q_B \quad ; \quad [P_k,Q_A] = \tfrac{1}{2}iR^{-1}(\gamma_o\gamma_k)_{AB}Q_B \qquad A,B = 1,\dots,4$$
$$\{Q_A,Q_B\} = (\gamma^k)_{AB}P_k + R^{-1}(\sigma^{ij}\gamma_o)_{AB}X_{ij}$$

where $\sigma^{ij}=\frac{1}{2}i[\gamma^i,\gamma^j]$ γ^μ forms the fourdimensional real realization of the relation $\{\gamma^\mu,\gamma^\nu\}=2g^{\mu\nu}$ ($g^{\mu\nu}=(-1,1,1,1)$).

Let us define

$$(15) \qquad L_k = \tfrac{1}{2}\epsilon_{kij}X_{ij} \quad ; \quad S_\alpha = \begin{pmatrix} Q_1-iQ_3 \\ Q_2-iQ_4 \end{pmatrix}$$

then in the limit $R\rightarrow\infty$ we get the Euclidean superalgebra in the form

-bosonic sector:

$$(16) \qquad o(3) \text{ rotations: } [L_i,L_j] = i\epsilon_{ijk}L_k$$

$$(17) \qquad \text{three translations: } [P_i,P_j]=0 \quad ; \quad [L_i,P_j] = i\epsilon_{ijk}P_k$$

-fermionic sector:

$$(18) \qquad \{S_\alpha,S^\beta\} = -2(\sigma_k)_\alpha{}^\beta P_k \quad ; \quad \{S_{\dot\alpha},S^{\dot\beta}\} = -2(\sigma_k)_{\dot\alpha}{}^{\dot\beta}P_k \quad ; \quad \{S_\alpha,S_{\dot\beta}\}=0$$

-covariance relations:

$$(19) \qquad [L_k,S_\alpha] = \tfrac{1}{2}(\sigma_k)_\alpha{}^\beta S_\beta \quad ; \quad [L_k,S_{\dot\alpha}] = -\tfrac{1}{2}(\sigma_k)_{\dot\alpha}{}^{\dot\beta}S_{\dot\beta} \quad ; \quad [P_k,S_\alpha] = [P_k,S_{\dot\alpha}] = 0$$

where $S_{\dot\alpha} = (S_\alpha)^*$, $(\sigma_k)_{\dot\alpha}{}^{\dot\beta} = ((\sigma_k)_\alpha{}^\beta)^*$, $S^\alpha = \epsilon^{\alpha\beta} S_\beta$.

Therefore, one can see that S_α transforms like SU(2) Weyl spinor. One can also notice that this superalgebra decomposes onto two complex conjugated to each other subalgebras generated by $E=(L_k,P_k,S_\alpha)$ and $E^* =(L_k,P_k,S_{\dot\alpha})$. The similar decomposition holds for D=4 Euclidean superalgebra [5] . In both cases, this fact allows us to introduce non-selfconjugate superfield formulation of D=3,4 Euclidean supersymmetry.

Now, let us consider the contraction (13b) yielding three-vectorial supercharges.

The fundamental realization of $sl(1,1;H)$ in terms of 2×2 quaternionic matrices has the form:

- bosonic sector

(20) O(4) rotations: $$M(\omega_1,\omega_2) = \begin{pmatrix} \omega_1 & 0 \\ 0 & \omega_2 \end{pmatrix} \qquad \bar\omega_1 = -\omega_1, \quad \bar\omega_2 = -\omega_2$$

(21) O(1,1) noncompact generator $$A = RM = \begin{pmatrix} 1 & 0 \\ 0 & 1 \end{pmatrix}$$

- fermionic sector:

(22) $$Q(\xi) = R^{-\frac{1}{2}} \xi_A Q_A = \begin{pmatrix} 0 & \xi \\ \bar\xi & 0 \end{pmatrix} \quad ; \quad S(\eta) = R^{-\frac{1}{2}} \eta_A S_A = \begin{pmatrix} 0 & \eta \\ -\bar\eta & 0 \end{pmatrix} \quad A = 1,\ldots,4$$

If we introduce the angular momentum generators L_k and "translation" generators P_k in curved Euclidean space-time as follows

(23) $$M(\omega,\omega) = 2\omega_k L_k \quad ; \quad M(\omega,-\omega) = 2R\omega_k P_k$$

then in the limit $R \to \infty$ we get the following superalgebra:

- bosonic sector

(24) $$[L_i,L_j] = \epsilon_{ijk} L_k \quad ; \quad [L_i,P_j] = \epsilon_{ijk} P_k$$
$$[P_i,P_j] = 0 \quad ; \quad [M,L_k] = [M,P_k] = 0$$

- fermionic sector

(25) $$\{Q_A,Q_B\} = 2\delta_{AB} M \quad ; \quad \{Q_i,S_j\} = 4\epsilon_{ijk} P_k$$
$$\{S_A,S_B\} = -2\delta_{AB} M \quad ; \quad \{Q_4,S_A\} = \{S_4,Q_A\} = 0$$

- covariance relations

(26) $$[L_i,Q_j] = \epsilon_{ijk} Q_k \quad ; \quad [P_k,Q_A] = 0 \quad ; \quad [M,Q_A] = 0 \quad ; \quad [L_k,Q_4] = 0$$
$$[L_i,S_j] = \epsilon_{ijk} S_k \quad ; \quad [P_k,S_A] = 0 \quad ; \quad [M,S_A] = 0 \quad ; \quad [L_k,S_4] = 0$$

We see that this superalgebra contains D=3 Euclidean superalgebra $E = (L_k, P_k, M; Q_k, S_k)$ with the real supercharges Q_k, S_k transforming like O(3) vector. The central charge M plays the role of the mass. It is interesting fact, that D=3 Euclidean supersymmetry can be realized or by the Weyl spinorial supercharges or by O(3) vectorial ones.

In D=4 only the spinorial charges are possible (spin-statistics theorem); in D=3 also vectorial charges are allowed. The supersymmetric D=3 theories with spinorial charges can be obtained in nonrelativistic limit from D=4 supersymmetry. The supersymmetric D=3 theories with vectorial charges describe three-dimensional nonrelativistic model without relativistic extension. It is interesting to find a physical model which would realize the second possibility.

Finally we would like to mention that more detailed discussion of D=3 Euclidean superalgebras and their superspace realization will be given elsewhere.

[1] S.Ferrara, Phys.Lett. 69B(1977),48 ;
[2] J.Lukierski,A.Nowicki,Phys.Lett. 27B(1983),40;
[3] J.Lukierski,A.Nowicki, Fortschr.Phys.30(1982),75;
[4] T.Kugo,P.Townsend,Nucl.Phys. B22 (1983),357;
[5] J.Lukierski,A.Nowicki,Trieste preprint SISA 34/82/EP(June 1982).

GAUGE THEORIES IN HIGHER DIMENSIONS: LINEAR RELATIONS FOR GAUGE FIELDS, INTEGRABILITY CONDITIONS, SPHERICAL SYMMETRY IN EIGHT DIMENSIONS

J. NUYTS
University of MONS
7000 MONS, BELGIUM

ABSTRACT : As a first approach to the study of classical solutions for gauge theories in dimensions higher than four we study linear relations among the fields which ensure the equation of motion as a consequence of the Bianchi identities. We show how these relations can be obtained as integrability conditions on linear covariant first order equations. Using spherical symmetry we exhibit specific solutions of these relations. We also hint at an algebraic approach of the ADHM type to obtain more general solutions at least in principle.

1. INTRODUCTION

With the hope of obtaining physically interesting theories via dimensional reduction, we have started [1,2,3] the study of gauge fields in dimensions greater than four. We have followed the road of generalizing the linear relations among the gauge fields in euclidian dimension d which, in four dimension, are usually called self-duality (or antiself-duality) relations. In section 2, we introduce the "secular equation" [1] for the gauge fields. This equation implies the equations of motion as a result of the Bianchi identities and its solutions generalize the concept of self duality. We then show the essential role of the stability group of the secular equation, a subgroup of the d-dimensional rotation group. A few examples are touched upon in section 3, in particular the interesting eight-dimensional case when the subgroup is $\tilde{SO}(7)$. In section 4, we show briefly that the linear relation can arise trivially as integrability conditions for systems of gauge covariant first order equations [2]. This generalizes the Belavin-Zakharov [4] four-dimensional equations. In section 5, we show that we can obtain $\tilde{SO}(7)$ spherically symmetric solutions (SO(7) or SO(8) gauge groups) in eight dimensions [3]: an obvious generalization of 't Hooft-Polyakov instantons in four dimensions. Finally in section 6 we draw our conclusions and suggest that further solutions can be obtained at least in principle, by generalizing the ADHM method [5] which transforms the problem into a purely algebraic one.

2. SECULAR EQUATION FOR d-DIMENSIONAL GAUGE FIELDS

Let $F_{\mu\nu}$ be the fields of a gauge theory for an arbitrary gauge group in euclidian space-time of dimension d.

It is obvious that if $F_{\mu\nu}$ satisfies the secular equation

$$\lambda F_{\mu\nu} = T_{\mu\nu\rho\sigma} F_{\rho\sigma} \tag{2.1}$$

where T is a completely antisymmetrical, non zero constant tensor and λ a non zero eigenvalue the equations of motion $D_\mu F_{\mu\nu} = o$ are satisfied in virtue of the Bianchi identities $D_\mu \wedge F_{\rho\sigma} = o$.

We now list briefly a few properties of the secular equation

a) it is an eigenvalue equation for the symmetric matrix $T_{\mu\nu\rho\sigma}$ (lines indexed by $\mu\nu$ and columns by $\rho\sigma$). T is obviously traceless, hence the sum of the eigenvalues is zero.

b) Distinct eigenvalues correspond to orthogonal subspaces of the $F_{\mu\nu}$ space.

c) Under the SO(d) group T behaves as the basis of a C_4^d dimensional representation, F of a d(d-1)/2 representation, irreducible except when d=4 in which case F splits into two representations [G = (3,1)+(1,3)] and d=8 where T splits into [70 = 35_s+35_a] where s and a refer to eight-dimensional self duality or antiself duality.

d) Every T has a little group or stability group L(SO(d) and F splits under L is such a way that F and hF ($h \in L$) in (2.1) correspond to the same eigenvalue.

e) The problem of the canonical forms of the secular equations is related to the classification of the points of the abstract T space under the action of SO(d) and its subgroup.

For example, points on the same orbit (or even stratum) in T space under the action of SO(d) have isomorphic stability groups and hence are canonically equivalent. Points not on the same stratum are in general not equivalent and correspond to widely different types of linear relations.

f) In four dimensions there is (up to a scale) only one $T_{\mu\nu\rho\sigma}$ i.e. $\varepsilon_{\mu\nu\rho\sigma}/2$. The eigenvalues +1 and -1 correspond to self duality and anti-self duality.

3. EXAMPLES IN DIMENSION 8.

Since the stability group of T plays a crucial role we have studied all cases in SO(8) which have a T invariant under a maximal subgroup of SO(8). Four maximal subgroups, namely $SU(3)/Z_3$, $SO(5)\otimes SO(3)$, $SO(6)\otimes SO(2)$ and SO(7) can have no invariant T i.e. the decomposition of the 35_s+35_d representations under the subgroup cannot contain a singlet piece. Four other maximal subgroup $SU(2) \otimes Sp(4)/Z_2$, $SO(4) \otimes SO(4)$, $SU(4) \otimes U(1)/Z_4$

and $\widetilde{SO}(7)$ have an allowed invariant T tensor. We describe briefly the results corresponding to the last allowed case. Here $\widetilde{SO}(7)$ (the covering of SO(7)) is embedded in SO(8) in a democratic way such that the following decompositions hold [$8 \to 8$, $28 \to 21+7$, $35_s \to 1+7+27$, $35_a \to 35$], we thus see that there exists essentially one canonical self dual T and that there are two eigenvalues $\lambda_7 = -3$ and $\lambda_{21} = 1$ (normalizing T and using the property (2.8))

By an explicit choice of T in canonical position on finds

(3.1) $F_{81}+F_{72}+F_{45}+F_{36} = 0$ (3.2) $F_{82}+F_{17}+F_{35}+F_{64} = 0$

(3.3) $F_{83}+F_{74}+F_{52}+F_{61} = 0$ (3.4) $F_{84}+F_{37}+F_{51}+F_{26} = 0$

(3.5) $F_{85}+F_{76}+F_{14}+F_{23} = 0$ (3.6) $F_{86}+F_{57}+F_{13}+F_{42} = 0$

(3.7) $F_{87}+F_{65}+F_{43}+F_{21} = 0$

as the 7 relations defining the 21 dimensional subspace for F.

The 21 relations defining the 7 dimensional subspace for F can be obtained by equating the terms of each line of (3.1-7) i.e.

(3.8) $$F_{12} = F_{34} = F_{56} = F_{78}, \text{ etc.,}$$

Numerous properties of the two sets of relations are obviously related to spinors in SO(8) and to octonions and hence seem a natural generalization of the four dimensional case where quaternions play the crucial role.

For later use it is useful to define the $\Lambda^A_{\mu\nu}$ matrices such that (3.1-3.7) become $(A = 1,\dots,7)$ $[8x8 \to 7 : \Lambda^A_{\mu\nu} F_{\mu\nu} = 0]$. $\Lambda^A_{\mu\nu} = -\Lambda^A_{\nu\mu}$.
The antisymmetric matrix is defined from eq. (3. A) as having zeroes everywhere except four times + 1 when needed by (3. A) et four times -1 by antisymmetry.

Define also

(3.9) $$8x8 \to 21 : \Omega^{AB}_{\mu\nu} = \frac{1}{2}(\Lambda^A_{\mu\gamma}\Lambda^B_{\gamma\nu} - \Lambda^B_{\mu\gamma}\Lambda^A_{\gamma\nu})$$

(3.10) $$7x21 \to 35 : \Sigma^{D,AB}_{\rho\sigma} = \frac{1}{2}((\Lambda^D_{\sigma\gamma}\Omega^{AB}_{\gamma\rho} + (\sigma\rho)) - Tr(\rho\sigma))$$

(3.11) $$21x21 \to 35 : \Phi^{AB,CD}_{\rho\sigma} = \frac{1}{2}((\Omega^{AB}_{\sigma\gamma}\Omega^{CD}_{\gamma\rho} + (\rho\sigma)) - Tr(\rho\sigma)).$$

All are obvious Clebsch-Gordan coefficients in SO(7).

4. THE LINEAR RELATIONS AS INTEGRATION CONDITIONS.

For the four-dimensional case there exist the Belavin-Zakharov equations whose integration conditions are exactly the self duality relations (see (2.e)). The following set of linear equations possesses the same property in the general case. Let $G_{\mu\nu}$ be a general antisymmetric tensor spanning the space of all the eigenvalues of (2.1) except one, say λ. The linear differential equations for $G_{\mu\nu}$, $D_\mu G_{\mu\nu} = 0$, imply, as integrability condition for all G, that $F_{\mu\nu}$ belongs to the space of the missing eigenvalue λ. Indeed they imply :

$$D_\nu D_\mu G_{\mu\nu} = \frac{1}{2} F_{\nu\mu} G_{\mu\nu} = 0. \tag{4.1}$$

i.e that $F_{\mu\nu}$ must be orthogonal to the space spanned by $G_{\mu\nu}$ and hence the result (see 2.b).

5. $\tilde{SO}(7)$ SPHERICALLY SYMMETRIC SOLUTIONS

Consider the potentials $A_\mu^{\alpha\beta} = -A_\mu^{\beta\alpha}$ $(\alpha,\beta,\mu=1,\dots,8)$ for an SO(8) gauge theory in eight dimensions. We look for $\tilde{SO}(7)$ spherically symmetric solutions of the equations. The most general form for $A_\mu^{\alpha\beta}$ is, with $F(x^2)$, $G(x^2)$ two arbitrary functions of x^2,

$$A_\mu^{\alpha\beta} = F\ \Omega_{\alpha\beta}^{CD}\ \Omega_{\rho\mu}^{CD}\ x^\rho + G\ \Lambda_{\alpha\beta}^{C}\ \Lambda_{\rho\mu}^{C}\ x^\rho \tag{5.1}$$

which in turn leads to (with $X_{\rho\sigma} = x_\rho x_\sigma - \delta_{\rho\sigma}x^2/8$)

$$\begin{aligned} F_{\mu\nu}^{\alpha\beta} = {} & \Omega_{\alpha\beta}^{AB}\Omega_{\mu\nu}^{AB}(2F+1/2F'x^2-5F^2-1/4G^2x^2) \quad +\Lambda_{\alpha\beta}^{A}\Lambda_{\mu\nu}^{A}(2G+1/2G'x^2-6Fx^2) \\ & +X_{\rho\sigma}(\Omega_{\alpha\beta}^{AB}\Omega_{\mu\nu}^{CD}\Phi_{\rho\sigma}^{AB,CD})(-1/4F'-1/2F^2-1/2G^2) \\ & +X_{\rho\sigma}(\Omega_{\alpha\beta}^{AB}\Lambda_{\mu\nu}^{C}\Sigma_{\rho\sigma}^{C,AB})(-1/2F'-3F^2+G^2) \\ & +X_{\rho\sigma}(\Lambda_{\alpha\beta}^{C}\Omega_{\mu\nu}^{AB}\Sigma_{\rho\sigma}^{C,AB})(-1/4G'-2FG) \end{aligned} \tag{5.2}$$

where the prime denotes differentiation with respect to x^2.

The equations (3.1-7) (i.e.F belongs to 21) admit two obvious solutions I $[F = \frac{1}{6x^2+a}$, $G = 0]$ and II $[F = \frac{1}{4x^2}$, $G = \pm\frac{1}{4x^2}]$ of the instanton type. The first one corresponds to an effective SO(7) gauge group. No such solution exists when F is restricted to the 7 dimensional space (3. 8).

6. CONCLUSIONS.

Finding linear relations which ensure that the equations of motion are satisfied by virtue of the Bianchi identities seems to be an interesting first approach to gauge theories in higher dimensions. In collaboration with D. FAIRLIE we are now looking into algebraic methods of ADHM type to generate new solutions, at least in principle.

ACKNOWLEDGMENTS

The author wishes to thank E. CORRIGAN, C. DEVCHAND and especially D. FAIRLIE. The results of this paper were obtained in collaboration with them. He also thanks A. MARTIN for a careful reading of the manuscript.

REFERENCES

[1] E. Corrigan, C. Devchand, D.B. Fairlie and J. Nuyts, Nuc. Phys. B 14 (1983) 452.

[2] D.B. Fairlie and J.Nuyts, Integration conditions for first order differential linear equations in higher dimensional gauge theories.

[3] D.B. Fairlie and J. Nuyts, Spherically symmetric solutions of gauge theories in eight dimensions.

[4] A. Belavin and V. Zakharov, Phys. Lett. 73B (1978) 53.

[5] D.B. Fairlie and J. Nuyts : in preparation.

QUANTUM VORTICES AND Diff ($\mathbb{R}^3$)

Mario Rasetti* and Tullio Regge**

*Dipartimento di Fisica, Politecnico di Torino, Torino, Italy

**CERN, Geneva

1. Introduction

Vortices in superfluids can be considered as a somewhat unique example of macroscopic quantum object. The construction of a canonical formalism to describe its dynamics leads to showing that the quantization of a vortex is indeed equivalent to constructing the unitary representations of Diff ($\mathbb{R}^3$), the group of diffeomorphism on $\mathbb{R}^3$.

We discuss as well the conjecture that the topological invariants of the vortex considered as a knot play a role analogous to that of the Casimir operators in Diff ($\mathbb{R}^3$).

From the physical point of view such a theoretical structure gives a rigorous frame for implementing - in a globally consistent way - Feynman's intuitive description of a macroscopic vortex as bearing a strong similarity to a ferromagnetic sheet of magnetic dipoles, where the magnetic moments are however replaced by the momenta of roton-like elementary excitations.

2. Canonical formalism and Current Algebra

A classical vortex is a Jordan curve Γ in $\mathbb{R}^3$, and we assign its configuration in terms of a map $\mu: S^1 \to \mathbb{R}^3$ by the set of functions $x^i(\sigma)$, $i=1,2,3$ where σ is a parameter ranging over a compact closed domain.

The description of the vortex is completed by the velocity potential $\varphi(\bar{x})$, such that $\bar{v}(\bar{x}) = -\mathrm{grad}\,\varphi$. The latter relation holds only locally, and in order to make it hold globally one needs the further requirement that continuing $\varphi(\bar{x})$ along a closed loop Λ , it changes by an integer multiple (the linking number of Λ and Γ) the vorticity κ .

Then $\bar{v}(\bar{x})$ is well defined over $\mathbb{R}^3 - \Gamma$, where $\mathrm{curl}\,\bar{v} = 0$; on Γ , $|\mathrm{curl}\,\bar{v}| = \kappa$.

The Lagrangian, reads /1/

$$\mathcal{L} = \frac{1}{3}\kappa\rho \int_\Gamma \varepsilon_{ijl}\, x^i(\sigma)\, dx^j \frac{dx^l}{dt} - \frac{1}{2}\rho \int_{\mathbb{R}^3} \bar{v}^2\, d^{(3)}\bar{x} \qquad (2.1)$$

where ρ is the mass density of the fluid, and the first term obviously describes the kinetics of the vortex whereas the second describes the global motion of the fluid.

By variation of $\varphi(\bar{x})$, one cheeks that $\mathcal{L}$ describes indeed an incompressible fluid, div $\bar{v} = 0$, so that one can write $\bar{v}(\bar{x})$ = curl $\bar{A}(\bar{x})$.

With the gauge choice such that div $\bar{A} = 0$, it is straight forward to get (no retardation effects are included)

$$\bar{A}(\bar{x}) = \frac{\kappa}{4\pi} \int_{\Gamma} \frac{1}{|\bar{x}-\bar{y}|} d\bar{y} \tag{2.2}$$

whereby the Lagrangian becomes

$$\mathcal{L} = \frac{1}{3}\kappa\rho \int_{\Gamma} \varepsilon_{ijl}\, x^i \frac{dx^l}{dt} dx^j - \frac{1}{8\pi}\kappa^2\rho \iint_{\Gamma\times\Gamma} \frac{1}{|\bar{x}-\bar{y}|} d\bar{x}\cdot d\bar{y} \tag{2.3}$$

The latter two formulas show explicitly where the difficulties come from when one tries to develop a canonical formalism: $\bar{A}(\bar{x})$, and hence $\bar{v}(\bar{x})$ are functio nals of the vortex configuration Γ , or-differently stated- $\mathcal{L}$ is degenerate in that linear in the velocities. The conjugate momenta to the $x^i(\sigma)$ are there fore functions of the x^i's themselves

$$P_i(\sigma) = \frac{\partial\mathcal{L}}{\partial \frac{dx^i}{dt}} = \frac{1}{3}\kappa\rho\, \varepsilon_{ijl}\, x^j \frac{dx^l}{d\sigma} \tag{2.4}$$

and this is incosistent with the Poisson bracket

$$\{x^i(\sigma), P_j(\sigma')\} = \delta^i_j\, \delta(\sigma-\sigma') \tag{2.5}$$

The way out such difficulties is using Dirac's formalism for constrained systems.[2]

A through application of it shows[3] that one has to introduce an additional wariable $\chi(\sigma)$ which is but the generator of reparametrization transformations of σ , and which - together with the $x^i(\sigma)$, $i=1,2,3$ generates all other dynami cal variables. In particular

$$P_i(\sigma) = \frac{1}{3}\kappa\rho\, \varepsilon_{ijl}\, x^j(\sigma) \frac{dx^l}{d\sigma} + \frac{\chi(\sigma)}{\left|\frac{d\bar{x}}{d\sigma}\right|^2} \frac{dx^i}{d\sigma} \tag{2.6}$$

The Dirac brackets are:

$$\{x_i(\sigma), x_j(\sigma')\}_D = \frac{1}{\kappa\rho\left|\frac{d\bar{x}}{d\sigma}\right|^2} \varepsilon_{ijl} \frac{dx^l}{d\sigma}\, \delta(\sigma-\sigma') + S_{ij}(\sigma,\sigma')\,\delta'(\sigma-\sigma') \tag{2.7-a}$$

$$\{x^i(\sigma), \chi(\sigma')\}_D = \frac{dx^i}{d\sigma} \delta(\sigma-\sigma') \qquad (2.7\text{-b})$$

$$\{\chi(\sigma), \chi(\sigma')\}_D = [\chi(\sigma) + \chi(\sigma')] \, \delta'(\sigma-\sigma') \qquad (2.7\text{-c})$$

where $S_{ij}(\sigma,\sigma') = C_{ij}(\sigma,\sigma')\chi(\sigma) + C'_{ij}(\sigma,\sigma')\chi(\sigma')$, C_{ij} and C'_{ij} denoting the Schwinger terms, which are indeed weakly zero.

The next step in order to have a consistent canonical theory-to be quantized in the customary way-is of course to get rid of $\chi(\sigma)$. This can be done in two different ways.

i) One may make a special choice of the gauge (e.g. equating σ to the arc-length on Γ). $\chi(\sigma)$ becomes then a second class constraint, and Dirac's should be further modified to be consistent with both the gauge fixing and the requirement $\chi = 0$ (strongly).

The procedure is lengthy and cumbersome, and can be easily worked out only in special cases[4]. The interesting results one gets from the latter are the following:

a) the vortex has one oscillator-like degree of freedom at each point.

b) it is impossible, quantum mechanically, to define sharply the configuration of Γ : if its projection on a plane is assigned, then the position of the points of Γ in the direction perpendicular to the plane is completely uncertain;

c) the scale of uncertainly in the definition of the vortex equals the average atomic dimension.

ii) One can build up the theory utilizing only reparametrization invariant quantities (i.e. variables which commute with χ), whose algebra is independent on the gauge fixing.

The second procedure is implemented by considering the functionals

$$J(\bar{b},\Gamma) = \int_\Gamma \bar{B}(\bar{x}) \cdot d\bar{x} \qquad (2.8)$$

where $\bar{B}$ is some classical field whose curl, $\bar{b}$ acts as a label for the dynamical variables on the phase space.

The J's have Dirac brackets - which one can obtain from (1.7) - given by

$$\{J(\bar{b}), J(\bar{c})\}_D = -\frac{1}{\kappa\rho} J(\mathrm{curl}(\bar{b}\times\bar{c})) \tag{2.9}$$

to which correspond the quantum commutators

$$[J(\bar{b}), J(\bar{c})] = i\frac{\Omega}{2\pi} J(\bar{c}\cdot\nabla\bar{b} - \bar{b}\cdot\nabla\bar{c}) \tag{2.10}$$

Ω denotes the specific volume per atom.

The $J(\bar{b},\Gamma)$ form an infinite Lie algebra $\mathcal{A}$, isomorphic with the algebra of divergenceless tangent vectors in $\mathbb{R}^3$.

Γ is classically almost everywhere determined by the set of all $J(\bar{b},\Gamma)$: the latter can therefore be regarded as a complete set of observables for the vortex configuration. The construction of a quantum theory for Γ turns then out to be equivalent to finding the unitary irreducible representations of the group $\mathcal{G}$ induced by $\mathcal{A}$. $\mathcal{G}$ is isomorphic to Diff ($\mathbb{R}^3$, ν), the group of measure preserving diffeomorphism of $\mathbb{R}^3$ (ν denotes the Lebesgue measure on $\mathbb{R}^3$).

3. Vortices and knots

Let $\mathcal{A}^{(n)}$ denote the subset of $\mathcal{A}$ consisting of all the $J(\bar{b})$'s for which $\bar{b}(\bar{x})$ is a homogeneous polynomial of degree n in $\bar{x}$.
$\mathcal{A}^{(0)}$ has three independent elements, which are but the components of the linear momentum along some given direction $\bar{u}$ (notice that the momentum is proportional to the area of the projection of the loop Γ on a plane orthogonal to $\bar{u}$). $\mathcal{A}^{(1)}$ has eigth elements, which form the subalgebra of $\mathcal{A}$ isomorphic to $SL(3,\mathbb{R})$, generating all affine volume preserving transformations of $\mathbb{R}^3$. The angular momentum belongs to $\mathcal{A}^{(1)}$.

In general one has

$$[\mathcal{A}^{(n)}, \mathcal{A}^{(m)}] \subset \mathcal{A}^{(n+m-1)} \tag{3.1}$$

so that the set of $\mathcal{A}^{(n)}$'s with $n\leq p$ is a subalgebra of $\mathcal{A}$ only if $p=1$.

On the other hand the set

$$\bigcup_{n=p}^{\infty} \mathcal{A}^{(n)} \doteq \mathfrak{a}(p) \tag{3.2}$$

is a subalgebra of $\mathcal{A}$, and one can define the factor algebra

$$\mathcal{T}(p) \doteq \mathcal{A}(1)/\mathcal{A}(p+1) \tag{3.2}$$

Turning now to representations, one representation of $\mathcal{A}$ is given by

$$\mathcal{D} : J(\bar{b}) \longrightarrow \frac{\Omega}{2\pi i} \, b^j(\bar{x}) \frac{\partial}{\partial x^j} \tag{3.3}$$

Its exponentiation gives the unitary representation induced by measure preserving flows. Indeed, if $\Psi(\bar{x})$ is a well behaved function on $\mathbb{R}^3$, $\Psi(\bar{x}) \in L^2(\mathbb{R}^3)$ and $\mathcal{F}$ is an incompressible flow on $\mathbb{R}^3$, the transformation $\mathcal{U}(\mathcal{F}) : L^2 \to L^2$

$$[\mathcal{U}(\mathcal{F})\Psi](\bar{x}) = \Psi(\mathcal{F}^{-1}\bar{x}) \tag{3.4}$$

defines a unitary operator on L^2. $\mathcal{D}$ is the so called single particle representation of the current algebra $\mathcal{A}$. Other representations of $\mathcal{A}$ can be of course obtained by tensor product of p copies of $\mathcal{D}$, $p \in \mathbb{Z}$ (the symmetric product giving the well known Fock space p-particle representation of $\mathcal{A}$).

However the most interesting generalization of $\mathcal{D}$ is the following.

Let $\{T_j^{i_1 \cdots i_k}\}_{k=1}^{p}$ be a representation of $\mathcal{T}(p)$, where $T_j^{i_1 \cdots i_k}$ are the infinitesimal generators of the stabilizer subgroup $\mathcal{H}$ of $\mathcal{G}$, leaving the point $\bar{x}$ fixed,

$$[T_j^{k_1 \cdots k_n}, T_q^{k_{n+1} \cdots k_\ell}] = \delta_j^{k_s} T_q^{k_1 \cdots k_n k_{n+1} \cdots \hat{k}_s \cdots k_\ell} - \delta_q^{k_t} T_j^{k_1 \cdots \hat{k}_t \cdots k_n k_{n+1} \cdots k_\ell} \tag{3.5}$$

$$n \geq 1 \; ; \; \ell \geq n+1 \; ; \; n+1 \leq s \leq \ell \; ; \; 1 \leq t \leq n$$

(indices with hat to be deleted).

A new representation of $\mathcal{A}$ is given by

$$\mathcal{D}^{(p)} : J(\bar{b}) \longrightarrow \frac{\Omega}{2\pi i} \left\{ b^j \frac{\partial}{\partial x^j} + \sum_{k=1}^{p} \frac{1}{k!} \frac{\partial^k b^j}{\partial x^{i_1} \cdots \partial x^{i_k}} T_j^{i_1 \cdots i_k} \right\} \tag{3.6}$$

$\mathcal{D}^{(p)}$ can be thought of as an induced representation with a nontrivial fibre transitive under the action of $\mathcal{G}/\mathcal{H}$.

Upon denoting by $\mathcal{T}^{(n)}$ the subset of all the $T_j^{k_1 \cdots k_n}$'s , $n > 0$; we have

$$[\mathcal{T}^{(n)}, \mathcal{T}^{(m)}] \subset \mathcal{T}^{(n+m-1)} \tag{3.7}$$

Notice that because $\mathcal{T}^{(0)}$ does not exist, it is possible to set $\mathcal{T}^{(n)} = 0$ for all

$n > p$, and to obtain a finite algebra. Intuitively, in such a case the $\mathcal{T}^{(n)}$ generate all the coordinate transformations leaving the point $\bar{x}$ fixed, when one neglects terms of order higher than p in the local power series expansion of $\bar{b}$.

Indeed writing, by Stokes theorem,

$$J(\bar{b}) = \int_S b^i \, dS_i \qquad , \quad \partial S = \Gamma \tag{3.8}$$

if the loop is small enough and $\bar{x}$ is a point on S, expanding $\bar{b}$ around $\bar{x}$ gives

$$J(\bar{b}) = b^i(\bar{x}) \int_S dS_i + \frac{\partial b^i}{\partial x^j} \int_S (y^j - x^j)\, dS_i + \cdots \tag{3.9}$$

where one can recognize in the first term at the r.h.s. the linear momentum - corresponding to the first factor in (3.6). The remaining terms in (3.9) correspond to the other factors in (3.6), showing how the representation in terms of $\{ T_j^{\; i_1 \dots i_k} \}_{k=1}^{p}$ is equivalent to describing the shape of the loop by its moments up to the finite order p.

The subset of the $T_j^{\{i\}}$'s corresponding to $n = l = 1$ in (3.5) satisfies the commutation relations of angular momentum, and in fact in the representation $\mathcal{D}$ its presence endows the vortex with spin.

Now the quantum vortex can be described in terms of a complex field $\psi(\bar{x})$, with commutation relations:

$$[\psi(\bar{x}), \psi^*(\bar{y})] = \delta(\bar{x} - \bar{y}) \tag{3.10}$$

The definition

$$J(\bar{b}) = \frac{\Omega}{2\pi} \int_{\mathbb{R}^3} b^k(\bar{x})\, j_k(\bar{x})\, d^{(3)}\bar{x} \tag{3.11}$$

where $j(\bar{x})$ is the current associated with the field $\psi(\bar{x})$,

$$j_k(\bar{x}) = \frac{1}{2i} \left[\psi^*(\bar{x}) \frac{\partial \psi}{\partial x^k} - \psi(\bar{x}) \frac{\partial \psi^*}{\partial x^k} \right] \tag{3.12}$$

is indeed consistent with the commutation relations (2.10),

$$[J(\bar{b}), J(\bar{c})] = i \frac{\Omega}{2\pi} J(\operatorname{curl} \bar{b} \times \bar{c}) \tag{3.13}$$

and shows as well that the total particle number commutes with all $J(\bar{b})$'s .

The definition of $\Psi(\bar{x})$ is better realized in terms of Clebsch potentials[5]. In such adescription a macroscopic vortex corresponds to states with a very large occupation number. In the N-particle sector the hamiltonian H turns out to be the sum of N self-energy terms and two body interactions,which are but the quadrupolar interactions between pairs of the elementary excitations one can associate to the currents in $\mathfrak{D}$, $J(\bar{b}) = \frac{\Omega}{2\pi i} b^{\kappa} \frac{\partial}{\partial x^{\kappa}}$ .The latter are in a way the smallest possible vortices, with only translational degrees of freedom, and very closely resemble the rotonic excitations introduced by Feynman[6].

In other words a large vortex is thus described as the bound state of a large number of elementary vortices.

A detailed analysis, which requires the introduction of a suitable cut-off factor to account for the finite size r_0 of the vortex filament, shows that- upon writing

$$H = \sum_{\alpha=1}^{N} \varepsilon_{\alpha} + \frac{1}{2} \sum_{\alpha,\beta} w_{\alpha\beta} \tag{3.14}$$

- one has

$$\varepsilon_{\alpha} = \frac{p_{\alpha}^{2}}{2M} \quad ; \quad w_{\alpha\beta} = \begin{cases} \frac{\bar{p}_{\alpha}\bar{p}_{\beta}}{M} & , r_{\alpha\beta} < r_0 \\ -\frac{1}{2M}\left(\frac{r_0}{r_{\alpha\beta}}\right)^3 \left[\bar{p}_{\alpha}\cdot\bar{p}_{\beta} - 3(\bar{n}\cdot\bar{p}_{\alpha})(\bar{n}\cdot p_{\beta})\right] & , r_{\alpha\beta} > r_0 \end{cases} \tag{3.15}$$

There follows that in a stable configuration a vortex is composed of several elementary excitations of momenta $\bar{p}_{\alpha}$ (and mass M). The momenta $\bar{p}_{\alpha}$ are approximately parallel for nearby excitations, and moreover:

i) the vortex is required to be the boundary of a smooth surface with a well defined normal,

ii) for $r_{\alpha\beta} < r_0$ the pair α, β behaves as a single excitation with mass M and momentum ($\bar{p}_{\alpha} + \bar{p}_{\beta}$). This acts as a repulsive core.

iv) for $r_{\alpha\beta} > r_0$ the quadrupolar interaction becomes attrattive (and decreasing with the distance) as $\bar{p}_{\alpha}$, $\bar{p}_{\beta}$ tend to become orthogonal to $\bar{n} = \frac{\bar{r}_{\alpha\beta}}{r_{\alpha\beta}}$.

In other words an equilibrium configuration is somewhat similar to a ferromagnetic lamina,(in which though the role of the magnetic moments is played by the momenta of the rotonic excitations) where the elementary excitations are bond with mutual spacing somewhat larger than r_0 .

Quantum mechanics would of course require the construction of all irreducible unitary representations of Diff ($\mathbb{R}^3$).

First one should construct a complete set of commuting observables.

Following the ideas of Cartan on Lie algebras, we consider first the abelian subalgebra $\mathfrak{h}$ of $\mathcal{A}$ generated by the current operators which perform translations along a fixed direction. Assuming the latter as z-axis, $\mathfrak{h}$ is generated by

$$J(V) = \int_{\Gamma} \left(\frac{\partial V}{\partial y} dx - \frac{\partial V}{\partial x} dy \right) = \int_{S} \left(\frac{\partial^2 V}{\partial x^2} + \frac{\partial^2 V}{\partial y^2} \right) dx\, dy \tag{3.16}$$

where $V = V(x,y)$ is C^2.

$J(V)$ induces a translation by the amount ΔV ; and of course the projection $\Gamma_{\perp}$ of Γ on the (x,y) - plane is completely determined once the eigen-values of $J(V)$ are known, for all V .

The algebra $\mathfrak{h}$, on the other hand can be studied by looking at its representation in the N-particle sector

$$\mathcal{D}_N(\mathfrak{h}) : J(V) \longrightarrow \sum_{\alpha=1}^{N} \frac{\Omega}{2\pi i} \Delta_\alpha V(x_\alpha, y_\alpha) \frac{\partial}{\partial z_\alpha} \tag{3.17}$$

Upon denoting by $\mathcal{S}_N$ the symmetrizing operator in the N-particle sector, one can check that the corresponding eigenvectors have the form

$$\Phi_\lambda^{(N)} = \mathcal{S}_N \prod_{\alpha=1}^{N} e^{i\lambda_\alpha z_\alpha} \delta(x_\alpha - \xi_\alpha)\, \delta(y_\alpha - \eta_\alpha) \tag{3.18}$$

with eigenvalue

$$\lambda = \frac{\Omega}{2\pi} \sum_{\alpha=1}^{N} \lambda_\alpha \Delta_\alpha V(\xi_\alpha, \eta_\alpha) \tag{3.19}$$

To complete the information provided by the latter we need as well the elements from the enveloping algebra of $\mathcal{A}$ commuting with $\mathfrak{h}$; and in particular an explicit construction of the center of the enveloping algebra (Casimir operators).

In fact one can expect, besides $\mathfrak{h}$, an infinite ascending sequence $\mathfrak{h}^{(p)}$ of subalgebras of degree $p, p \in \mathbb{Z}$ in the J's, all mutually commuting and it is plausible (no way is as yet known to construct the Casimir operators) that the operators in the center cannot be expressed as polynomials of finite degree in $\mathcal{A}$, and the whole set $\bigcup_{p=2}^{\infty} \mathfrak{h}^{(p)}$ is required.

However the role of the Casimir operators turns out to be closely related to the invariants of the vortex support Γ , considered as a knot.

In other words, both Dehn's diagram and Alexander's polynomials [7] can be retrieved from the algebras $\mathfrak{h}^{(p)}$, also for a quantum vortex, even though the three-dimensional configuration is not sharply defined.

For example the abelian subalgebra $\mathfrak{h}^{(2)}$ of quadratic elements in the enveloping algebra of $\mathcal{A}$ is generated by operators of the form:

$$J(F,\bar{u}) = \lim_{\substack{\varepsilon \to 0 \\ \eta \to 0}} \iint_{\Gamma \times \Gamma} \delta\left[x(\sigma) - x(\sigma') - \eta u^1\right] \cdot \delta\left[y(\sigma) - y(\sigma') - \eta u^2\right] F_{ij}\left[z(\sigma) - z(\sigma') - \varepsilon\right] dx^i dx^j \tag{3.20}$$

where $\bar{u}$ is a unit vector in the (x,y)-plane.

With the choice $F_{ij}(z) = \varepsilon_{ij} F_0(z)$, $F_0(-z) = -F_0(z)$, i.e. if one restricts to variables invariant under Euclidean motions in the (x,y) plane,

$$J(F) = \sum_{\gamma} \Xi(\gamma)\, F_0(z_\gamma - z'_\gamma) \tag{3.21}$$

where γ labels the self intersections of $\Gamma_\perp$, and

$$\Xi(\gamma) = \left.\frac{dx\,dy' - dx'dy}{|dx\,dy' - dx'dy|}\right|_\gamma \tag{3.22}$$

If all the $J(F)$'s had been measured the ($z_\gamma - z'_\gamma$) could be derived.

Multiple self-crossings require $\mathfrak{h}^{(p)}$ with $p > 2$ in order to remove the degeneracies and reconstruct the knot.

REFERENCES

/1/ J.J. Thompson, On the Motion of Vortex Rings, Adam Prize Essay, London (1883).
W. Thompson (Lord Kelvin), Mathematical and Physical Papers, vol.4, Cambridge University Press, Cambridge 1910; pag.152.

/2/ P.A.M. Dirac, Lectures on Quantum Mechanics, Belfer Graduate School of Science, Yeshiva University, New York 1965.
P.A.M. Dirac, Canad. J. Math. 2, 129 (1950); Proc. Roy. Soc. A246, 326 and 333 (1958); Phys. Rev. 114, 924 (1959).

/3/ M. Rasetti and T. Regge, Physica 80A, 217 (1975).

/4/ M. Rasetti and T. Regge, Quantum Vortices, in "Highlights on Condensed Matter Physics", M. Tosi ed., Academic Press, New York 1983.

/5/ A. Clebsch, J. Reine Angew. Math. 56, 1 (1859).
Y. Nambu, Phys. Lett. 92B, 327 (1980).

/6/ R.P. Feynman, in "Progress Low Temperature Physics", vol.1, C.G. Goerter ed., North Holland Publ. Co., Amsterdam 1955; chapter II.

/7/ R.H. Crowell and R.H. Fox, Introduction to knot Theory, Ginn and Co., Boston 1963.

ATOMIC AND NUCLEAR PHYSICS

THE TIME DEPENDENT Sp(2,ℝ) MODEL FOR THE BREATHING MODE

F. Arickx, J. Broeckhove, M. Buysse, P. Van Leuven
Universiteit Antwerpen
Rijksuniversitair Centrum
Groenenborgerlaan 171, 2020 Antwerpen, Belgium

In the study of nuclear systems the time-dependent methods are being fully recognized. In order to emphasize some particular degrees of freedom of the system, one can parametrize the A-particle wave-function accordingly. Assuming then the time evolution to be fully given through the parameters of the wave-function, one can apply the TDVP on the manifold of parametrized A-particle states [1]. The time dependence of the parameters is obtained via first-order differential equations. Provided the manifold and its underlying parameter space have appropriate geometric properties, one may develop the equations into a formalism of Hamiltonian mechanics in a generalized phase space. Parametrizing the manifold via the construct of coherent states of some Lie group provides the demanded geometric properties.

In the study of the nature of the nuclear breathing mode (or giant monopole excitation) the Sp(2,ℝ) group was used to provide a simple, yet accurate, method with clear-cut physical interpretation. However, this model was confined to a stationary-state formalism only. It is the purpose of this contribution to demonstrate that also in a time-dependent formalism the Sp(2,ℝ) group is of outstanding importance to describe the breathing mode phenomenon.

In order to derive the formal structures appropriate to the application of the TDVP, we turn to the Sp(2,ℝ) group whose algebra, sp(2,ℝ) is spanned by e.g.

$$M = \frac{1}{2}\sum \bar{r}_j^2 \quad ; \quad D = -\frac{1}{2\hbar}\sum (\bar{r}_j.\bar{p}_j+\bar{p}_j.\bar{r}_j) \quad ; \quad C = \frac{1}{2}\sum (\bar{p}_j^2+\bar{r}_j^2)$$

where M is the monopole operator, D the dilation operator and C, apart from a factor, the oscillator Hamiltonian. The latter is the weight operator in terms of the Cartan basis, written in terms of the familiar oscillator creation and annihilation operators:

$$A^+ = \frac{1}{2}\sum \bar{a}_j^+.\bar{a}_j^+ \quad ; \quad A = \frac{1}{2}\sum \bar{a}_j.\bar{a}_j \quad ; \quad C = \frac{1}{2}\sum (\bar{a}_j^+.\bar{a}_j+\bar{a}_j.\bar{a}_j^+)$$

with commutation rules

$$[C,A^+] = A^+ \quad ; \quad [C,A] = -A \quad ; \quad [A,A^+] = 2C$$

The irreducible representations (irrep) of Sp(2,$\mathbb{R}$), relevant to the breathing mode problem, can be constructed in the oscillator shell model framework using the concept of a lowest weight state $|0\rangle$:

$$A|0\rangle = 0 \qquad \text{and} \qquad C|0\rangle = k|0\rangle$$

Any such state fixes the irrep with label k whose basis is generated by repetitive action of the raising operator A^+. We will consider only the irrep built on the oscillator groundstate $\psi_0(\bar{r}_1,\ldots,\bar{r}_A)$ which is easily seen to be a lowest weight state.

Choosing ψ_0 as a reference state in the construction of Perelomov coherent states [2], and considering the Iwasawa decomposition for the Sp(2,$\mathbb{R}$) operator [3]:

$$T(\theta,\sigma,\varphi) = e^{i\sigma M}\, e^{i\theta D}\, e^{i\varphi C}$$

one obtains for the coherent states

$$T(\theta,\sigma,\varphi)|0\rangle = e^{ik\varphi}|\theta,\sigma\rangle$$

As the phase factor is irrelevant, the manifold of coherent states is isomorphic to the coset space (θ,σ) of Sp(2,$\mathbb{R}$)/SO(2). One can readily evaluate that the coherent states are

$$\langle\bar{r}_1,\ldots,\bar{r}_A|\theta,\sigma\rangle = \psi_{(\theta,\sigma)}(\bar{r}_1\ldots\bar{r}_A) = e^{i(\frac{\sigma}{2}\sum \bar{r}_j^2)}\,[e^{-\frac{3}{2}A\theta}\,\psi_0(e^{-\theta}\bar{r}_1,\ldots,e^{-\theta}\bar{r}_A)]$$

with $\theta=\theta(t)$ the regulator of instantaneous size, and $\sigma=\sigma(t)$ the boost that induces the velocity field associated with the change in time of the scale factor. This velocity field which is defined in the usual way through density and current density

$$\bar{v}(\bar{r},t) = \bar{j}(\bar{r},t)/\rho(\bar{r},t) = \frac{\hbar}{m}\,\sigma(t)\bar{r}$$

is seen to be linear and radial. Thus the variational manifold of coherent states is made up of states describing a scaling type oscillation with both dilational and boost coordinates.

The implementation of the TDVP is now a matter of group-theoretic calculations. Sp(2,$\mathbb{R}$)/SO(2) is found to be a curved space with a Poisson bracket of the form

$$\{F,G\} = (2k\hbar b^2 e^{2\theta})^{-1}\,[\partial_\theta F\partial_\sigma G - \partial_\theta G\partial_\sigma F]$$

with b the width parameter used in the oscillator shell model. By using dimensionless parameters $Q = e^\theta$ and $P = (2kb^2)e^\theta\sigma$, and applying TDVP to the $|P,Q\rangle$ manifold, one obtains $\dot{Q} = \{Q,\mathcal{H}\}$; $\dot{P} = \{P,\mathcal{H}\}$ where $\mathcal{H}(Q,P) = \langle Q,P|H|Q,P\rangle = \frac{1}{2}(\hbar\omega/2k)P^2 + U(Q)$ and

$U = \langle Q,P|V|Q,P\rangle$ can be calculated analytically, due to the explicit knowledge of $\langle\vec{r}|QP\rangle$.

We have used for V various Skyrme-type interactions. The solution of the Hamiltonian equations for spherical nuclei and a typical force (SkM) yield the time evolution shown in figures 1 and 2.

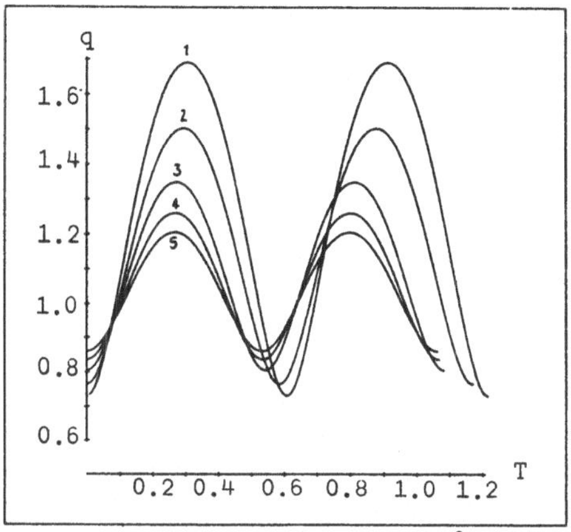

Fig. 1: Time evolution of $q=e^{2\theta}$.

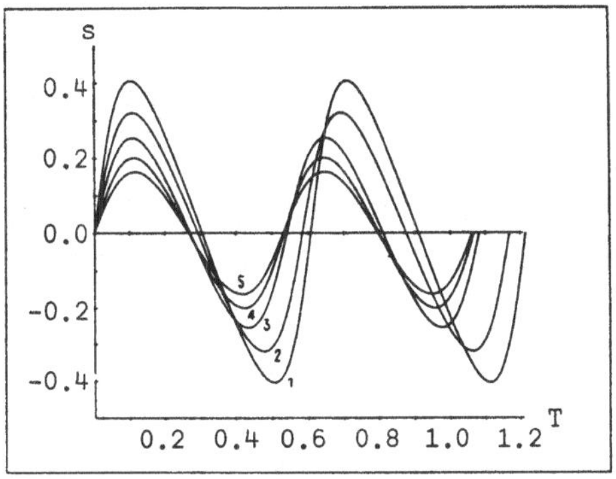

Fig. 2: Time evolution of s.
1: ^{40}Ca, 2: ^{56}Ni, 3: ^{90}Zr, 4: ^{140}Ce, 5: ^{208}Pb

Bound state energies of the system can be calculated by quantization in the phase-space of parameters by using the Gauge Invariant Periodic Quantization which demands [4], $I(E) = \oint \langle Q,P|i\hbar\partial_t|Q,P\rangle = n(2\pi\hbar)$ Best agreement of these results is obtained with the SkM force.

Nucleus:		^{56}Ni	^{90}Zr	^{140}Ce	^{208}Pb
This work: SkM		20.5	17.8	15.4	13.5
Experiment:	p	19.8±0.5	17.5±0.5		13.4±0.5
	^3He	17.1±0.2	16.4±0.3	14.8±0.2	13.2±0.3
	α	20.0±0.5	17.0±0.5		13.0±0.4

Table 1: Comparison of theoretical and experimental breathing mode energies (MeV). Data from ref. [5]

References

[1] P. Kramer and M. Saraceno, Lecture Notes in Physics 140 (Springer, 1981).
[2] A.M. Perelomov, Comm. Math. Phys. 261 (1972) 222
[3] S. Lang, $Sl_2(\mathbb{R})$ (Addison-Wesley, 1975)
[4] K. Goeke et al., Phys. Lett. 118B (1982) 1
[5] F.E. Bertrand et al., Phys. Lett. 80B (1979) 198
F.E. Bertrand et al., Phys. Rev. C22 (1980) 1832
M. Buenerd et al., Phys. Rev. Lett. 45 (1980) 1667
D. Lebrun et al., Phys. Lett. 97B (1980) 258
M. Buenerd et al., Phys. Lett. 84B (1980) 305

THE QUARK STRUCTURE OF NUCLEI FROM A GROUP THEORETICAL VIEWPOINT

K. Bleuler
Institut für Theoretische Kernphysik der Universität Bonn
Nußallee 14-16, D-5300 Bonn, BRD

Summary

Spherical nuclei are treated from a quark theoretical viewpoint by introducing a so-called overall bag structure. Under the assumption of a direct interaction between the uncorrelated quarks (determined, however, uniquely from general principles) it is shown (with the help of rather extended group theoretical methods) that this model accounts in a satisfactory way for the characteristic nuclear features usually interpreted within the framework of conventional shell structure.

I. Introduction

So far nuclear structure was understood in terms of various hadrons ($n, p, \Delta, \pi, \rho, \omega \ldots$) with their mutual interaction terms. (Recent detailed calculations of boson theory of nuclear matter showed that virtual Δ-excitations of the nucleons as well as a direct s-wave interaction between pions are responsible for the main part of nuclear binding[1)]). This means that a rather extended (and ill defined) phenomenological input (various masses, coupling schemes with corresponding form factors and coupling constants) has to be used. On the other hand Quantum--Chromodynamics (by now successfully checked in the framework of the Salam-Weinberg theory of electroweak interaction) suggests that all hadrons are to be considered as systems of quarks bound by the fundamental gauge field whose coupling scheme is uniquely determined through the general gauge principle. This viewpoint suggests that all phenomenological properties of hadrons (in particular those to be used in conventional nuclear theory) are - in principle - to be based on gauge theory. So far, various lattice theoretical approaches indicate that the characteristic properties of hadrons (masses and extension in space) may follow from this basic viewpoint[2)]. In view of the unsurmountable mathematical difficulties one has, however, to look for some simplifying approaches: In fact, the various well-known bag models for light hadrons should be visualized as rough approximation schemes for the general quantum chromodynamical structure, to some extent in the same way as atomic shell structure approximates the general N-electron problem.

Realizing the characteristic intrinsic quark properties of hadrons nuclear structure appears in fact to be a most complicated inter-correlated system of quarks: In view of the extension in space of the various hadrons there exists an appreciable overlaping as well as a relatively strong internal excitation of the individual structures. (This fact is empirically seen from the recent Cern Myon Experiments[6] as well as from conventional theory of boson exchange which automatically yields, as mentioned above, a strong virtual Δ-excitation[7]). Under these circumstances the question arises whether the overall structure of a nucleus might be considered from the outset as a general system of quarks to be treated by the principles of quantum chromodynamics, or, in other words: Whether the nucleus should just be viewed as a heavy hadron. This question can, of course, only be answered in the framework of a suitable approximation scheme. In view of the fact that the bag model prescriptions stand for an extremely simplified representation on a QCD-system, we here propose, within the framework of a first approximation scheme, to represent the nucleus just to be one single, but extended bag defined by a natural, practically unique generalization of the original assumptions[4]. (One might guess that the actual structure of nuclei lies somewhere in between this extreme model of uncorrelated quarks corresponding to completely dissolved nucleons on the one hand side, and the fully correlated system representing a system of unchanged nucleons on the other side).

The purpose of this paper is to show that, strangely enough, our extreme viewpoint leads, under the assumption of a special 2-body interaction between the independent quarks, directly to the interpretation of some fundamental features of spherical nuclei, i.e. the properties usually interpreted in the framework of conventional nucleonic shell structure[10]. This work is entirely due to H.R. Petry (Bonn) who will publish the explicit calculations as well as the mathematical arguments in full details[3].

II. The new assumptions about the quark structure of spherical nuclei

We thus apply the general bag model prescriptions in an extended form to the overall quark structure of spherical nuclei. This amounts to introduce a spherical bag whose diameter corresponds to the extension of the nucleus: The well-known rules of the MIT-bag model then read as follows[4]: The quarks are described by the Dirac-field $\psi_{k\tau}$ which contains the indexes for colour k = 1,2,3 as well as for flavour τ = 1,2 (up and down quark). In conventional notation the eigenstates ψ are determined through the following system:

$$\text{(1a)} \qquad (\vec{\alpha}\vec{p}+\vec{\beta}m)\psi_{k\tau} = \varepsilon\psi_{k\tau}$$

$$\text{(1b)} \qquad \frac{(\vec{\alpha}\vec{x})}{|x|}\psi_{k\tau} = i\beta\psi_{k\tau} \, , \quad \text{for} \quad |x| = R \text{ (nuclear radius).}$$

This means that there are independent quarks described by the free Dirac equation satisfying the characteristic boundary condition of the bag model. In addition, one has to assume the typical outer pressure B due to the degenerate QCD-vacuum state as well as the fundamental condition about vanishing total colour of the system.

In spite of the extreme simplicity of this model, it is appealing to realize that two fundamental nuclear properties are direct consequences of these assumptions[3]:

1. Introducing an appropriate value of the outer pressure B which differs only slightly from the one used in the MIT-bag, one obtains automatically the well-known nuclear radius law through the equilibrium condition between the inner pressure due to the occupied single particle levels and the outer pressure B. (At the same time the saturation property of the binding energy is approximately satisfied).

2. Looking in the details of the single quark level scheme (ε) defined by (1a) and (1b) and using an appropriate (but generally assumed) value for the quark mass, it turns out that the well-known Mayer-Jensen single particle level scheme of conventional shell structure[10] is approximately reproduced, i.e. the corresponding characteristic spin-orbit splitting usually described by a special (partly phenomenological) $(\vec{l}\vec{s})$-coupling term appears now as a relativistic effect due to the boundary condition (1b). (The relativistic wave-functions are characterized by j and parity from which the orbital quantum number l

may be determined).

This last fact shows that the well-known shell closures (i.e. the magic numbers) follow in a natural way from this model (the Coulomb interaction has so far been neglected). On the other hand,it may be immediately realized that the open quark shells lead to an extremely large degeneracy (due to the additional degrees of freedom with respect to the conventional case) of the various levels, i.e. a fact which at first sight appears to be in perfect contradiction to experimental data. Therefore, the question arises whether the introduction of a suitable direct interaction between the quarks might lead to an appropriate splitting, i.e. to a level-scheme in which the lowest term corresponds to the characteristic degeneracy of a system of independent nucleons. (This interaction should, in principle, follow from QCD; in the bag models one introduces, however, half-phenomenological expressions). The main scope of this paper is to show that a suitable direct interaction term is (apart from an overall coupling-constant) practically uniquely determined through the following assumptions:
1. The interaction is SU(3)-invariant.
2. It has the general features of a pairing-force, i.e. a property which follows from the assumption of a relatively short range with respect to the nuclear diameter. It is well-known that the pairing approximation scheme is most successful in various N-fermion systems. In our case, however, we can solve the corresponding enlarged eigenvalue problem in a rigorous way by introducing appropriate group-theoretical methods. It will turn out that the strength of this interaction has to be of the same order as the one introduced in the MIT-bag model.

III. The group theoretical method

The explicit form of our pairing interaction which in our model acts only between the quarks of a given (open) shell (corresponding to the angular momentum quantum number j) reads as follows: Within the framework of the second quantization scheme we introduce the fundamental emission and absorption operators (a^* and a) characterized by the indices k (colour), τ (flavour) and m (magnetic quantum number) of the single particle states of the shell considered, i.e.,

(1) $$a_{k\tau m} \qquad a^*_{k\tau m}$$

$$k = 1,2,3 \qquad \tau = 1,2 \qquad m = -j.....+j$$

The last two indices (τ and m) may be lumped together

(2) $$m,\tau \rightarrow \alpha \quad \text{taking now} \quad N = 2(2j+1) \quad \text{different values.}$$

With this notation our pairing operator P_j containing the free coupling constant g_j reads (applying the summation convention and introducing the totally skew-symmetric ε-symbol):

(3a) $$P_j = -g_j A_i^* A_i$$

(3b) $$A_i = \frac{1}{2} \sum_{m=-j}^{+j} \varepsilon_{ikl} (a_{k,m,1/2}\, a_{l,-m,-1/2} - a_{k,m,-1/2}\, a_{l,-m,+1/2}) (-1)^{j-m}$$

This expression is (apart from a single different, but inappropriate expression) determined uniquely by SU(3)-invariance (expressed by the invariance of the ε-symbol) and the pairing-principle (expressed by the occurrence of opposite values for m). Using the simplified notation (2) and the definition of an appropriate symmetric matrix $\eta^{\alpha\beta} = \eta^{\beta\alpha}$, one may write (3b) in a much simpler form:

(4) $$A_i = \varepsilon_{ikl} a_{k\alpha} \eta^{\alpha\beta} a_{l\beta} \, .$$

(The symmetry of η appears in characteristic contrast to the nuclear case, i.e. conventional pairing theory where an antisymmetric [simplectic] metric occurs; this difference is due to the additional antisymmetrization with respect to the flavour index τ enforced here by the needed antisymmetry with respect to the colour index k).

Our task now consists in an exact diagonalization of the corresponding total Hamiltonian H including also the energies of the free quarks and satisfying the fundamental condition of vanishing total colour. In the framework of our group theoretical method we construct in a first, but decisive step the larged possible (maximal) invariance group of H; we then determine the corresponding irreducible representations which, as we will see, characterize the various eigenstates of H in an unique way.

In order to reach that goal we proceed stepwise by considering first of all the (maximal) invariance group of the free Hamiltonian H_o of independent quarks in our given shell which reads:

$$H_o = \varepsilon a^*_{k\alpha} a_{k\alpha} \quad . \tag{5}$$

The corresponding eigenstates containing N_j-quarks in our open shell corresponding to the eigenvalue $E = \varepsilon N_j$ are, of course, given by

$$|\psi\rangle = \prod_{r=1}^{N_j} a^*_{k_r \alpha_r} |0\rangle \quad . \tag{6}$$

The characteristic (maximal) invariance group, i.e. SU(3N) may be characterized through its Lie elements given by a 3N-dimensional trace free and antihermitian matrix $D_{i\alpha;k\beta}$ (the double indices which characterize the various states take, in fact, 3N values). The corresponding actions on H_o and on the eigenvectors (6) are given by the following operators $\hat{D}$:

$$\hat{D} = D_{i\alpha;k\beta} a^*_{i\alpha} a_{k\beta} \ ; \qquad D^+ = -D \qquad \text{Trace } D = 0 \tag{7}$$

These operators which, of course, commute with H_o induce, when acting on the various eigenspaces (6) characterized by N_j a representation of SU(3N). According to the general principles about the properties of totally antisymmetric tensors (as represented by (6)) it is immediately clear that these eigenspaces just constitute the representation spaces of the well-known <u>irreducible</u> representations of SU(3N) which, in this way, characterize the different eigenstates of H_o in an unique way. (In this connection it might be observed that general theorems about representation theory take, within the second quantization scheme, a most natural form).

In the next step we have now to select the states which satisfy the fundamental condition of vanishing total colour: Such states are easily obtained by a contraction with help of the ε-symbol and correspond always to a number of quarks which is a multiple of 3, i.e. $N_j = 3n_j$ where n_j represents the number of nucleons:

$$|\psi\rangle = \prod_{r=1}^{n_j} A^*(\alpha_r \beta_r \gamma_r) |0\rangle \tag{8a}$$

with

$$A^*(\alpha,\beta,\gamma) = \varepsilon_{ikl} a^*_{i\alpha} a^*_{k\beta} a^*_{l\gamma} \tag{8b}$$

It may be proved, however[5)], that in this way all colourless states are obtained. This means that vanishing colour enforces automatically a factorization (i.e. clustering) in characteristic factors of 3 quarks which, according (8b) correspond to entire charges and will, as we shall see, represent the nucleons in their ground or excited states. The new reduced invariance group is by now given by SU(N) and the corresponding action which is obtained from (7) by contraction reads:

$$(9) \qquad \hat{d} = d_{\alpha\beta} a^{*}_{i\alpha} a_{i\beta} \qquad d^{+} = -d, \text{ i.e. } \quad d_{\alpha\beta} = -\bar{d}_{\beta\alpha} \qquad d_{\alpha\alpha} = 0$$

It is, by now, an important fact that (according H. Weyl[5)]) these colourless states given by (8) again generate irreducible representations of our reduced invariance group SU(N). (This fact exhibits a certain simularity with the methods used in the well-known Russel--Saunders coupling scheme). The condition of vanishing total colour thus leads to a first reduction of the (maximal) symmetry SU(3N) to SU(N).

In the third, decisive, step we now have to consider the additional reduction of this symmetry due to the introduction of the pairing term P_j given by (3). As seen from its structure, it is immediately realized that the additional condition reads

$$(10) \qquad d_{\alpha\gamma} \eta^{\gamma\beta} = -\eta^{\alpha\gamma} d_{\beta\gamma}$$

i.e. the metric η has to remain invariant under the action of our group, or, in other words our unitary group SU(N) is reduced to the orthogonal group SO(N). Our scheme thus reads: $SU(3N) \supset SU(N) \supset SO(N)$.

Within the framework of our program we now have to consider the irreducible representation of the orthogonal group SO(N) which will characterize the eigenstates of the full Hamiltonian. The states according (8) given above will under the action of the operation (9), still lead to a representation of this smaller group but, in general, to a reducible one and the question arises how to construct explicitly the right 'linear combinations' which correspond to the various irreducible representations of SO(N) contained in the eigenspaces of H_o. The decomposition of this space will, of course, correspond to the splitting of the eigenvalues of H_o through the interaction P_j. Our goal, i.e. the explicit construction of the linear combinations can be achieved according to a general method given by H. Weyl[8)] which consists in the introduction of the various contractions with help of the metric η within the expressions (8). The simplest example is given by

$$(11) \qquad |\psi^{o}\rangle = \prod_{\rho=1}^{n_j} B^{*}(\alpha_{\rho})|0\rangle \quad \text{with} \quad B^{*}(\alpha) = \varepsilon_{ikl} a^{*}_{i\alpha} a^{*}_{k\beta} \eta^{\beta\gamma} a^{*}_{l\gamma}$$

i.e. by introducing the maximum number of contractions. In the next case one single contractions has to be left out; this reads, using the notation (8b):

$$(12) \qquad |\psi^{1}\rangle = A^{*}(\alpha_r \beta_r \gamma_r) \prod_{\rho \neq r} B^{*}(\alpha_{\rho})|0\rangle$$

Taking from $|\psi^{1}\rangle$ the orthogonal complement $|\bar{\psi}^{1}\rangle$ with respect to ψ^{o}

$$(13) \qquad |\bar{\psi}^{1}\rangle \perp |\psi^{o}\rangle$$

one obtains the next irreducible representation of SO(N) contained in the eigenspace (8). This procedure must be iterated by leaving out stepwise more, i.e. L contractions:

$$(14) \qquad L = 0, \ldots\ldots, k \;, \qquad k = \min(n_j, N-n_j)$$

From the rules of tensor calculus it is immediately clear that this construction leads to representations of SO(N) but it is an important but by no means trivial fact that this method leads to <u>all irreducible</u> representations[8)] of SO(N).

IV. The exact diagonalization

True to our program, it may be hoped that these states corresponding to irreducible representations of our (maximal) invariance group SO(N) might be at the same time eigenstates of the total Hamiltonian. This is in fact the case; in fact, it will be shown that the Hamiltonian can be expressed (apart from trivial terms) by the Casimir operators C_1 of SU(N) and C_2 of SO(N) as well as the particle number operator $N_{op} = a^{*}_{k\alpha} a_{k\alpha}$:

$$(15) \qquad P_j = -g_j\left[(N-2)C_1 - NC_2 + \frac{1}{2N} N_{op}^2 - \frac{1}{2} N_{op}\right]$$

In order to prove this relation one has to consider the explicit expressions for the Killing form of the two groups and to insert the representations of the Lie-elements given by (9) and (10). In order to obtain, eventually, an explicit expression for the eigenvalue spectrum of H one has just to determine the eigenvalues of the two Casimir-operators for the various irreducible representations constructed above. For this purpose one has to know the characteristic weights of these representations. Following as before the method of H.R. Petry one constructs explicitly the 'maximal vectors' within the various eigen-spaces from which the weights may be readily read off. For the (lowest) representation space of SO(N) given by (11) this vector reads:

$$(16) \qquad |\psi\rangle = \prod_{\alpha=1}^{n_j} \varepsilon_{ikl} a^*_{i\alpha} a^*_{k\beta} \eta^{\beta\gamma} a^*_{l\gamma} |0\rangle$$

whereas in the next step corresponding to (12) and (13) one obtains

$$(17) \qquad |\psi\rangle = \varepsilon_{ikl} a^*_{i1} a^*_{k1} a^*_{l1} \prod_{\alpha=2}^{n_j} \varepsilon_{ikl} a^*_{i\alpha} a^*_{k\beta} \eta^{\beta\gamma} a^*_{l\gamma} |0\rangle \ .$$

Using the well-known expressions for the values of the Casimiroperators as functions of the various weights[9], one finally obtains for the energy eigenvalue of our (open) shell corresponding to the angular momentum j:

$$(18) \qquad E = 3n_j \cdot \varepsilon - g_j [n_j (N+3-n_j) - L(N+3-L)]$$

where $N = 2(2j+1)$, n_j the number of nucleons and L the integer defined by (14) characterizing the different representations of SO(N) contained in the eigenspaces of H_o. The last and most important term thus exhibits the splitting of the degenerate eigenvalue of H_o and indicates that the states given by (11) for L = 0 are, in fact, the lowest ones.

V. The results

Although our group theoretical deductions appear to be rather abstract (and make use in an essential way of several fundamental theorems), it turns out that our results admit a most natural physical interpretation: From the explicit form of the lowest states of our system which are in fact given by the expression (11) it is readily seen that each factor (containing three emission operators) appears to have the same quantum numbers (τ,m abbreviated by α) as an independent nucleon within the same open shell. One might thus visualize each factor to represent a strongly deformed nucleon containing a quark pair (with spin and isospin 0) and a 'valence quark' with the quantum numbers of the corresponding nucleon. On the other hand, a closer analysis of our result shows that the next excited levels with L = 1 stand for a system of (n_j-1)-nucleons completed by a Δ-resonance state. This interpretation also leads to an empirical determination of our overall coupling constant g_j which thus appears to be of the same order of strength as the quark interaction within the conventional quark model of the nucleon. A similar interpretation might be given for all higher excited states of our quark-level scheme. The occurrence of an inner excitation of nucleons embedded in nuclear matter is, of course, a natural consequence of our quark theoretical viewpoint and corresponds, at the same time, to experimental facts.

Our model thus reproduces the properties of conventional nuclear shell structure predicting a characteristic deformation of the embedded nucleons: Although such deformations are experimentally found through the Myon scattering experiments[6)] mentioned above, a numerical check shows that this model predicts too large an effect. It is hoped, however, that more detailed calculations (the influence of the direct interactions is considered here only in pairing approximation and leaves out some characteristic effects due to their finite range) will lead automatically to a stronger correlation between the three quarks representing so far the nucleon; at the same time, it may be realized that higher terms will also lead to additional correlations between our 3-quark clusters which replace, so to speak, the nucleon-nucleon force of the conventional scheme. It thus seems that realistic nuclear structure might, in fact, be obtained by starting from our extreme assumptions of independent quarks, i.e., the starting point opposite to the conventional one supposing unchanged nucleon structure within nuclear matter. As a final remark it should be emphasized that our enlargement of the characteristic assumptions used so far only in the case of light hadrons (and hopefully to be understood in the framework of general QCD), namely the characteristic vacuum pressure B and the condition about vanishing total colour, appears within the framework of our model directly related to fundamental nuclear properties.

References

1) Compare the article on Boson-Exchange by R. Machleidt (Bonn), to be published in "Quarks and Nuclear Structure", Springer Tracts in Physics (1984)

2) Compare for example: M. Bander, Phys. Rep. 75 (1981) 205

3) H.R. Petry (Bonn), 'Nuclear Shells from a quark theoretical Viewpoint', to be published in "Quarks and Nuclear Structure", Springer Tracts in Physics (1984)

   Further details are, however, found in:
   K. Bleuler, H. Hofestädt and H.R. Petry, Z. Naturforsch. 38a (1983) 705

   H. Hofestädt, S. Merk and H.R. Petry: "Ein Schalenmodell für Atomkerne", Bonn (1982), Preprint to be ordered at: Institut für Theoretische Kernphysik, Universität Bonn, Nußallee 14-16, D-5300 Bonn

   K. Bleuler, Proc. of 3rd Int. Conf. on Nuclear Reaction Mechanisms, Varenna 1982, Ricerca Scientifica, Milano 1982 (ed. by E. Gadioli)

   K. Bleuler, "Gauge Theory and Nuclear Structure", Report in the 12th Conference on "Differential Geometric Methods in Theoretical Physics", to be published by D. Reidel Publ. Company, Dordrecht, Holland

4) A. Chodos et al., Phys. Rev. D9 (1974) 3471 and T. de Grand et al., Phys. Rev. D12 (1975) 2060

5) Compare H. Weyl, Group Theory and Quantum Mechanics, Chapter V

6) J. Aubert et al., "The ratio of nuclear structure functions..." preprint, CERN-EP/83-84

7) M.R. Anastasio et al., Nucl. Phys. A322 (1979) 369

8) Compare H. Weyl, The Classical Groups, Chapter V

9) Compare: N. Jacobson, Lie Algebras, John Wiley, 1962

10) Compare for example: A. de Shalit and H. Feshbach, Theoretical Nuclear Physics, John Wiley (1974)

GROUP THEORETIC APPROACH TO SPHERICAL ANHARMONIC OSCILLATOR

Barnana Ghosh
and
Raj Kumar Roychoudhury
Electronics Unit
Indian Statistical Institute
Calcutta 700 035
INDIA

Abstract

We use the radial functions of the three dimensional isotropic harmonic oscillator, which form basis for unitary representation of O(2,1), to study the problem of spherical anharmonic oscillator (SAHO) of the form $V = r^2/2+\lambda r^4$. A variable scaling method gives a single formula for eigen values for both large and small coupling constants. The numerical results are found to be in good agreement with the exact numerical results quoted by Seetharaman et. al.

Some time ago Armstrong used a basis of O(2,1) to calculate matrix elements of r^s taken between spherical harmonic oscillator wave functions. In this note we used these results to find out an approximate analytical formula for the eigen values of SAHO, of the form $V=r^2/2+\lambda r^4$, valid for all n and l. The O(2,1)algebra and the radial harmonic oscillator wave functions are discussed in detail in ref 1-3 (also see (Miller[4])). Hence we give here only the essential steps needed for our calculations. The generator J_+, J_-, J_3 of O(2,1) satisfy the following commutation relations.

$$[J_3, J_\pm] = \pm J_\pm \; ; \; [J_+, J_-] = 2J_3 \qquad \ldots (2)$$

In a two dimensional space[4], the operators take the form

$$J_\pm = e^{\pm it}\left(z\frac{\partial}{\partial z} \mp i\frac{\partial}{\partial t} \mp \frac{z}{2}\right) \qquad \ldots (3)$$

$$J_3 = -i\frac{\partial}{\partial t} \qquad \ldots (4)$$

A basis for an irreducible representation of the above algebra is given by the states f_{ab} defined as follows

$$f_{ab} = \left[\frac{(l+\frac{1}{2})}{4\pi\,\rho^{\frac{1}{2}}}\right]^{\frac{1}{2}} e^{i(n+\frac{1}{2})t}\, z^{\frac{1}{4}}\, R_{nl}(z) \qquad \ldots (5)$$

where $R_{nl}(z)$ is the radial wave function for the spherical harmonic oscillator (SHO) and $\beta = mw/\hbar$, w being the classical frequency of oscillation.

Using equations (1) to (5) it can be shown that

$$J_3 f_{ab} = \frac{1}{2}(n + \frac{1}{2}) f_{ab} \quad \ldots \quad (6)$$

$$J_\pm f_{ab} = \pm (b \mp a)(b \pm a \pm 1) f_{ab} \quad \ldots \quad (7)$$

where
$$b = \frac{n + \frac{1}{2}}{2}, \quad a = \frac{l - 1/2}{2} \quad \ldots \quad (8)$$

R_{nl} is same as defined in Ref. 5.

For SAHO (with a quartic term λr^4) the radial part of the wave function satisfies the following differential equation (for the sake of simplicity we use the units in which $\hbar = m = 1$)

$$[-\frac{1}{2}\frac{d^2}{dr^2} + r^2/2 + \lambda r^4 + \frac{l(l+1)}{2r^2}] \bar{R}_{nl} = E \bar{R}_{nl} \quad \ldots \quad (9)$$

where $\bar{R}_{nl}(\beta r^2) = R_{nl}(\beta r^2)/r$

We treat this as a perturbation problem when the unperturbed Hamiltonian (after a suitable scaling) can be written as

$$H_o = -\frac{1}{2}\frac{d^2}{dr^2} + \frac{w^2 r^2}{2} \quad \ldots \quad (10)$$

then (9) can be written as

$$(H_o + V) \bar{R}_{nl} = E_{nl} \bar{R}_{nl} \quad \ldots \quad (11)$$

where
$$V = - w^2 r^2/2 + r^2/2 + \lambda r^4 \quad \ldots \quad (12)$$

and w can be treated as a sort of Ritz parameter.

In the basis given by (1) and (2)

$$V = \frac{(1-w^2)}{2w}[J_- e^{it} - J_+ e^{-it} + 2J_3] + \frac{\lambda}{w^2}[J_-^2 e^{2it} + J_+^2 e^{-2it} + 6J_3^2 - 2J^2 + 4e^{it} J_- J_3 - 4e^{-it} J_+ J_3 - 2e^{-it} J_+ - 2e^{it} J_-] \quad \ldots \quad (13)$$

If we take only the diagonal values of V(which can be expressed completely in terms of the Casimir operators of O(2,1)) and denote it by V_d, then

$$V_d = \frac{(1-w^2)}{w} J_3 + \frac{\lambda}{\beta^2}[4J_3^2 + 2J_3 - 2J_+ J_-] \quad \ldots \quad (14)$$

and also $(H_o)_d = 2J_3/w \quad \ldots \quad (15)$

Now it can easily be shown that

$$((H_o)_d + V_d) \bar{R}_{nl} = E_{nl}^0 \bar{R}_{nl} \quad \ldots \quad (16)$$

where $E^{o}_{n,l} = (n + \frac{1}{2})\, w + \frac{(1-w^2)}{w}\ \frac{(n + \frac{1}{2})}{2} + \frac{\lambda}{2w^2}[3(n + \frac{1}{2})^2$

$- (l - \frac{1}{2})(l + \frac{3}{2})]$ where $n = 2n_r + l+1$ ... (17)

w is fixed by the requirement that[7] E^o should be minimum, i.e. $\frac{\partial E^o}{\partial w} = 0$ and $\frac{\partial^2 E^o}{\partial w^2} > 0$. Then $w_n = w$ satisfies the following equation.

$$(n+ \frac{1}{2})w^3 - (n+ \frac{1}{2})w-\lambda\, [3(n + \frac{1}{2})^2 - (l- \frac{1}{2})(l + \frac{3}{2})] = 0 \qquad \text{... (18)}$$

(17) and (18) together give the eigen values $E_{n,l}$ for any n and l to the lowest order.

Second order corrections can be added using the formula

$$E^2_{n,l} = \sum_{m\neq n} \frac{\langle n \mid V \mid m \rangle^2}{E_{n,l} - E_{m,l}}$$

Where $E_{m,l}$ is calculated using $w' = w_n$ in (17) for all m. In Table 1 we compare our results with the accurate numerical results of Mathews et. al. (for λ = .5)(see ref. 6) for some selected values of n and λ .

Table 1

n	l	$E^o_{n,l}+E^2_{n,l}$	E (ref. 6)
50	0	214.937	213.991
	50	187.531	187.530
10	0	30.172	30.065
	4	29.597	29.510
	10	27.094	27.092
0	0	2.326	2.324

It is seen from the above table that the present method provides an analytical formula which is in good agreements with the accurate numerical results. The agreement is specially good when n -l is small.

References

1. M. Moshinsky, Harmonic Oscillator in Modern Physics (Gordon and Breach, New York 1963).
2. S. Goshen and H.J. Lipkin, Ann. Phys (N.Y) 6 301 (1959).
3. Lloyd Armstrong, JR., J. Math. Phys. 12 953 (1971)
4. W. Miller, Lie Theory and Special functions (Academic Press, New York, 1968).
5. P.M. Morse and H. Feshbach, Methods of Theoretical Physics (McGraw Hill, New York, 1953) Vol. II.
6. M. Seetharaman, Sekhar Raghavan and S.S. Vasan J. Phys. A : Math. Gen 15 1537 (1982).
7. I.D. Feranchuk and L.I. Komarov, Phys. Lett. 88A 211 (1982).

OPERATOR AVERAGES AND ORTHOGONALITIES

B.R. Judd
Physics Department, The Johns Hopkins University
Baltimore, Maryland 21218, USA

1. Introduction

Everyone working in theoretical spectroscopy has noticed that the splittings produced in a degenerate energy level by perturbing terms often leave the center of gravity of the level untouched. For example, the spin-orbit interaction $\lambda\vec{S}\cdot\vec{L}$ splits the atomic term characterized by S and L into levels whose energies are given by

$$E_J = \tfrac{1}{2}\lambda[J(J + 1) - S(S + 1) - L(L + 1)], \quad (J = S + L, S + L - 1,\ldots, |S - L|)$$

for which $\Sigma_J(2J + 1)E_J = 0$. On the other hand, the Coulomb interaction between the electrons of an atom splits a configuration into terms: but the center of gravity in this case is shifted. If an operator H is scalar with respect to S and L (like the Coulomb interaction), the condition for no shift is

$$\Sigma(2S_\psi + 1)(2L_\psi + 1)\langle \psi|H|\psi \rangle = 0, \tag{1}$$

where the sum runs over all the SL terms ψ of a configuration. If H possesses the form H_1H_2, where H_1 and H_2 are also scalar with respect to S and L, then closure over the complete set of terms ψ' yields

$$\sum_{\psi,\psi'} (2S_\psi + 1)(2L_\psi + 1) \langle \psi|H_1|\psi' \rangle \langle \psi'|H_2|\psi \rangle = 0 \tag{2}$$

It is in this sense that the idea of operator orthogonality has been introduced.[1] The fitting of atomic energy levels by orthogonal operators has a number of advantages: the addition of a new operator entails minimal adjustments to the strengths of those already brought into play, and the mean errors are smaller than they otherwise would be. Any conclusions that we can draw about the validity of (2) will be reflected in conditions on H_1 and H_2 when $H = H_1H_2$.

2. Theorem and Application

Some years ago, it was noticed that electric potentials described by harmonic components of the type Y_{2m} or Y_{4m} failed to shift the center of gravity of degenerate atomic levels produced by an icosahedral crystal field, while those of the type Y_{6m} did so.[2] This property was shown to depend on the fact that the representations D_2 and D_4 of O(3) do not provide the identity representation Γ_1 of the icosahedral group I under the reduction I → O(3), while D_6 does. This result is easily generalized to states ϕ_i that span a representation Γ of a compact Lie group G. If an operator H' belongs to Γ' of G, then

$$\sum_i \langle\phi_i|H'|\phi_i\rangle = 0 \tag{3}$$

unless Γ' contains in its reduction an irreducible scalar part Γ_1. Equation (3) is equivalent to (1) when the degeneracies of the terms are taken into account.

This principle can be at once applied to the spin-orbit interaction. We take $G = O_S(3) \times O_L(3)$. An SL term provides a basis for the irreducible representation $D_S \times D_L$. However, $\vec{S}\cdot\vec{L}$ belongs to $D_1 \times D_1$, and this does not contain the identity representation $D_0 \times D_0$. Thus (3) is satisfied and the spin-orbit interaction does not shift the center of gravity of a term.

3. Coulomb Interaction

The states of the atomic configuration ℓ^N form a basis for the irreducible representation [11...10...0] (in which N ones appear) of $U(4\ell + 2)$ whose generators are the $(4\ell + 2)^2$ products $a^+_\xi a_\eta$, where a^+_ξ creates the electronic state ξ and a_η annihilates the state η. The creation operators for an $n\ell$ electron belong to the irreducible representation of $U(4\ell + 2)$ with highest weight [10...0], while the annihilation operators correspond to [0...0 - 1].

A two-electron interaction such as the Coulomb interaction is a sum of operators of the type $a^+_\xi a^+_\eta a_\mu a_\nu$. The two creation operators belong to [110...0], the two annihilation operators to [0...0 - 1 - 1]. Putting the two parts together corresponds to forming the Kronecker product of the two representations, which decomposes as follows:

$$[110...0] \times [0...0 - 1 - 1] = [0...0] + [10...0 - 1] + [110...0 - 1 - 1]. \quad (4)$$

We have only to note that the Coulomb interaction transforms like $D_0 \times D_0$ of $O_S(3) \times O_L(3)$ and that this representation occurs in the identity [0...0] to conclude that equation (3) does not necessarily hold. So the Coulomb interaction can be expected to shift the center of gravity of an atomic configuration.

4. Orthogonalities

The representation $D_0 \times D_0$ occurs not only in [0...0] but also in [110...0 - 1 - 1]. The total number of occurrences is equal to the number of SL terms in ℓ^2, namely $(2\ell + 1)$. The corresponding operators H_i can be constructed to be mutually orthogonal in ℓ^2 in the sense of (2), but it would be highly convenient if the orthogonality could be guaranteed for all ℓ^N. An argument that this follows automatically[1] is flawed, as a counterexample by Uylings[3] has demonstrated. Uylings picks orthogonal operators H_{SL} that vanish in d^2 except for the term SL. On calculating their values in d^3 it is found that H_{00} possesses the single non-vanishing value $\langle {}^2D_1|H_{00}|{}^2D_1\rangle$. However, all other operators H_{SL} yield diagonal contributions to ${}^2D_1$ and so they cannot be orthogonal to H_{00}.

Group theory is of great help here. The product [0...0] × [110...0 - 1 - 1] does not contain [0...0], so every H_i belonging to [110...0 - 1 - 1] is orthogonal to the single total scalar H_1 belonging to [0...0]. However [0...0] occurs in $[110...0-1-1]^2$, so there is the possibility that two operators H_i and H_j belonging to [110...0 - 1 - 1] might not always be orthogonal for ℓ^N even though they are for ℓ^2. This can be discounted, however, because [0...0] occurs only once in $[110...0 - 1 - 1]^2$, and the existence of products H_i^2 for which (1) does not hold when we put $H = H_j^2$ shows that the [0...0] part of $[110...0-1-1]^2$ would be detected in ℓ^2. Thus the orthogonality of the operators for all ℓ^N is assured provided one operator (H_1) belongs to [0...0] and all the others belong to [110...0 - 1 - 1] and are separately (and arbitrarily) orthogonalized in ℓ^2. Uylings' counterexample amounts to the construction of mixed operators belonging to [0...0] + [110...0 - 1 - 1] for which the scalar parts of $[0...0]^2$ and $[110...0-1-1]^2$ exactly cancel in ℓ^2 but not for the general ℓ^N.

5. Developments

The above ideas are being applied to the atomic f shell, where the standard Racah operators e_0, e_1, e_2 and e_3, in terms of which the Coulomb interaction is traditionally expressed, can be very easily orthogonalized merely by replacing e_1 by $e_1 - 9e_0/13$. Small corrections provided by configuration interaction are added in such orthogonalized forms as $e_3 + 5\Omega$, where Ω is the classic operator defined by Racah.[4]

Other efforts on the $p^N d$ configurations are being pursued in collaboration with Dr. J.E. Hansen of the Zeeman Laboratorium. For such configurations it is useful to take

$$G = O_S^p(3) \times O_L^p(3) \times O_S^d(3) \times O_L^d(3).$$

The group G is not sufficient to give a unique classification of the three-electron operators that allow for configuration interaction, and it is necessary to include Sp(6), the symplectic group acting in the six-dimensional spin-orbital space of a p electron, to provide unambiguous labels for our orthogonal operators.

It is a pleasure to thank Mr. P. Uylings and Dr. J.E. Hansen for many valuable discussions. The work reported above was partially supported by the United States National Science Foundation under Grant No. PHY-8215320.

References

1. B.R. Judd, J.E. Hansen & A.J.J. Raassen, J. Phys. B: At. Mol. Phys. 15, 1457 (1982).
2. B.R. Judd, Proc. Roy. Soc. (London) A241, 122 (1957).
3. P. Uylings, private communication (1983).
4. G. Racah, Phys. Rev. 76, 1352 (1949).

ADVANCES IN THE THEORY OF COLLECTIVE MOTION IN NUCLEI*

P. Kramer
Institut für Theoretische Physik
der Univeristät Tübingen
D-7400 Tübingen, FRG

1. Introduction

The theory of collective motion in nuclei has as its geometric origin the comparison of certain nuclear phenomena with properties of a liquid drop. Bohr and Mottelson (1952,53) [1,2] introduced the idea of a irrotational flow and explained a variety of collective phenomena by the deformations and vibrations of a nuclear fluid. The nuclear shell theory succeeded later on in the representation of collective excitation by coherent superpositions of many single-particle excitations. Elliott (1958) [3] showed that collective levels in the shell theory can be connected to a group $SU(3)$. Independently of the shell theory, various attempts were made to develop the geometric ideas implicit in the Bohr-Mottelson model. Weaver, Biedenharn and Cusson [4,5,6] introduced the group $SL(3,\mathbb{R})$ of volume-preserving deformations into the collective theory. With this group, they connected kinematical transformations of the system of A nucleons, the vortex spin, and a spectrum generating algebra. Inclusion of the mass quadrupole tensor leads to a natural extension of this group which was also studied by Rowe, Rosensteel and collaborators [7,8]. In the geometric models, it is the final goal to explain collective phenomena from the point of view of many-body dynamics. Therefore one has to link the collective coordinates to the single-particle coordinates. This program was already started by Lipkin (1955) [9] and by Villars (1957) [10]. Whereas these authors tried to keep the single-particle coordinates, new viewpoints were developed later by Zickendraht [11], by Dzyublik et al. [12], and by Buck Biedenharn and Cusson [13] by use of the orthogonal intrinsic group $SO(n,\mathbb{R})$ acting on the particle indices and commuting with $SO(3,\mathbb{R})$. Rowe and Rosensteel (1980) [14] were the first to analyze this scheme through an orbit analysis in configuration space. Vanagas (1977) [15] pointed out the close relation of the group $SO(n,\mathbb{R})$ to the symmetric group of orbital permutations and proposed the group $SO(n,\mathbb{R})$ as a symmetry group of the collective hamiltonian. In the following sections, the dynamical implications of this proposal will be analyzed, based on work done with Z. Papadopolos, M. Saraceno and W. Schweizer.

*Work supported by Deutsche Forschungsgemeinschaft.

Before starting the next section it should be mentioned that the symplectic group $Sp(6, \mathbb{R})$ or rather $Sp(2r, \mathbb{R})$ (r = 1,2,3 is the dimension of the space) as a phenomonological group for collective motion was proposed by Goshen and Lipkin for r = 1 [16] and 2 [17] and studied by Rosensteel and Rowe [18] for r = 3.

2. Group action on phase space and the moment map

Consider the phase space M of A particles and use as a chart the n = A-1 relative Jacobi coordinates ξ_{is} and momenta π_{is}, i = 1,2,3, s = 1,2,..,n. A point p of M can be associated with a 6 x n matrix as

$$p \in M : p \rightarrow \begin{bmatrix} \xi_{is} \\ -\pi_{is} \end{bmatrix}$$

We consider the orthogonal group $G = SO(n, \mathbb{R})$ which acts on the index s and maps $M \rightarrow M$ according to $g'' \in G : \quad p \rightarrow pg''$.

Now we introduce the moment map m due to Souriau [19] from M to the dual L^* of the Lie algebra $L = so(n, \mathbb{R})$ by $m : p \rightarrow T(p) = {}^t p\, Kp$, $K = \begin{bmatrix} 0 & I \\ -I & 0 \end{bmatrix}$. Here ${}^t p$ is the transposed of p taken as a matrix, and T is an antisymmetric n x n matrix with elements

$$T_{st}(p) = - \sum_{i=1}^{3} (\xi_{is}\, \pi_{it} - \xi_{it}\, \pi_{is})$$

The action of G on a function f(p) on M is generated by the Poisson action [20]

$$\{ \sum_{s,t} \alpha_{ts}\, T_{st}, f\}_{PB}(p)$$

of the moments T via the standard Poisson brackets. On the dual L^*, the coadjoint action Ad^* for an element $g'' \in G$ becomes

$$Ad^*_{g''}(T) = {}^t g''\, Tg''$$

Under Ad^*, L^* can be decomposed into orbits which carry a symplectic structure and corresponding generalized Poisson bracket. For the G-actions on M and L^* we have the commutative diagram

$$\begin{array}{ccc} M & \xrightarrow{g''} & M \\ m \downarrow & & \downarrow m \\ L^* & \xrightarrow{Ad^*_{g''}} & L^* \end{array}$$

as can be seen from the computation

$T(pg'') = {}^t(pg'')K(pg'') = {}^t g''\, T(p)g''$.

3. The reduced phase space for a hamiltonian invariant under SO(n, ℝ)

We shall adopt the characterization of the collective hamiltonian H_{coll} due to Vanagas [15] that H_{coll} be invariant under G = SO(n, ℝ). The consequences of this idea are described in the following sections. The analysis of Marsden and Weinstein [21], see Arnold [20], shows that for a hamiltonian with symmetry group G there exists a reduced phase space F with a generalized Poisson bracket {,}. For construct F we note that the moments m(p) of G have vanishing Poisson brackets with $H = H_{coll}$ so that fixed values of $T = (T_{st})$ can be chosen as integrals of motion. For fixed T, the inverse images of T under m form a subset $M_T = m^{-1}(T)$ of M. The stability group H_T of T on L^* acts on M_T. The reduced phase space is the projection $F_T = \pi(M_T)$ obtained by factorizing out the action of H_T from M_T. This subset F_T is shown to carry under appropriate condition a non-degenerate symplectic structure and generalized Poisson bracket [20,21].

To implement the reduced phase space, it proves useful to construct the G-orbits on L^* and on M and to relate them. From its definition, the reduced phase space contains points from the transversal to the G-action on M which are mapped into the same G-orbit on L^*. To analyze these transversals we now introduce the collective group $G = Sp(6, \mathbb{R})$ whose elements g' act on M according to

$$g' \in G \;:\; p \rightarrow (g')^{-1}\, p.$$

We shall also employ a complex setting defined by

$$p_c = Rp = \begin{bmatrix} z_{is} \\ \bar{z}_{is} \end{bmatrix}, \quad R = \sqrt{\frac{1}{2}} \begin{bmatrix} I & iI \\ I & -iI \end{bmatrix},$$

and the corresponding complex form of G,

$$R\, Sp(6, \mathbb{R}) R^{-1} = Sp(6, \mathbb{C}) \cap U(3,3)$$

For G we get a moment map $m : G \rightarrow L^*$, $L = sp(6, \mathbb{R})$ as

$$m : p \rightarrow T'(p) = p^t p\, K,$$

$$p_c \rightarrow T(p_c) = p_c(p_c^+)M, \qquad M = \begin{bmatrix} I & 0 \\ 0 & -I \end{bmatrix}$$

$$= i\, R\, T'(p) R^{-1}$$

The 6 x 6 matrices T' and T have the 3 x 3 block form

$$T' = \begin{bmatrix} A & Q \\ -P & -{}^tA \end{bmatrix} \qquad T = \begin{bmatrix} C & -2K_+ \\ 2K_- & -C^+ \end{bmatrix}$$

$$A_{ij} = \sum_{s=1}^{n} \xi_{is}\,\pi_{js}, \qquad C_{ij} = \frac{1}{2}\sum_{s=1}^{n}(z_{is}\,\bar{z}_{js} + z_{js}\,\bar{z}_{is}),$$

$$Q_{ij} = \sum_{s=1}^{n} \xi_{is}\,\xi_{js}, \qquad K_+ = \frac{1}{2}\sum_{s=1}^{n} z_{is}\,z_{js},$$

$$P_{ij} = \sum_{s=1}^{n} \pi_{is}\,\pi_{js}, \qquad K_- = \frac{1}{2}\sum_{s=1}^{n} \bar{z}_{is}\,\bar{z}_{js}.$$

The elements of $T'(p)$ and $T(p)$ respectively generate a Poisson action of G on M which commutes with the action of G. The coadjoint action of G on L^* is obtained as

$$\mathrm{Ad}^*_{g'}(T') = (g')^{-1}\,T'\,g'$$

$$\mathrm{Ad}^*_{g}(T) = g^{-1}\,T\,g, \qquad g = Rg'\,R^{-1}$$

In [22, 23, 24], we studied the orbit analysis for the group G, G and $G \times$ G acting on L^*, L*and M respectively. In the matrix notation introduced above, an orbit on M under $Sp(6,\mathbb{R}) \times SO(n,\mathbb{R})$ has the matrix form

$$p = c^{-1}\,\overset{o}{p}\,c''$$

where c is a matrix from a coset $Sp(6,\mathbb{R})/H'$ and c'' is a matrix from a coset $SO(n,\mathbb{R})/H''$. The representative point $\overset{o}{p}$ determines corresponding points on L^* and L* by the moment maps $\overset{o}{T} = T(\overset{o}{p})$ and $\overset{o}{\mathrm{T}} = \mathrm{T}(\overset{o}{p})$ respectively. Suppose that we can choose c'' to be the appropriate coset for the orbit of G on L* at $\overset{o}{\mathrm{T}}$. Then the reduced phase space $F_{\overset{o}{T}}$ becomes the coset $Sp(6,\mathbb{R})/H'$ on L^* or on $M_{\overset{o}{T}}$ respectively.

3.1 Proposition: The reduced phase space $F_{\overset{o}{T}}$ for the group $SO(n,\mathbb{R})$ acting on M is an orbit on the dual L^* of $sp(6,\mathbb{R})$ with the representative point $\overset{o}{T} = T(\overset{o}{p})$. It carries the standard symplectic structure and generalized Poisson bracket for this orbit.

In the next sections we shall consider particular cases of this construction.

4. Hyperbolic orbits on L^* and collective degrees of freedom for collective dynamics of closed and open shells

To study the orbits on L^* under $Sp(6,\mathbb{R})$ we look for standard forms of the matrix T' or T under symplectic transformations. For the complex setting we define adjoint matrices **for** the groups $Sp(6,\mathbb{C})$ and $U(3,3)$ by

$$q \to q^{\S} = -K^t q K, \qquad q \to q^* = M q^+ M$$

For the matrix T one finds $T^{\S} = T$ and $T^* = -T$. Moreover from the moment map one infers that $M\ T \geq 0$. We call the part of L^* with $M\ T > 0$ hyperbolic since it generalizes hyperbolic orbits under SU(1,1).

4.1 Proposition [22,25]: Under $Sp(6, \mathbb{R})$ the hyperbolic matrix T, $M\ T > 0$ has the standard form

$$\overset{o}{T} = \begin{bmatrix} \sigma & 0 \\ 0 & -\sigma \end{bmatrix}, \quad \sigma = (\sigma_i\ \delta_{ij}), \quad \sigma > 0$$

For different degen of the matrix σ, we obtain the following stability group H and dimensions of the coset space $Sp(6, \mathbb{R})/H$:

σ	H	$\dim(Sp(6, \mathbb{R})/H)$
$\sigma_1 > \sigma_2 > \sigma_3$	$U(1) \times U(1) \times U(1)$	18
$\sigma_1 > \sigma_2 = \sigma_3$ $\sigma_1 = \sigma_2 > \sigma_3$	$U(1) \times U(2)$	16
$\sigma_1 = \sigma_2 = \sigma_3 = \sigma_o$	$U(3)$	12

We consider in more detail the case $\sigma = \sigma_o I$. The coset $Sp(6, \mathbb{R})/U(3)$ may be parametrized by a symplectic matrix

$$c = \begin{bmatrix} I & -B \\ 0 & I \end{bmatrix}, \quad {}^tB = B, \quad I - BB^+ > 0$$

In these complex parameters the form of T is

$$T(B,B^+) = \sigma_o \begin{bmatrix} (I+BB^+)(I-BB^+)^{-1} & -2(I-BB^+)^{-1}B \\ 2B^+(I-BB^+)^{-1} & -(I-B^+B)^{-1}(I-B^+B) \end{bmatrix}$$

The fundamental generalized Poisson brackets become

$$\{b_{ij}, b_{lk}\} = 0 \qquad \{\overline{b_{ij}}, \overline{b_{lk}}\} = 0$$

$$\{\overline{b_{ij}}, b_{lk}\} = \sigma_o^{-1} \left[(I-BB^+)_{il}(I-BB^+)_{jk} + (I-BB^+)_{ik}(I-BB^+)_{jl} \right]$$

Under $U(3)$, the elements of B transform according to the irreducible representation $D^{[2]}$ and hence represent an s- and five d-quasi particles. The interpretation of this orbits in collective theory is obtained through dequantization [23,26]. Then the degenerate matrix σ can be shown to correspond to the collective dynamics of closed oscillator shell configurations, and σ_o is given by

$$\sigma_o = M/3 + n/2$$

where M is the total excitation. For large mass number A, the quasi-

particles may be transformed into bosons, and the dynamics on these orbits is given by the interaction of these bosons [27].
For open-shell dynamics we have additional degrees of freedom which are under study.

5. Conal orbits and irrotational flow

If the condition $M\,T > 0$ is relaxed to $M\,T \geq 0$, there appears the possibility of Jordan decompositions of T, compare Brunet and Kramer [25]. The Jordan chains have a maximum length 2, and if there are three such chains, the standard form of T' is

$$\overset{o}{T}{}' = \begin{bmatrix} 0 & I \\ 0 & 0 \end{bmatrix}$$

We call the orbits with Jordan chains conal, again in analogy to the case of SU(1,1). In this case we use the real setting and get as the stability group the group $H = t(6) \wedge SO(3, \mathbb{R})$ with elements

$$\begin{bmatrix} I & \gamma \\ 0 & I \end{bmatrix} \begin{bmatrix} a & o \\ o & a \end{bmatrix} \quad , \quad {}^t\gamma = \gamma \quad , \quad {}^ta = a^{-1}$$

and 9 parameters. The coset $Sp(6, \mathbb{R})/H$ admits the parametrization

$$c = \begin{bmatrix} I & 0 \\ -Z & I \end{bmatrix} \begin{bmatrix} s & 0 \\ 0 & s^{-1} \end{bmatrix} \quad , \quad s = {}^ts \geq 0, \; Z = {}^tZ$$

The form of T' becomes

$$T'(s^2, Z) = \begin{bmatrix} s^2 Z & s^2 \\ -s^2 Z s^2 & -Z s^2 \end{bmatrix}$$

and the fundamental generalized Poisson brackets are

$$\{(s^2)_{ij}, (s^2)_{lk}\} = 0 \; , \qquad \{Z_{ij}, Z_{lk}\} = 0 \; ,$$

$$\{(s^2)_{ij}, Z_{lk}\} = \delta_{il}\,\delta_{jk} + \delta_{ik}\,\delta_{jl}$$

Since $s^2 = Q$, the matrix Z determines generalized momenta corresponding to Q. If now one computes the positions and momenta on this orbit one finds

$$\pi_{is} = \sum_{j=1}^{3} Z_{ij}\,\xi_{js} \qquad\qquad s = 1,2,\ldots,n$$

Since $Z = {}^tZ$, the momenta and (for local interactions) the velocities are linearly related to the positions, and the velocity field is irrotational. On these orbits we therefore discover a feature of collective motion which was postulated by Bohr and Mottelson [1,2].

6. Conal coordinates on hyperbolic orbits of L^* and collective motion of closed shells

We return to the standard form of $\overset{o}{T}'$ for hyperbolic orbits and $\sigma = \sigma_o I$ but transform the representative point to

$$\overset{o}{S}' = \begin{bmatrix} \lambda^{-1} & 0 \\ 0 & \lambda \end{bmatrix} \begin{bmatrix} 0 & \sigma_o I \\ -\sigma_o I & 0 \end{bmatrix} \begin{bmatrix} \lambda & 0 \\ 0 & \lambda^{-1} \end{bmatrix}, \quad \lambda = ((\sigma_o)^{1/2}\,\delta_{ij})$$

$$= \begin{bmatrix} 0 & I \\ -\sigma_o^2 I & 0 \end{bmatrix}$$

The stability group H of $\overset{o}{S}'$ is

$$\begin{bmatrix} \lambda^{-1} & 0 \\ 0 & \lambda \end{bmatrix} R^{-1} \begin{bmatrix} u & 0 \\ 0 & \bar{u} \end{bmatrix} R \begin{bmatrix} \lambda & 0 \\ 0 & \lambda^{-1} \end{bmatrix}$$

$$= \begin{bmatrix} u + \bar{u} & i\,\sigma_o^{-1}(u-\bar{u}) \\ -i\,\sigma_o(u-\bar{u}) & u + \bar{u} \end{bmatrix}$$

A group contraction with $\sigma_o \to o$ but finite $i\,\sigma_o^{-1}(u-\bar{u})$ yields $\lim H = t(6) \wedge SO(3, \mathbb{R})$, the stability group of the conal orbit of section 5 . But without taking this limit one can use the parameters s^2, Z of the conal orbit and define T' by

$$T' = \begin{bmatrix} s^2 Z & s^2 \\ -s^2 Z s^2 - \sigma_o^2\, s^{-2} & -Z s^2 \end{bmatrix}$$

An explicit computation shows that the generalized Poisson bracket reproduces the Lie algebra of $Sp(6, \mathbb{R})$! We consider now a collective hamiltonian on this orbit. From

$$H_{coll} = \frac{1}{2m} \sum_{i=1}^{3} P_{ii} + V(\mathrm{trace}\, Q,\ \mathrm{trace}\, Q^2,\ \mathrm{trace}\, Q^3)$$

one finds for the kinetic energy

$$T_{kin} = \frac{1}{2m}\,\mathrm{trace}\,(ZQZ) + \frac{1}{2m}\,\sigma_o^2\,\mathrm{trace}(Q^{-1})$$

Transforming Q to diagonal form $\mu = (\mu_i\,\delta_{ij})$ one finds [2

$$T_{kin} = m^{-1} \sum_{l=1}^{3} 2\mu_l(\pi_l)^2 + (2m)^{-1} \sum_{l<k} (\mu_l+\mu_k)(\mu_l-\mu_k)^{-2}(L_{lk})^2$$

$$+ (2m)^{-1}\,\sigma_o^2 \sum_{l=1}^{3} (\mu_l)^{-1}.$$

The momenta π_l are related to the eigenvalues of Q and hence to collective vibrations. The L_{1k} are the angular momenta in the body-fixed system. As an example of collective dynamics assume that the potential V depends on $\rho = \mu_1 + \mu_2 + \mu_3$ only and consider collective motion without vibration, $\dot{\pi}_1 = 0$, $\dot{\mu}_1 = 0$. The dynamical equation for the lowest energy and fixed angular momentum require that the energy ellipsoid touches the angular momentum in one point. The energy surface can be represented as a function of the quantities $y_i = \rho^{-1} \mu_i$ in triangular coordinates. The system rotates with a pseudo-rigid tensor of inertia. Its deformation is determined by the balance between angular momentum part which favours deformation and the term proportional to σ_o^2 which favours a spherical shape, compare Figs. 1 and 2.

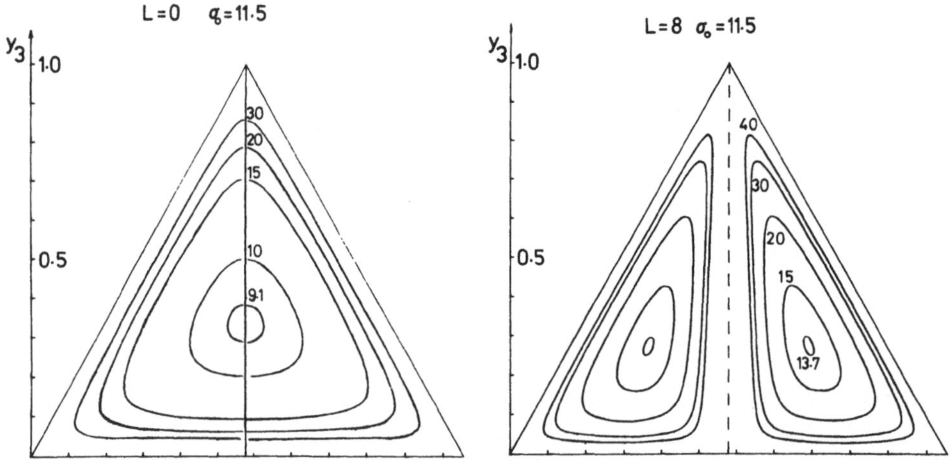

Figs. 1, 2: Lines of constant energy as a function of the coordinates $y_i = \rho^{-1} \mu_i$ for $\sigma_o = 11,5$, L = 0 and L = 8 respectively.

If the potential contains additional terms in μ_i, the lines of constant energy may change and favour other deformations as in the Bohr-Mottelson theory. For open-shell dynamics which has not yet been treated, we expect deformation already in the lowest energy state.

References

[1] A. Bohr, Dan.Mat.Fys.Medd. 26 (1952) 14

[2] A. Bohr and B. Mottelson, Dan.Mat.Fys.Medd. 27 (1953) No. 16

[3] J.P. Elliott, Proc.Roy.Soc. A245 (1958) 562

[4] R.Y. Cusson, Nucl.Phys. A114 (1968) 289

[5] L. Weaver, L.C. Biedenharn and R.Y. Cusson, Ann.Phys. 77 (1973) 250

[6] O.L. Weaver, R.J. Cusson and L.C. Biedenharn, Ann.Phys. 102 (1976) 493

[7] P. Gulshani and D.J. Rowe, Canad.J.Phys. 54 (1976) 970

[8] G. Rosensteel and D.J. Rowe, Ann.Phys. 123 (1979) 36

[9] H.J. Lipkin, A. DeShalit and I. Talmi, Nuovo Cimento II (1955) 773

[10] F. Villars, Nucl.Phys. 3 (1957) 240

[11] W. Zickendraht, J.Math.Phys. 12 (1971) 1663

[12] A.Ya. Dzyublik, V.I. Ovcharenko, A.I. Steshenko and G.F. Filippov, Sov.J.Nucl.Phys. 15 (1972) 487

[13] B. Buck, L.C. Biedenharn and R.Y. Cusson, Nucl.Phys. A317 (1979) 205

[14] D.J. Rowe and G. Rosensteel, Ann.Phys. 126 (1980) 198

[15] V. Vanagas, The microscopic theory, Lecture notes, University of Toronto, 1977

[16] S. Goshen and H.J. Lipkin, Ann.Phys. (N.Y.) 6 (1959) 301

[17] S. Goshen and H.J. Lipkin in Spectroscopic and Group Theoretical Methods in Physics, ed. F. Block, North-Holland, Amsterdam 1968

[18] G. Rosensteel and D.J. Rowe, Ann.Phys. (N.Y.) 126 (1980) 343

[19] J.M. Souriau, Structure des systèmes dynamiques, Dunod, Paris 1970

[20] V.I. Arnold, Mathematical Methods of Classical Mechanics, Springer, Berlin 1978

[21] J. Marsden and A. Weinstein, Rep.Math.Phys. 5 (1974) 121

[22] P. Kramer, Ann.Phys. (N.Y.) 141 (1982) 254

[23] P. Kramer, Ann.Phys. (N.Y.) 141 (1982) 269

[24] P. Kramer, Z. Papadopolos and W. Schweizer, Ann.Phys. (N.Y.) 149 (1983) 44

[25] M. Brunet and P. Kramer, Rep.Math.Phys. 15 (1979) 287

[26] P. Kramer and M. Saraceno, Lecture Notes in Physics, Vol. 140, Berlin 1981

[27] A. Arima and F. Iachello, Phys.Lett. 35 (1975) 1069

QUANTUM EFFECTS IN CLASSICAL PHASE SPACE: SYMPLECTIC STRUCTURES ASSOCIATED TO THE SCATTERING OF NUCLEAR FRAGMENTS

P. Kramer
Institut für Theoretische Physik, Universität Tübingen

M. Saraceno
Departamento de Física, Comisión Nacional de Energía Atómica, Buenos Aires

Recently[1] we have studied the geometry of the Time Dependent Variational Principle (TDVP) in quantum mechanics. We have shown that to a parametrized trial state there is associated a symplectic structure which derives from the overlap of the wave function. The TDVP is seen to provide a dequantization procedure which maps quantum observables into functions on a (generalized) phase space, commutators into Poisson brackets and time evolution in state space into classical evolution along a trajectory. The dequantization, however, retains some of the quantum features which were present in the trial wave function but depicts them in a classical phase space. The case when the parametrization is obtained via the coordinates of a Lie Group leads to the well known symplectic structures on orbits of its coadjoint representation studied by Kostant[2] and Souriau[3].

Physical applications of this method using various coset spaces associated to the group Sp(6,R) have been studied in relation to collective motion in nuclei[4,5].

Another line of research[6,7] that we want to report here is related to the scattering of composite nuclei. We consider three types of trial wave functions, all parametrized by a complex vector $\underset{\sim}{S}$.

$$|\underset{\sim}{S}\rangle = \exp(\underset{\sim}{S}.\underset{\sim}{B}^{+})|0\rangle \tag{1}$$

$$|\underset{\sim}{S},f\rangle = C^{f}|\underset{\sim}{S}\rangle \tag{2}$$

$$|\underset{\sim}{S},f,\pm\rangle = \sqrt{\tfrac{1}{2}}\,\{|\underset{\sim}{S},f\rangle \pm |-\underset{\sim}{S},f\rangle\} \tag{3}$$

Here $|0\rangle$ is a product of the oscillator shell model states $|\phi_i\rangle$ which represent the two scattering fragments. We assume S=T=0 for their spin and isospin. $\underset{\sim}{B}^{+}$ is the creation operator for quanta of relative motion

$$\underset{\sim}{B}^{+} = (\mu/N_2)^{1/2}\sum_{i\varepsilon 2}\underset{\sim}{a}_i^{+} - (\mu/N_1)^{1/2}\sum_{i\varepsilon 1}\underset{\sim}{a}_i^{+} \tag{4}$$

while $\underset{\sim}{S}$ is the complex vector

$$\underset{\sim}{S} = (\mu/2)^{1/2}\underset{\sim}{R}/b - ib\underset{\sim}{K}(2\mu)^{-1/2} \tag{5}$$

$\underset{\sim}{R}$ and $\underset{\sim}{K}$ are the relative position and wave number of the centers of mass of the fragments, N_i their mass numbers, $\mu = N_1N_2/(N_1+N_2)$, and $b = (\hbar/(m\omega))^{1/2}$ is the single particle oscillator length.

The states (1), (2), (3) are many body wave packets parametrized by $\underset{\sim}{S}$ (or equivalently by $\underset{\sim}{R},\underset{\sim}{K}$). State (1) is a product wave function of two fragments translated and boosted towards each other. In the relative coordinate $\underset{\sim}{B}^{+}$ it is a Weyl coherent state. Its properties are very simple and well known. However, the Pauli principle is violated as (1) does not have a definite permutation symmetry under particle exchange. This defect is remedied in (2) by the projection operator C^{f}. The partition f appropriate for S=T=0 is [4,4...4]. In (3) we have also projected to states of definite parity.

The symplectic structures associated to the above states are derived from the norm overlaps[1)] by

$$g_{ij} = \partial^{2}/\partial S_{i}\partial S_{j}\ \ln\langle \underset{\sim}{S}|\bar{\underset{\sim}{S}}\rangle \tag{6}$$

The fact that $g_{ij}\neq\delta_{ij}$ in general means that the coordinates $\underset{\sim}{S}$, $\bar{\underset{\sim}{S}}$ are not canonical. However, due to Darboux' theorem one can always find canonical coordinates on any symplectic manifold. In our case they are given explicitly by

$$\underset{\sim}{\omega} = u^{1/2}\,\underset{\sim}{S} \tag{7}$$

where $u = \partial/\partial(\underset{\sim}{S}.\bar{\underset{\sim}{S}})\ \ln\langle \underset{\sim}{S}|\underset{\sim}{S}\rangle$.

Analytic expressions for the overlaps are given by

$$N_{0} = \langle \underset{\sim}{S}|\underset{\sim}{S}\rangle = \exp(\underset{\sim}{S}.\bar{\underset{\sim}{S}}) \tag{8}$$

$$N = \langle \underset{\sim}{S},f|\underset{\sim}{S},f\rangle = N_{0}\left[1-(1+\underset{\sim}{S}.\bar{\underset{\sim}{S}}/\mu)\exp(-\underset{\sim}{S}.\bar{\underset{\sim}{S}}/\mu)\right]^{4} \tag{9}$$

$$N^{\pm} = N(\underset{\sim}{S}.\bar{\underset{\sim}{S}})\pm N(-\underset{\sim}{S}.\bar{\underset{\sim}{S}}) \tag{10}$$

(8) and (10) are general while (9) is, for definiteness, written for the special case of α-^{16}O scattering. The calculation of these overlaps is detailed in [8)].

Using (7) the Darboux transformation can be implemented explicitly. To study its features it is convenient to look at the function

$$\underset{\sim}{\omega}.\bar{\underset{\sim}{\omega}} = \underset{\sim}{S}.\bar{\underset{\sim}{S}}\ u(\underset{\sim}{S}.\bar{\underset{\sim}{S}}) \tag{11}$$

This function is plotted in Fig. 1 for some SU(3) scalar fragments with states of type (2). The general pattern that emerges is the following. For large $\underset{\sim}{S}.\bar{\underset{\sim}{S}}$ the $\underset{\sim}{S}$ and $\underset{\sim}{\omega}$ coordinates are equal and the symmetry projection unimportant. For small $\underset{\sim}{S}.\bar{\underset{\sim}{S}}$, $\omega.\bar{\underset{\sim}{\omega}}$ tends to a definite non-zero limit N_0. Thus the canonical phase space can be divided into three regions: a) $\underset{\sim}{\omega}.\bar{\underset{\sim}{\omega}} < N_0$ is forbidden. This region corresponds in the quantum treatment to states of relative motion forbidden by the Pauli principle; b) $N_0<\underset{\sim}{\omega}.\bar{\underset{\sim}{\omega}}<N_f$ where N_f is some value where $\underset{\sim}{\omega}$ and $\underset{\sim}{S}$ can be considered equal. In this region $\underset{\sim}{\omega}$ differs significantly from $\underset{\sim}{S}$ and the Pauli principle has dynamical effects; c) finally for $\underset{\sim}{\omega}.\bar{\underset{\sim}{\omega}}>N_f$ the effects of the projection C^{f} have disappeared. In Fig. 2 we show the additional features brought about by the projection of parity in the α-^{16}O case. The forbidden region is larger for odd states ($N_0^{+}=8$ and $N_0^{-}=9$). The effects of the projection die off as $\underset{\sim}{S}.\bar{\underset{\sim}{S}}$ increase although they do so much sooner than the antisymmetrization.

The boundaries discussed above can be looked at in radial phase space considering that from (5)

$$\underset{\sim}{\omega}.\bar{\underset{\sim}{\omega}} = 1/2u[\mu R^2/b^2+b^2K^2/\mu+b^2l^2/(\mu R^2)] \qquad (12)$$

For constant $\omega.\bar{\omega}$=N these are closed curves that depend on N and L (see Fig. 3). They divide the phase space in regions where different quantum effects are important.

Once having taken into account the structure of the phase space it is possible to evaluate bound states, resonances and phase shifts using WKB formulas. This has been done[6] in the α-α and α-^{16}O case and compared successfully with the results of Resonating Group Method (RGM) calculations. In synthesis we find that, depending on the trial state used in the TDVP, the dequantization procedure treats selected quantum features classically and displays their effects in phase space.

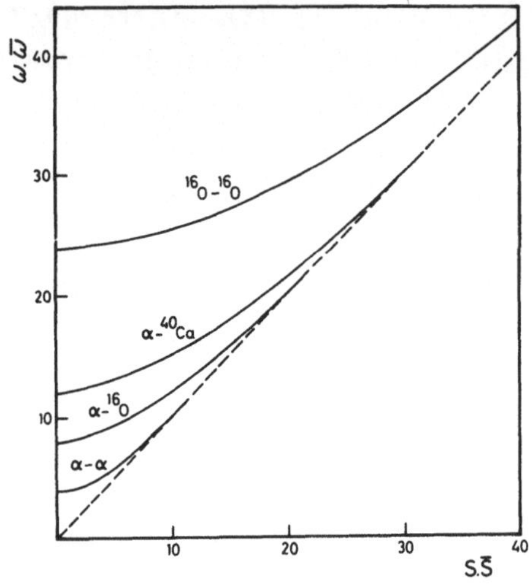

Fig. 1

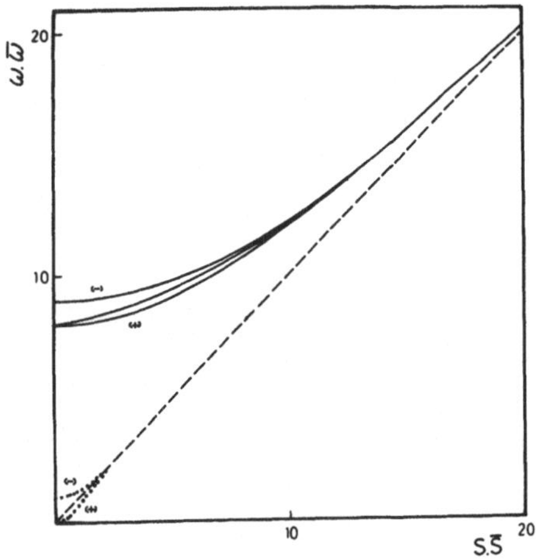

Fig. 2

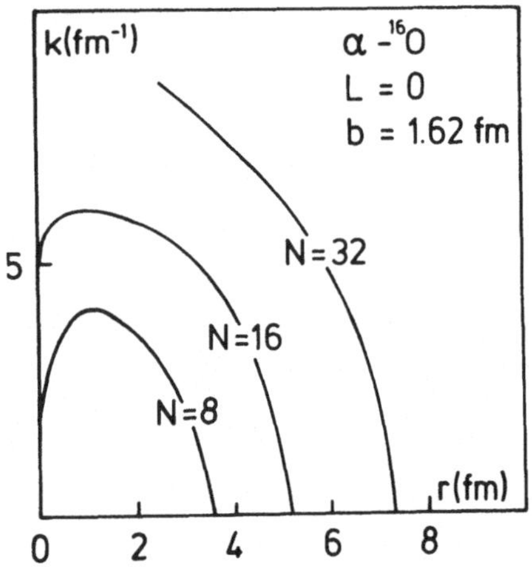

Fig. 3

1) P.Kramer, M.Saraceno, "Geometry of the Time Dependent Variational Principle in Quantum Mechanics", Lect.Notes in Phys. 140, Springer, Berlin, 1981.
2) B.Kostant, "Quantization and Unitary Representations" Lect.Notes in Math. 170, Springer, Berlin, 1970.
3) J.M.Souriau, "Structure des Systemes Dynamiques", Dunod, Paris, 1970.
4) P.Kramer, Ann.Phys. (N.Y.) 141 (1982) 254; Ann.Phys. (N.Y.) 141 (1982) 269.
5) P.Kramer, Z.Papadopolos, W.Schweizer, to be published in Ann.Phys.(N.Y.).
6) M.Saraceno, P.Kramer, F.Fernández, Nucl.Phys. A 405, (1983) 88.
7) M.Saraceno, Contribution to the V Workshop on Nuclear Physics in Brasil, Itatiaia (1982), Rev.Bras.Fis. (Vol.Especial) Sept. 1982.
8) P.Kramer, G.John, D.Schenzle, "Group Theory and the Interaction of Composite Nucleon Systems", Vieweg, Braunschweig, 1981.

GAMOW STATES IN MOMENTUM REPRESENTATION

A. Mondragón and E. Hernández
Instituto de Física UNAM
Apdo. Postal 20-364, 01000 México D.F.

It has been appreciated for many years that there are distinct advantages to performing nuclear scattering calculations in momentum space, since many physical effects are then readily expressed and evaluated. We intend to show that it is possible to work with Gamow states in momentum representation[1].

Gamow states in momentum representation are defined as right solutions of a homogeneous Lippmann-Schwinger equation for purely outgoing particles. The Lippmann-Schwinger equation for the partial wave $u_{n\ell}(q\,;\,k_n)$ is

$$u_{n\ell}(q\,;\,k_n) = \left\{ \int_0^\infty K_\ell^{(+)}(q,q';k)\,u_{n\ell}(q';k)\,dq' \right\}_{k_n} \tag{1}$$

with the kernel

$$K_\ell^{(+)}(q,q';k) = \int_0^\infty \frac{\delta(q,q'')}{k^2-q''^2}\,U_\ell(q'',q')\,dq'' \tag{2}$$

with Im $k > 0$, $q = \frac{1}{\hbar}p$ is the momentum variable, $k = \left(\frac{2mE}{\hbar^2}\right)^{1/2}$ is the wave number and $U_\ell(q,q') = \frac{2m}{\hbar^2}V_\ell(q,q')$ is the potential, which, for spinless particle is real and symmetric. In order to have outgoing state solutions, the integral in (1) is defined with Im $k > 0$, and, in order to have outgoing particle states of energy E_n, (Im $E_n < 0$), that is Gamow states, the integral is analytically continued to k_n.

Solutions of (1) exist, even when the kernel is infinite at one point[2,3], provided that the Fredholm determinant of $(\underline{1} - \underline{K}^{(+)}(k))$ vanishes

$$\Delta_\ell^{(+)}(k) = \det\left(\underline{1} - K_\ell^{(+)}(k)\right) = 0 \tag{3}$$

and the integrals $\int_0^\infty \int_0^\infty |U_\ell\ (q, q')|^2\, dq\, dq'$, $\int_0^\infty\ |U_\ell\ (q,q')|^2\, dq'$ with $U_\ell\ (q,q') = U_\ell(q',q)$ exist, and the latter regarded as function of q remains below a fixed bound. The energy eigenvalues E_n are the roots of equation (3).

The kernel $K_\ell^{(+)}(q,q';\ k)$ is not hermitean, but the causality condition implies that

$$K_\ell^{(+)}\ (q,q';\ k) \ = \ K_\ell^{(+)*}\ (q',q;\ -k^*) \tag{4}$$

Therefore, the adjoint (left solution of the L-S equation) $u_{n\ell}(q,\ k_n)$ of $u_{n\ell}(q,k_n)$ is $u^*_{n\ell}(q,\ -k^*_n)$. Since the potential is real and symmetric $u^*_{n\ell}(q,\ -k^*_n)$ is equal to $u_{n\ell}(q,\ k_n)$. Hence, the adjoint of the Gamow function is the same function

$$\tilde{u}_{n\ell}(q,\ k_n) \ = \ u_{n\ell}(q,\ k_n) \tag{5}$$

and bound and resonant states form a biorthogonal set with their adjoints

$$\left\{ \int_0^\infty u_{n\ell}(q;\ k)\ u_{m\ell}(q;\ k')\ dq \right\}_{\substack{k \to k_n \\ k' \to k_m}} \ = \ \delta_{nm}\left\{ \int_0^\infty u^2_{n\ell}(q;\ k)\ dq \right\}_{k_n} \tag{6}$$

the integrals in (6) are defined with Im $k > 0$, Im $k' > 0$, then they are analytically continued to k_n and k_m . From (4), it follows that

$$\Delta_\ell^{(+)}\ (k_n) \ = \ \Delta_\ell^{(+)*}\ (-k^*_n\ ) \tag{7}$$

Hence, the zeroes of the Fredholm determinant $\Delta_\ell^{(+)}\ (k)$ are located on the imaginary axis and in the lower half of the k plane, including the real axis.

In the case of potentials satisfying the condition stated above, the momentum representation of the restriction to the ℓth wave of the resolvent of the hamiltonian, $\langle q|\frac{1}{E^+ - H^\ell}|q'\rangle$, and the collision matrix, S_ℓ , have poles in the k plane located precisely where $\Delta_\ell^{(+)}\ (\ k\ )$ has zeroes[4]. Since resonant states correspond to poles of the transition matrix (t matrix) in unphysical sheets of the complex energy plane[5], a convenient way to relate Gamow states with processes of physical interest is provided by the study of matrix elements of the resolvent between arbitrary states. It is found that

$$\lim_{E \to E_n} (E - E_n)\langle q|\frac{1}{E^+ - H_\ell}|q'\rangle = u_{n\ell}(q;\, k_n)\, \frac{1}{\{\int_0^\infty u_{n\ell}^2(q;\, k)\,dq\,\}}\, u_{n\ell}(q';\, k_n) \tag{8}$$

and

$$\lim_{E \to E_n} (\, E - E_n\,)\langle \phi|\theta_\ell \frac{1}{E^+ - H_\ell} Q_\ell|\chi\rangle$$

$$= \frac{\{\, \int_0^\infty \phi^*(q)\theta_\ell\,(q,q')u_{n\ell}(q';\, k)\,\, dq\,\, dq'\,\}_{k_n} \{\, \int_0^\infty u_{n\ell}(q;\, k)Q_\ell(q,q')\chi(q')dqdq\}_{k_n}}{\{\, \int_0^\infty u_n^2\,(q;\, k)\,\, dq\,\}_{k_n}} \tag{9}$$

Therefore, the matrix elements of quantum mechanical operators between properly normalized resonant states and an arbitrary state are

$$\langle\phi|\theta_\ell|u_{n\ell}\rangle = \frac{\{\, \int_0^\infty \phi^*\,(q)\theta_\ell\,(q;\, q')u_{n\ell}(q';\, k)\,\, dq\,\, dq'\,\}_{k_n}}{\{\, \int_0^\infty u_{n\ell}^2\,(q;\, k)\,\, dq\,\,\}_{k_n}^{1/2}} \tag{10}$$

and

$$\langle u_{n\ell}|Q_\ell|\chi\rangle = \frac{\{\, \int_0^\infty u_{n\ell}\,(q;\, k)Q_\ell(q;\, q')\chi(q')\,\, dq\,\, dq'\,\}_{k_n}}{\{\, \int_0^\infty u_{n\ell}^2\,(q;\, k)\,\, dq\,\}_{k_n}^{1/2}} \tag{11}$$

The integrals are defined with Im $k > 0$, and analytically continued to k_n. From (10) and (11), it follows that the normalization condition for Gamow states in momentum space is

$$\{\, \int_0^\infty u_{n\ell}^2(q;\, k)\,\, dq\,\,\}_{k_n} = 1$$

This rule simplifies the orthogonality condition and makes the set of bound and resonant states an orthonormal set

We have also shown that the Gamow state function in configuration space is obtained from the Gamow state function in momentum space as the Fourier transform of $u_{n\ell}(q;\ k)$, defined with Im $k > 0$ and analytically continued to k_n. The momentum representation of Gamow states is obtained from its position representation in a similar way. Then, it is readily verified that the norm of a Gamow state is independent of the representation, in fact, it was shown that the norm of resonant states defined in this work is equal to the norm defined by Peierls[6], Hokkyo[7] and Romo[8] in configuration space.

Although Gamow state functions in momentum space are square integrable, Gamow states are unnormalizable in terms of the integral of the modulus squared of the state function. It is possible to define a normalized momentum probability density in terms of $|u_{n\ell}(q;k_n)|^2 dq$. However, since Im $k_n < 0$, the expectation values obtained with this probability density correspond to purely incoming particle states rather than to Gamow states. When we try to define the normalization integral in terms of $\int_0^\infty |u_{n\ell}(q;\ k)|^2\, dq$ with Im $k > 0$, it corresponds to outgoing particle states, but when the integral is analytically continued to k_n, it vanishes for all eigenfunctions belonging to an eigenvalue E_n such that Im $E_n < 0$. Therefore, when we try to calculate expectation values of quantum mechanical operators as the analytic continuation to k_n of the integral

$$\int_0^\infty \int_0^\infty u^*_{n\ell}(q;\ k)\,\theta_\ell(q,\ q')\,u_{n\ell}(q';\ k)\ dq\ dq'$$

defined with Im $k > 0$, the results are finite but cannot be normalized.

Finally, following a standard procedure, it was shown that a square integrable function may be expanded in terms of a set containing bound and resonant states and a continuum of scattering functions of complex wave number.

References.

1. E. Hernández and A. Mondragón
"Resonant states in momentum representation"
Preprint IFUNAM 82 - 18, submitted for publication.
See also
E. Hernández and A. Mondragón
Notas de Física 6, (2) (1983)

2. R. Courant and D. Hilbert
"Methods of Mathematical Physics"
Interscience Publishers Inc. New York (1953)
Ch. III

3. S. Weinberg
"Quasiparticles and Perturbation theory"
in "Lectures on Particles and Field Theory"
Brandeis Summer Institute in Theoretical Physics
S. Deser and K. W. Ford Editors. Vol II. Ch. 5

4. G. C. Ghirardi, V. Gorini and G. Parravicini
IL Nuovo Cimento 57A, 1, (1980)

5. F. J. Siegert
Phys Rev. 56, 750 (1939)

6. R. E. Peierls
Proc. Cambridge Phys. Soc. 44, 242 (1948)

7. N. Hokkyo
Prog. Theor. Phys. 33, 1116 (1965)

8. W. Romo
Nucl. Phys. A116, 618 (1968)

GEOMETRY OF NUCLEAR COLLECTIVE MOTIONS

M. Moshinsky
Instituto de Física, UNAM, Apdo. Postal 20-364
México, D.F. 01000 México

We discuss the relation between collective behavior of many body systems and geometrical concepts. Specifically we shall reexamine the microscopic description of nuclear collective motions and their relation with the symplectic geometry of the A-nucleon system.

The first widely successful way of introducing collective degrees of freedom in the nucleus was the liquid drop model of Niels Bohr[1]. In it the surface of the drop was given in spherical coordinates by $r=f(\Theta,\phi)$, and the right hand side was developed in spherical harmonics with the coefficients becoming the collective degrees of freedom. In the more systematic approach developed years later by Aage Bohr and Ben Mottelson[2] the drop is restricted to quadrupole deformations and the fluid is taken as incompressible so that the equation of the surface becomes

$$r=r_o[1+\sum_m \alpha^m Y_{2m}(\Theta,\phi)] \;, \tag{1}$$

with $\alpha^m=(-1)^m\alpha_{-m}$; $m=2,1,0,-1,-2$ being the collective degrees of freedom. The α_m are implicitly related[2),3)] with the mass quadrupole of the many nucleon system defined by

$$q_m= \sum_{k,k'} \langle 1k,1k'|2m\rangle \sum_{s=1}^{n} x_{ks}x_{k's} \tag{2}$$

where $\langle|\rangle$ is a Clebsch-Gordan coefficient and x_{ks}; $k=1,0,-1$; $s=1,2,\ldots n$ are the spherical components of the $n=A-1$ relative Jacobi vectors of the A nucleon system as, from the beginning, we wish to eliminate the center of mass coordinate which is irrelevant for the analysis. It is immediately clear that the q_m defined by (2) is a scalar with respect to the $O(n)$ group of rotations in the space associated with the index $s=1,2,\ldots n$. This point will be central to much of the following discussion.

In view of the great success of the Bohr-Mottelson model in explaining many of the features of nuclear structure a very extensive literature developed trying to justify it microscopically. From the standpoint of this note we are concerned only with that part of the literature based on group theoretical considerations, which we shall proceed to review briefly so as to place our approach in the context of the work done previously in this field.

Probably the first publication that develops the point of view in which we are interested is that of Goshen and Lipkin[4]. In the one that appeared in 1959, they consider an n-body system in one dimensional space and indicate how from the symplectic 2n dimensional group Sp(2n) of the problem, one can pass to the groupSp(2)xO(n), where the latter is the orthogonal group in n-dimensions. Collective excitations are then related with the Sp(2) group and, in particular, the collective Hamiltonian is associated with the O(2) subgroup of the latter. In several respects the paper of Goshen and Lipkin provides the foundations on which many of the group theoretical microscopic collective theories were based, including the one to be presented in this note. They did not provide though a detailed formalism for carrying out nuclear structure calculations.

By the early seventies a considerable push was given to the development of group theoretical microscopic collective models through the work of Zickendraht[5] and Dzublik et al.[6] in which they introduced a transformation of coordinates that take us from the 3n variables x_{is}; i=1,2,3; s=1,...n, to six collective degrees of freedom (three deformation parameters along the principal axis and three Euler angles) and 3n-6 variables associated with single particle excitations. Almost immediately the group of Filippov[7] in Kiev and of Vanagas[8] in Vilnus, realized that these transformations allowed one to discuss collective effects by restricting the states to a definite irreducible representation (irrep) of the O(n) group which could be derived from shell model considerations[7,8].

A problem of 3n degrees of freedom can be associated with a definite irrep of a dynamical group Sp(6n), which among its subgroups has Sp(6)xO(n). It was shown by Moshinsky and Quesne[9] in 1971, that the irrep of the group O(n) determines that of Sp(6), i.e. that they are "complementary"[9], and thus the procedure of Filippov and Vanagas restricted to a specific irrep of O(n) implied also a restriction to a definite irrep of Sp(6), as these authors realized in their later publications[10,11].

By the middle seventies Rosensteel and Rowe[12] in Toronto and

Biedenharn, Buck, Cusson and Weaver[13] at Duke, initiated an approach to the problem by first identifying the desired collective motions and then determining the operators that generate these motions as well as the Lie algebra that they satisfy. Thus appeared the collective motion Lie algebra[12,13] known as cm(3) as well as the explicit determination of the vortex spin operator which plays a very important role in coupling rotational motion with internal dynamics[14].

Rowe and Rosensteel[15] and their collaborators quickly realized that the CM(3) group was a subgroup of Sp(6)(which they call SP(3,R)) and thus the basis for the irreps of the latter were fundamental for implementing their views on the microscopic origin of collective motions. A natural chain for discussing these basis is $Sp(6)\supset U(3)$ where the latter is the unitary group in three dimensions introduced by Elliott[16]. Rowe and Rosensteel then proceeded to find the basis for the irreps $Sp(6)\supset U(3)$ both in an abstract fashion and as shell model states, and to carry an extensive program of calculations[17] using these basis as well as discussing the implications for the states characterized by irreps in the chain $Sp(6)\supset CM(3)$[12,13].

The remark made previously[9] about the relation between the irreps of Sp(6) and O(n), when Sp(6)xO(n) is considered as a subgroup of a definite irrep of Sp(6n), immediately relate the work done in Toronto and Duke, with that carried out in Vilnus and Kiev as was quickly clear to all the authors involved[10,11,14,15]. Thus the Sp(6) group became the paramount one for the microscopic analysis of nuclear collective motions, and it seemed that the discussion of the previously mentioned authors and, in particular, the very extensive analysis of Rowe and Rosensteel[15], provided all the background necessary for calculations in nuclear structure.

The reason that, in this note and in a series of papers submitted for publication in J. Math. Phys., the author and his collaborators are looking again at the Sp(6) problem, stems from their experience[18,19,20] both in the Bohr-Mottelson model[2] and the Interacting Boson Approximation (IBA)[21] introduced by Arima and Iachello. In particular in the latter, where pairs of nucleons are associated with s and d bosons, the fundamental group is U(6) and it admits, among others, the chain of subgroup $U(6)\supset U(3)\supset O(3)$ and $U(6)\supset U(5)\supset O(5)$ associated respectively with rotational and vibrational limits[21]. Now Deenen and Quesne[22] showed that for closed shells or, equivalently, scalar representation of O(n), the states that are basis for irreps of $Sp(6)\supset U(3)$ can be put into one to one correspondance with basis states characterized by irreps of

$U(6) \supset U(3)$ in the boson approximation. A similar result was proved for open shells, where now there is vortex spin, by Castaños and Frank[23], but only in the limit when the number of nucleons $A \to \infty$. Thus the procedures of Rosensteel and Rowe, are closely correlated with the $U(6) \supset U(3)$ chain in the boson approximation, as they themselves have stressed[24].

Is there something equivalent in the Sp(6) model of the $U(6) \supset U(5) \supset O(3)$ chain in the boson model? We showed in recent papers[25,26] that this is the case if we consider the $Sp(6) \supset Sp(2) \times O(3)$ chain which, to our knowledge, has not been fully discussed before. Thus it becomes interesting to analyze the basis for the irreps in the chain $Sp(6) \supset Sp(2) \times O(3)$ as well as the matrix element of the generators of Sp(6) in this basis. This will be the main objective of this note.

Before presenting a summary of our analysis we would like to stress that the developments outlined above for a group theoretical approach to a microscopic theory of collective motions in nuclei are not contradictory but complementary. As we indicated, all of them can be related with the Sp(6) group, though some analysis emphasize the complementary group O(n),[10,11] while others stress particular subgroups of Sp(6) whose generators are relevant to collective operators.[12,14,15]. Thus one could affirm that the bare bones ideas introduced almost a quarter of century ago[4] have been fleshed out in a variety of ways that illuminate the microscopic nature of nuclear collective motions and help us also to understand better macroscopic collective models such as the Bohr-Mottelson one and the Interacting Boson Approximation.

Before outlining the main ideas of our approach we would like to indicate a useful analogy. If we look at the three dimensional Euclidean space of coordinates (x,y,z) we have in it the simple geometrical concepts and theorems discussed in Euclids book. If we introduce a <u>constraint</u> in this space by the relation $x^2+y^2+(z-a)^2=a^2$ we have now the more complex geometry of a sphere despite the fact that we are dealing with a two rather than a three dimensional manifold. We recover the simple Euclidean geometry, but now in two rather than three dimensions,for the points of the sphere near the origin when $a \to \infty$, as then that part of the sphere becomes essentially the x-y plane.

In a many body problem we have a 6n dimensional phase space of coordinates x_{is} and momenta p_{is}, $i=1,2,3$; $s=1,\ldots n$. The dynamical group of this problem is Sp(6n) as it is also the dynamical group[9] of the 3n dimensional harmonic oscillator that provides a complete set of states for the many body problem.

The 3n(6n+1) generators of Sp(6n) are given by $x_{is}x_{js}$, $x_{is}p_{js}+p_{js}x_{is}$, $p_{is}p_{js}$ and clearly a many body Hamiltonian, which for simplicity we assume spin and isospin independent, can be written in terms of elements in the enveloping algebra of Sp(6n). Formulated in this way the problem is just as difficult as any other approach to many body theory. The mass quadrupole q_m of (2) indicates though that collective effects are associated with operators that are scalar in O(n). This suggests that collective Hamiltonians can be constrained to invariant expressions in O(n) i.e. functions of elements in the enveloping algebra of Sp(6) in which the 21 generators of this group have the form

$$B^+_{ij}=\sum_{s=1}^{n}\eta_{is}\eta_{js},\quad C_{ij}=(\tfrac{1}{2})\sum_{s=1}^{n}(\eta_{is}\xi_{js}+\xi_{js}\eta_{is}),\quad B_{ij}=\sum_{s=1}^{n}\xi_{is}\xi_{js} \qquad (3)$$

where, in appropriate units, $\eta_{is}=(1/\sqrt{2})(x_{is}-ip_{is})$, $\xi_{is}=(1/\sqrt{2})(x_{is}+ip_{is})$ are creation and annihilation operators.

It is clear then that the eigenstates of these collective Hamiltonians will be constrained to a definite irreducible representation (irrep) of O(n). How can we obtain this irrep for a given nucleus? One way is from shell model considerations, by filling compactly with nucleons the levels of an harmonic oscillator common potential. Thus, for example[11,25], the partitions characterizing the irreps of O(n) for ^4He, ^{16}O and ^{20}Ne are respectively (0,0,0), (4,4,4) and (12,4,4). Alternative procedures for medium and heavy nuclei have also been proposed[27),10)]

The constraints indicated above, in a fashion similar to the analogy mentioned for Euclidean space, drastically reduce the size of our Hilbert space[26] but complicate the geometry of our problem. As now our collective state corresponds to a given irrep $(\omega_1\omega_2\omega_3)$ of O(n), the complementarity relation mentioned previously[29] indicates that it is also characterized by the irrep[26]

$$[(n/2)+\omega_3,\ (n/2)+\omega_2,\ (n/2)+\omega_1] \qquad (4)$$

of Sp(6). Our program then is to determine the matrix elements of our collective Hamiltonian with respect to states characterized by the irrep (4) of Sp(6) as well as by irreps of subgroups of this group.

To implement this program we need, as indicated above, first to get the matrix elements of the generators of Sp(6) with respect to states characterized by the chain of groups $Sp(6)\supset Sp(2)\times O(3)$ where the generators of O(3) are $L_{ij}=C_{ij}-C_{ji}$ while those of Sp(2) become

$$I_+ = (\tfrac{1}{2})\sum_{i=1}^{3} B^+_{ii} \equiv (\tfrac{1}{2})\overline{B}^+, \quad I_o = (\tfrac{1}{2})\sum_{i=1}^{3} C_{ii}, \quad I_- = (\tfrac{1}{2})\sum_{i=1}^{3} B_{ii} \qquad (5)$$

with $2I_o$ being the Hamiltonian of a 3n dimensional oscillator.

To get the states we start by considering the irrep (4) of Sp(6) corresponding to the scalar representation of O(n) i.e. $(\omega_1\omega_2\omega_3)=(000)$. In this case the state of highest weight L in O(3) and lowest weight Λ in Sp(2) can be expressed as a polynomial $P_{\Lambda L}$ in the B^+_{ij} of (3), of degree $\Lambda-(3n/4)$ in these variables, and acting on the ground state $|0)$ of the 3n dimensional oscillator, that satisfies the equations

$$I_o P_{\Lambda L}|0) = \Lambda P_{\Lambda L}|0), \quad I_- P_{\Lambda L}|0) = 0 \qquad (6)$$

$$L_o = P_{\Lambda L}|0) = L P_{\Lambda L}|0), \quad L_+ P_{\Lambda L}|0) = 0 \qquad (7)$$

where $L_0 = L_{12}$, $L_+ = L_{23} + iL_{31}$. This state is the more general needed as for matrix elements involving arbitrary weights in O(3) and Sp(2)=SU(1,1) we can use the Wigner Eckart theorem for these groups, as their Clebsch-Gordan coefficients are well known.

The determination of the $P_{\Lambda L}$ can then be carried out[28] in an entirely parallel fashion to the procedure followed for the exact evaluation of the states in the Bohr-Mottelson vibrational Hamiltonian.[18,19] By Dragts theorem[29], the polynomials $P_{\Lambda L}(B^+_{ij})$ can be transformed into $P_{\Lambda L}(q_{ij})\exp(-q/2)$ where $q_{ij} = \sum_{s=1}^{n} x_{is}x_{js}$, and $q = \sum_{i=1}^{3} q_{ii}$, and so matrix elements of q_{ij} with respect to our states can be determined, from which those of the other generators of Sp(6) follow immediately.[26]

Our program can then be implemented for the irrep [(n/2), (n/2), (n/2)] of Sp(6). To carry it out for the general irrep (4) of this group we need to apply $P_{\Lambda L}$ not the ground state $|0)$ of the harmonic oscillator but to the intrinsic state associated with the Slater determinant for the nucleus in question that is a lowest weight state of Sp(6). This is trivial for doubly closed shell nuclei and more involved in the open shell case where an appropriate algorithm is being developed.

We note also[25,26] that the formalism simplifies drastically when $n\to\infty$, similarly to what happens in our Euclidean space analogy when $a\to\infty$. Thus in this case the Casimir operator of the Sp(2) group and that of its O(2) subgroup given by $\overline{C}$ of (4), can be correlated respectively with the Bohr-Mottelson vibrational Hamiltonian and the linear Casimir operator of U(6) in IBA.

Thus the geometry of collective motions in nuclei, along the lines of the Sp(6)$\supset$Sp(2)xO(3) chain of groups, is at a stage in which significant applications to nuclei can be expected in the near future. These could be compared both to experiment as well as to microscopic models related to different subgroups[10,11,15,17] of the fundamental Sp(6) group.

REFERENCES

1. N. Bohr, Nature 137, 344 (1936)
2. A. Bohr and B. Mottelson, K. Dan. Vidensk. Selsk. Mat. Fys. Med. 27, No. 16 (1953)
3. J.M. Eisenberg and W. Greiner, Nuclear Models (North Holland Publishing Co., Amsterdam 1970) p. 56
4. S. Goshen and H.J. Lipkin, Ann. Phys. (N.Y.) 6, 301 (1959)
5. W. Zickendraht, J. Math. Phys. 12, 1663 (1971)
6. A.Ya. Dzublik, V.I. Ovcharenko, A.I. Steshenko and G.F. Filippov, Yad. Fiz. 15, 869 (1972), Sov. J. Nucl. Phys. 15, 487 (1972)
7. G.F. Filippov, A.I. Steshenko and V.I. Ovcharenko, Preprint ITF-72-66 (Institute of Theoretical Physics, Ukranian Academy of Sciences 1972), also in Isv. Akad. Nak (Seria Fizika) 37, 1613 (1973)
8. V. Vanagas and R.K. Kalinauskas, Yad. Fiz 18, 768 (1973); V. Vanagas, E. Nadjakov, P. Raychev, Bulg. J. Phys. 6 559 (1975)
9. M. Moshinsky and C. Quesne, J. Math. Phys. 12, 1772 (1971)
10. G.F. Filippov and V.I. Ovcharenko, Yad. Fiz. 30, 646 (1979); V.C. Vasilevsky, Ya.F. Smirnov and G.F. Filippov, Yad. Fiz. 32, 987 (1980); G.F. Filippov, V.I. Ovcharenko and Ya.F. Smirnov, Microscopic Theory of Collective Excitations in Nuclei (in Russian) (Naukova Dumka, Kiev, 1981). Further references are given in this book
11. V. Vanagas, "The Microscopic Nuclear Theory", Lecture Notes in Physics (University of Toronto Press, Toronto, 1977); V. Vanagas, "The Microscopic Theory of the Collective Motions in Nuclei" in Group Theory and its Applications in Physics - 1980 edited by T.H. Seligman (AIP, New York 1980). Further references are given in these books.
12. G. Rosensteel and D.J. Rowe, Ann. Phys. (N.Y.) 96, 1 (1976), 123, 36 (1978); 126, 198 (1980)
13. O.L. Weaver, R.Y. Cusson and L.C. Biedenharn, Ann. Phys. (N.Y.) 102, 493 (1976)
14. B. Buck, L.C. Biedenharn and R.Y. Cusson, Nucl. Phys. A317, 205 (1979)
15. G. Rosensteel and D.J. Rowe, Phys. Rev. Lett. 38, 10 (1977); Ann. Phys. (N.Y.) 126, 343 (1980); G. Rosensteel, J. Math. Phys. 21, 924 (1980)
16. J.P. Elliott, Proc. Roy. Soc. (London) A245, 128 (1958)
17. J. Carvalho, P. Park, D.J. Rowe and G. Rosensteel, Phys. Lett. 119B, 249 (1982); M. Vassanji and D.J. Rowe, Phys. Lett. 115B, 77 (1982) and two other papers in press
18. E. Chacón, M. Moshinsky and R.T. Sharp, J. Math. Phys. 17, 668 (1976)
19. E. Chacón and M. Moshinsky, J. Math. Phys. 18, 870 (1977)
20. O. Castaños, E. Chacón, A. Frank and M. Moshinsky, J. Math. Phys. 20, 35 (1979)
21. A. Arima and F. Iachello, Ann. Phys. (N.Y.) 94, 253 (1976). 111, 201 (1978); A. Arima, F. Iachello and D. Scholten, Ann. Phys. (N.Y.) 115, 325 (1978)
22. J. Deenen and C. Quesne, J. Math. Phys. 23, 878, 2004 (1982)

23. O. Castaños and A. Frank, J. Math. Phys. (in press)
24. G. Rosensteel and D.J. Rowe, Phys. Rev. Lett. 47, 223 (1981); D.J. Rowe and G. Rosensteel, Phys. Rev. C. 25, 3236 (1982)
25. O. Castaños, E. Chacón, A. Frank, P. Hess and M. Moshinsky, J. Math. Phys. 23, 2537 (1982)
26. M. Moshinsky, J. Math. Phys. Submitted for publication
27. L. Sabaliauskas, Liet. Fiz. Rink. 19, 5 (1979)
28. O. Castaños, E. Chacón and M. Moshinsky, J. Math. Phys. Submitted for publication
29. A.J. Dragt, J. Math. Phys. 6, 533 (1965)

IS IT POSSIBLE TO SEPARATE THE KINETIC ENERGY AND THE VELOCITY FIELD INTO A COLLECTIVE AND AN INTRINSIC PART W.R.T. THE $GL_+(3,\mathbb{R})$ COLLECTIVITY? *

Z. Papadopolos** and P. Kramer
Institut für Theoretische Physik, Universität Tübingen

In the nuclear theory of collective motion the problem of separation of collective effects from intrinsic ones has been studied making use of different coordinate transformations (c.t.) which introduced some collective and intrinsic coordinates. The c.t. we are interested in was defined in ref. 1 and 2. The result of this c.t. was also obtained in ref. 3, 4, 5, and 6 and discussed from the point of view of the microscopic realization of cm(3) (or MQC) model. These authors introduced the c.t.

$$(\xi_{is} | i=1,2,3\ s=1,\dots,n) \to (\omega_\mu, \xi_i, \beta_\tau | \mu,\ i=1,2,3\ \tau=1,\dots,3n-6)$$

where ξ_{is} are 3n coordinates for n relative vectors (n≡A-1, where A is the number of nucleons and the center of mass is considered separated) in configuration space, and the c.t. is defined as $\xi_{is} = \sum_{k=1}^{3} o_{ik}(\omega)\, \mathring{\xi}_k o'_{ks}(\beta)$ where

$$o(\omega)\in SO(3);\ \mathring{\xi}_1 > \mathring{\xi}_2 > \mathring{\xi}_3 > 0;\ o'(\beta) \in SO(n)/SO(n-3) \equiv R_C$$

the right coset space. This c.t. in our opinion[7)] is uniquely defined by the orbit analysis of the configuration space for n relative vectors w.r.t. the $SO(3)\times SO(n)$ group. The quadratic part in the momenta of the expression for the kinetic energy in new coordinates is[7)]

$$T_{(Q)} = \frac{1}{2m} \sum_{g<v}^{3} \frac{\mathring{\xi}_g^2 + \mathring{\xi}_v^2}{(\mathring{\xi}_g^2 - \mathring{\xi}_v^2)^2} \hat{L}_{gv} + \frac{1}{2m} \sum_{g<v}^{3} \frac{\mathring{\xi}_g^2 + \mathring{\xi}_v^2}{(\mathring{\xi}_g^2 - \mathring{\xi}_v^2)^2} \hat{\mathcal{L}}_{gv}$$

$$- \frac{1}{2m} \sum_{g<v}^{3} \frac{4\mathring{\xi}_g \mathring{\xi}_v}{(\mathring{\xi}_g^2 - \mathring{\xi}_v^2)^2} \hat{L}_{gv} \hat{\mathcal{L}}_{gv} + \frac{1}{2m} \sum_{v=1}^{3} \hat{t}_v + \frac{1}{2m} \sum_{g=1}^{3} \sum_{v=4}^{n} \mathring{\xi}_g^{-2} \hat{\mathcal{Q}}_{gv}^2$$

where $\hat{L}_{gv} = i\hbar \sum_{\mu=1}^{3} {}^R\Theta_{gv,\mu}(\omega) \frac{\partial}{\partial \omega_\mu}$ are angular momenta of the whole system referred to the body[8)], $\hat{\mathcal{L}}_{gv} = i\hbar \sum_{\tau=1}^{3n-6} {}^L\Theta_{gv,\tau}(\beta) \frac{\partial}{\partial \beta_\tau}$ g<v=1,2,3 is the vortex spin[9)], $\hat{t}_v = -i\hbar \frac{\partial}{\partial \mathring{\xi}_v}$ are the vibrational momenta and $\hat{\mathcal{Q}}_{gv} = i\hbar \sum_{\tau=1}^{3n-6} {}^L\Theta_{gv,\tau}(\beta) \frac{\partial}{\partial \beta_\tau}$ g=1,2,3 v=4,...,n are the 3n-9 "intrinsic" momenta. The functions Θ's are the derivatives of the structure function of a group w.r.t. their parameters[10,7)]. The letters L and R denote left or right action on groups or coset spaces[7)]. This result is identical to those given

* Work supported by Deutsche Forschungsgemeinschaft
** Permanent adress: Institute of Physics, Belgrade, Jugoslavia

in ref. 1-6. From the point of view of the $GL_+(3, \mathbb{R})$ collectivity or cm(3) (or MQC) model it is claimed in ref. 3-6 that the sum of the first four terms in the kinetic energy is the collective part and the last term is the intrinsic one. Even if we agree that certain three coset parameters of ${}^{R}C$ can be chosen to be the collective SO(3) group parameters (SO(3) < SO(n)) which complete the $GL_+(3, \mathbb{R})$ manifold and on which (SO(3)) the vortex spin is defined as the left action generators, the 3n-9 "intrinsic" $\hat{\mathcal{Q}}$ differential operators still contain the partial derivatives w.r.t. all 3n-6 ${}^{R}C$ coordinates. In order to introduce the intrinsic momenta $\hat{\tilde{\mathcal{Q}}}$ as the differential operators which do not contain partial derivatives w.r.t. the $GL_+(3, \mathbb{R})$ collective coordinates, we propose the following c.t. defined by the orbit analysis w.r.t. the $GL_+(3, \mathbb{R})\times SO(n)$ group[7]:

$$(\xi_{is}|i=1,2,3\ s=1,\dots,n) \to (\omega_\mu, \mathring{\xi}_i, \omega'_{\mu'}, \beta_\tau | \mu, i, \mu'=1,2,3\ \tau=1,\dots,3n-9)\ .$$

This c.t. is explicitly defined by $\xi_{is} = \sum_{k,l}^{3} o_{ik}(\omega)\mathring{\xi}_k\, o'_{kl}(\omega')o''_{ls}(\beta)$, where $o(\omega)$ and $\mathring{\xi}_k$ are the same as in the previous case, $o'(\omega') \in SO(3) < GL_+(3, \mathbb{R})$ and $o''(\beta) \in SO(n)/SO(3)\times SO(n-3)$, the right coset space. The quadratic part in the momenta of the kinetic energy now takes the form

$$T_{(Q)} = \frac{1}{2m}\sum_{g<v}^{3} \frac{\mathring{\xi}_g^2+\mathring{\xi}_v^2}{(\mathring{\xi}_g^2-\mathring{\xi}_v^2)^2}\hat{L}_{gv}^2 + \frac{1}{2m}\sum_{g<v}^{3} \frac{\mathring{\xi}_g^2+\mathring{\xi}_v^2}{(\mathring{\xi}_g^2-\mathring{\xi}_v^2)^2}\hat{\mathcal{L}}_{gv}^2$$

$$- \frac{1}{2m}\sum_{g<v}^{3} \frac{4\mathring{\xi}_g\mathring{\xi}_v}{(\mathring{\xi}_g^2-\mathring{\xi}_v^2)^2}\hat{L}_{gv}\hat{\mathcal{L}}_{gv} + \frac{1}{2m}\sum_{v=1}^{3}\hat{t}_v^2$$

$$+ \frac{1}{2m}\sum_{i}^{3}\sum_{k=4}^{n}\sum_{g<v}^{3} 2\mathring{\xi}_i^{-2}\, {}^{L}\Theta^{b(\omega')}_{ik,gv}(\beta)\ \hat{\tilde{\mathcal{L}}}\ \hat{\tilde{\mathcal{Q}}}_{ik}$$

$$+ \frac{1}{2m}\sum_{i}^{3}\sum_{k=4}^{n}\sum_{g<v}\sum_{\tilde{g}<\tilde{v}} \mathring{\xi}_i^{-2}\, {}^{L}\Theta^{b(\omega')}_{ik,gv}(\beta)\, {}^{L}\Theta^{b(\omega')}_{ik,\tilde{g}\tilde{v}}(\beta)\ \hat{\tilde{\mathcal{L}}}_{gv}\ \hat{\tilde{\mathcal{L}}}_{\tilde{g}\tilde{v}}$$

$$+ \frac{1}{2m}\sum_{i=1}^{3}\sum_{k=4}^{n} \mathring{\xi}_i^{-2}\hat{\tilde{\mathcal{Q}}}_{ik}^2$$

where $$\hat{\tilde{\mathcal{L}}}_{gv} = i\hbar\sum_{\mu'=1}^{3} {}^{L}\Theta_{gv,\mu'}(\omega')\frac{\partial}{\partial\omega'_{\mu'}} \qquad g<v=1,2,3$$

$$\hat{\tilde{\mathcal{Q}}}_{ik} := i\hbar\sum_{m=1}^{3} o'_{im}(\omega')\sum_{\tau=1}^{3n-9} {}^{L}\Theta_{mk,\tau}(\beta)\frac{\partial}{\partial\beta_\tau} \qquad i=1,2,3\quad k=4,\dots,n$$

$${}^{L}\Theta^{b(\omega')}_{ik,gv}(\beta) := \sum_{m,l,a}^{3} o'_{im}(\omega')o'_{gl}(\omega')o'_{va}(\omega')\,{}^{L}\tilde{\Theta}_{mk,la}(\beta)$$

$i=1,2,3 \qquad g<v=1,2,3 \qquad k=4,\ldots,n$.

A similar conclusion one obtains for the velocity field[7] in the sense of ref. 11. The separation could be obtained if ${}^{L}\Theta^{b(\omega')}_{ik,gv}(\beta)=0$, which is equivalent to the statement that ${}^{L}\Theta_{2\,1}(\beta)=0$ or ${}^{L}\psi_{2\,1}(\beta)=0$ ($2\equiv ik$, $1\equiv gv$ for the same range of indices). We define only ${}^{L}\Theta_{2\,1}(\beta)=\frac{\partial}{\partial\beta_2}\,\phi_1(\tilde{\beta},\delta\beta)|\tilde{\beta}=0,\delta=0$ where ϕ is the structure function of SO(n)[10,7]. The result is not that surprising because the constraint ${}^{L}\psi_{2\,1}(\beta)=0$ can be shown to be identical to the nonholonomic constraint postulated in ref. 11 from the very beginning. In this reference the complete separation of T and the velocity field was achieved. But, the constraint ${}^{L}\psi_{2\,1}(\beta)=0$ does not follow from the properties of the group SO(n, $\mathbb{R}$) and we conclude that the total separation of the collective kinetic energy and the velocity field w.r.t. $GL_{+}(3,\mathbb{R})$ is not obtained.

1) Zickendraht, W., J. Math. Phys. 12 (1971) 1663.
2) Dzyublik, A.Ya, Ovcharenko, V.I., Steshenko, A.I. and Filippov, G.F., Sov. J. Nucl. Phys. 15 (1972) 487.
3) Buck, B., Biedenharn, L.C. and Cusson, R.Y., Nucl. Phys. A 317 (1979) 205.
4) Rowe, D.J. and Rosensteel, G., J. Math. Phys. 20 (1979) 465.
5) Rowe, D.J. and Rosensteel, G., Ann. Phys. (N.Y.) 126 (1980) 198.
6) Rowe, D.J., "The Microscopic Realization of Nuclear Collective Models", in Group Theory and its Applications in Physics, ed. T.H. Seligman (AIP Conference Proceedings 71, New York, 1981).
7) Papadopolos, Z., Dissertation, Universität Tübingen, 1983.
8) Biedenharn, L.C. and Brussaard, P.J., "Coulomb Excitation", Clarendon Press, Oxford, 1965.
9) Weaver, O.L., Cusson, R.Y. and Biedenharn, L.C., Ann. Phys. (N.Y.) 102 (1976) 493.
10) Gilmore, R., "Lie Groups, Lie Aögebras and Some of Their Applications", J.Wiley & Sons, New York, 1974.
11) Gulshani, P. and Rowe, D.J., Canad. J. Phys. 54 (1976) 970.

SYMMETRIES IN CONDENSED MATTER PHYSICS
AND
STATISTICAL MECHANICS

COMPUTER GENERATED CLEBSCH-GORDAN (C-G) COEFFICIENTS FOR SPACE GROUPS

B.L. Davies[1] and R. Dirl[2]

[1]School of Mathematics and Computer Science, University College of North Wales BANGOR LL57 2UW, Wales, U.K.

[2]Institut für Theoretische Physik, TU Wien, A-1040 WIEN, Karlsplatz 13; Austria

We have shown, by computer, that for *all* Kronecker products $\Lambda^{(\kappa,q)} \otimes \Lambda^{(\kappa',q')}$ of (Miller and Love[1]) induced matrix unirreps $\Lambda^{(\kappa'',q'')} = \Gamma^{(\kappa'',q'')}\uparrow G$ in *all* 230 space groups G, *special solutions of the multiplicity problem exist*, so that *all* the elements of *every* unitary C-G matrix $C^{(\kappa,q)(\kappa',q')}$ can be computed using a single, explicit formula[2] in terms of only the allowed matrix unirreps $\Gamma^{(\kappa'',q'')}$ of the little groups $G^{q''}$; $q'' = q,q',q_o$ occurring in the Kronecker product and its C-G series decomposition:

$$(C^{(\kappa,q)(\kappa',q')})^{\dagger} (\Lambda^{(\kappa,q)} \otimes \Lambda^{(\kappa',q')}) C^{(\kappa,q)(\kappa',q')} =$$

$$= \sum_{\kappa_o,(\underline{\sigma},\underline{\sigma}')q_o} \oplus\, m^{(\underline{\sigma},\underline{\sigma}')}_{(\kappa,q)(\kappa',q');(\kappa_o,q_o)} \Lambda^{(\kappa_o,q_o)} \tag{1}$$

Full details of the algorithm on which the computer program is based will be published elsewhere (Davies[3]).

Let T denote the abelian, invariant subgroup of translations of G then $P \simeq G/T$, $P^q \simeq G^q/T$. The dimension of $\Lambda^{(\kappa,q)}$ is $n_\kappa|P:P^q|$ where n_κ is the dimension of $\Gamma^{(\kappa,q)}$. The C-G coefficients are the elements of the C-G matrix:

$$C^{(\kappa,q)(\kappa',q')}_{\underline{\tau}d,\underline{\tau}'d';(\kappa_o,(\underline{\sigma},\underline{\sigma}')q_o)\; w\, \underline{\sigma}_o\, j} \tag{2}$$

where the rows are indexed lexicographically by $\underline{\tau}d,\underline{\tau}'d'$ such that $\underline{\tau} \in P:P^q$, $d = 1,2, .. n_\kappa$, $\underline{\tau}' \in P:P^{q'}$, $d' = 1,2, .. n_{\kappa'}$ and the columns are indexed by $(\kappa_o,(\underline{\sigma},\underline{\sigma}')q_o)\; w\, \underline{\sigma}_o\, j$ for those $(\kappa_o,(\underline{\sigma},\underline{\sigma}')q_o)$ such that in Eq.(1) $m^{\cdot\cdot}_{\cdot\cdot;\cdot} > 0$, the *multiplicity index* $w = 1,2, .. m^{(\underline{\sigma},\underline{\sigma}')}_{(\kappa,q)(\kappa',q');(\kappa_o,q_o)}$, (N.B. For a given Kronecker product, the left coset representatives $\underline{\tau} \in P:P^q$, $\underline{\tau}' \in P:P^{q'}$ and $\underline{\sigma}_o \in P:P^{q_o}$ are chosen at the outset and *must remain fixed*.) For a given Kronecker product, the solution of the multiplicity problem consists in identifying the multiplicity index w with special column indices $(\underline{\sigma}_v,c_v;\underline{\sigma}'_v,c'_v)$ of the Kronecker product so that

$$w = (\underline{\sigma}_v,c_v;\underline{\sigma}'_v,c'_v) \;;\quad v = 1,2, .. m^{(\underline{\sigma},\underline{\sigma}')}_{(\kappa,q)(\kappa',q');(\kappa_o,q_o)}. \tag{3}$$

Example: We take as our example the Kronecker square $\Lambda^{(3,P)} \otimes \Lambda^{(3,P)}$ in the non-symmorphic, body-centred cubic space group G = Ia3d (230). The allowed matrix unirreps, etc. may be found in Ref.1. The allowed matrix unirrep $\Gamma^{(3,P)}$ of G^P is 4-dimensional and the index of G^P in G is 2 so that $\Lambda^{(3,P)} \otimes \Lambda^{(3,P)}$ is of dimension $(2\times4)^2 = 64$.

The C-G series decomposition is[4]:

$$\Lambda^{(3,P)} \otimes \Lambda^{(3,P)} \simeq \Lambda^{(1\pm,GM)} \oplus \Lambda^{(2\pm,GM)} \oplus \Lambda^{(3\pm,GM)} \oplus 2*\Lambda^{(4\pm,GM)} \oplus 2*\Lambda^{(5\pm,GM)} \oplus$$
$$\oplus\, 2*\Lambda^{(1,H)} \oplus \Lambda^{(2,H)} \oplus \Lambda^{(3,H)} \oplus 4*\Lambda^{(4,H)} \qquad (4)$$

where q_o takes on the values: q_o = GM = 0 and q_o = H, and $G^H = G^{GM} = G$. The pairs of fixed left coset representatives $(\underline{\sigma},\underline{\sigma}')$ associated with q_o are $(\underline{1},\underline{13})$ for q_o = GM and $(\underline{1},\underline{1})$ for q_o = H (see Ref.4).

The special solutions of the multiplicity problem are given in Table 1. Using Table 1 and Eq.(I.9) of Dirl[2], the complete C-G matrix was computed and is displayed in Table 2.

A notable feature of this example is the (for space groups) high value of the multiplicity (m = 4) occurring in the C-G series decomposition (4). Even for this high multiplicity, a special solution of the multiplicity problem has been found so that *all* the elements of the unitary C-G matrix in Table 2 are computed using a *single, explicit formula* in terms of only the (Miller and Love) *allowed matrix unirreps*[1]. Furthermore, the allowed matrix unirreps in this Kronecker product are not all *monomial* (i.e. not all rows and columns contain only one non-zero entry). Finally, the C-G matrix in Table 2 is a relatively sparse matrix with less than 10% of the entries being non-zero.

References:

1. A.P. Cracknell, B.L. Davies, S.C. Miller, W.F. Love; Kronecker Product Tables.vol.I (Plenum Press, New York, 1979)
2. R. Dirl, J.Math.Phys.,20,671(1979)
3. B.L. Davies (to be published)
4, B.L. Davies, A.P. Cracknell; Kronecker Product Tables. vol.II (Plenum Press, New York, 1979)

Table 1: *Special solutions of the multiplicity problem*

κ_o	n_{κ_o}	$(\underline{\sigma},\underline{\sigma}')q_o$	m	v	$(\underline{\sigma}_v,c_v;\underline{\sigma}'_v,c'_v)$	κ_o	n_{κ_o}	$(\underline{\sigma},\underline{\sigma}')q_o$	m	v	$(\underline{\sigma}_v,c_v;\underline{\sigma}'_v,c'_v)$
1±	1	(1,13)GM	1	1	(1,1;13,1)	1	2	(1,1)H	2	1	(1,1;1,3)
2±	1	(1,13)GM	1	1	(1,1;13,1)	1	2	(1,1)H	2	2	(1,3;1,1)
3±	2	(1,13)GM	1	1	(1,1;13,4)	2	2	(1,1)H	1	1	(1,1;1,2)
4±	3	(1,13)GM	2	1	(1,1;13,1)	3	2	(1,1)H	1	1	(1,3;1,4)
4±	3	(1,13)GM	2	2	(1,1;13,4)	4	6	(1,1)H	4	1	(1,1;1,1)
5±	3	(1,13)GM	2	1	(1,1;13,1)	4	6	(1,1)H	4	2	(1,1;1,4)
5±	3	(1,13)GM	2	2	(1,1;13,4)	4	6	(1,1)H	4	3	(1,3;1,2)
						4	6	(1,1)H	4	4	(1,3;1,3)

Using the special solutions in Table 1, the following C-G matrix was constructed: (see Table 2)

Table 2: Clebsch-Gordan matrix $C^{(3,P)(3,P)}$

$(\underline{\sigma},\underline{\sigma}')\vec{q}_0$				$(\underline{1},\underline{13})\vec{GM}$																																$(\underline{1},\underline{1})\vec{H}$																															
			κ_0	1+	2+	3+		4+			4+			5+			5+			1−	2−	3−		4−			4−			5−			5−			1		1		2		3		4						4						4						4					
			v	1	1	1		1			2			1			2			1	1	1		1			2			1			2			1		2		1		1		1						2						3						4					
			$\underline{\sigma}_0$	$\underline{1}$	$\underline{1}$	$\underline{1}$		$\underline{1}$			$\underline{1}$			$\underline{1}$			$\underline{1}$			$\underline{1}$	$\underline{1}$	$\underline{1}$		$\underline{1}$			$\underline{1}$			$\underline{1}$			$\underline{1}$			$\underline{1}$		$\underline{1}$		$\underline{1}$		$\underline{1}$		$\underline{1}$						$\underline{1}$						$\underline{1}$						$\underline{1}$					
			j	1	1	1	2	1	2	3	1	2	3	1	2	3	1	2	3	1	1	1	2	1	2	3	1	2	3	1	2	3	1	2	3	1	2	1	2	1	2	1	2	1	2	3	4	5	6	1	2	3	4	5	6	1	2	3	4	5	6	1	2	3	4	5	6
$\underline{1}$	1	$\underline{1}$	1	.	.	.	.	.	.	.	.	.	.	.	.	.	.	.	.	.	.	.	.	.	.	.	.	.	.	.	.	.	.	.	.	.	.	.	.	.	.	.	.	1	.	U^*	.	.	.	.	.	.	.	.	.	.	.	.	.	.	.	.	.	.	$\bar{1}$	.	$\bar{U}$
			2	.	.	.	.	.	.	.	.	.	.	.	.	.	.	.	.	.	.	.	.	.	.	.	.	.	.	.	.	.	.	.	.	.	.	.	.	1	.	.	$\bar{1}$	.	U	.	.	.	.	.	.	.	.	.	.	.	.	.	.	.	.	.	.	.	.	W^*	.
			3	.	.	.	.	.	.	.	.	.	.	.	.	.	.	.	.	.	.	.	.	.	.	.	.	.	.	.	.	.	.	.	.	1	.	.	$\bar{J}$	.	.	.	.	.	.	.	.	.	.	.	1	.	.	.	.	.	.	.	.	$\bar{1}$	.	.	.	.	.	.	.
			4	.	.	.	.	.	.	.	.	.	.	.	.	.	.	.	.	.	.	.	.	.	.	.	.	.	.	.	.	.	.	.	.	.	.	.	.	.	.	.	.	.	.	.	.	.	.	1	.	J	.	.	.	.	.	.	$\bar{J}$	.	$\bar{1}$	.	.	.	.	.	.
		$\underline{13}$	1	1	1	.	.	1	.	.	.	.	.	1	.	.	.	.	.	1	1	.	.	1	.	.	.	.	.	1	.	.	.	.	.	.	.	.	.	.	.	.	.	.	.	.	.	.	.	.	.	.	.	.	.	.	.	.	.	.	.	.	.	.	.	.	.
			2	.	.	.	.	.	J	1	.	.	.	.	J	1	.	.	.	.	.	.	.	.	J	1	.	.	.	.	J	1	.	.	.	.	.	.	.	.	.	.	.	.	.	.	.	.	.	.	.	.	.	.	.	.	.	.	.	.	.	.	.	.	.	.	.
			3	.	.	.	.	.	.	.	.	$\bar{W}$	U^*	.	.	.	.	$\bar{W}$	U^*	.	.	.	.	.	.	.	.	$\bar{W}$	U^*	.	.	.	.	$\bar{W}$	U^*	.	.	.	.	.	.	.	.	.	.	.	.	.	.	.	.	.	.	.	.	.	.	.	.	.	.	.	.	.	.	.	.
			4	.	.	1	.	.	.	.	1	.	.	.	.	.	1	.	.	.	.	1	.	.	.	.	1	.	.	.	.	.	1	.	.	.	.	.	.	.	.	.	.	.	.	.	.	.	.	.	.	.	.	.	.	.	.	.	.	.	.	.	.	.	.	.	.
	2	$\underline{1}$	1	.	.	.	.	.	.	.	.	.	.	.	.	.	.	.	.	.	.	.	.	.	.	.	.	.	.	.	.	.	.	.	.	.	.	.	.	$\bar{1}$	.	.	1	.	U	.	.	.	.	.	.	.	.	.	.	.	.	.	.	.	.	.	.	.	.	W^*	.
			2	.	.	.	.	.	.	.	.	.	.	.	.	.	.	.	.	.	.	.	.	.	.	.	.	.	.	.	.	.	.	.	.	.	.	.	.	.	.	.	.	1	.	$\bar{U}^*$	.	.	.	.	.	.	.	.	.	.	.	.	.	.	.	.	.	.	1	.	$\bar{U}$
			3	.	.	.	.	.	.	.	.	.	.	.	.	.	.	.	.	.	.	.	.	.	.	.	.	.	.	.	.	.	.	.	.	.	.	.	.	.	.	.	.	.	.	.	.	.	.	J	.	1	.	.	.	.	.	.	$\bar{1}$	.	$\bar{J}$	.	.	.	.	.	.
			4	.	.	.	.	.	.	.	.	.	.	.	.	.	.	.	.	.	.	.	.	.	.	.	.	.	.	.	.	.	.	.	.	J	.	.	1	.	.	.	.	.	.	.	.	.	.	.	$\bar{J}$	.	.	.	.	.	.	.	.	J	.	.	.	.	.	.	.
		$\underline{13}$	1	.	.	.	.	.	1	J	.	.	.	.	1	J	.	.	.	.	.	.	.	.	1	J	.	.	.	.	1	J	.	.	.	.	.	.	.	.	.	.	.	.	.	.	.	.	.	.	.	.	.	.	.	.	.	.	.	.	.	.	.	.	.	.	.
			2	J	J	.	.	$\bar{J}$	.	.	.	.	.	$\bar{J}$	.	.	.	.	.	J	J	.	.	$\bar{J}$	.	.	.	.	.	$\bar{J}$	.	.	.	.	.	.	.	.	.	.	.	.	.	.	.	.	.	.	.	.	.	.	.	.	.	.	.	.	.	.	.	.	.	.	.	.	.
			3	.	.	$\bar{1}$	.	.	.	.	1	.	.	.	.	.	1	.	.	.	.	$\bar{1}$	.	.	.	.	1	.	.	.	.	.	1	.	.	.	.	.	.	.	.	.	.	.	.	.	.	.	.	.	.	.	.	.	.	.	.	.	.	.	.	.	.	.	.	.	.
			4	.	.	.	.	.	.	.	.	W	U^*	.	.	.	.	W	U^*	.	.	.	.	.	.	.	.	W	U^*	.	.	.	.	W	U^*	.	.	.	.	.	.	.	.	.	.	.	.	.	.	.	.	.	.	.	.	.	.	.	.	.	.	.	.	.	.	.	.
	3	$\underline{1}$	1	.	.	.	.	.	.	.	.	.	.	.	.	.	.	.	.	.	.	.	.	.	.	.	.	.	.	.	.	.	.	.	.	.	$\bar{J}$	$\bar{1}$	.	.	.	.	.	.	.	.	.	.	.	.	.	.	.	$\bar{J}$	.	.	$\bar{J}$	.	.	.	.	.	.	.	.	.	.
			2	.	.	.	.	.	.	.	.	.	.	.	.	.	.	.	.	.	.	.	.	.	.	.	.	.	.	.	.	.	.	.	.	.	.	.	.	.	.	.	.	.	.	.	.	.	.	.	.	.	$\bar{J}$	.	1	1	.	$\bar{J}$	.	.	.	.	.	.	.	.	.
			3	.	.	.	.	.	.	.	.	.	.	.	.	.	.	.	.	.	.	.	.	.	.	.	.	.	.	.	.	.	.	.	.	.	.	.	.	.	.	.	.	.	.	.	1	.	U^*	.	.	.	.	.	.	.	.	.	.	.	.	1	.	U	.	.	.
			4	.	.	.	.	.	.	.	.	.	.	.	.	.	.	.	.	.	.	.	.	.	.	.	.	.	.	.	.	.	.	.	.	.	.	.	.	.	1	1	.	.	.	.	.	U	.	.	.	.	.	.	.	.	.	.	.	.	.	.	$\bar{W}^*$	.	.	.	.
		$\underline{13}$	1	.	.	.	.	.	.	.	.	$\bar{U}^*$	$\bar{W}$	.	.	.	.	$\bar{U}^*$	$\bar{W}$	.	.	.	.	.	.	.	.	$\bar{U}^*$	$\bar{W}$	.	.	.	.	$\bar{U}^*$	$\bar{W}$	.	.	.	.	.	.	.	.	.	.	.	.	.	.	.	.	.	.	.	.	.	.	.	.	.	.	.	.	.	.	.	.
			2	.	.	.	1	.	.	.	$\bar{1}$	.	.	.	.	.	1	.	.	.	.	.	1	.	.	.	1	.	.	.	.	.	$\bar{1}$	.	.	.	.	.	.	.	.	.	.	.	.	.	.	.	.	.	.	.	.	.	.	.	.	.	.	.	.	.	.	.	.	.	.
			3	J	$\bar{J}$	.	.	$\bar{J}$	.	.	.	.	.	J	.	.	.	.	.	$\bar{J}$	J	.	.	J	.	.	.	.	.	$\bar{J}$	.	.	.	.	.	.	.	.	.	.	.	.	.	.	.	.	.	.	.	.	.	.	.	.	.	.	.	.	.	.	.	.	.	.	.	.	.
			4	.	.	.	.	.	$\bar{J}$	1	.	.	.	.	J	$\bar{1}$	.	.	.	.	.	.	.	.	J	$\bar{1}$	.	.	.	.	$\bar{J}$	1	.	.	.	.	.	.	.	.	.	.	.	.	.	.	.	.	.	.	.	.	.	.	.	.	.	.	.	.	.	.	.	.	.	.	.
	4	$\underline{1}$	1	.	.	.	.	.	.	.	.	.	.	.	.	.	.	.	.	.	.	.	.	.	.	.	.	.	.	.	.	.	.	.	.	.	.	.	.	.	.	.	.	.	.	.	.	.	.	.	.	.	1	.	$\bar{J}$	$\bar{J}$	.	1	.	.	.	.	.	.	.	.	.
			2	.	.	.	.	.	.	.	.	.	.	.	.	.	.	.	.	.	.	.	.	.	.	.	.	.	.	.	.	.	.	.	.	.	1	J	.	.	.	.	.	.	.	.	.	.	.	.	.	.	.	$\bar{1}$	.	.	$\bar{1}$	.	.	.	.	.	.	.	.	.	.
			3	.	.	.	.	.	.	.	.	.	.	.	.	.	.	.	.	.	.	.	.	.	.	.	.	.	.	.	.	.	.	.	.	.	.	.	.	.	$\bar{1}$	$\bar{1}$	.	.	.	.	.	U	.	.	.	.	.	.	.	.	.	.	.	.	.	.	$\bar{W}^*$	.	.	.	.
			4	.	.	.	.	.	.	.	.	.	.	.	.	.	.	.	.	.	.	.	.	.	.	.	.	.	.	.	.	.	.	.	.	.	.	.	.	.	.	.	.	.	.	.	1	.	$\bar{U}^*$	.	.	.	.	.	.	.	.	.	.	.	.	$\bar{1}$	.	U	.	.	.
		$\underline{13}$	1	.	.	.	$\bar{1}$	.	.	.	$\bar{1}$	.	.	.	.	.	1	.	.	.	.	.	$\bar{1}$	.	.	.	1	.	.	.	.	.	$\bar{1}$	.	.	.	.	.	.	.	.	.	.	.	.	.	.	.	.	.	.	.	.	.	.	.	.	.	.	.	.	.	.	.	.	.	.
			2	.	.	.	.	.	.	.	.	$\bar{U}^*$	$\bar{W}$	.	.	.	.	U^*	W	.	.	.	.	.	.	.	.	U^*	W	.	.	.	.	$\bar{U}^*$	$\bar{W}$	.	.	.	.	.	.	.	.	.	.	.	.	.	.	.	.	.	.	.	.	.	.	.	.	.	.	.	.	.	.	.	.
			3	.	.	.	.	.	1	$\bar{J}$	.	.	.	.	$\bar{1}$	J	.	.	.	.	.	.	.	.	$\bar{1}$	J	.	.	.	.	1	$\bar{J}$	.	.	.	.	.	.	.	.	.	.	.	.	.	.	.	.	.	.	.	.	.	.	.	.	.	.	.	.	.	.	.	.	.	.	.
			4	$\bar{1}$	1	.	.	$\bar{1}$	.	.	.	.	.	1	.	.	.	.	.	1	$\bar{1}$	.	.	1	.	.	.	.	.	$\bar{1}$	.	.	.	.	.	.	.	.	.	.	.	.	.	.	.	.	.	.	.	.	.	.	.	.	.	.	.	.	.	.	.	.	.	.	.	.	.
$\underline{13}$	1	$\underline{1}$	1	1	$\bar{1}$	.	.	$\bar{1}$	.	.	.	.	.	1	.	.	.	.	.	1	$\bar{1}$	.	.	$\bar{1}$	.	.	.	.	.	1	.	.	.	.	.	.	.	.	.	.	.	.	.	.	.	.	.	.	.	.	.	.	.	.	.	.	.	.	.	.	.	.	.	.	.	.	.
			2	.	.	.	.	.	$\bar{1}$	$\bar{J}$	.	.	.	.	1	J	.	.	.	.	.	.	.	.	$\bar{1}$	$\bar{J}$	.	.	.	.	1	J	.	.	.	.	.	.	.	.	.	.	.	.	.	.	.	.	.	.	.	.	.	.	.	.	.	.	.	.	.	.	.	.	.	.	.
			3	.	.	.	.	.	.	.	.	$\bar{U}^*$	W	.	.	.	.	U^*	$\bar{W}$	.	.	.	.	.	.	.	.	$\bar{U}^*$	W	.	.	.	.	U^*	$\bar{W}$	.	.	.	.	.	.	.	.	.	.	.	.	.	.	.	.	.	.	.	.	.	.	.	.	.	.	.	.	.	.	.	.
			4	.	.	.	1	.	.	.	$\bar{1}$	.	.	.	.	.	1	.	.	.	.	.	$\bar{1}$	.	.	.	$\bar{1}$	.	.	.	.	.	1	.	.	.	.	.	.	.	.	.	.	.	.	.	.	.	.	.	.	.	.	.	.	.	.	.	.	.	.	.	.	.	.	.	.
		$\underline{13}$	1	.	.	.	.	.	.	.	.	.	.	.	.	.	.	.	.	.	.	.	.	.	.	.	.	.	.	.	.	.	.	.	.	.	.	.	.	.	.	.	.	.	.	.	$\bar{1}$	.	U^*	.	.	.	.	.	.	.	.	.	.	.	.	$\bar{1}$	.	$\bar{U}$	.	.	.
			2	.	.	.	.	.	.	.	.	.	.	.	.	.	.	.	.	.	.	.	.	.	.	.	.	.	.	.	.	.	.	.	.	.	.	.	.	.	$\bar{1}$	1	.	.	.	.	.	U	.	.	.	.	.	.	.	.	.	.	.	.	.	.	W^*	.	.	.	.
			3	.	.	.	.	.	.	.	.	.	.	.	.	.	.	.	.	.	.	.	.	.	.	.	.	.	.	.	.	.	.	.	.	.	$\bar{1}$	J	.	.	.	.	.	.	.	.	.	.	.	.	.	.	.	1	.	.	$\bar{1}$	.	.	.	.	.	.	.	.	.	.
			4	.	.	.	.	.	.	.	.	.	.	.	.	.	.	.	.	.	.	.	.	.	.	.	.	.	.	.	.	.	.	.	.	.	.	.	.	.	.	.	.	.	.	.	.	.	.	.	.	.	1	.	J	$\bar{J}$	.	$\bar{1}$	.	.	.	.	.	.	.	.	.
	2	$\underline{1}$	1	.	.	.	.	.	$\bar{J}$	$\bar{1}$	.	.	.	.	J	1	.	.	.	.	.	.	.	.	$\bar{J}$	$\bar{1}$	.	.	.	.	J	1	.	.	.	.	.	.	.	.	.	.	.	.	.	.	.	.	.	.	.	.	.	.	.	.	.	.	.	.	.	.	.	.	.	.	.
			2	J	$\bar{J}$	.	.	J	.	.	.	.	.	$\bar{J}$	.	.	.	.	.	J	$\bar{J}$	.	.	J	.	.	.	.	.	$\bar{J}$	.	.	.	.	.	.	.	.	.	.	.	.	.	.	.	.	.	.	.	.	.	.	.	.	.	.	.	.	.	.	.	.	.	.	.	.	.
			3	.	.	.	$\bar{1}$	.	.	.	$\bar{1}$	.	.	.	.	.	1	.	.	.	.	.	1	.	.	.	$\bar{1}$	.	.	.	.	.	1	.	.	.	.	.	.	.	.	.	.	.	.	.	.	.	.	.	.	.	.	.	.	.	.	.	.	.	.	.	.	.	.	.	.
			4	.	.	.	.	.	.	.	.	$\bar{U}^*$	$\bar{W}$	.	.	.	.	U^*	W	.	.	.	.	.	.	.	.	$\bar{U}^*$	$\bar{W}$	.	.	.	.	U^*	W	.	.	.	.	.	.	.	.	.	.	.	.	.	.	.	.	.	.	.	.	.	.	.	.	.	.	.	.	.	.	.	.
		$\underline{13}$	1	.	.	.	.	.	.	.	.	.	.	.	.	.	.	.	.	.	.	.	.	.	.	.	.	.	.	.	.	.	.	.	.	.	.	.	.	.	1	$\bar{1}$	.	.	.	.	.	U	.	.	.	.	.	.	.	.	.	.	.	.	.	.	W^*	.	.	.	.
			2	.	.	.	.	.	.	.	.	.	.	.	.	.	.	.	.	.	.	.	.	.	.	.	.	.	.	.	.	.	.	.	.	.	.	.	.	.	.	.	.	.	.	.	1	.	$\bar{U}^*$	.	.	.	.	.	.	.	.	.	.	.	.	1	.	$\bar{U}$	.	.	.
			3	.	.	.	.	.	.	.	.	.	.	.	.	.	.	.	.	.	.	.	.	.	.	.	.	.	.	.	.	.	.	.	.	.	.	.	.	.	.	.	.	.	.	.	.	.	.	.	.	.	J	.	1	$\bar{1}$	.	$\bar{J}$	.	.	.	.	.	.	.	.	.
			4	.	.	.	.	.	.	.	.	.	.	.	.	.	.	.	.	.	.	.	.	.	.	.	.	.	.	.	.	.	.	.	.	.	$\bar{J}$	$\bar{1}$	.	.	.	.	.	.	.	.	.	.	.	.	.	.	.	$\bar{J}$	.	.	J	.	.	.	.	.	.	.	.	.	.
	3	$\underline{1}$	1	.	.	.	.	.	.	.	.	$\bar{W}$	U^*	.	.	.	.	$\bar{W}$	U^*	.	.	.	.	.	.	.	.	$\bar{W}$	$\bar{U}^*$	.	.	.	.	$\bar{W}$	$\bar{U}^*$	.	.	.	.	.	.	.	.	.	.	.	.	.	.	.	.	.	.	.	.	.	.	.	.	.	.	.	.	.	.	.	.
			2	.	.	1	.	.	.	.	1	.	.	.	.	.	1	.	.	.	.	$\bar{1}$	.	.	.	.	$\bar{1}$	.	.	.	.	.	$\bar{1}$	.	.	.	.	.	.	.	.	.	.	.	.	.	.	.	.	.	.	.	.	.	.	.	.	.	.	.	.	.	.	.	.	.	.
			3	J	J	.	.	J	.	.	.	.	.	J	.	.	.	.	.	$\bar{J}$	$\bar{J}$	.	.	$\bar{J}$	.	.	.	.	.	$\bar{J}$	.	.	.	.	.	.	.	.	.	.	.	.	.	.	.	.	.	.	.	.	.	.	.	.	.	.	.	.	.	.	.	.	.	.	.	.	.
			4	.	.	.	.	.	$\bar{1}$	J	.	.	.	.	$\bar{1}$	J	.	.	.	.	.	.	.	.	1	$\bar{J}$	.	.	.	.	1	$\bar{J}$	.	.	.	.	.	.	.	.	.	.	.	.	.	.	.	.	.	.	.	.	.	.	.	.	.	.	.	.	.	.	.	.	.	.	.
		$\underline{13}$	1	.	.	.	.	.	.	.	.	.	.	.	.	.	.	.	.	.	.	.	.	.	.	.	.	.	.	.	.	.	.	.	.	J	.	.	$\bar{1}$	.	.	.	.	.	.	.	.	.	.	.	$\bar{J}$	.	.	.	.	.	.	.	.	$\bar{J}$	.	.	.	.	.	.	.
			2	.	.	.	.	.	.	.	.	.	.	.	.	.	.	.	.	.	.	.	.	.	.	.	.	.	.	.	.	.	.	.	.	.	.	.	.	.	.	.	.	.	.	.	.	.	.	$\bar{J}$	.	1	.	.	.	.	.	.	1	.	$\bar{J}$	.	.	.	.	.	.
			3	.	.	.	.	.	.	.	.	.	.	.	.	.	.	.	.	.	.	.	.	.	.	.	.	.	.	.	.	.	.	.	.	.	.	.	.	.	.	.	.	1	.	U^*	.	.	.	.	.	.	.	.	.	.	.	.	.	.	.	.	.	.	1	.	U
			4	.	.	.	.	.	.	.	.	.	.	.	.	.	.	.	.	.	.	.	.	.	.	.	.	.	.	.	.	.	.	.	.	.	.	.	.	$\bar{1}$	.	.	$\bar{1}$	.	U	.	.	.	.	.	.	.	.	.	.	.	.	.	.	.	.	.	.	.	.	$\bar{W}^*$	.
	4	$\underline{1}$	1	.	.	$\bar{1}$	.	.	.	.	1	.	.	.	.	.	1	.	.	.	.	1	.	.	.	.	$\bar{1}$	.	.	.	.	.	$\bar{1}$	.	.	.	.	.	.	.	.	.	.	.	.	.	.	.	.	.	.	.	.	.	.	.	.	.	.	.	.	.	.	.	.	.	.
			2	.	.	.	.	.	.	.	.	W	U^*	.	.	.	.	W	U^*	.	.	.	.	.	.	.	.	$\bar{W}$	$\bar{U}^*$	.	.	.	.	$\bar{W}$	$\bar{U}^*$	.	.	.	.	.	.	.	.	.	.	.	.	.	.	.	.	.	.	.	.	.	.	.	.	.	.	.	.	.	.	.	.
			3	.	.	.	.	.	J	$\bar{1}$	.	.	.	.	J	$\bar{1}$	.	.	.	.	.	.	.	.	$\bar{J}$	1	.	.	.	.	$\bar{J}$	1	.	.	.	.	.	.	.	.	.	.	.	.	.	.	.	.	.	.	.	.	.	.	.	.	.	.	.	.	.	.	.	.	.	.	.
			4	$\bar{1}$	$\bar{1}$	.	.	1	.	.	.	.	.	1	.	.	.	.	.	1	1	.	.	$\bar{1}$	.	.	.	.	.	$\bar{1}$	.	.	.	.	.	.	.	.	.	.	.	.	.	.	.	.	.	.	.	.	.	.	.	.	.	.	.	.	.	.	.	.	.	.	.	.	.
		$\underline{13}$	1	.	.	.	.	.	.	.	.	.	.	.	.	.	.	.	.	.	.	.	.	.	.	.	.	.	.	.	.	.	.	.	.	.	.	.	.	.	.	.	.	.	.	.	.	.	.	$\bar{1}$	.	$\bar{J}$	.	.	.	.	.	.	$\bar{J}$	.	1	.	.	.	.	.	.
			2	.	.	.	.	.	.	.	.	.	.	.	.	.	.	.	.	.	.	.	.	.	.	.	.	.	.	.	.	.	.	.	.	$\bar{1}$	.	.	$\bar{J}$	.	.	.	.	.	.	.	.	.	.	.	$\bar{1}$	.	.	.	.	.	.	.	.	$\bar{1}$	.	.	.	.	.	.	.
			3	.	.	.	.	.	.	.	.	.	.	.	.	.	.	.	.	.	.	.	.	.	.	.	.	.	.	.	.	.	.	.	.	.	.	.	.	1	.	.	1	.	U	.	.	.	.	.	.	.	.	.	.	.	.	.	.	.	.	.	.	.	.	$\bar{W}^*$	.
			4	.	.	.	.	.	.	.	.	.	.	.	.	.	.	.	.	.	.	.	.	.	.	.	.	.	.	.	.	.	.	.	.	.	.	.	.	.	.	.	.	1	.	$\bar{U}^*$	.	.	.	.	.	.	.	.	.	.	.	.	.	.	.	.	.	.	$\bar{1}$	.	U
$\underline{\sigma}$	d	$\underline{\sigma}'$	d'	b	b	a	a	b	b	b	b	b	b	b	b	b	b	b	b	b	b	a	a	b	b	b	b	b	b	b	b	b	b	b	b	a	a	a	a	a	a	a	a	a	a	a	a	a	a	a	a	a	a	a	a	a	a	a	a	a	a	a	a	a	a	a	a

N.B. $J = \exp(i\pi/2)$, $U = \exp(i\pi/6)$, $W = \exp(i2\pi/3)$; $a = 2$, $b = \sqrt{8}$.

The number at the base of each column divides each element of that column.

$\bar{A} = -A$, for $A = 1, J, U, W$, and the asterisk * denotes the complex conjugate.

AUTOMORPHISM SYMMETRIES OF SPACE GROUP REPRESENTATIONS

R. Dirl

Institut für Theoretische Physik, TU Wien
A-1040 Wien, Karlsplatz 13; Austria

Equivalence properties of space group unirreps arising from their automorphism groups are discussed. *Alternative* space group unirreps are introduced and their merits for some physical applications are outlined.

1. General remarks

In many problems of solid state physics (and many other fields) it is necessary to determine relationships between the symmetry group of the system and some of its subgroups or factor groups as well as between their representations[1,2]. On the other hand physical properties must not depend on a particular form of the used representations. Taking this aspect into account there have been made in solid state physics obvious and succesful attempts to unify space group unirreps (to be more precise as regards the representations of the *little space groups* $G(\mathbf{k})$[3]). Nevertheless special forms of space group unirreps may simplify substantially computations. For instance space group C-G coefficients are now available in a systematic way for *all* Kronecker products for *all* of the 230 space groups[4], since among others the peculiar form of the Miller and Love matrices[5] for the allowed representations of $G(\mathbf{k})$ has been exploited. Or structural phase transitions in solids can be described in a more natural way, if corresponding space group unirreps are suitably adapted. For this and other reasons the structure of space group representations is revisited with particular regard to their automorphism groups.

2. Automorphism groups of space groups

The automorphism group $A(G)$ of a given space group G is isomorphic to the factor group of the *affine normalizer* $N(G)$ with respect to the *affine centralizer* $C(G)$ [6]. For convenience we assume that the *centre* $Z(G)$ of G and $C(G)$ are trivial in order to be able to embed G as normal subgroup in $N(G) = A(G)$. Fortunately, this situation is realized for many space groups, as can be seen from Table 3 of Ref.6. In this connection it is worth to mention that even though the automorphism groups of space groups are well known, they still require further investigations, since only gross classifications thereof have been published[6,7].

In order to get more insight into this complex task let us write down some persuasive examples:

$$A(P23) = \ldots = A(Pm3m) = A(Pn3n) = \ldots = Im3m \tag{1}$$

$$A(P222_a) = \ldots = A(Fm3m_a) = A(Fm3c_a) = Pm3m_{a/2} \tag{2}$$

In the case of (1) G as well as A(G) have the same lattice constant a, whereas in the case (2) G has a lattice constant that is twice of A(G). For convenience we also give for some examples corresponding coset decompositions of A(G) with respect to G, where the coset representatives are chosen so that, if possible, they form a group.

$$G = Pm3m: \qquad A(G) = G + (I|\mathbf{b}_o)\, G \tag{3}$$

$$G = Pn3n: \qquad A(G) = G + (I|\mathbf{0})\, G \tag{4}$$

$$G = Fm3m_a: \qquad A(G) = G + (I|\mathbf{b}_o)\, G \tag{5}$$

Note in particular that in the decomposition (3) as well as in (5) $\mathbf{b}_o$ is the same primitive translation of $Pm3m_{a/2}$ or of $Im3m_a$. Of course also chains of normal subgroups can be considered, like

$$Fm3m_a \triangleleft Pm3m_{a/2} \triangleleft Im3m_{a/2} \, . \tag{6}$$

3. Equivalence properties of G-unirreps with respect to A(G)

Space group unirreps, when constructed by induction[8] and when using partly matrix notation, take the form

$$D^K_{\underline{R},\underline{S}}(R|\mathbf{t}) = \Delta^{\mathbf{k}}(\underline{R},\underline{RS})\; e^{-i\underline{R}\mathbf{k}.\mathbf{t}}\; D^{\alpha}(\underline{R}^{-1}\underline{RS}) \tag{7}$$

$$D^K_{\underline{R},\underline{S}}(R|\mathbf{n}(R)+\mathbf{t}) = \Delta^{\mathbf{k}}(\underline{R},\underline{RS})\; \xi^{\mathbf{k}}_{\underline{R},\underline{S}}(R)\; e^{-i\underline{R}\mathbf{k}.(\mathbf{n}(R)+\mathbf{t})}\; \mathbf{D}^{\alpha}(\underline{R}^{-1}\underline{RS}) \tag{8}$$

where (7) refers to symmorphic and (8) to non-symmorphic space groups respectively. For brevity the equivalence classes are in both cases denoted by $K = (\mathbf{k},\alpha)\uparrow G \in A_G$ (*index set*) or for short $K = (\mathbf{k},\alpha)$ if there is no confusion, where $\mathbf{k} \in \Delta BZ_G$ (*representation domain*[8]). The coset representatives $\underline{R},\underline{S} \in P:P(\mathbf{k})$ (where $P \simeq G/T$ and $P(\mathbf{k}) \simeq G(\mathbf{k})/T$) are *uniquely* fixed. The symbol $\Delta^{\mathbf{k}}(\underline{R},\underline{RS})$ indicates the generalized permutation structure of the unirreps. Finally D^{α} forms in the case of a symmorphic space group a n_{α}-dimensional vector unirrep of $P(\mathbf{k})$, whereas $\mathbf{D}^{\alpha}$ turns out to be a projective matrix unirrep of $P(\mathbf{k})$ with well-defined factor system, if G is a non-symmorphic space group. The quantities $\mathbf{n}(R)$ are non-primitive lattice translations and $\xi^{\mathbf{k}}_{\underline{R},\underline{S}}(R)$ are well-defined unimodular factors.

Since due to our assumption G is a normal subgroup of A(G), each element $a = (S_a|\mathbf{v}_a) \in A(G)$ defines an equivalenve relation, i.e.

$$D^K(a(g)) = Z^K(a)\ D^{a(K)}(g)\ Z^K(a)^\dagger \qquad g \in G \tag{9}$$

where K, $a(K) \in A_G$; $a(g) = a\ g\ a^{-1}$ and space group elements are denoted by g. As usual we call $A(G)^K = \{a \in A(G)|a(K) = K\}$ the corresponding *little group*.

One goal of this contribution is to specify the mappings $a(K) = K' \in A_G$ for each non-trivial coset representative of A(G) with respect to $A(G)^K$. These mappings have been called in Ref.3 *translations*. We investigate this problem for several reasons. One reason is that by virtue of the choice

$$Z^K(a) = 1_K \qquad a \in A(G):A(G)^K \tag{10}$$

Eq.(9) can be exploited to construct space group unirreps out of known ones. This has been partly utilized in Ref.8 (Theorem 5.5.2) to define space group unirreps. A second aspect is that (9,10) allow to derive a new class of symmetry and generating relations for space group C-G coefficients[9] as well as symmetry relations for reduced matrix elements of irreducible tensor operators[10]. These useful correlations between space group C-G coefficients shall be exploited to reduce the space needed for their tabulation in a systematic and reasonable way[11].

The following examples refer to *Pm3m* and *Pn3n* respectively. We call $\mathbf{k}_g$ a *general point* of ΔBZ_G , if $P(\mathbf{k}_g)$ is trivial. A simple group theoretical argument yields

$$A(Pm3m)^K = A(Pn3n)^K = Im3m \tag{11}$$

for $K = (\mathbf{k}_g,\alpha_1)\uparrow G$; G = *Pm3m* or *Pn3n*, where α_1 denotes the trivial unirrep of $P(\mathbf{k}_g)$. Considering now the symmetry point *R* of ΔBZ_G; G = *Pm3m* or *Pn3n* we obtain

$$\begin{array}{ll} A(Pm3m)^K = Pm3m & \text{if } K = (\mathbf{k}_R;v,s)\uparrow Pm3m\ ;\ v = 1,2,3,4,5\ ;\ s = \pm \\ a = (I|\mathbf{b}_o)\ : & a((\mathbf{k}_R;v,+)\uparrow Pm3m) = (\mathbf{k}_R;v,-)\uparrow Pm3m\ ;\ v = 1,2,3,4,5 \end{array} \tag{12}$$

$$\begin{array}{ll} A(Pn3n)^K = \begin{cases} Pn3n & \text{if } K = (\mathbf{k}_R;w)\uparrow Pn3n\ ;\ w = 2,3 \\ Im3m & \text{if } K = (\mathbf{k}_R;w)\uparrow Pn3n\ ;\ w = 1,4 \end{cases} & \\ a = (I|\mathbf{0})\ : \quad a((\mathbf{k}_R;2)\uparrow Pn3n) = (\mathbf{k}_R;3)\uparrow Pn3n & \end{array} \tag{13}$$

For α = v,s or w we adopted the notation of Ref.5.

4. *Alternative* space group unirreps

On account of $G \lhd A(G)$ it is obvious to construct $A(G)$-unirreps by means of induction out of G-unirreps. Now if $A(G) = G^*$ is a space group, which is often the case (see Table 3 of Ref.6), then G^*-unirreps shall be constructed by induction:

$$\{D^K \uparrow A(G)^K\} \uparrow A(G) = D^{K\uparrow G^*} \tag{14}$$

where $K = (\mathbf{k},\alpha)\uparrow G$; $\mathbf{k} \in \Delta BZ_G$ and $\alpha \in A_{P(\mathbf{k})}$. Of course usual G^*-unirreps D^{K^*}, where $K^* = (\mathbf{k}^*,\alpha^*)\uparrow G^*$; $\mathbf{k}^* \in \Delta BZ_{G^*}$ and $\alpha^* \in A_{P(\mathbf{k}^*)}$ are (in general) only equivalent to $D^{K\uparrow G^*}$. For distinction we call G^*-unirreps of the type (14) as *alternative*.

In order to be able to estimate the merits of *alternative* space group unirreps let us consider at first some persuasive examples. We choose $A(Pm3m) = A(Pn3n) = Im3m = G^*$. For $K = (\mathbf{k}_g,\alpha_1)\uparrow Pm3m$ we found $A(Pm3m)^K = G^*$. A simple manipulation yields

$$Z^K_{\underline{R},\underline{S}}(I|\mathbf{b}_o) = \delta_{\underline{R},I\underline{S}}\; e^{-i\underline{R}\mathbf{k}_g\cdot\mathbf{b}_o} \qquad \underline{R},\underline{S} \in O_h \tag{15}$$

which satifies (9). Hence corresponding *alternative* $Im3m$-unirreps are given by

$$D^{(K,\omega)\uparrow G^*}(R|\mathbf{t}) = D^K(R|\mathbf{t}) \tag{16}$$

$$D^{(K,\omega)\uparrow G^*}((I|\mathbf{b}_o)(R|\mathbf{t})) = (-1)^\omega\, Z^K(I|\mathbf{b}_o)\, D^K(R|\mathbf{t}) \qquad (R|\mathbf{t}) \in G \tag{17}$$

where $\omega = 0,1$. Now it is suggestive to ask which G^*-unirreps D^{K^*}; $K^* = (\mathbf{k}^*,\alpha^*)\uparrow G^*$ with $\mathbf{k}^* \in \Delta BZ_{G^*}$ and $\alpha^* \in A_{P(\mathbf{k}^*)}$ are equivalent to (16,17). It is readily verified that $\omega = 0$ corresponds to $K^* = (\mathbf{k}^*,\alpha^*)\uparrow G^*$ with $\mathbf{k}^* = \mathbf{k}_g$ and $\alpha^* = \alpha_1$. For $\omega = 1$ it holds $K^* = (\mathbf{k}'_g,\alpha_1)\uparrow G^*$ where $\mathbf{k}'_g$ is the mirror point of $\mathbf{k}_g$ reflected on the plane *RXM* of ΔBZ_G. This implies a drastic reduction of ΔBZ_{G^*} to ΔBZ_G, i.e.

$$\Delta BZ_{Im3m} \rightarrow \Delta BZ_{Pm3m} \tag{18}$$

where the seize of the latter is only half of the former! In the case of $K = (\mathbf{k}_R;v,+)\uparrow Pm3m$ we obtain because of (12) the following induced G^*-unirreps

$$D^{K\uparrow G^*}(R|\mathbf{t}) = \begin{vmatrix} D^K(R|\mathbf{t}) & \mathbf{0} \\ \mathbf{0} & D^K(a(R|\mathbf{t})) \end{vmatrix} \qquad a = (I|\mathbf{b}_o) = a^{-1} \tag{19}$$

$$D^{K\uparrow G^*}(I|\mathbf{b}_o) = \begin{vmatrix} \mathbf{0} & D^K(I|\mathbf{b}_o) \\ D^K(I|\mathbf{b}_o) & \mathbf{0} \end{vmatrix} \;;\quad D^{K\uparrow G^*}((I|\mathbf{b}_o)(R|\mathbf{t})) = D^{K\uparrow G^*}(I|\mathbf{b}_o)D^{K\uparrow G^*}(R|\mathbf{t}) \tag{20}$$

where (12) has to be taken into account. Of course analogous expressions can be derived for $A(Pn3n) = Im3m$. For instance if $K = (\mathbf{k}_g;\alpha_1)\uparrow Pn3n$ we have

$$Z^{K}_{\underline{R},\underline{S}}(I|\mathbf{0}) = \delta_{\underline{R},I\underline{S}} \qquad \underline{R},\underline{S} \in O_h \tag{21}$$

which defines G*-unirreps that are equivalent to (16,17). In the case of K = $= (\mathbf{k}_R; w)\uparrow Pn3n$; w = 1,2,3,4 we obtain by induction G*-unirreps that are only equivalent to (19,20). For instance if K = $(\mathbf{k}_R;4)\uparrow Pn3n$ the corresponding G*-unirrep $D^{K\uparrow G^*}$ must have because of (13) the same dimension as the G-unirrep D^K. This is an essential difference to the G*-unirreps $D^{K\uparrow G^*}$ with K = $(\mathbf{k}_R;v,+)\uparrow Pm3m$; v = 4,5. Note that in the usual approach we have for $\mathbf{k}^* = \mathbf{k}_P^*$ two 2-dimensional, one 4-dimensional and two 6-dimensional vector unirreps of G* = *Im3m*, since the order of the *star* of $\mathbf{k}_P^*$ is two. This of course coincides with the results (19,20) or the corresponding expressions for $D^{K\uparrow G^*}$ when K = $(\mathbf{k}_R;w)\uparrow Pn3n$. For details of the present material and the following brief discussion the reader is referred to Ref.12.

An obvious question is now: What could be the merit of *alternative* space group unirreps? Firstly each G*-unirrep $D^{K\uparrow G^*}$ subduce immediately to a direct sum of G-unirreps. This might be advantageous for some applications, like a more natural description of phase transitions in solids. Of course, if necessary, *alternative* space group unirreps can also be constructed that are adapted to chains of space groups, like for (6). Secondly this new classification of space group unirreps reduces drastically in some cases (like (3,4,5)) the representation domain ΔBZ_{G^*} to ΔBZ_G, which could give rise to computational simplifications in *k-integrations* or an *alternative classifications of energy bands*. Thirdly in each case where $D^{K\uparrow G^*}\downarrow G$ is a G-unirrep or where $A(G)^K = G$ it suffices to consider G instead of G*. Finally if $D^{K\uparrow G^*}\downarrow G = D^K$, corresponding G*-C-G coefficients follow immediately from G-C-G coefficients[11].

Concluding this discussion it should be noted that the tendency of the present approach conflicts to some extent to the unification attempts of space group unirreps[3].

References

1. D.B. Litvin, J.N. Kotzev, J.L. Birman; Phys.Rev., B26, 6947 (1982)
2. M. Sutton, R.L. Armstrong; Phys.Rev., B25, 1813 (1982)
3. L. Michel, J. Morzymas; Lecture Notes in Physics, 180, 292 (1983)
4. B.L. Davies, R. Dirl: "Computer generated Clebsch-Gordan coefficients for space groups" (preprint, August 1983)
5. A.P. Cracknell, B.L. Davies, S.C. Miller, W.F. Love; Kronecker Product Tables. vol.I (Plenum Press, New York, 1979)
6. H. Burzlaff, H. Zimmermann; Z.Krist., 153, 151 (1980)
7. E. Koch, W. Fischer; Acta Cryst., A31, 88 (1975)
8. C.J. Bradley, A.P. Cracknell; The Mathematical Theory of Symmetry in Solids (Clarendon, Oxford, 1972)
9. R. Dirl; J.Math.Phys., 24, 1935 (1983)
10. R. Dirl; Lecture Notes in Physics, 180, 313 (1983)
11. R. Dirl: "Automorphism Symmetries of Space Group C-G Coefficients" (in preparation)
12. R. Dirl: "Representation Theory of Space Groups revisited" (in preparation)

LATTICES OF SYMMETRIC GROUPS S_5 AND S_6 AND EXOMORPHISM OF GROUP-SUBGROUP RELATIONS UP TO INDEX 6

Jiří Fuksa and Vojtěch Kopský

Institute of Physics, Czechoslovak Academy of Sciences
Na Slovance 2, POB 24, 180 40 Praha 8, Czechoslovakia

There are several reasons, for which investigation of group-subgroup relations is of interest. The most practical field of applications is the structural phase transition theory, another field provides the theory of colour groups. Generally, it is of interest to know, how a given group F can be embedded into various supergroups. The key to the theory of group-subgroup relations from several viewpoints is given by the definition of "exomorphism" /1/ which, in its most suitable form, says that two group-subgroup realtions $G\downarrow F$ and $G'\downarrow F'$ belong to the same exomorphic type just if the permutation representations of G, G' on cosets of F, F' in G, G', respectively, are equivalent as permutation groups. The indices $[G:F]$ and $[G':F']$ are equal and we call them the index of group-subgroup relation.
A straightforward way for determination of group-subgroup relations of a given index q is given by the

Theorem (generalized Jordan-Cayley theorem): For any group-subgroup relation $G\downarrow F$ of index q there exists a homomorphism σ : $G \longrightarrow S_q$ such, that $\sigma(G)\downarrow\sigma(F)$ is exomorphic to $G\downarrow F$.

The kernel of σ is the group $H = \ker\sigma = \text{core } F = \bigcap_i F_i$, where F_i is the set of conjugate subgroups in G, the image $\sigma(G) = \text{Im}\,\sigma$ is a transitive group on q symbols, and $\sigma(F)$ is its subgroup of index q, fixing one or several symbols.

Determination of exomorphic classes of group-subgroup relations of a given index q is as well intriguing as, for example, the problem of groups of a given finite order and it is closely related to the problem of the structure of lattices of subgroups of symmetric groups S_q. Any subgroup R of S_q defines a partition $\{q_i\}$ of the set $\mathcal{Q} = \{1,2,\ldots\ldots,q\}$ into orbits of R. To each partition there corresponds a direct product $\bigotimes_i S(q_i)$ and a set of transitive constituents $R(q_i)$ of the group R, which arise by the restriction of the action of R on q_i's. The group R is then a subdirect product of the groups $R(q_i)$, "transitive" on q_i and $\bigotimes_i S(q_i)$ is the least direct product of symmetric groups, of which R is a subgroup. Each partition of $\mathcal{Q}$ defines a certain Young table. The partitions, Young tables and direct products of symmetric groups form isomorphic lattices, of which the last is the sublattice $\mathcal{L}_0(S_q)$ of the lattice $\mathcal{L}(S_q)$ - the lattice of the symmetric group S_q.

To determine exomorphic classes of index q, we have to calculate those subgroups of S_q, which are transitive on the whole set This can be performed with use of Sylow subgroups of the symmetric group /2/. Each set of conjugate transitive subgroups then defines just one exomorphic class of index q. If we now denote by G the transitive group itself, then F is just its subgroup fixing one of the symbols.

It is further of interest to know, how the subgroup F of G is connected with irreducible representations (ireps) of G. For this purpose, we can use the fact that permutation representations of G on cosets of F in G, considered as a linear representation, is just the induced representation $1_F \uparrow G$. On the other hand, the representation of the symmetric groups S_q in which the q symbols are considered as a basis of a linear space of dimension q is the representation $[(q) \oplus (q-1,1)]$ (S_q). Since the permutation representation of G on cosets of F as a linear representation is obtained by restricting the permutations to the group G, we obtain the relation $1_F \uparrow G =$ $= [(q) \oplus (q-1,1)](S_q) \downarrow G$. Combining this result with Frobenius reciprocity theorem /3/, we obtain easily the subduction coefficients $s_\alpha(F)$, giving the number of times an irep of G of label α subduces the identity irep of F.

It is also of interest to notice, that the direct products of symmetric groups of lower degrees in S_q are epikernels of the irep $(q-1,1)(S_q)$.

We have carried so far the systematic investigation up to q=6. Up to q=4 there are actually no problems, because the groups S_2, S_3, S_4 are isomorphic to groups C_2, D_3, O, well known from crystal physics. The numbers and types of exomorphic relations for q = 2,3,4 are respectively: 1: $(C_2 \downarrow C_1)$; 2: $(C_3 \downarrow C_1, D_3 \downarrow C_{2i})$; 5: $(D_2 \downarrow C_1,$ $C_4 \downarrow C_1, D_4 \downarrow C_{2i}, T \downarrow C_{3j}, O \downarrow D_{3j})$ in crystallographic notation. It is also worth mentioning, that the existence of two exomorphic relations of index three has been used for the proof of the original Landau conjecture concerning impossibility of continuous phase transitions into subgroups of index three.

The last help in our calculation is given by the icosahedral group, which is isomorphic with the alternating group A_5. We have found complete lattices of groups S_5 and S_6 with use of the transitive groups on five and six symbols. There were found 5 exomorphic classes of index 5 and 16 classes of index 6.

References:

/1/ V. Kopský, Phys. Lett., 69 A (1978), 82

/2/ C.A. Miller, H.F. Blichfelt and L.E. Dickson, Theory and applications of finite groups. Wiley, New York. 1916
M. Hall, The Theory of Groups. McMillan, New York. 1959

/3/ F.G. Frobenius, Sber. preuss. Akad. Wiss. (1898), 501

A DIRECT-EXPANSION METHOD FOR TENSOR PROPERTIES OF CRYSTALS

F.G. Fumi and C. Ripamonti
Istituto di Scienze Fisiche, Università di Genova (Italy)
and GNSM-CNR, Unità di Genova (Italy)

Abstract

We present a direct-expansion method for a general tensorial set and a criterion for an optimal choice of an expansion basis in a rotational symmetry group. The expansion relations are determined in two stages: one imposes first the permutational constraints arising from the permutational multiplicity of the tensorial products; one imposes next the rotational constraints by using a numerical vector representation of the tensorial set in the group. The expansion relations for a tensor property possessing also permutational symmetry are obtained from the general expansion relations by removing the pertinent permutational multiplicities. This procedure drastically reduces the algebra with respect to the traditional methods for the symmetry axes of order 3, 6 and ∞.

1. Basic ideas

In the physics of materials rather complex objects have to be introduced: for instance, to account for the elastic properties it has proved necessary to consider 4th order products of the (six) strain components, amounting to a total of 126 (besides ordering rearrangements)[1]. More generally, to specify the directional behaviour of the physical properties of a material, 3^N Nth-order products of the 3 components of a vector(Nth-rank tensorial set) have to be introduced. Accordingly, the specification of materials' properties may be cumbersome even for smallish N.

Structural symmetry of the material reduces the complexity by setting rotational constraints which allow to express the whole tensorial set in terms of a (minimal) subset (expansion relations). These relations are naturally simple for symmetry axes of order 1, 2 or 4 (direct-inspection method) but not for symmetry axes of order 3, 6 or ∞. In the latter cases, the complexity of the expansion is well known: it requires both the specification of a large number of fairly involved expansion relations, and the solution of a large number of equations to obtain these relations. It is clearly desirable to achieve simplicity also in these cases.

Most of the complexity of a general tensorial set without rotational symmetry is of trivial permutational nature, due to the many different orderings of a given product (tensorial-product type): in fact, the same is true also for a general tensorial set with rotational symmetry, provided a few ordering permutations are excluded.

This permits one to set permutational constraints on the possible forms of the expansion relations in terms of an optimal expansion basis, thus achieving a high degree of permutational multiplicity of the expansion itself: this implies both a drastic reduction in the number of expansion relations to be specified, and a great formal simplicity of the relations. The optimal basis consists of appropriate sets of tensor products closed under the maximum possible number of ordering permutations (permutational "multiplets") and can be chosen by correspondence with a complete set of tensorial invariants of the rotational symmetry group.

The formal simplicity of the permutationally-constrained expansion relations motivates a direct-expansion method which simply determines the expansion coefficients with respect to an assigned basis. In this way the formal simplicity of the expansion causes a simplification of the expansion procedure. Indeed the use of rotational constraints is kept to a minimum since it is only necessary to determine the few coefficients which are free from permutational constraints.

The rotational constraints are imposed via a numerical vector representation of the tensorial products in the rotational symmetry group: the representation is provided by the coefficients in any complete set of invariant linear combinations of tensorial products. The replacement of the tensorial products in an expansion relation by these numerical vectors gives directly the vector equation for the (few) permutationally-free coefficients.

The expansion relations are thus derived in two independent stages, using first the permutational constraints and then the rotational constraints. With rising rank the relative importance of the permutational constraints increases, as the permutational multiplicity of tensorial products is a rapidly increasing function of the rank.

Tensorial sets of interest in materials' physics usually have permutational symmetry in addition to rotational symmetry (e.g. the dielectric tensor is symmetric in the permutation of its two indices). The pertinent expansion relations can be obtained simply by removing the pertinent permutational multiplicities from the most general expansion relations. As a result, the form of these expansion relations may be less simple.

The procedure just described should be contrasted to the traditional methods with special reference to the symmetry axes of order 3, 6 and ∞. In essence these methods (i) impose rotational constraints on the individual tensorial products, thus obtaining a (highly) redundant system of equations; or (ii) put equal to zero a (minimal) set of tensorial non-invariants, which have, however, to be constructed; or (iii) express directly each tensorial product in terms of tensorial invariants, which must then be eliminated. The natural evolution of these traditional methods in going to higher ranks - specifically to the 4th-order elastic tensor (a particular 8th-rank tensor) - has been to recur to electronic computers, preparing appropriate programs, which could of course be adapted to treat different tensors[2].

The method described here[3] can properly be characterized among the existing methods as the only one based on the direct expansion of a (whole) tensorial set in terms of an assigned subset. The criterion given here[3] to choose an optimal expansion basis is to take suitable permutational "multiplets". The combination of the two ideas drastically reduces the algebra with respect to the traditional methods for the symmetry axes of order 3, 6 and ∞, and does not at all induce one to recur to an electronic computer even in treating the general 8th-rank tensor[3].

2. Applications*

We consider general tensorial products of rank 6 in x and y even in x and y in group ∞ (∞_z) (with the z axis parallel to the cylindrical axis). We adopt the tensorial invariants Re(+++---)-, 10 with

$$+ = x + iy \ , \quad - = x - iy \tag{1}$$

where the round bracket means that we consider the (distinct) permutations of the symbols + and - inside the bracket: the number of these permutations is 10.

I) Determination of the optimal expansion basis

The optimal basis is chosen by correspondence with the tensorial invariants: the correspondence used is + $\leftrightarrow$ x, - $\leftrightarrow$ y, on all indices but the last (left free to fix the pertinent parity). The basis is

---

* For technical details see reference 3.

thus (xxxyy)x , 10.

II) Determination of the expansion relations

a) Expansion of x^6

α) Form of the expansion relation*

The permutational constraints simplify the general expansion relation

$$x^6 = \sum_i^{10} c_i\,(xxxyy)_i x \tag{2}$$

to

$$x^6 = c \sum_i^{10} (xxxyy)_i x \tag{3}$$

β) Coefficients of the expansion relation

The rotational constraints are imposed via a numerical vector representation of the tensorial products taken from the invariant Re+++---. The representatives are:

$$x^6 \leftrightarrow [1]\ ;\quad xxxyyx,\ xyyxxx,\ yxyxxx,\ yyxxxx \leftrightarrow [-1]$$

$$xxyxyx,\ xyxxyx,\ yxxxyx,\ xxyyxx,\ xyxyxx,\ yxxyxx \leftrightarrow [1]$$

The permutationally-constrained expansion relation can then be rewritten as

$$[1] = c\,[6-4] = c\,[2] \tag{4}$$

and thus c = ½.

Finally
$$x^6 = \tfrac{1}{2} \sum_i^{10} (xxxyy)_i x \tag{5}$$

γ) Expansion relation for a tensor property with permutational symmetry

For the 3rd-order elastic tensor which has the permutational symmetry ijk = jik = kji (i,j,k = 1,2,3,4,5,6 with the usual convention 1 = xx, 2 = yy, 3 = zz, 4 = yz, 5 = zx, 6 = xy) the removal of the pertinent permutational multiplicities yields

$$111 = \tfrac{1}{2}(2\cdot 112 + 8\cdot 166) = 112 + 4\cdot 166 \tag{6}$$

b) Expansion of x^4y^2

α) Form of the expansion relation

The permutational constraints simplify the general expansion re-

---

* A more compact (but perhaps less transparent) notation is used in reference 3.

lation

$$x^4y^2 = \sum_i^{10} c_i(xxxyy)_i x \qquad (7)$$

to

$$x^4y^2 = c_1 \sum_i^{4} (xxxy)_i yx + c_2 \sum_i^{6} (xxyy)_i xx \qquad (8)$$

β) <u>Coefficients of the expansion relation</u>

The rotational constraints are imposed via the numerical vector representation taken from the invariants Re+++--- and Re++--+-:

$x^4y^2 \leftrightarrow \begin{bmatrix}-1\\1\end{bmatrix}$; $xxxyyx \leftrightarrow \begin{bmatrix}-1\\1\end{bmatrix}$; $xxyxyx \leftrightarrow \begin{bmatrix}1\\1\end{bmatrix}$; $xyxxyx$, $yxxxyx \leftrightarrow \begin{bmatrix}1\\-1\end{bmatrix}$

$xxyyxx \leftrightarrow \begin{bmatrix}1\\-1\end{bmatrix}$; $yyxxxx \leftrightarrow \begin{bmatrix}-1\\-1\end{bmatrix}$; $xyxyxx, yxxyxx \leftrightarrow \begin{bmatrix}1\\1\end{bmatrix}$; $xyyxxx, yxyxxx \leftrightarrow \begin{bmatrix}-1\\1\end{bmatrix}$

The permutationally-constrained expansion relation can then be rewritten as

$$\begin{bmatrix}-1\\1\end{bmatrix} = c_1\begin{bmatrix}3-1\\2-2\end{bmatrix} + c_2\begin{bmatrix}3-3\\4-2\end{bmatrix} = c_1\begin{bmatrix}2\\0\end{bmatrix} + c_2\begin{bmatrix}0\\2\end{bmatrix} \qquad (9)$$

and thus $c_1 = -½$, $c_2 = ½$.

Finally

$$x^4y^2 = -½ \sum_i^{4} (xxxy)_i yx + ½ \sum_i^{6} (xxyy)_i xx \qquad (10)$$

γ) <u>Expansion relation for a tensor property with permutational symmetry</u>

For the 3rd-order elastic tensor (see a)γ)) the removal of the pertinent permutational multiplicities yields only an identity:

$$112 = -½(4\cdot 166) + ½(2\cdot 112 + 4\cdot 166) = 112 .$$

<u>References</u>

1. See e.g. X. Markenscoff, J. Appl. Phys. <u>48</u>, 3752/55 (1977) and <u>50</u>, 1325/27 (1979).
2. See e.g. R. Brendel, Acta Cryst. <u>A35</u>, 525/33 (1979).
3. F.G. Fumi & C. Ripamonti, Acta Cryst. <u>A36</u>, 535/51 (1980) and <u>A39</u>, 245/51 (1983).

Table of contents

ISOTROPY GROUPS OF SPACE GROUPS - A SIMPLE METHOD FOR THEIR DETERMINATION

Dorian M. Hatch

Department of Physics and Astronomy
Brigham Young University
Provo, Utah 84602 USA

1. INTRODUCTION

The most widely used description of continuous phase transitions was introduced by Landau nearly forty five years ago.[1] More recent renormalization group methods due to Wilson[2] extend the Landau theory and numerous systems have been studied using these methods. Within the context of the above methods we here consider a physical system in equilibrium having a space group structure G. At the transition a vector $\vec{\psi}$ of an n-dimensional irreducible real representation of G minimizes the free energy of the system. Methods have been introduced[3] to obtain the solutions corresponding to the extrema of nonlinear polynomial forms by exploiting the isotropy groups (cokernels) of the irreducible representation carried by basis vectors $\{\psi_i | i=1,2,\ldots n\}$. Here we present a simple method of obtaining isotropy groups of G corresponding to a representation $D^{(*\vec{k})}$. The method has been used by the author[4] to specify the possible phases of $R\bar{3}c$ (D^6_{3d}), $R\bar{3}m$ (D^5_{3d}), and Immn (D^{25}_{2h}). In approach it differs from both References [5] and [6] but in philosophy has several points in common with Reference [6].

2. SUBGROUPS

The following versions of two important theorems concerning spacegroups were recently discussed by Senechal.[7]

Theorem 1 - Hermann's Theorem

Let G be an μ dimensional space group with arithmetic class S and translation group T. If H is a subgroup of G and of finite index k then

(i) $T' = H \cap T$ is a maximal abelian normal subgroup of H.

(ii) H/T' is isomorphic to a subgroup S' of S and thus

$$H = \bigcup_i T'\{s'_i | \vec{\tau}'_i\}$$

Theorem 2 - The Frobenius Subgroup Theorem.

Let $G = \bigcup_{i=1}^{m} T\{s_i|\vec{\tau}_i\}$ be a space group. Then each subgroup H of finite index k of G is determined by the pair of subgroups $T'\subset T$ and $S'\subset S$ where $[S:S']\cdot[T:T'] = k = \mu\Delta$.

Furthermore
$$H = \bigcup_{i=1}^{m/\mu} T'\{s_i'|\vec{t}_i+\vec{\tau}_i'\}$$
where $\vec{t}_i \in T$ and not in T' and $s_i' \in S'$. Also S' and T' are constrained by the conditions

(i) $S' \in \text{Aut } T'$ (consistency conditions)

(ii) $\vec{t}_i + \vec{\tau}_i + s_i'\vec{t}_j + s_i'\vec{\tau}_j = \vec{t}_\ell + \vec{\tau}_\ell$ (ModT') (congruency conditions)

3. ISOTROPY SUBGROUPS

To construct representations of a space group we necessarily have a specific $\vec{k}$ -vector in mind. A representation of the space group is then induced from a representation of the little group $G^{\vec{k}}$.

If the coset elements of $G^{\vec{k}}$ are represented by pxp dimensional matrices obtained from the little cogroup $G^{\vec{k}}$ and if there are q inequivalent arms of the star then the induced representation will be (pq)x(pq) dimensional. The character for any element of the space group corresponding to the induced representation $D^{(*\vec{k})}$ is then
$$\chi(\{s|\vec{t} + \vec{\tau}\}) = \sum_{\sigma=1}^{q} e^{-i\vec{k}_\sigma\cdot\vec{t}} \chi_\sigma(\{s|\vec{\tau}\}).$$

We denote the basis vectors of the representation as $\{\psi_i;i = 1,2,3 \ldots, pq\}$.

An isotropy group is a subgroup of G which leaves all vectors of a subspace of the representation space invariant (fixed). For a subgroup H to be an isotropy group a necessary and sufficient condition is that the subduction frequency i(H),

$$i(H) = \frac{1}{|H|}\sum_{\vec{t}'\in T'}\sum_{s'\in S'}\sum_{\sigma=1}^{q} e^{-i\vec{k}_\sigma\cdot\vec{t}'} \chi_\sigma(\{s'|\vec{\tau}'\}) \qquad (1)$$

be equal to a positive integer. The integer indicates the dimensionality of the invariant subspace.

The isotropy groups of G corresponding to $D^{(*\vec{k})}$ are constructed by first looking for a translation subgroup $T'\subset T$ which yield

$$\sum_{\vec{t}'\in T'} \frac{e^{-i\vec{k}_\sigma\cdot\vec{t}'}}{|T'|} = 1 . \qquad (2)$$

Different sublattices T' are obtained by allowing successively more arms of the star to contribute in i(H), i.e., the sum (2) is one for more values of σ.

The cosets $\{s_i \mid \vec{t}_i + \vec{\tau}_i\}$ are from the original coset representatives of G together with a primitive $\vec{t}_i$ which is a fractional of T'. The collection $\{s_i\}$ are to form a subgroup $S' \subset S$. The selection of S' and T' are such that i(H) = n (integer) and are motivated by theorems 1 and 2 above. The consistency conditions must be checked, i.e. the point transformations $s_i \in S'$ *must be an automorphism of* the T' lattice. Also congruency conditions must be checked for the selections of the translation $\vec{t}_i + \vec{\tau}_i$.

For a given selection of $T' \subset T$ and $S' \subset S$ the coset representatives $\{s_i' \mid \vec{t}_i + \vec{\tau}_i\}$ and the translations $\vec{t}' \in \vec{T}'$ are, to this point, expressed in terms of the original lattice primatives and coset representatives. If the consistency conditions and the conditions of congruence are satisfied we are assured we have obtained a subgroup H. However a correspondence to a "conventional" setting and origin is useful particularly for a conventional labeling of the group. This often necessatates a new selection of primatives (or conventionals), i.e., a rotation of basis vectors, as well as a translation of origin.

Since the process was carried out with a particular representation initially chosen the transformation properties of the order parameter are explicit. Thus for any connection with lattice dynamics or microscopic forces the restriction to order parameter symmetry is given in conjunction with the obtained subgroup. Additionaly since the subduction frequency i(H) is known the subduction condition is obviously satisfied and the chain criterion can be applied to a collection of subgroups. If one starts with a space group G in a conventional setting the expression of H in a conventional setting and the relation between them is useful for the details of a microscopic correspondence, i.e. for specific selection of basis atoms.

4. EXAMPLE

Consider the space group D_{2h}^{25}, the $\vec{R}$-point (1/2,0,0), and irreducible representation $D_2^{(*\vec{R})}$. For one arm contributing equation (2) is satisfied by basis vectors $\vec{t}_1' = (-2,-1,-1)$, $\vec{t}_2' = (0,1,0)$, and $\vec{t}_3' = (0,0,-1)$ as expressed in terms of the original primatives.[8] Equation (1) yields i(H) = 1 for the cosets $\{E|000\}$, $\{C_{2y}|000\}$, $\{I|100\}$, $\{\sigma_y|100\}$, i.e. $S' = C_{2h}$. The consistency and congruency conditions are easily seen to be satisfied. A translation of origin (-1/2,-1/4,1/4) (in terms of new primatives) yields C_{2h}^6. This result together with the results for the two arms $\{\vec{R}, \vec{R}'\}$ contributing are listed below for representation $D_2^{(*\vec{R})}$.

Table 1. Isotropy groups of D_{2h}^{25} corresponding to representation $D_2^{(*\vec{R})}$. The associated translation of origin is indicated.

	Primative Cell		Conventional Cell	
Space Group	New Basis	Origin	New Basis	Origin
C_{2h}^{6}	-2,-1,-1;010;0,0,-1	-1/2,-1/4,1/4	-1,0,1;1,0,1;0,1,0	0,1/4,1/4
D_{2h}^{21}	0,1,-1;2,1,1;0,-1,-1	0,0,0	0,0,2;2,0,0;0,1,0	1/4,0,0
D_{2h}^{21}	0,1,-1;2,1,1;0,-1,-1	-1/2,0,0	0,0,2;2,0,0;0,1,0	0,1/4,0
C_{2h}^{4}	-2,0,-2;0,1,-1;0,-1,-1	-1/2,0,0	-1,0,1;2,0,0,;0,1,0	0,1/4,0

ACKNOWLEDGEMENTS

For helpful conversations I would like to thank Harold Stokes, Marjorie Senechal, and Marko Jaric.

REFERENCES

1. L.D. Landau, Z. Phys. 11, 546 (1937)
2. K.G. Wilson, Phys. Rev. B4, 3174, 3184 (1971)
3. M.V. Jaric, Phys. Rev. Lett. 48, 1641 (1982)
4. D.M. Hatch and L. Merrill, Phys. Rev. B 23, 368 (1981); D.M. Hatch, Phys. Rev. B 23, 2346 (1981): D.M. Hatch, Phys. Stat. Sol. (b) 106, 473, (1981). The consideration of Immn is to be published.
5. S. Deonarine and J.L. Birman, Phys. Rev. B 27, 4261 (1983).
6. M.V. Jaric to be published.
7. M. Senechal, Acta. Cryst. A 36, 845 (1980)
8. The notation and convention used here is that of C.J. Bradley and A.P. Cracknell, "The Mathematical Theory of symmetry in Solids" (Clarendon, Oxford, 1972).

LANDAU'S THEORY OF CRYSTALLINE PHASE TRANSITIONS IN A SUPERSPACE FORMULATION

A.Janner and T. Janssen, Institute for Theoretical Physics,
University of Nijmegen. Toernooiveld, 6525 ED Nijmegen, The Netherlands.
and
J.C. Toledano, Centre National d'Etudes des Télécomunications,
92220 Bagneux, France.

1. Introduction

We assume that the reader is familiar with the superspace group approach and with Landau's theory for continuous structural phase transitions [1],[2]. This Landau theory assumes that the high temperature crystal phase has a space group G_o as symmetry group and that there is a group-subgroup relation between the symmetry groups above and below the phase transition. Normally such a subgroup is a space group also. This type of crystal state can be destabilized by the occurrence of a so called Lifshitz term, leading to a phase which is incommensurate and lacks 3-dimensional lattice symmetry. The superspace approach has been developed precisely for recovering the space group symmetry of such phases by adding an internal space (see [3] and references therein). The problem is that those space groups have a dimension higher than three. In order to retain the group-subgroup relation (essential in a Landau theory) the high temperature phase is embedded as well (trivially) in a superspace of the same dimension (as that needed for the incommensurate phase) and has accordingly as symmetry group the direct product group $G_o \times E^d$, with E^d the euclidean group of the d-dimensional internal space. A preliminary paper on this subject has already appeared [4], and a more detailed one will be published elsewhere [5].

2. The Landau free energy of the supercrystal

The high temperature crystal density $\rho_s(\vec{r})$ is embedded as a supercrystal in the (3+d)-dimensional superspace according to

$$\rho_{os}(\vec{r}_s) = \sum_{\vec{k} \in \Lambda^*} \hat{\rho}(\vec{k}) \exp(i\vec{k}_s\vec{r}_s) \tag{1}$$

with $\hat{\rho}(\vec{k})$ the Fourier components of ρ_o, Λ^* the reciprocal lattice of the 3-dimensional space group G_o (symmetry of ρ_o), $\vec{r}_s = (\vec{r}, \vec{t}) \in R^{3+d}$ and $\vec{k}_s = (\vec{k},0)$. Accordingly the symmetry of ρ_s is $G_o \times E^d$ as mentioned above. Structural variations are considered :

$$\rho_{os} \rightarrow \rho_s = \rho_{os} + \delta\rho_s \tag{2}$$

which satisfy the matter conservation condition, so that each section $\vec{t}$ = constant defines a possible crystal configuration. Associated there is a variational free energy $F_s(T,P;\rho_s)$ whose minimum above T_c provides ρ_{os} and thus ρ_o. One requires invariance of F_s with respect to $G_o \times E^d$, expands F_s in powers of $\delta\rho_s$ and expresses the latter in terms of irreducible representations D^α of $G_o \times E^d$ with basis functions $\phi_j^\alpha(\vec{r}_s)$:

$$\delta\rho_s = \sum_{i,\alpha} c_i^\alpha \phi_i^\alpha(\vec{r}_s) \tag{3}$$

Keeping these basis functions fixed, F_s becomes a function of the variational constants c_i^α and is called Landau free energy $F_s = F_s(T,P,c_i^\alpha)$. The arguments of Landau's theory can be applied here also and lead to consider for a continuous phase transition the variation of one single irreducible representation.

3. Physical equivalence of low-temperature supercrystal states

For simplicity we restrict here considerations to $G_o = P\bar{1}$ and $d = 1$. This is enough for expressing the ideas. The irreducible representations of $G_o \times E^1$ are labeled by the set of four vectors $(\pm\vec{q},\pm\vec{Q})$ with $\vec{q}$ in the first Brillouin zone of Λ^*, and $\vec{Q}$ an arbitrary element of the internal space. One can write :

$$\delta\rho_s(\vec{r}_s) = c_1\exp(i\vec{q}\vec{r} + i\vec{Q}\vec{t}) + c_2\exp(i\vec{q}\vec{r} - i\vec{Q}\vec{t}) + c.c. \tag{4}$$

Putting $c_1 = \rho_1 \exp(i\phi_1)$ and $c_2 = \rho_2 \exp(i\phi_2)$ the form of the Landau free energy for $\vec{q}$ incommensurate with Λ^* is up to 4th order :

$$F_s = \frac{\alpha}{2}(\rho_1^2 + \rho_2^2) + \frac{\beta}{4}(\rho_1^2 + \rho_2^2)^2 + \frac{\gamma}{4}\rho_1^2\rho_2^2 \tag{5}$$

which admits three sets of non trivial minima for $T < T_c$: $\rho_1 \neq 0, \rho_2 = 0$, $\rho_1 = 0, \rho_2 \neq 0$, and $\rho_1 = \rho_2 \neq 0$. These solutions define different superspace equilibrium densities $\langle\rho_s(\vec{r}_s)\rangle$ with different symmetry groups in the superspace, which only in the first and second case are superspace groups. The physical crystal structure is that of a $\vec{t}$ = constant section :

$$\langle\rho_s(\vec{r}, \vec{t}=0)\rangle = \langle\rho(\vec{r})\rangle \tag{6}$$

and this defines an equivalence relation among superspace equilibrium densities. One finds that solutions involving the same $\vec{q}$ component are physically equivalent. The arbitrariness in the supercrystal configuration and of its symmetry is analogous to that of a potential with respect to the field.

4. Lifshitz criterion

The stability of a commensurate crystal state labeled by $\vec{k}_o$ with respect to those defined by neighbouring $\vec{k}$'s (in general incommensurate with Λ^*) is investigated by considering expansions with as coefficients slowly varying functions of the space coordinates:

$$\delta\rho = \sum_i c_i(\vec{r})\phi^i_{k_o}(\vec{r}) \tag{7}$$

The free energy is then written as integral of a free energy density $f = f(P,T,c_i(\vec{r}),\vec{\partial}c_i,\phi^i_{k_o})$ for fixed function ϕ_{k_o}. The lowest order term in a polynomial expansion involving derivatives and not giving rise to "surface" contributions only, has the form of a Lifshitz term :

$$\sum_{i,j} b_{ij}\,(c_i\vec{\partial}c_j - c_j\vec{\partial}c_i) \tag{8}$$

If such a term is different from zero, the Euler-Lagrange equations for $f(c_i,\vec{\partial}c_i)$ lead to an extremal value for F_s and to a possible minimum for non-constant c_i's. This implies the instability of the state $\vec{k}_o$, and eventually a transition to an incommensurate low-temperature phase labeled by $\vec{k}$, with symmetry requiring an internal space.

Consider now a high-temperature phase already incommensurate. On can repeat the discussion of the stability of a state labeled by a $\vec{k}_{os}$ commensurate with respect to Σ^*, the reciprocal lattice of the high temperature superspace group G_{os}, and consider variational expansions :

$$\delta\rho_s(\vec{r}_s) = \sum_i c_i(\vec{r})\phi^i_{k_{os}}(\vec{r}_s) \tag{9}$$

The possibility of limiting the coefficients to functions of the space coordinate only, is connected with the physical equivalency discussed above. For fixed $\phi^i_{k_{os}}(\vec{r}_s)$ one gets a free energy density f as above and the same Lifshitz criterion for the instability of the state $\vec{k}_{os}$. This indicates that the existence of a Lifshitz invariant is required in continuous phase transitions between crysstal phases of different internal dimensions.

References

[1] L.D. Landau and E.M. Lifshitz, Statistical Physics (Pergamon, London, 1968) 2nd ed., Chap XIV.

[2] G.Y. Lyubarskii, The application of Group Theory in Physics (Pergamon, London, 1960), Chap. VII.

[3] A. Janner, Physica 114A (1982) 614-616.

[4] A. Janner and T. Janssen, Helv. Phys. Acta 56 (1983) 665-675.

[5] A. Janner, T. Janssen and J.C. Toledano, to be published.

SYMMETRY BREAKING IN SOLID STATE AND PARTICLE PHYSICS

Marko V. JARIĆ

Department of Physics, Montana State University,
Bozeman, Montana 59717, U.S.A.
and Department of Nuclear Physics,
Weizmann Institute of Science, Rehovot 76100, Israel.

Abstract. Spontaneous symmetry breaking in the Landau theory of phase transitions in crystals and in the Higgs mechanism of gauge theories is reviewed from a single point of view: It is shown that in both cases the problem can be reduced to the one concerning a finite group. Subsequently, the results regarding minimization of a G-invariant polynomial, where G is a finite group, are reviewed.

I. INTRODUCTION. In both, the Landau theory of phase transitions[1,2] and in the gauge theories with the Higgs fields[3,4], a physical symmetry group G is broken in a similar way: the symmetry is determined by an absolute minimum of a G-invariant polynomial, the Landau free energy in the former case or the Higgs potential in the latter. To be sure, since in the first case G is typically an infinite, countable group, a crystallographic space group, while in the second case it is a compact Lie group, important differences do arise. Nevertheless, one has to face the same problems in both cases: First, given an orthogonal representation $R(G) \leq O(n)$ of G over a real vector space $\mathcal{R}^n$ find the most general fourth degree, G-invariant polynomial F: $\mathcal{R}^n \to \mathcal{R}$ (the restriction to fourth degree is dictated in the Landau theory by the requirement that the transition be a simple continuous one and, in the Higgs mechanism, by the renormalizability requirement). Second, for any given F determine the points at which it reaches absolute minima and find their symmetries, that is isotropy groups, in G.

In this contribution an overview of recent developments in solving the above problems will be presented. Following the pioneer work of Michel[5] the emphasis will be on treating the Landau and the Higgs problems from a single perspective. Furthermore, an attempt will be made to convince the reader that the above problems may be reduced for the Lie groups as well as for the crystallographic space groups to the case where G is a finite group.

It will be assumed in the following that the reader is fairly familiar with the basic notions and results for a compact group action. They have been presented, for example, by Michel[5-7] on several occasions. Only a brief summary is given in the appendix A. Similarly, the appendix B contains a summary of the basic results of the invariant theory. Sections II and III demonstrate how a finite group arises in the Higgs and in the Landau problems, respectively. Sections IV and V review carrier space and, respectively, orbit space approaches to minimization of a G-invariant potential when G is finite.

Although an extensive list of references will be given, it will only reflect the author's preference and his sources.

II. CASE OF A COMPACT LIE GROUP. It is assumed in this section that G is a compact Lie group acting orthogonally on R^n. The goal of this section is to show how the above G-action corresponds to a finite group action on a subspace of R^n and how this is used in the minimization problem. The main idea is to consider the smallest subspace which contains the fundamental region for the G-action. The reader is expected to be familiar with the contents of the appendix A and parts of the appendix B. The reader is furthermore strongly urged to become familiar with Ref. 18, especially its appendix.

It follows from Theorem A.5 of the appendix that all the information about a G-invariant function $F:R^n \to R$ is contained in its restriction to any global slice. Furthermore, since the points of the generic stratum have the smallest isotropy groups their orbits and the associated tangent planes will be the largest (i.e. have the largest cardinality). Consequently, a generic global slice is the smallest global slice and it is most beneficial to consider the restriction of F to that slice. We will denote a point from the generic stratum by x_o and its global slice (= its invariant slice ≡ generic slice, c.f. Corollary A.4.1) by $N(x_o)=N_o(x_o)$. Since $N_o(x_o)$ is not an invariant subspace under the action of G, the action of $N(N_o(x_o)) \leq G$ must be considered. Although $N(N_o(x_o))$ does not act effectively on $N_o(x_o)$ the quotient group $W(x_o) = N(N_o(x_o))/C(N_o(x_o)) = N(N_o(x_o))/L(x_o)$ does [it follows from Corollary A.4.1 that $C(N_o(x_o)) = L(x_o)$]. Since $N_o(x_o)$ is perpendicular to $\Omega_G(x_o)$ and using Theorem A.4 one concludes that in the neighborhood of $\Omega_G(x_o)$ the generic slice cuts the orbits in isolated points. Thus one arrives at the most important conclusion of this section: W is a finite group[5].

Therefore, instead of constructing a general (fourth degree) G-invariant polynomial $F:R^n \to R$ and minimizing it on R^n, one can construct a general (fourth degree) W-invariant polynomial $F_o:N_o(x_o) \to R$ and minimize it on $N_o(x_o)$. Note, however, that although every W-orbit will be contained in the intersection of a G-orbit by $N_o(x_o)$, the converse is not true: in general $\Omega_G \cap N_o(x_o) = \cup \Omega_W$ will contain (a continuum of) different W-orbits. Therefore, a W-invariant polynomial F_o is not in general the restriction of a G-invariant polynomial, and the additional condition that F_o be constant at the intersection of any G-orbit with $N_o(x_o)$ must be imposed. The special and favorable case does occur when the intersection of each G-orbit with $N_o(x_o)$ is a single W-orbit and, thus, when each W-invariant function is the restriction of a G-invariant function (the generic slice is then called[19] the "meridian section"). This is the case, for example, in the adjoint representation of a connected, compact, semisimple Lie group when $N_o(x_o)$ is a Cartan subalgebra and W is the Weyl group. A simple example is offered by SO(3) acting on R^3 where $N_o(x_o)=\{\lambda x_o, \lambda\epsilon R\}$, $L(x_o)=SO(2)=C_{oo}$, $N(N_o(x_o))= SO(2)xZ_2=\mathbf{C}_{ooh}$, $W=Z_2$.

Let us now return to the general case [which can be exemplified by the action of SO(2) on R^4 (vector $\oplus$ vector)]. The problem to be addressed is the following: Given the ring P_o of W-invariant polynomials (c.f. Appendix B) how does one determine the set P_o^* of W-invariant polynomials which are constant at any $\Omega_G \cap N_o(x_o)$ and what is the structure of P_o^*.

It is easy to see that just as the W-invariant polynomials form the ring P_o, the W-invariant polynomials which are constant at certain W-invariant subsets of $N_o(x_o)$ [e.g. on $\Omega_G \cap N_o(x_o)$] form a ring [e.g. P_o^*]. Clearly $P_o^* \leq P_o$. Furthermore, just as P_o has a natural grading, $P_o = \oplus_{m=o}^{oo} P_o^{(m)}$, so does P_o^*, $P_o^* = \oplus_{m=o}^{oo} P_o^{*(m)}$, in the subspaces of homogeneous polynomials of degree m. This follows from the remark that for every $F_o^* \epsilon P_o^*$ and every Ω_G, $\Omega_W' < \Omega_G$ and $\Omega_W'' < \Omega_G$ one has $F_o^*(\Omega_W') = F_o^*(\Omega_W'')$ if and only if $F_o^*(\lambda\Omega_W') = F_o^*(\lambda\Omega_W'')$ for every $\lambda\epsilon R$. Therefore, it is proposed here that the ring P_o^* (or the vector space $P_o^{*(m)}$) is generated by the basic polynomials $\{F_i^*\}_{i\epsilon I}$ obtained from the basic polynomials $\{F_i\}_{i\epsilon I}$ which generate P_o (or the vector space $P_o^{(m)}$) by

$$\{F_i^*(x)\}_{i\epsilon I} = \{\int d\mu(g) F_i(\mathbb{P}\cdot g\cdot x)\}_{i\epsilon I} \ , \ x\epsilon N_o(x_o), \tag{1}$$

where $d\mu(g)$ is the G-invariant (Haar) measure and $\mathbb{P}$ is the canonical projector $\mathbb{P}: R^n \to N_o(x_o)$. Note, however, that not all $F_i^*(x)$, $i\epsilon I$, will be independent in general (we will denote by I* the index set associated with the independent ones). Furthermore, the transformation Eq.(1) will change a basis $\{F_i^*\}_{i\epsilon I^*}$ to an in general different basis.

To justify the above proposition let us consider $z=x\oplus y\epsilon R^n$, $x\epsilon N_o(x_o)$, $y\epsilon T(x_o)$ and a function $\phi(z) \equiv F(\mathbb{P}\cdot z)$. By group averaging this function one obtains a G-invariant function $\int d\mu(g) F(\mathbb{P}\cdot g\cdot z)$. Finally, restricting this G-invariant to $N_o(x_o)$ amounts to replacing z by $\mathbb{P}\cdot z=x$ and one arrives at the result Eq.(1).

Once a W-invariant function F* constant at each $\Omega_G \cap N_o(x_o)$ is determined it can be minimized using the techniques developed in sections IV and V.

III. CASE OF CRYSTAL GROUPS. This section gives basic notions of space groups[8] and demonstrates that the symmetry group relevant in crystal-to-crystal transitions is a finite group.

The symmetry group G of a crystal in d-dimensions, a d-dimensional space group, is a denumerable subgroup of the Euclidean group $E(d)=R^d \rtimes O(d)$ which is an extension of a group of lattice translations T_G, $Z^d \sim T_G < R^d$ by a finite point group $P_G < O(d)$ such that the mapping $P_G \to Aut(T_G)$ is an injection ($\rtimes$ denotes the semi-direct product). Each element of E(d) can be uniquely written as a product of an element from R^d and an element from O(d). Therefore, one writes each element $g\epsilon G$ in the form $g=tfp$ where $t\epsilon T_G$, $p\epsilon P_G$ and a unique coset representative $f\epsilon[R^d:T_G]$ (a fractional translation) can be associated with each p. Closure under the multiplication requires that whenever $p_1, p_2, p_3 \equiv p_1 p_2 \epsilon P_G$ the following congruences are satisfied:

$$f_1+p_1 \cdot f_2-f_3 \epsilon T_G \qquad (2)$$

where f_i is associated with p_i, i=1,2,3.

Some basic facts about irreducible representation of G need to be recalled. Irreducible characters of a group of lattice translations form its dual group which is a d-torus. Among physicists it is better known as the first Brillouin zone (=Wigner-Seitz cell or Dirichlet region) of the lattice of reciprocal translations. Thus, an irreducible representation of T_G, say $R_k(T_G)$, is labelled by a "wave vector" k from the first Brillouin zone and for any $t\epsilon T_G$, $R_k(t)=\exp[2\pi i(k,t)]$, where (,) denotes the usual scalar product in R^d.

An irreducible representation R of G is in general reducible when restricted to T_G:

$$R(T_G)=\ell \cdot \bigoplus_{i=1}^{s} R_{k_i}(T_G) \ , \ \ell > 0 \qquad (3)$$

The set ("star") $\{k_1,\ldots,k_s\}$ is uniquely determined by R. The vectors k_i are in one-to-one correspondence with the left cosets of $[P_G:L(k_1)]$ where $L(k_1)$ is the isotropy (=little) group of k_1 in P_G.

Our task in the Landau theory is to find the isotropy group associated with an absolute minimum of an R(G)-invariant thermodynamic potential. An important theorem permits reduction of this problem to a problem involving a finite group:

Theorem III.1.[8] A group of lattice translations is the group of lattice translations of an isotropy subgroup of G if and only if its reciprocal lattice is generated by the reciprocal lattice of T_G and by a subset of the star $\{k_1,\ldots k_s\}$.

Therefore, whether the transition is crystal-to-crystal or crystal-to-incommensurate-crystal, one ends up with a finite group. In the first case the components of the wave vectors of the star are rational so that the kernel K of the representation R is a d-dimensional space group and the relevant group $R(G)\sim G/K$ is a finite group. In the second case, although R(G) is not finite, its isotropy groups are. Only the first case, crystal-to-crystal transitions, will be considered in the following.

Therefore, the remainder of this contribution deals with a finite group G acting effectively, $G\sim R(G)$, and orthogonally, $R(G)<O(n)$, on a carrier space R^n. It will be furthermore assumed that the G-invariant polynomial $F:R^n \to R$, which is to be minimized, is given (c.f. Section II and Appendix B).

IV. CARRIER SPACE APPROACH.[20] Absolute minima of a G-invariant polynomial F are a special kind of its critical points. Therefore, they are to be found among the points at which the gradient of F, ∇F, vanishes. Since ∇F is a G-covariant vector field, that is, for every $g \in G$ and every $x \in \mathbb{R}^n$,

$$\nabla F(g \cdot x) = g \cdot \nabla F(x), \qquad (4)$$

we know that $L(\nabla F(x)) \geq L(x)$. Consequently, if $L \leq G$ is an isotropy group then for every $x \in \mathrm{Fix}L$, $\nabla F(x)$ will be contained in FixL. This is just a generalization of a well known fact that a vector field cannot "cross" a reflection hyperplane or a rotation axis. Therefore, at FixL the [n-dim(FixL)] components of F perpendicular to FixL automatically satisfy $\nabla F=0$.

It is a straightforward observation that for any two isotropy groups L_1 and L_2 $\mathrm{Fix}L_1$ contains $\mathrm{Fix}L_2$ if and only if L_2 contains L_1. Thus, an efficient scheme in determining all the solutions of $\nabla F=0$ is to proceed by solving $\nabla F=0$ restricted to FixL's of a successively higher dimension. In the process, previously found solutions (in FixL's of lower dimensions) are removed from the equations. In this fashion one solves equations in systematically more unknowns, but the equations of lower degree then the original equation $\nabla F=0$.

To be more specific, let $\mathbb{P}_L$ be the projector $\mathbb{P}_L : \mathbb{R}^n \to \mathrm{Fix}L$ so that $(1-\mathbb{P}_L)\cdot x=0$ gives [n-dim(FixL)] independent linear equations of the plane FixL. The remaining dim(FixL) equations for $\nabla F=0$ at that plane are

$$\mathbb{P}_L \cdot \nabla F = 0 \qquad (5)$$

Let $L'>L$ be another isotropy group so that $\mathrm{Fix}L' < \mathrm{Fix}L$. Then, the solutions from FixL' are contained in Eq.(5) and they satisfy $(1-\mathbb{P}_{L'})\cdot x = (1-\mathbb{P}_{L'})\cdot \mathbb{P}_L \cdot x = 0$. Whenever dim(FixL)-dim(FixL')=1 these solutions may be factored out of Eq.(5),

$$(1-\mathbb{P}_{L'})\cdot \mathbb{P}_L \cdot \nabla F(x) \equiv f'(x)\,[(1-\mathbb{P}_{L'})\cdot \mathbb{P}_L \cdot x] \qquad (6)$$

to give an equation of a lower degree:

$$f'(x) = 0 \qquad (7)$$

Each $L'>L$, dim(FixL)-dim(FixL')=1, gives such an equation. When the number of such L' is greater than dim(fixL) one first uses precisely dim(FixL) of them to obtain a corresponding number of equations of the type Eq.(7). The solutions associated with the remaining L' must be contained in those equations. They may be removed using the same technique now applied to the action of N(L)=N(FixL) on FixL. One only needs to observe that the functions $f'(x)$ transform under this action as the permutation representation of $N(L) \cap N(L')$ in N(L).

Of course, in order to use the above method it is first necessary to be able to determine all the isotropy groups for the given G-action. For finite images R(G) Theorems A.1 and A.2 (i.e. the "chain criterion"[21]) suffice. Nonetheless, the case when G is a space group is particularly intricate and needed a separate treatment which led to an efficient algorithm.[22,8]. Furthermore, Ref. 12 contains the isotropy groups for all $R(G)\leq O(n)$, n=2,3.

The above technique has been recently[23] used with great success to find the first counterexamples to the Ascher-Michel maximality conjecture[24,5] and to its extension, the Gell-Mann-Slansky conjecture.[25,26] This is reported in another contribution to this conference.

V. ORBIT SPACE APPROACHES.[9,12,26-28] Several results of the invariant theory which are summarized in Appendix B will be used in this section. A motivation for the approach presented here lies in the fact that a physical problem which possesses a symmetry G always has "degenerate" solutions which fall on a G-orbit. Therefore, it is sometimes advantageous to formulate the problem in terms of invariants which are constant on any orbit. In this section $G\sim R(G)$ is a finite group acting orthogonally on $\mathbb{R}^n$.

Let us denote the map $\omega:\mathbb{R}^n\to\mathbb{R}^{n+k}$ defined by the integrity basis, $\omega(x)=[\theta_1(x),\dots,\theta_n(x),\phi_1(x),\dots,\phi_k(x)]$. Usefulness of ω arises from the fact that ω is constant on any given G-orbit. It also separates orbits, that is $\omega(x)=\omega(y)$ if and only if x and y are on the same G-orbit. An elementary constructive proof may be found in the second citation of Ref. 28. The subset $\omega(\mathbb{R}^n)<\mathbb{R}^{n+k}$, which will be called the orbit space, is an n-dimensional, semi-algebraic surface. Its determination is an important problem. Namely, since every G-invariant polynomial is a polynomial in the integrity basis its absolute minima may be determined directly in the orbit space. Corresponding orbits in $\mathbb{R}^n$, when necessary, can be determined by inverting the map ω.

There are essentially two methods for determining $\omega(\mathbb{R}^n)$. In the first method[28], given an isotropy group L, one constructs linear subspace FixL

$$\mathrm{Fix}L=\{x\,|\,(1-\mathbb{P}_L)\cdot x=0\}. \tag{8}$$

Next, one constructs a G-invariant equation for the closure of the stratum $\Sigma[L]$,

$$\bigcup_{L\in[L]}\mathrm{Fix}L=\{x\,|\,\prod_{L\in[L]}||(1-\mathbb{P}_L)\cdot x||^2=0\}. \tag{9}$$

Since $\Pi||(1-\mathbb{P}_L)\cdot x||^2\equiv p(x)$ is a G-invariant polynomial it can be expressed as a polynomial in $\omega(x)$, $p(x)\equiv\hat{p}(\omega(x))$. Thus, the image of FixL in the orbit space is given by the equation $\hat{p}(\omega)=0$. By considering all conjugacy classes of isotropy groups one can construct the whole orbit space and the images of the strata. Note, however, that some caution must be exercised since this procedure will, in general, lead to some spurious, but easily identifiable, pieces outside of $\omega(\mathbb{R}^n)$.[28]

The second method[9,27] for determining $\omega(\mathbb{R}^n)$ is based on Theorem B.2. It follows from that theorem that at any $x\in\mathbb{R}^n$ the rank of the $(n+k)\times n$ matrix $\nabla\omega(x)$

is equal to the dimension of FixL(x). It is also equal to the rank of the $(n+k)\times(n+k)$ Grammian matrix $M(x)\equiv(\nabla\omega(x),\nabla\omega(x))$, where $(\ ,\ )$ is the usual scalar product in R^n. Since the scalar product is G-invariant, the components of M(x) are G-invariant polynomials and, thus, polynomials in the integrity basis $\omega(x)$, $M(x)\equiv\hat{M}(\omega(x))$. Therefore, the following conclusion is reached: Image of a connected component of a stratum of dimension m is entirely contained in a connected component of the semi-algebraic surface given by rank $\hat{M}(\omega)=m$, $\hat{M}(\omega)\geq 0$. The second condition is imposed since M(x) is clearly positive semi-definite. This surface may also be defined[28] as

$$Sp_i\hat{M}(\omega)\begin{cases} >0 : i\leq m \\ =0 : i>m \end{cases}, \tag{10}$$

where Sp_i is the sum of principal minors of order i. In particular, m=n gives $\omega(R^n)$ but it may, just as in the first method, also give some spurious pieces outside $\omega(R^n)$.

Once the orbit space $\omega(R^n)$ and its stratification are determined one can proceed to minimize a G-invariant polynomial F(x) [which is also a polynomial in the integrity basis, $F(x)\equiv\hat{F}(\omega(x))$]. Of course, to determine the symmetry breaking direction, it suffices to minimize $\hat{F}(\omega)$ on the hyperplane $\delta\equiv\{\omega\in R^{n+k},\ \omega_1\equiv\theta_1=1\}$ which corresponds to the unit sphere, $||x||^2=1$, in R^n. In the special case when R is irreducible and F(x) is of fourth degree it is clear that $\hat{F}(\omega)=$ const defines a hyperplane in δ. Therefore, $\hat{F}(\omega)$ reaches its minimum on $\omega(R^n)$ at a point where a hyperplane $\hat{F}(\omega)=$const, $\omega\in\delta$, is the tangent plane to $\delta\cap\omega(R^n)$. This geometric picture[26] may be useful when ω contains a small number (≤ 4) of third and fourth degree invariants, but it is of limited value otherwise.

Generally, the equation $\nabla F(x)=0$ can be replaced, c.f. Theorem B.2, by the (n+k) equations $(\nabla F(x),\nabla\omega(x))=\hat{M}(\omega)\partial_\omega\hat{F}(\omega)=0$. Of course, at a stratum of dimension m one only needs m equations, e.g.

$$(\nabla F(x),\nabla Sp_iM(x))=[\partial_\omega\hat{F}(\omega)]\hat{M}(\omega)[\partial_\omega Sp_i\hat{M}(\omega)]=0,\ i=1,\dots,m\leq n. \tag{11}$$

Solving these equations one obtains the solutions to the minimization problem directly in the orbit space.

This procedure has been generalized to the determination of zeros of a general G-covariant vector field.[12] As stated in Appendix B such a field, $v:R^n\to R^n, v\cdot g=g\cdot v$, can be expressed in terms of nk basic fields $e_j(x)$, $v(x)=\Sigma_{j=1}^{nk}\hat{q}_j(\omega)e_j(x)$ where $\hat{q}_j$ are the polynomials in the invariants $\theta_i(x)$. The essential point is to determine m basic fields, say $e_j(x)$, j=1,...,m, which are <u>linearly</u> independent on (a part of) a given m-dimensional stratum. The invariant equations for the zeros v(x)=0 at that (part of the) stratum are

$$(v(x),e_j(x))=\sum_{i=1}^{nk}\hat{q}_i(\omega)\hat{M}_{ij}(\omega)=0,\ j=1,\dots,m. \tag{12}$$

The matrix $\hat{M}_{ij}(\omega) \equiv M_{ij}(x) \equiv (e_i(x), e_j(x))$ is a generalization of the previously defined matrix $\hat{M}$.

VI. CONCLUSION. The main goal of this contribution was to demonstrate that a problem of minimizing a G-invariant function over a representation space R^n can be reduced in cases when G is either a compact Lie group or a space group to a similar problem involving only a finite group. Furthermore, carrier and orbit space approaches to minimization of G-invariant polynomials (when G is finite) where reviewed. This author's opinion is that while the orbit space approaches may be helpful in deriving some general results, the carrier space approach is more suitable in practical applications. Finally, it appears that the methods for finding critical points of Landau-Higgs potentials are fairly complete and most of the remaining problems rest on the isolation and further specification of absolute minima.

APPENDIX A: GROUP ACTION. Orthogonal action of a compact Lie group (finite group being a special case) on R^n is considered in this section. Most of the contents of this section can be found in Ref's. 5 and 7 where a more general group action is considered.

Relative to a G-action one defines various subgroups of G. (a) The isotropy group (=little group) of a vector $x \in R^n$ is the largest subgroup $L(x) \leq G$ such that x is fixed by L(x). (b) The centralizer of a subspace $V \leq R^n$ is the largest subgroup $C(V) \leq G$ such that V is pointwise fixed by C(V). (c) The normalizer of a subspace $V \leq R^n$ is the largest subgroup $N(V) \leq G$ such that V is invariant under N(V).

Furthermore, one defines various geometric structures in R^n. (d) A G-orbit through x is $\Omega_G(x) = \{g \cdot x, g \in G\}$. Since isotropy groups of points of an orbit are conjugated, the orbit types are defined by conjugacy classes [L] of isotropy groups and one defines (e) the stratum $\Sigma[L]$ as the union of orbits of the same type. (f) FixL is the largest subspace of R^n which is pointwise fixed by a subgroup $L \leq G$. (g) The tangent plane $T(x)$ to $\Omega_G(x)$ at x is the plane $\{\tau \cdot x, \tau \in \text{Lie algebra of } G\}$. (h) The normal plane (=global slice) $N(x)$ to $\Omega_G(x)$ at x is the orthogonal complement to $T(x)$ in R^n, $R^n = T(x) \oplus N(x)$. (i) The invariant slice through $x \in R^n$ is $N_0(x) = N(x) \cap \text{FixL}(x)$ (the term "invariant" is somewhat misleading, it refers to the action of L(x) on $N(x)$).

The following results will be quoted without the proofs which can be found in the references. However, quoted references do not necessarily give the first or the most simple proofs.

Theorem A.1.[8] A subgroup $C \leq G$ is the centralizer of a subspace of R^n if and only if for every L, $C < L \leq G$, $\dim(\text{FixL}) < \dim(\text{FixC})$.

Theorem A.2.[8] The isotropy group of any $x \in R^n$ is the centralizer of some V,

$\{\lambda x, \lambda\varepsilon R\} \leq V \leq R^n$. Conversely, a centralizer $C(V)$, $V \leq R^n$, is the isotropy group of some $x\varepsilon \mathrm{Fix} C(V) \leq R^n$ whenever the set of isotropy groups which contain $C(V)$ is countable.

Theorem A.3.[5] There is a unique minimal conjugacy class of isotropy subgroups and corresponding stratum called the generic stratum is open and dense (the partial order is defined by $[L_1]<[L_2]$ iff $\exists L_1\varepsilon[L_1]$ and $\exists L_2\varepsilon[L_2]$ such that $L_1<L_2$).

Theorem A.4.[5] For every $\Omega_G(x)$ there is a G-invariant neighborhood such that for every x' in that neighborhood $[L(x')] \leq [L(x)]$.

Corollary A.4.1.[5] If $x_o\varepsilon R^n$ is a generic point, i.e. $[L(x_o)]$ is minimal, then $N(x_o)=N_o(x_o)$.

Lemma A.1. For every $x\varepsilon R^n$, $\mathrm{Fix}L(x)=N(x)\cap \mathrm{Fix}L(x) \oplus T(x)\cap \mathrm{Fix}L(x)$.

Theorem A.5.[5] Each global slice $N(x)$ cuts every G-orbit in at least one point. The references 5 and 7 as well as relevant chapters of Ref. 9 are recommended for further reading.

APPENDIX B: INVARIANT THEORY. In this appendix some of the relevant results of the invariant theory are briefly summarized. A mathematical review of the topic may be found in Ref. 10. An excellent review, readable to physicists, may be found in Ref. 11. The subject is also treated and summarized in several places in physics literature.[5,9,12,13] Henceforth, G and its image R(G) are identified and G is considered a finite subgroup of O(n).

The infinite dimensional vector space P of all polynomials in $x\varepsilon R^n$ is also a ring. It can be decomposed as a direct sum

$$P = \bigoplus_{m=o}^{\infty} P^{(m)} \tag{B.1}$$

where $P^{(m)}$ is the $\binom{n+m-1}{m}$-dimensional space of homogeneous polynomials of degree m. Since the action of G on the polynomials preserves the degree, each $P^{(m)}$ carries a linear representation of G. This representation and the space $P^{(m)}$ can be reduced into a sum of isotypical components $P^{(m)}=\oplus P^{(m)}_\alpha$, where α labels different equivalence classes of irreducible representations of G. The multiplicity c^m_α of an irreducible representation whose characters are χ_α is given by the generating function

$$M_\alpha(t) \equiv \sum_{m=o}^{\infty} c^m_\alpha t^m = \frac{1}{|G|} \sum_{g\varepsilon G} \chi^*_\alpha(g)\det(1-gt)^{-1} \tag{B.2}$$

when α=0, which will be reserved for the trivial representation, $M_o(t)$ is called the Molien function, while for general α it is called by mathematicians the Poincaré series.

An important result of the invariant theory is that the ring of invariant polynomials $P_o = \oplus_m P_o^{(m)}$ is finitely generated. More specifically, there are precisely n algebraically independent, homogeneous polynomials $\theta_i(x)$, i=1,...,n, and certain number k of other, algebraically dependent, homogeneous polynomials $\phi_j(x)$, j=1,...k, $[\phi_1(x) \equiv 1]$ such that every G-invariant polynomial can be uniquely written as a linear form in ϕ's with polynomials in θ's as coefficients. One says that P_o is a free module over the ring $R[\theta_1,\ldots,\theta_n]$ of polynomials in θ's. The invariant polynomials θ and ϕ are called the integrity basis of P_o. Its structure is reflected in the form of the Molien function:

$$M_o(t) = \left(\sum_{j=1}^{k} t^{\delta_j}\right) / \prod_{i=1}^{n} (1-t^{d_i}), \tag{B.3}$$

where d_i=degree of θ_i and δ_j=degree of ϕ_j.

Therefore, a knowledge of the Molien function is very useful in constructing the integrity basis. For the point groups, n=3, the Molien functions can be found in several publications[13-16]. A method for calculating the Molien functions for images of space groups is also available[17]. Actual construction of the invariant polynomials can be accomplished by projecting the invariant part $p_o(x)$ out of a polynomial $p(x)$:

$$p_o(x) = |G|^{-1} \sum_{g \in G} p(g \cdot x). \tag{B.4}$$

The module structure of P_o is, in fact, a consequence of a much more general theorem.

<u>Theorem B.1.</u>[11] P is a free $R[\theta_1,\ldots,\theta_n]$-module and its quotient by $R[\theta_1,\ldots\theta_n]$ carries k-times the regular representation of G.

In particular, for the case of interest in this paper, there are kn independent (but not linearly!) basic polynomial G-covariant vector fields such that every polynomial G-covariant vector field can be uniquely expressed as a linear combination of the basic fields with coefficients from $R[\theta_1,\ldots,\theta_n]$. Specifically, gradients of the integrity basis are G-covariant vector fields. Their importance lies in the following theorem

<u>Theorem B.2.</u>[12] The gradients $\nabla\theta_i(x)$, i=1,...,n, $\nabla\phi_j(x)$, j=1,...,k, span FixL(x).

<u>REFERENCES.</u>

1) L. D. LANDAU, Zh. Eksp. Teor. Fiz. 7, 19 (1937); 7, 627 (1937); L. D. LANDAU and E. M. LIFSHITZ, *Statistical Physics* (Pergamon, New York, 1968).
2) For a review of group theoretical results until 1981 see M. V. JARIC, Physica 114A, 550 (1982).
3) P. HIGGS, Phys. Rev. Lett. 13, 508 (1964); Phys. Rev. 145, 1156 (1966).
4) For a review see L. O'RAIFEARTAIGH, Rep. Prog. Phys. 42. 159 (1979).

5) L. MICHEL, preprint no. TH-2716-CERN (1979).
6) L. MICHEL, Rev. Mod. Phys. 52, 617 (1980).
7) L. MICHEL in *Statistical Mechanics and Field Theory*, ed. R. N. SEN and C. WEIL (Israel University Press, Jerusalem, 1972) p. 133.
8) M. V. JARIC and M. SENECHAL (to be published).
9) M. ABUD and G. SARTORI (to be published).
10) J. DIEUDONNE and J. B. CARRELL, *Invariant Theory, Old and New* (Academic Press, New York, 1970).
11) R. P. STANLEY, Bull. Am. Math. Soc. 1, 475 (1979).
12) M. V. JARIĆ, L. MICHEL and R. T. SHARP (to be published); Lec. Notes Phys. 180, 317 (1983); preprint no. IHES/P/83/23.
13) L. MICHEL in *Group Theoretical Methods in Physics*, ed. R. T. SHARP and B. KOLMAN (Academic Press, New York, 1977) p. 75.
14) B. MEYER, Can. J. Math. 6, 135 (1953).
15) J. KILLINGBECK, J. Phys. C5, 2497 (1972).
16) V. KOPSKY, J. Phys. C8, 3251 (1975).
17) M. V. JARIC and J. L. BIRMAN, J. Math. Phys. 18, 1459 (1977).
18) L. MICHEL, Proc. of the Primersko Summer School (1981) (to appear in Lecture Notes in Mathematics).
19) L. MICHEL (private communication).
20) M. V. JARIĆ, Phys. Rev. Lett. 48, 164 (1982).
21) M. V. JARIĆ, Phys. Rev. B23, 3460 (1981).
22) M. V. JARIĆ, preprint IHES/P/82/17 (1982); J. Math. Phys. (to be published).
23) D. MUKAMEL and M. V. JARIĆ, preprint WIS-83/16 April-Ph (1983); M. V. JARIĆ (to be published); proceedings of this conference.
24) E. ASCHER, Helv. Phys. Acta 39, 40 (1966); 466 (1966); Phys. Lett. 20, 352 (1962).
25) R. SLANSKY, Phys. Rep. 79C, 1 (1981).
26) J. S. KIM, Nuc. Phys. B196, 285 (1982); B197, 174 (1982); S. FRAUTSCHI and J. S. KIM, Nuc. Phys. B196, 301 (1982).
27) M. ABUD and G. SARTORI, Phys. Lett. 104B, 147 (1981); G. SARTORI, J. Math. Phys. 24, 765 (1983).
28) M. V. JARIĆ, J. Math. Phys. 24, 917 (1983); Lec. Notes Phys. 135, 12 (1980).

ACKNOWLEDGEMENT. The author wishes to use this opportunity to express his gratitude to Professor L. Michel for providing over the last several years an inexhaustible source of knowledge and intellectual stimulation.

The author acknowledges the hospitality and financial support from the Einstein Center for Theoretical Physics at the Weizmann Institute of Science.

COUNTEREXAMPLES TO THE MAXIMALITY CONJECTURE OF LANDAU-HIGGS MODELS

Marko V. JARIĆ

Physics Department, Montana State University,
Bozeman, Montana 59717, USA, and
Department of Nuclear Physics, Weizmann Institute of Science,
Rehovot 76100, Israel

Abstract. Two counterexamples to the Ascher-Michel maximality conjecture are presented. One of the models is extended to the case of reducible order parameter/Higgs field to provide a counterexample to the Gell-Mann-Slansky conjecture. Physical consequences of these counterexamples are discussed in the context of phase transitions.

I. Introduction

For almost fifty years the Landau theory[1] of phase transitions has been one of the most successful theories used to predict phase diagrams in various systems. Starting from a high symmetry group G, the low symmetry group is determined as the symmetry of the equilibrium value of an order parameter $x \in R^n$. This value is obtained by minimizing a G-invariant thermodynamic potential $F: R^n \to R$. Such minimization is often a nontrivial task which has been confronted in both the Landau theory as well as in the Higgs mechanism of gauge field theories.[2]

In 1962 Asher[3] introduced an essentially empirical conjecture which, if correct, would remove in an elegant stroke most of the difficulties of the minimization. The conjecture stated that in a simple continuous transition the low symmetry phase must be a maximal isotropy (strict) subgroup of G. The reader is reminded that R^n spans an irreducible (on R^n) G-representation $R(G) \leq O(n)$ and that an isotropy group of a vector $x \in R^n$ is the largest subgroup of G which fixes x.

Although the above conjecture withstood the test of time for over twenty years, it became clear[4,5] that the relevant group is not G, or equivalently R(G), but a group $R^*(G) \leq O(n)$, the symmetry of the R(G)-invariant potential F. The group $R^*(G)$ contains R(G) and may contain it strictly. The reason is that a requirement that the transition be a simple continuous one dictates that F be a polynomial of a definite (fourth) degree. Such a particular restriction may have as a consequence that the true symmetry of F in O(n) is a group $R^*(G)$ strictly larger than R(G). In such a case, although the low symmetry group might be a maximal isotropy group in $R^*(G)$, its intersection with R(G), which is the relevant group for the G-action, need not be a maximal isotropy group in R(G). Precisely this occurred in an example of TOLÉDANO.[6] Therefore, MICHEL[5] has reformulated the conjecture and extended it to compact groups: A fourth degree polynomial $F: R^n \to R$ which is bounded from below, has maximum at the origin and whose symmetry group in O(n), $R \leq O(n)$, is compact and irreducible on $\mathbb{R}^n$

has absolute minima only at maximal isotropy subgroups of R.

The ASCHER-MICHEL conjecture has not been disproved and it found fruitful ground in the Higgs problems. However, the Higgs fields must often be taken to belong to a reducible representation R and the conjecture must be extended. Let $R=S\oplus T$, $S\leq O(n)$, $T\leq O(m)$ be a reducible symmetry of the potential $F:(R^n\oplus R^m)\to R$. In such a case the GELL-MANN-SLANSKY conjecture[7,8] amounts to replacing the maximal isotropy subgroups by maxi-maximal subgroups. The maxi-maximal subgroups are defined in the following way. Consider a maximal isotropy subgroup L_S of R in the representation S. The representation T of R is in general reducible when restricted to L_S. For each of its irreducible components one can then construct a maximal isotropy subgroup, say $L_{ST}\leq L_S$. The subgroups L_{ST} (and L_{TS}) obtained in such a way are the maxi-maximal subgroups of R.

In the following sections we will discuss essential features of the first counterexamples to the conjectures mentioned above. More details will be published elsewhere.[9,10]

II. First example.[9]

Let the carrier space be $R^{2n}=R^n\oplus R^n$ and its vectors $x=y\oplus z$. Furthermore, let $R\leq O(2n)$ be the group generated by B_n, the symmetry of n-dimensional cube which acts simultaneously on both y and z, and by the elements μ_i, $i=1,\dots,n$ and η. The elements μ_i interchange i-th components of y and z while η changes the sign of z. Order of R is $2^{2n+1}n!$ and R is irreducible on R^{2n}. There are no third degree invariants and there are three linearly independent fourth degree invariants other than $||x||^4$: $I_1=\Sigma(y_i^4+z_i^4)$, $I_2=(\Sigma y_iz_i)^2$, $I_3=\Sigma y_i^2z_i^2$. The role of I_3 is not important in the following analysis and we will consider the anisotropic, quortic part of F, vI_1+wI_2. R is the complete symmetry of such F so that the conditions of the Ascher-Michel conjecture are met.

When the coefficient v is positive the first term is minimized when each y and z are along a body diagonal of the n-cube of B_n. On the other hand, when w is negative the second term is minimal when y and z are parallel while for w>o it is minimal when y and z are perpendicular. In the first case, v>o and w<o, the minimum is at x along a direction where both y and z are along the same body diagonal. However, in the second case, v>o and w>o, y and z try to be both along body diagonals and, at the same time, perpendicular to each other. The two tendencies can be reconciled only for even n. For odd n one expects and indeed finds that y and z compromise and y points along (a...ab..b) direction, while z points along (a..a-b...-b) direction. The ratio a^2/b^2 varies with w, from $a^2/b^2=\frac{n-m}{m}$ as $w\to+\infty$ to $a^2/b^2=1$ as $w\to o$ $(m=[n/2])$. The isotropy group L of such x is clearly nonmaximal since it is contained in both, the isotropy group L_a of x_a such that y=z=(a...ao..o) as well as in the isotropy group L_b of x_b such that y=-z=(o...ob...b).

As a counterexample to the GELL-MANN-SLANSKY conjecture one can consider the same group, a representation S the same as the representation defined above and a second representation T, a one dimensional representation where all the generators

B_n and μ_i are represented by 1 and η by -1. The maximal isotropy groups of S are the same as before while the maximal isotropy group of T is the group L_T generated by B_n and $\mu_i, i=i,\dots,n$. Since neither L_a nor L_b do contain η they are maxi-maximal subgroups (they occur in both L_{ST} and L_{TS} schemes) while L is not a maxi-maximal subgroup. Now, consider the simplest case when S and T are not coupled, i.e. $F(x)=F_S(x_S)+F_T(x_T)$. The isotropy group of an absolute minimum is the intersection of isotropy groups of corresponding x_S and x_T. Therefore, in a special case, this symmetry is $L\cap L_T=L$ which is not a maxi-maximal subgroup.

III. Second example[10]

In this example a group $R\leq O(4)$ acting on $x\epsilon R^4$ is considered. Due to the lack of space the reader is directed to Refs. 10 and 11 for details on the notation.

The group to be considered is $R=[\bar{D}_3/\bar{C}_1,\bar{O}/\bar{D}_2]$ where bars indicate corresponding inverse images under $SU(2)\to SO(3)$ and the traditional Schoenfils notation for subgroups of SO(3) is employed. There are two nonisotropic, linearly independent quartic R-invariants: $I_1=\Sigma x_i^4$ and $I_2=\Sigma\ [x_ix_j(x_k^2-x_1^2)-x_1x_i(x_j^2-x_k^2)]$ where the last sum is over cyclic permutations of (ijk)=(234). The group R is irreducible on R^4 and it is the complete symmetry of an R-invariant fourth degree polynomial containing vI_1+wI_2, therefore it fulfills the assumptions of the Ascher-Michel conjecture.

As a preliminary step one can set $\|x\|^2=1$ since the direction of symmetry breaking does not depend on $\|x\|^2$. The parametrization of the unit sphere can be set by $x_1=\cos\alpha_1$, $x_2=\sin\alpha_1\cos\alpha_2$, $x_3=\sin\alpha_1\sin\alpha_2\cos\alpha_3$ and $x_4=\sin\alpha_1\sin\alpha_2\sin\alpha_3$ with $\alpha_1,\alpha_2\epsilon[o,\pi)$ and $\alpha_3\epsilon[o,2\pi)$. Next, an (I_1,I_2) graph is plotted by evaluating $I_1(\alpha_1,\alpha_2,\alpha_3)$ and $I_2(\alpha_1,\alpha_2,\alpha_3)$ at several thousand randomly chosen points $(\alpha_1,\alpha_2,\alpha_3)$.[8] In such a way one obtains a fairly good idea of the domain over which the potential $F=vI_1+wI_2$ needs to be minimized as a function of I_1 and I_2 (Fig.1). Clearly, the minima (or maxima) occur at the points on the boundary which are such that the tangent <u>does not</u> cross the domain. Therefore, the minima can occur only along the thick line connecting the points A and B. On the other hand, a singularity (cusp) at the boundary of the domain must be associated with a maximal isotropy group. Thus, if A and B are indeed cusps then associated isotropy subgroups would strictly include the isotropy group associated with a point on the thick line between A and B and one would have a counterexample. Explicit calculations, like in the previous example, which use the method of Ref. 12, confirm the above. The isotropy groups associated with A and B are $[\bar{D}_3/C_1,\bar{D}_3/C_1]$ and $[\bar{D}_3/C_1,\bar{D}_3/C_1]'$ (not equivalent in R) whose strata are one dimensional. Isotropy group associated with the thick line is $[\bar{C}_3/C_1,\bar{C}_3/C_1]$ whose stratum is two-dimensional. For completeness we included the only other nontrivial, also two-dimensional stratum (thin line) of the isotropy group $[\bar{C}_2/C_1,\bar{C}_2/C_1]$.

IV. Conclusion

Two counterexamples to the Ascher-Michel conjecture and a counterexample to the Gell-Mann-Slansky conjecture have been demonstrated. It is important to emphasize the central role played in these discoveries by recent technical advances in the field.[2,12] At the same time, these counterexamples showed that the Landau theory can describe more complex physical situations. For example, whereas the Ascher-Michel conjecture would imply that a transition between two ordered phases is necessarily discontinuous the counterexamples show that a continuous transition between a maximal and a nonmaximal ordered phase is possible. Furthermore, in a nonmaximal phase the order parameter direction is never fixed and may vary with the temperature.

Acknowledgement. The author acknowledges the hospitality and financial support from the Einstein Center for Theoretical Physics at the Weizmann Institute of Science.

References

1. L.D. LANDAU, Zh. Eksp. Teor. Fiz. 7, 19 (1937); 627 (1937); L.D. LANDAU and E.M. LIFSHITZ, Statistical Physics (Pergamon, New York, 1968).
2. The reader is urged to read the review by M.V. JARIĆ in the same proceedings.
3. E. ASCHER, Helv. Phys. Acta 39, 40, 466 (1966); Phys. Lett. 20, 352 (1962).
4. M.V. JARIĆ, Phys. Rev. B18, 2237 (1978).
5. L. MICHEL, preprint TH-2716-CERN (1979).
6. J.C. TOLÉDANO and P. TOLÉDANO, J. de Phys. 41, 189 (1980).
7. R. SLANSKY, Phys. Rep. 79C, 1 (1981).
8. J.S. KIM, Nuc. Phys. B196, 285 (1982).
9. D. MUKAMEL and M.V. JARIĆ, preprint WIS-83/16-Ph (1983) and to be published.
10. M.V. JARIĆ (to be published).
11. L. MICHEL, J.C. TOLÉDANO and P. TOLÉDANO in Symmetries and Broken Symmetries in Condensed Matter Physics, ed. N. BOCCARA (IDSET, Paris 1981) p.263.
12. M.V. JARIĆ, Phys. Rev. Lett. 48, 164 (1982).

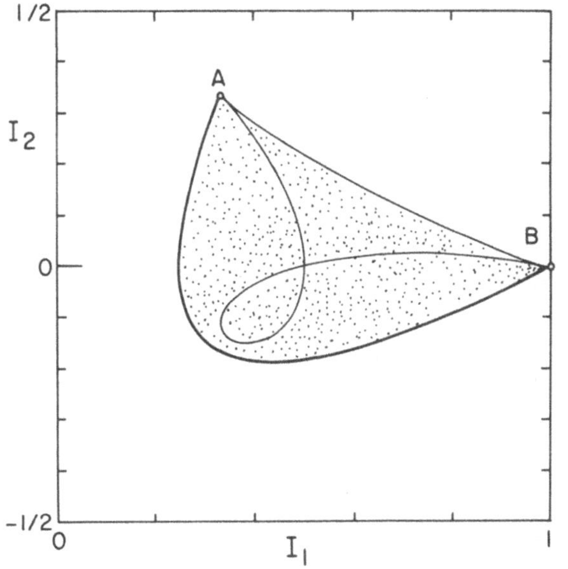

Figure 1

SOME MATHEMATICAL PROBLEMS IN RENORMALIZATION GROUP THEORY

G. Jona-Lasinio (*)
Laboratoire de Physique Théorique et
Hautes Energies - Université Paris VI

Summary

Under the name Renormalization Group (RG) one usually collects some techniques and points of view developed over a long period, but more intensely over the last fourteen years, to deal with systems of infinitely many degrees of freedom. Recently there have been also important applications in other directions like the long time behaviour of deterministic systems with sensitive dependence on initial conditions. Most of the problems raised by the RG fall outside the domain of the existing mathematics. A discipline which accomodates RG concepts in a most natural way is probability theory. Here the RG point of view has an obvious counterpart in the study of limit theorems. The new problem raised by the RG is the limit behaviour of sums of strongly dependent random variables for which the variance of a sum is not proportional to the number of addends but may grow with a power higher than one. Typically, strongly dependent variables have correlation functions which decay like a non integrable power law of the distance and appear at critical points in phase transitions.

The classification of the possible limit behaviours of sums of strongly dependent random variables corresponds in the physical language to the study of the fixed points of the so-called "block-spin" transformation and the probabilistic concept of domain of attraction corresponds to the physicist notion of universality class. Non trivial explicit examples can be constructed within such simple models as mean field theories.

There is a new symmetry which usually emerges in connection with strongly dependent random variables : scaling.

Scaling in this context is a symmetry different from the usual ones in physics which reflect an invariance group of the basic interaction.

(*) Permanent address : Dipartimento de Fisica
Universita di Roma "La Sapienza"

Scaling invariance is produced by laws of large numbers and is therefore of statistical origin.
There is a deep connection between various aspects of critical phenomena and long standing renormalization problems in quantum field theory. This connection has been most clearly elucidated in constructive field theory for which statistical mechanics constitutes a basic reference model. Very interesting results have been obtained recently in the study of ϕ^4 theories in four dimensions.
There is an important idea which has emerged over the years from scaling and the algorithmic practice of the RG : hierarchical structures provide a clue to understand complex systems. This idea can be studied in its pure form in certain special models where the hierarchical structure coincides with the recursive character of the algorithm that defines the model. It constitutes now a guideline of much rigorous work in quantum field theory.
Strictly tied with the hierarchical philosophy seems to be the effort of describing the typical configurations of complex systems through inductive procedures over different length scales. This point of view has given important results in the analysis of Coulomb systems and of disordered systems like the so called Anderson model of the metal insulator transition. This trend constitutes probably the beginning of the second generation of RG ideas. There are some remarkable examples in mathematics where a result is established on the basis of an inductive analysis of its truth over different scales. Two examples : the small denominator problem in classical mechanics and Kolmogorov theorem on the representation of a continuous function of several variables in terms of functions of only one variable. More recently Fefferman theorem on the almost everywhere convergence of Fourier series as well as his work on Schrödinger operators remind of the RG. The RG seems therefore to imply a more general line of thought and is likely to have far reaching consequences outside the restricted set of problems for which it was conceived.

References

The literature on the R.G. is enormous. Since there are many well known survey articles which emphasize the physical applications we shall restrict ourselves to references where the mathematical aspects are dominant. A bridge-article between physics and probability theory where the main references prior to 1978 are included is :

(1) - M. Cassandro, G. Jona-Lasinio - Adv. Phys. 27, 913 (1978)

An up to date review of several problems mentioned in the text can be found in :

(2) - J. Frölich, T. Spencer - seminaire Bourbaki, N° 586, 1981/82, Asterisque N° 92-93, 1982, page 159.

There is also a forthcoming book :

(3) - "Self-similarity and Scaling in Physics" - J. Frölich ed., Birkhäuser, Basel 1983.

For recent developments in ϕ^4 theory for $d \geq 4$, see (2). A different point of view is contained in :

(4) - G. Gallavotti, V. Rivasseau : "ϕ^4 Field Theory in Dimension 4 : A Modern Introduction to its Unsolved Problems", Preprint 1983.

The hierarchical approach to field theory was initiated by Gallavotti and is reviewed in :

(5) - G. Gallavotti, in "Quantum Fields-Algebras, Processes" - L. Streit ed., Springer, Wien 1980.

For more recent developments of this line see the contribution of K. Gavedsky and A. Kupiainen in ref. (3).

An account of the work by Frölich and Spencer on Coulomb systems can also be found in (3).

The connection between RG ideas and the small denominator problem is developed in the contribution of G. Gallavotti to (3).

Kolmogorov theorem on the representation of continuous functions is clearly exposed in :

(6) - V.M. Tikhomirov, Russ. Math. Surv. 18, N° 5, 51 (1963).

Fefferman's works mentioned in the text are :

(7) - C. Fefferman - Ann. Math. 98, 551 (1973)
C. Fefferman - Bull. Am.. Math. Soc. 9; 129 (1983)

For the application of the R.G. to the study of the behaviour deterministic systems we refer to the excellent book :

(8) - P. Collet, J.P. Eckmann - "Iterated Maps on the Interval as Dynamical Systems" - Birkhäuser, Basel 1980.

ON THE RACAH ALGEBRA FOR SHUBNIKOV MAGNETIC GROUPS

J.N.Kotzev, M.I. Aroyo and M.N.Angelova
Faculty of Physics, the University of Sofia,
Sofia 1126, Bulgaria

ABSTRACT

The generalization of the Racah algebra in the case of corepresentations of antiunitary Shubnikov black-and-white and grey groups is presented. The basic properties of the elements of the Racah algebra are discussed.

INTRODUCTION

The generalization of the method of irreducible tensorial sets or the Racah algebra for the case of corepresentations (coreps) [1] of the Shubnikov magnetic groups [2] is a wide programme which includes both the development of the theory and the making up of large tables. In this paper, which is a review of a series of our papers [3-9], we discuss the main results of our investigations along these lines, which lay the foundations of the Racah algebra for systems with magnetic symmetry.

ON THE THEORY OF COREPS

Schur Lemma. Using a corollary of the Schur lemma for reducible representation we give a new and strict proof of the Schur lemma for irreducible coreps [5], which contributes to the clarifying and the specifying of the similar results obtained by other authors [10, 11]. Of particular interest for the genralization of the Racah algebra is the Schur lemma for reducible coreps $D = \bigoplus_A (e_A \times D^A)$, where D^A are irreducible coreps, and e_A is an unit matrix with dim $e_A = (D|D_A)$, the multiplicity of every D^A in D. The matrices M_D, commuting with D, are of the type

$$M_D = \bigoplus_A (L_A \times M^A) \qquad (1)$$

where the unitary matrices M^A belong to the commutator algebra of the irreducible corep D^A [5], and only in the case of coreps the matrices $L_A = L_A^*$ are orthogonal, with dim L_A = dim e_A = $(D|D_A)$.

BASIC ELEMENTS OF THE RACAH ALGEBRA

(i) Clebsch-Gordan Coefficients (CGC) and Isoscalar Factors - for the definitions, properties, the methods for their calculation and full tables refer to [3-8].

(ii) Generalized CGC (GCGC). The matrix elemets of the unitary transformation $U^{A1A2...An}$, reducing the inner Kronecker product of n irreducible coreps $D^{A1} \times D^{A2} \times ... \times D^{An}$ into a block-diagonal form are called the GCGC.

$$(U^{A1...An})^{-1}(D^{A1} \times ... \times D^{An})(g)U^{A1...An(*)} = \bigoplus_{Ai} (e_{Ai}^{A1...An} \times D^{Ai}(g)) \ , \ g \in G \qquad (2)$$

where (*) means complex conjugation iff $g \in G$ is an antiunitary operator, and $e_{Ai}^{A1...An}$ is unit matrix with dim $e_{Ai}^{A1...An}$ = (A1... An|Ai), the multiplicity of D^{Ai} in $D^{A1} \times ... \times D^{An}$.

(iii) Racah coefficients for coreps. These are the matrix elements of the unitary transformation, which gives the transition between lineary dependent sets of basis functions of a system of n coupled moments, where every set of functions is obtained by different coupling schemes.

For example, for the matrix relating the bases obtained by the following coupling schemes for the corep D^{A123}

$$(A1A2,A3): [D^{A1} \times D^{A2}] \times D^{A3} \rightarrow D^{A12} \times D^{A3} \rightarrow D^{A12,3} = D^{A123} \qquad (3a)$$

$$(A1,A2A3): D^{A1} \times [D^{A2} \times D^{A3}] \rightarrow D^{A1} \times D^{A23} \rightarrow D^{A1,23} = D^{A123} \qquad (3b)$$

can be factorized as follows:

$$M(A1A2,A3;A1,A2A3) = \bigoplus_{A123} (L_{A123}^{A1,A2A3;A1A2A3} \times M^{A123}) =$$

$$= [\bigoplus_{A23} (e_{A23}^{A2A3} \times U^{A1A23})]^{-1} (E^{A1} \times U^{A2A3})^{-1} (U^{A1A2} \times E^{A3}) [\bigoplus_{A12} (e_{A12}^{A1A2} \times U^{A12A3})] \qquad (4)$$

The matrix elements of $L_{A123}^{A1,A2A3;A1A2,A3}$ are generalized Racah coefffici ents for coreps.

(iv) Symmetrized GCGC. The matrix elements of a rectangular matrix $V^{A1...An}$, consisting of those columns of the square matrix $U^{A1...An}$, which projects the n-fold direct product of coreps on the subspace of the identity (fully symmetric) coreps $D^{Ao} \in D^{A1} \times ... \times D^{An}$, where $D^{Ao} \equiv$ $\equiv D^{Aor_o}$, $r_o = 1, ...,$ (A1A2...An|Ao) = dim $e_{Ao}^{A1...An}$.

$$(V^{A1...An})^{-1}(D^{A1} \times ... \times D^{An})(g)V^{A1...An\ (*)} = (e_{Ao}^{A1...An} \times D^{Ao}(g)), \qquad g \in G \qquad (5)$$

are called the symmetrized GCGC, which for the case n=3 for systems with spherical symmetry are known as 3j-symbols of Wigner [1].

BASIC PROPERTIES OF THE RACAH ALGEBRA ELEMENTS

All the properties will be discussed for the case of GCGC. Similar results are obtained for the CGC [4], Isoscalar factors [6-8], symmetrized GCGC [9], Racah coefficients [9], etc.

(i) Transformation properties. The GCGC matrix into a new equivalent basis $\bar{U}^{A1..An}$ can be expressed by the GCGC matrix of the initial basis $U^{A1...An}$ and the unitary matrices S^{Ai}, $i = 1,...,n+1$, which transform the basis of the corresponding D^{Ai}:

$$\bar{U}^{A1...An} = (S^{A1} \times ... \times S^{An})^{-1} U^{A1...An} [\bigoplus_{An+1} (e^{A1..An}_{An+1} \times S^{An+1})] [\bigoplus_{An+1} (L^{A1...An}_{An+1} \times M^{An+1})] \quad (6)$$

For the case of coreps $L^{A1...An}_{An+1}$ is an orthogonal matrix, which is a corrollary of the Schur lemma (1)

(ii) Lemma of Derome and Sharp. This lemma gives the relation between the GCGC reducing $D^{A1} \times ... \times D^{An}$ and the corresponding GCGC $U^{A1^*... An^*}$ for the product of the complex conjugated coreps $D^{A1^*} \times ... \times D^{An^*}$

$$U^{A1^*... An^*} = (K^{A1} \times .. \times K^{An})^{-1} (U^{A1...An})^* [\bigoplus_{An+1} (e^{A1...An}_{An+1} \times K^{An+1})] [\bigoplus_{An+1} (L^{A1^*... An^*}_{An+1^*} \times M^{An+1^*})] \quad (7)$$

where $L^{A1^*... An^*}_{An+1^*}$ is an orthogonal matrix (1) and K^{Ai}, $i=1,...,n+1$ are unitary matrices transforming the complex conjugated bases into the standard ones (analogous to the Wigner metric tensor or 1j-symbols).

(iii) Racah lemma. The generalized Racah lemma gives the relation between the GCGC for the coreps $U^{A1... An}$ of a given group G_A and the corresponding GCGC $U^{B1...Bn}$ of its antiunitary subgroup $G_B \subset G_A$

$$\bigoplus_{B1... Bn} (e^{A1}_{B1} \times ... \times e^{An}_{Bn} \times U^{B1...Bn}) =$$

$$= (S^{A1} \times .. \times S^{An})^{-1} U^{A1...An} [\bigoplus_{An+1} (e^{A1...An}_{An+1} \times S^{An+1})] [\bigoplus_{Bn+1} (X^{A1...An,Bn+1} \times M^{Bn+1})]^{-1} \quad (8)$$

The matrix elements of the orthogonal matrices $X^{A1... An, Bn+1}$ are the so called isoscalar factors for coreps of a given group-subgroup chain. These factors are real for the case of coreps, which once again is a corollary of the Schur lemma (1). The matrices e^{Ai}_{Bi}, $i= 1,...,n$ are unit matrices with dim $e^{Ai}_{Bi} = (Ai|Bi)$, the subduction multiplicity of D^B in $D^A \downarrow G_B$. The lemma gives the possibility for building up the Racah algebra of an antiunitary subgroup G_B starting from the algebra of its supergroup G_A or vice versa.

(iv) Associative symmetry. Alongside with the permutational symmetry [4] we investigate a new type of symmetry of the elements of the generalized Racah algebra - symmetry under association of irreducible coreps. The coreps D^A and $D^{A'} = D^A \boxtimes D^I$ are associated if D^I is an one-dimensional corep of the same group. The relation between the GCGC $U^{A1...An}$ and the coefficients $U^{A1'A2''... An^{(n)}}$ reducing the direct product of coreps $D^{Ai^{(i)}}$, $i=1,...,n$, associated to D^{Ai} is of the type

$$U^{A1'\ldots An(n)} =$$

$$=(U^{A1I1} x \ldots x U^{AnIn})^{-1} (U^{A1\ldots An} \times U^{I1\ldots In}) [\bigoplus_{An+1} (e^{A1\ldots An}_{An+1} \times U^{An+1In+1})] [\bigoplus_{An+1} (L^{A1\ldots An}_{\substack{(n+1)\\An+1}} x M^{An+1}_{(n+1)})] \quad (9)$$

Here U^{AiIi} are CGC reducing D^{AixIi} and $L^{A1\ldots An}_{An+1}{}^{(n+1)}$ are orthogonal matrices (1)

THE WIGNER-ECKART THEOREM

The theorem is generalized for the case of coreps for arbitrary antiunitary magnetic groups (grey and black-and-white). Symmetrized CGC for coreps, i.e. 3D-symbols are used for the factorization of the matrix elements of quantum-mechanical operators:

$$\left\langle A3a3 \middle| T^{A1}_{a1} \middle| A2a2 \right\rangle = \sum_{ra3} \left\langle A3 \middle\| T^{A1} \middle\| A2 \right\rangle_r \begin{pmatrix} A3 & A3^* \\ a3 & a3^* \end{pmatrix}^* \begin{pmatrix} A3^* & A1 & A2 \\ a3^* & a1 & a2 \end{pmatrix}_r \quad (10a)$$

$$\left\langle A3 \middle\| T^{A1} \middle\| A2 \right\rangle^*_r = \left\langle A3 \middle\| T^{A1} \middle\| A2 \right\rangle_r \quad , \; r=1,\ldots, \; (A1A2A3 \mid Ao) \quad (10b)$$

The main advantages of this new form of the theorem are the following: (i) the reduced matrix elements are real in all cases. (ii) the matrix elements are completely factorized. (iii) the form of the theorem is simplified and it does not depend on the type of coreps.

REFERENCES

[1] Wigner E.P.: Group Theory. Academic Press, N.Y. 1959.
[2] Bradley C.J. and A.P.Cracknell: The Mathematical Theory of Symmetry in Solids. Clarendon Press, Oxford 1972.
[3] Kotzev J.N. and M.I. Aroyo: Comm.Jt.Inst.Nucl.Res.P17-10987, Dubna (1977).
[4] Kotzev J.N. and M.I.Aroyo: J.Phys.A:Math.Gen.13, 2275(1980), ibid.14, 1543(1981), ibid.15,711(1982), ibid. 15,725(1982).
[5] Kotzev J.N. and M.I.Aroyo: On the Coupling Coefficients for Systems with Antiunitary Symmetry, in Group Theoretical Methods in Physics, Proceedings of the International Seminar, Zvenigorod-1982 ed.M.A.Markov, Nauka, Moscow 1983.
[6] Kotzev J.N., M.I.Aroyo and M.N. Angelova: Physica 114A, 533(1982).
[7] Kotzev J.N., M.I.Aroyo and M.N.Angelova: Comm.Jt.Inst.Nucl. Res, E17-81-376, Dubna(1981).
[8] Kotzev J.N., M.I.Aroyo and M.N. Angelova: Comm.R oy.Soc. Edinburgh 19, 253 (1983).
[9] Aroyo M.I.: On the Racah Algebra for Systems with Magnetic Symmetry, Ph.D.Thesis, Sofia University, 1983 (unpublished).
[10] Van den Broek P.: J.Math.Phys. 20, 2028 (1979).
[11] Newmarch J.D. and R.M.Golding: J.Math.Phys. 22, 233 (1981).

On periodic and non-periodic space fillings of $\mathbb{E}^m$ obtained by projection

P.Kramer and R.Neri
Institut für Theoretische Physik
der Universität Tübingen
D-7400 Tübingen, FRG

For the two-dimensional non-periodic patterns introduced by Penrose [1], de Bruijn [2,3] gave an algebraic description in terms of pentagrids. We propose to treat these and similar types of patterns by projection from high-dimensional lattices.
Consider a n-dimensional cubic lattice. Its cell faces form an n-grid Y of n families of equidistant oriented hyperplanes

$$Y^i=\{ \underline{x} \mid \underline{x}\cdot\underline{b}_i= \tfrac{1}{2}k_i-\underline{\gamma}\cdot\underline{b}_i \ , \quad k_i=\pm1,\pm3,\pm5,\ldots \} \quad i=1,2\ldots,n$$

where $\underline{b}_i$ are the primitive vectors of the cubic lattice and $\underline{\gamma}$ is an arbitrary fixed vector of the $\mathbb{E}^n$.
Consider now a subgroup H of the point group $\Omega(n)$ of the cubic lattice. The representation D^n of H in $\mathbb{E}^n$ is in general reducible and this leads to

$$D^n(H) = D_1^m(H) + D_2^{n-m}(H) \quad 0<m<n$$

The corresponding decomposition of $\mathbb{E}^n$ is $\mathbb{E}^n= \mathbb{E}_1^m+ \mathbb{E}_2^{n-m}$, $\mathbb{E}^m \perp \mathbb{E}^{n-m}$
We define the projected n-grid 'Y as the intersection of Y with the subspace $\mathbb{E}^m$ which contains the point $\underline{\gamma}$. Denoting the projections of vectors on $\mathbb{E}_1$ and $\mathbb{E}_2$ by the indices 1 and 2, we get

$$'Y^i=\{ \underline{x} \mid \underline{x}\cdot\underline{b}_{i1}= \tfrac{1}{2}k_i-\underline{\gamma}\cdot\underline{b}_i \ , \quad k_i=\pm1,\pm3,\pm5,\ldots\} \quad i=1,2\ldots,n$$

'Y yields a division of $\mathbb{E}_1^m$ into cells each with an index system $(k_1,\ldots,k_n)$. We define the dual graph 'Z by its vertices

$$\{ \underline{k} \mid \underline{k} = \sum_i k_i\cdot\underline{b}_{i1} \ , \quad (k_1,\ldots,k_n) \text{ a cell index of 'Y } \}$$

and by connecting any pair of vertices with $k_i' = k_i''+\delta_{ij}, 1 \leqslant j \leqslant n$ by the vector $\underline{b}_{j1}$. The cells of the <u>dual space filling</u> have the vertices and edges of 'Z and their (m-1)-dimensional faces are perpendicular to the edges of the cells of 'Y. To any vertex of 'Y there corresponds a cell of 'Z. We call 'Y <u>subperiodic</u> if the translation group T of the cubic lattice in $\mathbb{E}^n$ has a non-trivial subgroup in $\mathbb{E}_1^m$. We restrict the discussion to the <u>regular</u> case: at most m hyperplanes in 'Y intersect in one point.

<u>Example</u> 1: Projection from $\mathbb{E}^3$ to $\mathbb{E}_1^2$ with the subgroup H = S(3). Start with the cubic lattice and space filling in $\mathbb{E}^3$. Choose the subgroup H = S(3) of the full octahedral group and decompose $\mathbb{E}^3$ as

$$\mathbb{E}^3 = \mathbb{E}_1^2 + \mathbb{E}_2^1 , \quad \underline{b}_{i1} = \underline{b}_i - \frac{1}{3}\sum_j \underline{b}_j , \quad \underline{b}_{i2} = \frac{1}{3}\sum_j \underline{b}_j$$

The 3-grid Y is projected into a 3-grid 'Y in a plane $\mathbb{E}_1^2$ perpendicular to $\underline{b}_{i2}$, the cells of 'Y have a shape that depends on $\underline{\gamma}\cdot\underline{b}_i$. The condition for subperiodicity is found to be

$$\underline{t} = \sum_i t_i \cdot \underline{b}_{i1} , \quad \sum_i t_i = 0$$

which yields a non-trivial translation subgroup of 'Y. The point group of 'Y is S(3), compare Fig.1.

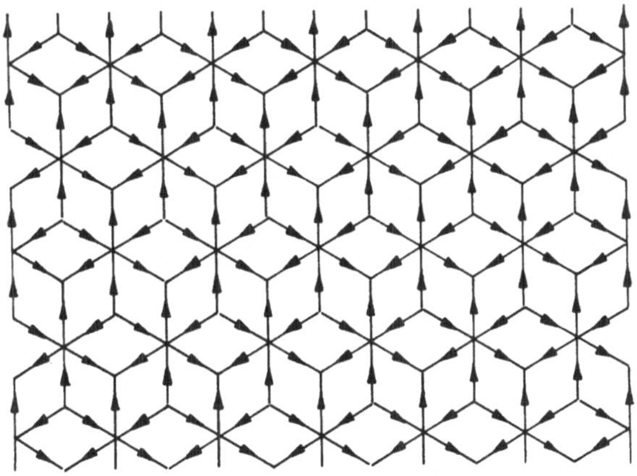

Fig.1 For regular 'Y, the dual space filling and the dual graph 'Z consist of a rhombus pattern with oriented edges $\underline{b}_{i1}$, i=1,2,3.

Example 2: Projection from $\mathbb{E}^{12}$ to $\mathbb{E}^{3}$ with the subgroup H = A(5). The hyperoctahedral group $\Omega(12)$ of the cubic lattice in $\mathbb{E}^{12}$ has the subgroup S(12). Consider the symmetry operations of the regular dodecahedron as permutations of its 12 faces. This yields an embedding A(5) in S(12) which we shall use. The representation D^{12} of A(5) in $\mathbb{E}^{12}$ reduces as

$$D^{12} = D_1^3 + D_2^9 \;, \quad D_2^9 = D_{1'}^3 + D_{2'}^1 + D_{3'}^5$$

where D_1^3 and $D_{1'}^3$ are inequivalent irreducible representations and D_1^3 is the usual one of A(5) in $\mathbb{E}^3$. The respective decomposition $\mathbb{E}^{12} = \mathbb{E}^3 + \mathbb{E}^9$ yields the 12 vectors $\underline{b}_{i1}$ perpendicular to the faces of the regular dodecahedron. By an appopriate choice of 6 components of $\underline{\gamma}$, the 12-grid in $\mathbb{E}_1^3$ can be reduced to a hexagrid

$$'Y^i = \{ \; \underline{x} \,|\, \underline{x}\cdot\underline{b}_{i1} = \tfrac{1}{2}k_i - \underline{\gamma}\cdot\underline{b}_i \;, \quad k_i = \pm 1, \pm 3, \pm 5, \ldots\} \; i = 1,2\ldots,6$$

whose planes are parallel to the faces of the regular dodecahedron. The orientations of the planes are equal for $k_i = +1,-3,+5,\ldots$ and opposite for $k_i = -1,+3,-5,\ldots$ Since three of the six parameters of $\underline{\gamma}$ can be changed by a translation of the hexagrid in $\mathbb{E}_1^3$, the internal structure of 'Y depends on three real parameters. The analysis of the projected grid 'Y yields no subperiodicity since no subgroup of the full translation group T exists in the subspace $\mathbb{E}_1^3$. The cell faces of the dual space filling do all have the same rhombus shape with angles α_1, α_2 given by $tg\alpha_{1,2} = \pm 2$. For a regular hexagrid, the rhombohedra derived by projection coincide with the ones introduced by Mackay [4] as a generalization of the Penrose pattern to three dimensions. The present projection method leads to a systematic derivation of these rhombohedra, yields results on the orientation of their edges which will be essentially connected with the question of non-periodicity, and provides an algebraic treatment in terms of three real parameters. The details will be given in [5].

[1] Penrose,R. Mathe. Intelligencer 2(1979)32
[2,3] de Bruijn,N.G. Nederl.Akad.Weten.Proc.Ser.A 43(1981)39 and 53
[4] Mackay,A.L. Sov.Phys.(Cryst.) 26(1981)517
[5] Kramer,P. and Neri,R. to be published

INVARIANTS FOR PHYSICALLY IRREDUCIBLE REPRESENTATIONS OF SPACE GROUPS

HERBERT WILLI KUNERT

Institute of Physics,Technical University,Piotrowo3,Poznań 60-965 Poland

We introduce a method calculating of invariants for so-called "physically" irreducible representations(PhIR's) of space groups.In particular,theory of coupling coefficients of the representation for space groups is adapted to the PhIR's.According to Landau's theory the continuous transition is induced by one real or physically irreducible representation of a given space group of the crystal that satisfies the Landau and Lifshitz criteria.Our method can be used to calculate the Clebsch-Gordan coefficients for the IR's contained in the Kronecker product of two PhIR's.

Introduction.

Recently much attention has been directed to group-theoretical methods in second-order phase transitions in crystals.
In this paper the theory of coupling coefficients or Clebsch-Gordan coefficients(CGcs)of the representations for space groups [1,2] is adapted to calculation of invariants for PhIRs of a given space group.

Theory

Let us consider the unitary matrix U which brings $D \otimes \overline{D}$ into the fully reduced matrix (see ref.1.eq.18.16)

$$U^{-1}\left[D(R) \otimes \overline{D}(R)\right]U = \Delta(R) \tag{1}$$

Usually $D(R)$ and $\overline{D}(R)$ are irreducible representations(IRs) of space group.We will use the short notation $D(R)\to D$.It is often convenient,but not necessary,to take D and $\overline{D}$ to be irreducible.Here we assume

$$D = D^1 \oplus D^2 \oplus,\ldots,\oplus D^n \quad \text{and} \quad \overline{D} = D^{\overline{1}} \oplus D^{\overline{2}} \oplus,\ldots,\oplus D^{\overline{k}} \tag{2}$$

Using of eq.1 we have:

$$U^{-1}\left[\left(D^1 \oplus,\ldots,\oplus D^n\right)\otimes\left(D^{\overline{1}} \oplus,\ldots,\oplus D^{\overline{k}}\right)\right]U = \Delta \tag{3}$$

but $D^1 \otimes D^{\overline{1}} = U_{1\overline{1}}\,\Delta_{1\overline{1}}\,U^{-1}_{1\overline{1}}$ and $\Delta_{1\overline{1}} = \bigoplus c_{1\overline{1},m_{1\overline{1}}}\,D^{m_{1\overline{1}}}$

$,\ldots\ldots,$ $\qquad$ $,\ldots\ldots,$

$,\ldots\ldots,$ $\qquad$ $,\ldots\ldots,$

$$D^n \otimes D^{\overline{k}} = U_{n\overline{k}}\,\Delta_{n\overline{k}}\,U^{-1}_{n\overline{k}} \quad \text{and} \quad \Delta_{n\overline{k}} = \bigoplus c_{n\overline{k},m_{n\overline{k}}}\,D^{m_{n\overline{k}}} \tag{4}$$

Substituting eq.4 to eq.3 we have

$$U^{-1}\left[U_{1\bar{1}}\Delta_{1\bar{1}}U^{-1}_{1\bar{1}}\oplus,\dots,\oplus U_{n\bar{k}}\Delta_{n\bar{k}}U^{-1}_{n\bar{k}}\right]U = \Delta \qquad (5)$$

and
$$U_{1\bar{1}}\Delta_{1\bar{1}}U^{-1}_{1\bar{1}}\oplus,\dots,\oplus U_{n\bar{k}}\Delta_{n\bar{k}}U^{-1}_{n\bar{k}} = U\Delta U^{-1} \qquad (6)$$

or in matrix form

$$\begin{bmatrix} U_{1\bar{1}}\Delta_{1\bar{1}}U^{-1}_{1\bar{1}} & & \\ & \ddots & \\ & & U_{n\bar{k}}\Delta_{n\bar{k}}U^{-1}_{n\bar{k}} \end{bmatrix} = \begin{bmatrix} U \end{bmatrix}\begin{bmatrix} \Delta_{1\bar{1}} & & \\ & \ddots & \\ & & \Delta_{n\bar{k}} \end{bmatrix}\begin{bmatrix} U^{-1} \end{bmatrix} \qquad (7)$$

The left side of eq.7 we write as

$$\begin{bmatrix} U_{1\bar{1}} & & \\ & \ddots & \\ & & U_{n\bar{k}} \end{bmatrix}\begin{bmatrix} \Delta_{1\bar{1}} & & \\ & \ddots & \\ & & \Delta_{n\bar{k}} \end{bmatrix}\begin{bmatrix} U^{-1}_{1\bar{1}} & & \\ & \ddots & \\ & & U^{-1}_{n\bar{k}} \end{bmatrix} = \begin{bmatrix} U \end{bmatrix}\begin{bmatrix} \Delta_{1\bar{1}} & & \\ & \ddots & \\ & & \Delta_{n\bar{k}} \end{bmatrix}\begin{bmatrix} U^{-1} \end{bmatrix} \qquad (8)$$

To satisfy this eqution we take

$$U = U_{1\bar{1}}\oplus U_{1\bar{2}}\oplus,\dots\oplus U_{1\bar{k}}\oplus U_{2\bar{1}}\oplus,\dots,\oplus U_{2\bar{k}}\oplus,\dots,\oplus U_{n\bar{k}} \qquad (9)$$

From this equation we can calculate the matrix U in eq.1.The submatrices $U_{1\bar{1}}, U_{1\bar{2}},\dots,U_{n\bar{k}}$ ineq.9. can be calculate by standart method [1,2]

Example

We take into consideration the PhIR $H_2\oplus H_2^{*} = H_2\oplus H_3$ at the corner point in Brillouin zone(BZ) of the garnet structure O_h^{10}(Ia3d)space group.We will calculate the invariants for the irreducible representations contained in Kronecker product $(H_2\oplus H_3)\otimes(H_2\oplus H_3)$.From CDML Tables [3] and [4] we have $(H_2\oplus H_3)\otimes(H_2\oplus H_3) = (H_2\otimes H_2)\oplus(H_2\otimes H_3)\oplus(H_3\otimes H_2)\oplus(H_3\otimes H_3)\sim\Gamma_{1-}\oplus\Gamma_{2-}\oplus\Gamma_{3+}\oplus\Gamma_{1+}\oplus\Gamma_{2+}\oplus\Gamma_{3-}\oplus\Gamma_{1+}\oplus\Gamma_{2+}\oplus\Gamma_{3-}\oplus\Gamma_{1-}\oplus\Gamma_{2-}\oplus\Gamma_{3+}$ .So,that $(H_2\otimes H_2)\sim\Gamma_{1-}\oplus\Gamma_{2-}\oplus\Gamma_{3+}$, $H_2\otimes H_3 = H_3\otimes H_2\sim\Gamma_{1+}\oplus\Gamma_{2+}\oplus\Gamma_{3-}$ and $H_3\otimes H_3\sim\Gamma_{1-}\oplus\Gamma_{2-}\oplus\Gamma_{3+}$. The eq.1.for this example has the form

$$\left(U^{H_2\otimes H_2}\right)^{-1}\left(H_2\otimes H_2\right)U^{H_2\otimes H_2} = \Gamma_{1-}\oplus\Gamma_{2-}\oplus\Gamma_{3+}$$

and like for $U^{H_2\otimes H_3}$, $U^{H_3\otimes H_2}$ and $U^{H_3\otimes H_3}$.

Now,using the standart method [1,2] we have calculated, that

$$U^{H_2 \otimes H_2} = \begin{bmatrix} 0 & 0 & 0 & 1 \\ a & a & 0 & 0 \\ a & -a & 0 & 0 \\ 0 & 0 & 1 & 0 \end{bmatrix}, \quad U^{H_2 \otimes H_3} = U^{H_3 \otimes H_2} = \begin{bmatrix} a & a & 0 & 0 \\ 0 & 0 & 0 & 1 \\ 0 & 0 & -1 & 0 \\ a & -a & 0 & 0 \end{bmatrix}$$

and

$$U^{H_3 \otimes H_3} = \begin{bmatrix} 0 & 0 & 1 & 0 \\ a & a & 0 & 0 \\ a & -a & 0 & 0 \\ 0 & 0 & 0 & 1 \end{bmatrix}$$

where $a = 1/\sqrt{2}$, So,using our method for the case $(H_2 \oplus H_3) \otimes (H_2 \oplus H_3)$ we have

$$U^{-1}\left[(H_2 \oplus H_3) \otimes (H_2 \oplus H_3)\right]U = \Delta = \Delta_{22} \oplus \Delta_{23} \oplus \Delta_{32} \oplus \Delta_{33} =$$

$$= (\Gamma_{1-} \oplus \Gamma_{2-} \oplus \Gamma_{3+}) \oplus (\Gamma_{1+} \oplus \Gamma_{2+} \oplus \Gamma_{3-}) \oplus (\Gamma_{1+} \oplus \Gamma_{2+} \oplus \Gamma_{3-}) \oplus (\Gamma_{1-} \oplus \Gamma_{2-} \oplus \Gamma_{3+})$$

and from eq.9. we have

$$U = U^{H_2 \otimes H_2} \oplus U^{H_2 \otimes H_3} \oplus U^{H_3 \otimes H_2} \oplus U^{H_3 \otimes H_3}$$

The elements of the matrix U are presented in Table I.

Conclusions

Recently we have investigated the allowed symmetries of soft modes in cubic crystals [5] for all cubic space groups O_h^i $(i = 1,2,\dots,10)$.For these cases we find two active physically irreducible representations $R_2 \oplus R_2^* = R_2 \oplus R_3$,for A-15 structure,$O_h^3$ (Pm3n)space group, and $H_2 \oplus H_2^*$ $= H_2 \oplus H_3$,for the garnet structure,O_h^{10} (Ia3d) space group .In both structures,the second order phase transition have been found [6,7] .

From Table I we have the following invariants,i.e. basis functions of the irreducible representations contained in the product $(H_2 \oplus H_3) \otimes$ $\otimes (H_2 \oplus H_3)$.

Invariants

$$\Phi_1^{\Gamma_{1-}} = \frac{1}{\sqrt{2}}\left\{\psi_1^{H_2}\varphi_2^{H_2} + \psi_2^{H_2}\varphi_1^{H_2}\right\}$$

$$\Phi_1^{\Gamma_{2-}} = \frac{1}{\sqrt{2}}\left\{\psi_1^{H_2}\varphi_2^{H_2} - \psi_2^{H_2}\varphi_1^{H_2}\right\}$$

$$\Phi_1^{\Gamma_{3+}} = \psi_2^{H_2}\varphi_2^{H_2}, \quad \Phi_2^{\Gamma_{3+}} = \psi_1^{H_2}\varphi_1^{H_2}$$

$$\Phi_1^{\Gamma_{1+}} = \frac{1}{\sqrt{2}}\left\{\Psi_1^{H_2}\varphi_1^{H_3} + \Psi_2^{H_2}\varphi_2^{H_3}\right\}$$

$$\Phi_1^{\Gamma_{2+}} = \frac{1}{\sqrt{2}}\left\{\Psi_1^{H_2}\varphi_1^{H_3} - \Psi_2^{H_2}\varphi_2^{H_3}\right\}$$

$$\Phi_1^{\Gamma_{3-}} = -\Psi_2^{H_2}\varphi_1^{H_3}, \quad \Phi_2^{\Gamma_{3-}} = \Psi_1^{H_2}\varphi_2^{H_3}$$

$$\Phi_1^{\Gamma_{1+}} = \frac{1}{\sqrt{2}}\left\{\Psi_1^{H_3}\varphi_1^{H_2} + \Psi_2^{H_3}\varphi_2^{H_2}\right\}$$

$$\Phi_1^{\Gamma_{2+}} = \frac{1}{\sqrt{2}}\left\{\Psi_1^{H_3}\varphi_1^{H_2} - \Psi_2^{H_3}\varphi_2^{H_2}\right\}$$

$$\Phi_1^{\Gamma_{3-}} = -\Psi_2^{H_3}\varphi_1^{H_2}, \quad \Phi_2^{\Gamma_{3-}} = \Psi_1^{H_3}\varphi_2^{H_2}$$

$$\Phi_1^{\Gamma_{1-}} = \frac{1}{\sqrt{2}}\left\{\Psi_1^{H_3}\varphi_2^{H_3} + \Psi_2^{H_3}\varphi_1^{H_3}\right\}$$

$$\Phi_1^{\Gamma_{2-}} = \frac{1}{\sqrt{2}}\left\{\Psi_1^{H_3}\varphi_2^{H_3} - \Psi_2^{H_3}\varphi_1^{H_3}\right\}$$

$$\Phi_1^{\Gamma_{3+}} = \Psi_1^{H_3}\varphi_1^{H_3}, \quad \Phi_2^{\Gamma_{3+}} = \Psi_2^{H_3}\varphi_2^{H_3}$$

References

[1] J.L.Birman,Theory of Crystal Space Groups and Infra-Red and Raman Lattice Processes of Insulating Crystals,in:Handbuch der Physik (Encyklopedia of Physics,vol.XXV/2b,Light and Matter Ib,S.Flügge, ed.Springer-Verlag,Berlin-Heidelberg-New York,)(1974).

[2] R.Berenson,I.Itzkan and J.L.,Birman.,J.Math.Phys. 16 (1975) 236 .

[3] A.P.Cracknell,B.L.Davies,S.C.Miller and Love, Kronecker Product Tables,vol.1-4, Plenum,New York,Washington,London, (1979).

[4] A.P.Cracknell,B.L.Davies,H.Kunert,J.Popenda and M.Suffczyński, Commun.Roy.Soc.Edinb.Phys.Sci. 1 (1979) 177 .

[5] H.Kunert and J.Zieliński,Allowed Symmetries of Soft Modes in Cubic Crystals, will be submited to J.de Physique, (1983).

[6] R.N.Bhatt and W.L.McMillan,Phys.Rev. B 14 (1976) 1007 .

[7] S.Geller and G.Balestrino,Phys.Rev. B 21 (1980) 4055 .

TABLE I

The Clebsch-Gordan coefficients for $(H_2 \oplus H_3) \otimes (H_2 \oplus H_3)$ of space group O_h^{10} (Ia3d) of the garnet structure.

$(H_2 \oplus H_3) \otimes (H_2 \oplus H_3) =$		$H_2 \otimes H_2$				$H_2 \otimes H_3$				$H_3 \otimes H_2$				$H_3 \otimes H_3 \leftarrow (i,j)$			
		Γ_{1-}	Γ_{2-}	Γ_{3+}		Γ_{1+}	Γ_{2+}	Γ_{3-}		Γ_{1+}	Γ_{2+}	Γ_{3-}		Γ_{1-}	Γ_{2-}	Γ_{3+} $\leftarrow q$	
α	α'	1	1	1	2	1	1	1	2	1	1	1	2	1	1	1	2 $\leftarrow \alpha''$
1	1	0	0	0	1												
1	2	a	a	0	0												
2	1	a	-a	0	0												
2	2	0	0	1	0												
1	1					a	a	0	0								
1	2					0	0	0	1								
2	1					0	0	-1	0								
2	2					a	-a	0	0								
1	1									a	a	0	0				
1	2									0	0	0	1				
2	1									0	0	-1	0				
2	2									a	-a	0	0				
1	1													0	0	1	0
1	2													a	a	0	0
2	1													a	-a	0	0
2	2													0	0	0	1

where $a = 1/\sqrt{2}$. From this Table and equation.10. [1] we can calculate the invariants. The empty places in this Table mean zeros.

The basis functions of the irreducible representations contained in the product $(H_2 \oplus H_3) \otimes (H_2 \oplus H_3)$ are [1,2,8,9]

$$\Phi_{\alpha''}^{\Gamma_q} = \sum U_{i\alpha j\alpha', q\alpha''} \Psi_{\alpha}^{H_i} \Psi_{\alpha'}^{H_j} \tag{10}$$

References (continued)

[8] H.Kunert and M.Suffczyński., J.de Physique 40 (1979) 199

[9] H.Kunert and M.Suffczyński., J.de Physique 41 (1980) 1361

ON SYMMETRY ASPECTS OF PHASE TRANSITIONS WITH COUPLED PARAMETERS

by

Jan Lorenc

Institute of Theoretical Physics, University of Wroclaw,

Cybulskiego 36, 50-205 Wroclaw, Poland.

ABSTRACT

We discuss symmetry aspects of (continuous) phase transitions with coupled parameters belonging to irreducible representations. It is stressed that the minimization procedure introduces essential restrictions on possible symmetry changes. The effects of critical fluctuations have also been discussed. The application to first order phase transitions is briefly reviewed.

1. INTRODUCTION

The Landau theory of symmetry changes at continuous (second order) phase transition has been applied to numerous transitions since the original work of Landau [1] and it is still in use [2,3]. What is more, new ideas have been put forward in an attempt to "simplify" the task of analysing symmetry changes at continuous phase transition [4]. All these concepts, however, are based on the original Landau assumption that the order parameter associated with a continuous phase transition, $\phi_1, \ldots, \phi_m$, forms a basis of a single irreducible (or physically irreducible) representation $D^{(n)}$ of the symmetry group G_o of the disordered phase. And only so-called active irreducible representations are considered [1-3].

There exist, however, phase transitions described by parameters transforming according to reducible or several irreducible representations of G_o [5]. The purpose of this article is to review critically symmetry aspects of such phase transitions with a particular emphasis on the minimization procedure and fluctuation effects.

2. SYMMETRY ASPECTS OF CONTINUOUS PHASE TRANSITIONS WITH COUPLED PARAMETERS

A change of the probability density function $\delta\rho(\bar{r})$ at continuous crystalline phase transitions, as introduced by Landau, has proved very useful in describing symmetry changes at such transitions [1,2]. And all possible ways of symmetry breaking could be obtained by inspection of minima of a Landau free energy functional $F=F(p;T;\delta\rho(\bar{r}))$ for constant pressure (p) and temperature (T) [3]. However, to obtain quantitative results Landau introduced two assumptions [1,2]:

(1) Only one irreducible representation was considered, i.e. $\delta\rho(\bar{r})=\sum_{i=1}^{m} c_i\varphi_i(\bar{r})$,

where c_i's can be chosen as a basis of a given n-dimensional irreducible representation of G_o (Note that $\varphi_i(\vec{r})$ can, in principle, be determined in the vicinity of a transition by a variational method [2]).

(2) A truncated expansion of F in C_i's up to fourth or sixth order terms was used. Now it is easy to see that (1) results from the assumption that a phase transition occurs along a line in the (p,T) phase diagram and from considering a (truncated) Gaussian (or quadratic) form of F. Therefore Landau[2] and others[6] concluded that reducible representations would be responsible for first order (discontinuous) phase transitions (only at isolated points of the phase diagram such transitions would be continuous). This result, however, appears not to be true when a non-Gaussian (or non-quadratic) form of F is taken into account.

2.1 LANDAU NECESSARY CONDITION (LNC) FOR SYMMETRY CHANGES

For the case of a single irreducible representation one can determine all possible lower-symmetry groups which might occur at a continuous phase transition by considering all non-equivalent forms of $\delta\rho(\vec{r})$. And this is just LNC[7]. It is easy to generalize this concept to reducible or several irreducible representations of G_o. First, let $\bar{c}_i = (c_i^1, \ldots, c_i^n)(i=1,2,\ldots)$ denote a basis of the irreducible representation $D_i^{(ni)}$ of G_o. Secondly, let us determine isotropy subgroups $G(\bar{c}_1), G(\bar{c}_2), \ldots$ for $\bar{c}_1, \bar{c}_2 \ldots$ directions in the representation space, respectively. Then, the symmetry group of a phase with non-zero $\bar{c}_i^2$ is just $G(\bar{c}_1, \bar{c}_2, \ldots) = G(\bar{c}_1) \cap G(\bar{c}_2) \cap \ldots$. Note that for not-equivalent representations this intersection is a (proper) subgroup of $G(\bar{c}_1)$ and $G(\bar{c}_2)$ and ...[8]. Therefore, by inspecting all non-equivalent (with respect to G_o) directions in $(\bar{c}_1, \bar{c}_2, \ldots)$ (for $\bar{c}_i^2 \neq 0; i=1,2,\ldots$) we obtain all possible lower-symmetry groups which might occur at a continuous phase transition to such phases. Of course, a minimization of $F=F(p,T;\bar{c}_1,\bar{c}_2,\ldots)$ will determine the actual symmetry breaking. (Note that parameters $\bar{c}_1, \bar{c}_2, \ldots$ will couple with one another in an invariant, with respect to G_o, expansion of F).

Remark 1 . Note, that possible lower-symmetry groups of a $(\bar{c}_1 \neq 0, \bar{c}_2 \neq 0, \ldots)$ phase were determined in Ref. 9 as $G(\bar{c}_1)$ or $G(\bar{c}_2)$ or ... . Here, however, a less restrictive type of necessary condition was used.

Remark 2 . Here we would like to clarify the idea of the so-called "maximality" property in crystalline phase transitions [4,7,10]. First we note that a conjecture that LNC[7] and maximality principle of Ascher[10] are equivalent is not correct. The reason is that LNC specifies for some direction $\bar{c}$ of a representation space the maximal subgroup of G_o (or isotropy subgroup for $\bar{c}$). And only if $\bar{c}$ affords an absolute minimum to F, then this subgroup would describe the symmetry breaking (up to conjugation in G_o). There are no physical reasons whatsoever to expect that maximal epikernels [10] will become lower-symmetry groups at continuous phase transitions. We conclude, therefore, that within the Landau theory of symmetry changes LNC will determine correctly all possible lower-symmetry groups. This conclusion is also consistent with the results of Refs. 11,12 and 13. We also note that the above discussion can be easily generalized to reducible or irreducible representations.

2.2 MINIMIZATION PROCEDURE

For the sake of simplicity let us consider two parameters $\bar{c}_1=(c_1^1,\dots,c_1^n)$ and $\bar{c}_2=(c_2^1,\dots,c_2^m)$ transforming under two irreducible (or physically irreducible) representations of G_o, $D_1^{(n)}$ and $D_2^{(m)}$, respectively. It is reasonable to expect that only lowest order (invariant) couplings between $\bar{c}_1$ and $\bar{c}_2$ would be important. Hence, three cases are to be distinguished.

a) D_2^{m} occurs in the decomposition of the direct square of $D_1^{(m)}$ -(invariant) couplings $\alpha_{ijK}\, c_1^i c_1^j c_2^k$ (i,j=1,...,n; k=1,...,m) shall appear[15].

b) $D_1^{(n)}$ and $D_2^{(m)}$ are non-equivalent - the (invariant) coupling $\bar{c}_1,\bar{c}_2$ shall appear[16] (see also Ref. 17).

c) $D_1^{(n)}$ and $D_2^{(m)}$ are equivalent (or the same) - the following (invariant) couplings shall appear[18]: $\alpha_{ij}\, c_1^i c_2^j$, $\alpha_{ijkl}\, c_1^i c_1^j c_1^k c_2^l$ and $\beta_{ijkl}\, c_1^i c_2^j c_2^k c_2^l$, and $\gamma_{ijkl}\, c_1^i c_1^j c_2^k c_2^l$.

Higher order couplings are possible but for simplicity they will be neglected[9,14]. For these three cases it was shown, using a usual minimization procedure, that continuous phase transitions are possible to ($\bar{c}_1\neq0$, $\bar{c}_2\neq0$) phases. Hence, the lower-symmetry group became $G(\bar{c}_1)\cap G(\bar{c}_2)$. (up to conjugation in G_o). Note that this result is nontrivial for the a) and b) cases. And again we see that the minimization procedure is more restrictive than LNC as far as symmetry changes are concerned.

3. FLUCTUATION EFFECTS

Now it is well-known that the Landau theory results and resulting phase diagrams can be drastically changed when critical fluctuations are taken into account [6,19]. And a similar picture can be expected for coupled parameters. The reason is that the exchange symmetry is broken: always there is a temperature-like variable (τ) and a symmetry breaking field (g). And this has led even for a single parameter (less dimensional parameter space), to very complicated phase diagrams revealing critical and/or first order behaviour, tricritical and higher-order points, critical end points ...[19,20]. For sufficiently large g the renormalization group analyses[21] have been performed for the cases a) and b), Refs. 16,17, respectively. (The case c) has not been analysed yet). And the result is that the critical fluctuations prefer a $\bar{c}_1^2 \neq 0$ phase <u>or</u> $\bar{c}_2^2 \neq 0$ phase to appear below the transion . Thus we expect that the symmetry breaking occurs to $G(\bar{c}_1)$ <u>or</u> $G(\bar{c}_2)$ but <u>not</u> to $G(\bar{c}_1) \cap G(\bar{c}_2)$.

When g is small one has to integrate differential recursion relations until the effective symmetry breaking field, g, becomes sufficiently large[19] and then one can proceed as in the previous case. To the author's knowledge, however, this has not been done yet. Neverthless one can expect a rather involved phase diagram revealing a complicated behaviour.

It is also possible to have g=0. This would correspond to multicritical point (bicritical in the case b),Ref. 17). And now the resulting symmetry is expected to become $G(\bar{c}_1) \cap G(\bar{c}_2)$.

Finally, we would like to mention a similar analysis performed for the case of UO_2 and some other magnetic structures.[23]

4. FIRST ORDER PHASE TRANSITIONS

It is believed that the Landau theory of symmetry changes could be applied to first order phase transitions provided that a group-subgroup relation is preserved and the change of the order parameter is sufficiently small. Then the result of sections 1 and 2 apply to first order phase transitions as well (sixth-order invariants and higher-order couplings have to be included). And indeed, there are experimentally found first order phase transitions where symmetry changes cannot be described by considering only a single (even multicomponent) parameter. These are, for example, improper ferroelectric transitions[15] and triggered phase transitions[8] (see also Ref. 24). And because the fluctuations are expected not to be important for first order transitions, we can conclude that all predictions concerning symmetry changes for transitions with coupled parameters (with the Section 3 modified accordingly) will remain valid in this case as well.

5. CONCLUSIONS

The above critical review of symmetry aspects of phase transitions with coupled parameters has been undertaken in view of the existing evidence that more than one parameter is responsible for some symmetry changes observed experimentally.

i) Following the Landau procedure we have to consider the minima of $F(p,T;\bar{c}_1,\bar{c}_2,\ldots)$ that determine the symmetry breakings. Thus it was possible to show explicitly that the lowering of symmetry due to coupled parameters corresponds to the intersection of symmetry changes due to the separate parameters and it is greater (for non-equivalent representations) than the lowering of symmetry connected with separate parameters only.

ii) However, the above picture is drastically changed for continuous transitions when critical fluctuations are included. We expect that the lowering of symmetry would be as it is for separate parameters.Only at multicritical points or for a small symmetry breaking field g the lowering of symmetry could be as predicted in i).

iii) The case of first order phase transitions is quite interesting for fluctuations will not mask the (nontrivial) lowering of symmetry (see i)).

In addition we have clarified (see Remark 2) the status of the "maximality" principle at continuous crystalline phase transitions.

Finally, I would like to stress that because of a vast number of available papers on this subject only very few where cited for reference purposes.

Acknowledgements

It is a pleasure to thank Jerzy Przystawa for discussions and the critical reading of the manuscript.

Discussions with Arthur Cracknell and Krzysztof Parlinski are also acknowledged.

REFERENCES

1. L.D.Landau, Phys.Z.Sovjet., II (1937)26,545 (translated in L.D. Landau "Collected Papers" edited by D.ter Haar, Pergamon, 1965).
2. E.M. Lifshitz and L.P. Pitaevskii, Landau and Lifshitz Course of Theoretical Physics, Vol. 5, Statistical Physics (Pergamon, Oxford, 1980).
3. G.Ya. Lyubarskii, The Application of Group Theory in Physics,(Pergamon,Oxford,1980).
4. See,e.g., E. Ascher, J.Phys. C, 10(1977) 1365; M.V. Jaric, Physica 114A (1982) 550; J.Przystawa , Physica 114A (1982) 557; M. Sutton and R.L.Armstrong, Phys.Rev. B, 25 (1982) 1813 and erratum in Phys. Rev. B, 27 (1983) 1937.
5. See, e.g., Yu.A.Izyumov, Sov. Phys. Uspekhi 23(1980)356 and references therein.
6. P. Bak,S.Krinsky and D. Mukamel, Phys.Rev.Lett. 36 (1976)52; D. Mukamel and S. Krinsky, Phys.Rev., 13 (1976) 5065.

7. J.Lorenc, J.Przystawa and A.P.Cracknell, J.Phys,C.,13 (1980) 1955.

8. J.Holakovsky, Phys.Stat.Sol.(b), 56(1973) 615.

9. See,e.g., J.O.Dimmock, Phys.Rev., 130 (1963)1337; O.V. Kovalev,Sov.Phys.- Solid State, 5(1964)2309; C.Haas, Phys.Rev., 140(1965)A863.

10. E.Ascher, Phys.Lett., 20 (1966)352;J.Phys.C.,10(1977)1365.

11. J.C.Toledano,P. Toledano,J.Phys.(France),41(1980)189.

12. D.Mukamel, M.V.Jaric (1983,to be published).

13. M.V.Jaric (1983,Proceedings:XII International Colloquium on Group Theoretical Methods in Physics, Trieste,Italy).

14. I.E. Dzyaloshinski, V.I.Man'ko, Sov.Phys. JEPT,19(1964)915.

15. See,e.g.,A.P.Levanyuk, D.G.Sannikov, Sov.Phys.JEPT, 28(1969)134; Sov.Phys. Uspekhi 17 (1974) 199.

16. Y.Imry, J.Phys., C8(1975)567; and references therein.

17. A slightly more general Ymry's model was considered by J.Lorenc, Phys. Rev. B17 (1978) 363.

18. G. Oleksy, J.Przystawa, Physica (1983) in press. (note that for n=m 1 additional invariants might appear).

19. See,e.g., E.Domany, D.Mukamel,M.E.Fisher,Phys.Rev.B,15(1977)543 and references therein.

20. See,e.g., M.Kerszberg, D.Mukamel, Phys.Rev.B., 23(1981)3943;3953; D.Blankschtein, D.Mukamel, Phys.Rev. B., 25 (1982) 6939.

21. K.G.Wilson, J.Kogut, Phys.Rep. C, 12 (1974) 75.

22. Y.Achian, Y.Imry, Phys.Rev.B, 12 (1975) 2768.

23. S.A. Brazovskii, I.E. Dzyaloshinskii, B.G. Kukharenko, Sv.Phys. JEPT,43 (1976) 1178; V.A. Alessandrini, A.P. Cracknell, J.A. Przystawa, Comm.on Physics, 1 (1976) 51; M.A. Savchenko, A.V. Stefanovich, Sov.Phys. JEPT, 47(1978)1195.

24. V.V. Gene, E.F. Dudnik, V.A. Podol'skii, L.Ya. Sodovskaya, Sov.Phys. - Solid State, 19 (1977) 1128; K. Parlinski, Phys. Stat. Sol. (a), 74 (1982) K37.

QUASISYMMETRY (P-SYMMETRY) IN CRYSTALS*

K. RAMA MOHANA RAO
Department of Applied Mathematics, A.U.P.G. Extension Centre
NUZVID 521201, Andhra Pradesh, INDIA

and

M. KONDALA RAO
Department of Applied Mathematics, Andhra University, Waltair

The concept of antisymmetry was introduced into the study of crystallographic point groups by Shubnikov (1951). The interpretation of antisymmetry as two-colour symmetry has led to the idea of polychromatic symmetry (Belov and Tarkhova, 1956). Generalizations of Shubnikov's antisymmetry and Belov's colour symmetry were enveloped by the concept of P-symmetry introduced by Zamorzaev (1967) and discussed in detail by Shubnikov and Koptsik (1974).

Zamorzaev (1967) classified groups as major, minor and intermediate. Let us recall the essentials of the so-called geometric approach to the classification of P-symmetries : To every point of a geometric figure we assign at least one of the indices 1, 2, ..., p and we use the term P-symmetry transformation to denote an isometric transformation of the figure, which transforms each point of the figure with index i into a point with index K_i, the permutation of the indices $\begin{pmatrix} 1 & 2 & \dots & p \\ K_1 & K_2 & \dots & K_p \end{pmatrix} \in P$. We shall call a group of P-symmetry transformations a full P-symmetry group if the group p_i of permutations of indices involved in the transformation of the group coincides with P. Such groups are divided into major, minor and intermediate groups when the subgroups of the permutations $Q = G \cap P$ coincide with P, consists of the identity transformation or are a non-trivial subgroup of P, respectively.

Whenever the generator group G can be expressed as $G = S \wedge T$ of two of its constituent subgroups S and T, Krishnamurty, Prasad and Rama Mohana Rao (1978a) have established

---

* The authors acknowledge the editor, Acta Crystallographica for his kind permission to discuss and submit the results of this paper in brief at this Colloquium

a general method of obtaining quasisymmetry (P-symmetry) groups as semi-direct products. In the case of those generators which can be written as a quasi semi-direct product i.e. G=H o G(mod H), similar results as in the case of semi-direct products were obtained (Krishnamurty, Prasad and Rama Mohana Rao, 1980).

The general method of obtaining P-symmetry groups as semi-direct products established by Krishnamurty, Prasad and Rama Mohana Rao (1978a) makes use of the concept of semi-direct products and the fundamental quasisymmetry theorem of Zamorzaev (1967). The proof is accomplished with the following three theorms, the statements of which are given below :

Theorem 1 : Let $G = S \wedge T$, S' be a full P-symmetry group with S as generator and T' be a full Q-symmetry group with T as generator, where $P \neq Q$, $Q \neq \{I\} \neq P$. If $G' = S' \wedge T'$ and if S' and T' are of the same category (both major or both minor or both intermediate), then G' is a full PQ-symmetry group of the same category with G as generator.

Theorem 2 : Under the conditions of the theorem 1 cited above, if $G' = S' \wedge T'$ and if S' and T' are of different categories, then G' is a full PQ-symmetry intermediate group with G as generator.

Remark 1 : If $Q = P \neq \{I\}$, then we have $PQ = P$, Then in this case some of the results of theorem 1 and theorem 2 cited above have to be modified as under :

i) If both S' and T' are intermediate groups and if $G' = S' \wedge T'$, then G' may be either a full intermediate group or a full major group with G as generator.

ii) If one of S' and T' is a major group and the other a minor group and if $G' = S' \wedge T'$, then G' is a full P-symmetry major group with G as generator.

iii) If one of S' and T' is a major group and the other an intermediate group, and if $G' = S' \wedge T'$, then G' is a full P-symmetry major group with G as generator.

Remark 2 : If S' is a full P-symmetry group with $P = \{I\}$, then S' is a major group as well as a minor group. Such a group is referred to as a major/minor group. In the case of a major/minor group, the following theorem can be established.

Theorem 3 : If one of the groups S' and T' is a major/minor group, and the other a major or a minor or an intermediate group, and if $G' = S' \wedge T'$, then G' is a major or minor or intermediate group respectively with G as generator.

If both S' and T' are major/minor groups and if $G' = S' \wedge T'$, then G' is also a major/minor group with G as generator.

In the present paper, the general method of construction of quasisymmetry groups as semi-direct and quasi semi-direct products (Krishnamurty et al, 1978a; 1980) is explored to construct minor quasisymmetry groups with Fedorov groups as generators. A connection between the semi-direct and quasi semi-direct products with those of the symmorphic and non-symmorphic groups is established. The consequent developments of the concept of P-symmetry, the different aspects on the applications of generalized colour symmetry and the position of the P-symmetry groups vis-a-vis the Wreath product groups is discussed elaborately. The method of construction of minor quasisymmetry space groups is exemplified in the case of both symmorphic and non-symmorphic cubic space groups for the chosen boundary condition $T_x^2 = T_y^2 = T_z^2 = E$, considering the symmorphic cubic space groups F23, F432 and the non-symmorphic ones Fd3, Fd3m. The minor quasisymmetry cubic space groups thus generated are associated with the appropriate 1-D complex and 2-D real IRs of the generator groups using the ideas of little groups and their 1-D allowable irreducible representations. The results obtained are discussed briefly and some suggestions have been made as to the possible studies in which the groups obtained in this paper can be applied.

As a connected account of the minor groups obtained as semi-direct and quasi semi-direct products and associated with the non-degenerate complex IRs and degenerate IRs of the space groups is not so readily available as far as our knowledge goes. the minor quasisymmetry cubic space groups obtained above are tabulated.

REFERENCES

Shubnikov, A.V. (1951) Symmetry and antisymmetry of finite figures, Moscow Ussr academy of Science.

Belov, N.V. and Torkhova, T.N. (1956) Kristallogr. 1 , 4-13 (1956 Sov. Phys. Crystallogr. 1, 5-11).

Zamdzaev, A.M. (1967) Kristallogr. 12 , 819-825 (1967 Sov. Phys.Cristallogr. 12, 717-722).

Shubnikov, A.V. and Koptsik, V.A. (1974) Symmetry in Science and Art, New York Plenum.

Krishnamurty, T.S.G., Prasad, L.S.R.K. and Rama Mohana Rao,K (1978 a) Jour.Phys. A: Math and Cren. 11, 805-811.

Krishnamurty, T.S.G., Prasad, L.S.R.K. and Rama Mohana Rao, K. (1980) Journ. Phys. A.: Math.and Cren, 13 , 1942-56.

BRAID GROUPS AND EUCLIDEAN LIE ALGEBRAS IN STATISTICAL MECHANICS OF SPIN SYSTEMS

Mario Rasetti

Dipartimento di Fisica del Politecnico di Torino
Torino, Italy

1. Introduction

A rich algebraic stucture has recently emerged in the analysis of two classes of models which at present play a major role in the statistical mechanics of lattice spin systems:the Ising model on three-dimensional lattices homogeneous under finitely presented groups /1/, and the Baxter vertex models /2/.

One of the unexpected results of such an analysis is the deep structural equivalence between the two sets of models,which crucially depends on what turns out to be for both the basic element of their combinatorial stucture:the braid group.

Another result,induced by the previous one,depends on the geometrical picture it gives rise to,which allows a far reaching generalization of the concept of integrability of the connected dynamical systems. It leads,in the case of finite,globally symmetric Ising systems,to exciting connections with the theory of simple finite groups /3/;in the case of vertex models,to the conjecture of solvability for a whole new class of models of much richer combinatorial structure,characterized by graded Lie algebras /4/.

In present note only the basic elements of the theory will be concisely reviewed,enphasizing the general philosophy of the method more than its technical details.

2. The Ising Model

Among the different procedures of attack of the Ising model,the richest in structure,and probably the most promising - in particular when the model is defined on a lattice L homogeneous under some finitely presented finite group G - is the so called Pfaffian (or dimer) method /5/.

Its interest for the 3-dimensional case stems out of the formulation that was recently given of it /6/,based on:

i) the relabeling of the positional degrees of freedom in terms of anticommuting Grassmann variables,associated in a one-to-one way with the elements of G,

ii) the successive extension of the group G to a group $\hat{G}$,whereby all the orientations of the lattice bonds compatible with the combinatorial requirements expressed by Kasteleyn's theorem /7/,and only those, might be obtained as the invariant set of configurations of the oriented graph matching L,

iii) the crucial theorem stating that the calculation of the partition function for the model reduces - if $\hat{G}$ exists - to the evaluation of a single Pfaffian,associated with the generalized incidence matrix $\hat{M}$ of L,extended with respect to $\hat{G}$ /1/:

$$Z(L) = \mathrm{Pf}\,\hat{M} \tag{2.1}$$

Since the regular representation of a finite group is the direct sum of its irreducible representations, each contained as many times as its dimension j, (2.1) can be naturally reduced to the evaluation of a finite number of finite determinants:

$$Z(L) = \prod_J \left(\det \hat{M}^{(J)} \right)^{j/2} \tag{2.2}$$

where $\hat{M}^{(J)}$ is a matrix of rank j.

There are severe constraints in the choice of L and $\hat{G}$ dictated by both combinatorial and topological limitations. The latter essentially consist in the requirement that L should be embeddable in a two-dimensional orientable compact surface A of arbitrarily high topological genus h.

It was shown in ref./1/ that the most general extension $\hat{G}$ of G satisfying all the requisites is of the form

$$\hat{G} = C \text{ wr } S_{2h} \tag{2.3}$$

where wr denotes the wreath product /8/, and S_{2h} is the permutation group on 2h objects. $C = (\text{Homeo } A/\text{Isot } A)/H$, is the mapping class group P of A modulo the subgroup H of those diffeomorphisms which preserve the isotopy class of any configuration in which the handles of A are cut by a system of h disjoint cycles c_i, i=1,...,h.

It should be noted that maps and spaces are to be thought of in the piecewise linear category.

The determination of the finite group of mapping classes /9/

$$P = \text{Homeo } A/\text{Isot } A = \pi_0 \text{ Diff}^+(A) \tag{2.4}$$

is indeed equivalent to the description of all finite extensions of the fundamental group of the surface A.

One can, with no loss of generality from the point of view of physical applications /1/, assume G to be a Fuchsian group, namely a discontinuous group of orientation preserving motions of the hyperbolic plane. A finitely generated Fuchsian group has presentation /10/

$$G = \langle d_1,\dots,d_n,t_1,u_1,\dots,t_h,u_h \mid d_i^{k_i}, i=1,\dots,n; d_1\cdots d_n \prod_{j=1}^{h} [t_j,u_j] \rangle$$

$$k_i \in \mathbf{Z}\ ,\quad k_i \geqslant 2\ ;\ i=1,\dots,n \tag{2.5}$$

The fundamental group of A, in this case, has presentation

$$\pi_1(A) = \langle t_1,u_1,\dots,t_h,u_h \mid \prod_{j=1}^{h} [t_j,u_j] \rangle \tag{2.6}$$

The analysis of the first cohomology group $H^1(G)$ of G, shows that G is finite if $\sum_i (1/k_i) > n - 2$.

The homeomorphism ext: $G \to \hat{G}$, acts locally mapping $d_i^{k_i}$ to $-d_i^{k_i}$, i=1,...,n and globally extending G by its fundamental group, i.e. mapping t_j,u_j to ± 1, j=1,...,h (namely $\pi_1(A)$ to $\mathbf{Z}_2$).

We have the exact sequence /9/:

$$\mathbf{Z} \to \mathbf{Z}^h \oplus B_{2h-1} \to H \to \pm S_h \to 0 \tag{2.7}$$

and this implies that H as well has finite presentation and therefore $\hat{G}$ is finitely presented.

In (2.7) B_{2h-1} is the pure braid group /11/ on (2h-1) strands, i.e.

the fundamental group of the space of unordered sets of (2h-1) distinct points in a plane;more precisely a representation of the fundamental group of the complement of the branching manifold in the symmetric group onto Artin's braid group /12/.

The latter (that we still denote B_n) has (n-1) generators $b_1,\dots,b_{n-1}$ and (n-1)(n-2)/2 relations

$$\begin{aligned} b_i b_j &= b_j b_i & |i-j| > 1 \\ b_i b_{i+1} b_i &= b_{i+1} b_i b_{i+1} & i=1,\dots,n-2 \end{aligned} \tag{2.8}$$

P is indeed generated by H and the set of elements supported locally by an homology exchange between two cycles c_i, c_j.

It follows that all the relations of P derive from relations supported in subsurfaces of A of genus at most 2. There exists therefore a finite matrix representation of $\hat{G}$ whenever G is finite.

Finally Z(L),as a function on the space of parameters of the model, can be expected to be an automorphic function of $\hat{G}$,and hence to depend only on the representations of H (namely of B_{2h-1}) and on the set of characters of S_{2h}.

3. The Baxter Vertex Models

Baxter's vertex models are special cases of a class of spin systems generalizing the Ising model,defined on a two-dimensional lattice L' (which,for simplicity of notation,we assume square) and characterized by an hamiltonian of the form

$$H = \sum_{f \in L'} E(s_i, s_j, s_k, s_m) \tag{3.1}$$

where f denotes any plaquette of L',whose vertex sites (named anticlockwise) are (i,j,k,m).

Baxter showed that upon splitting L' into 4 congruent 2-dimensional sublattices $L^{(k)}$,and 4 congruent 1-dimensinal sublattices $X^{(k)}$, k=1,..,4,such that

$$L' = \bigcup_{k=1}^{4} (L^{(k)} \cup X^{(k)}) , \qquad \partial L^{(k)} = X^{(k)} \cup X^{(k+1)} \quad (k \bmod 4)$$

$$L^{(k)} \cap L^{(k')} = \emptyset , \quad X^{(k)} \cap L^{(k')} = \emptyset , \quad X^{(k)} \cap X^{(k')} = \text{site } O$$

for all k,k';the corresponding partition function is given by /2/

$$Z(L') = \mathrm{Tr}(A_1 A_2 A_3 A_4) \tag{3.2}$$

where A_k,k=1,..,4,called "corner transfer matrices" are defined as

$$\begin{aligned} (A_k)_{ss'} &= \delta_{s_o s'_o} \sum_{\{s_j, j \in L^{(k)}\}} \prod_{f \in L^{(k)}} \omega(s_i, s_j, s_k, s_m) \\ \omega &= \exp(-\beta E) \; ; \\ s &= (s_{x_o}, s_{x_1}, \dots, s_{x_n} ; x_i \in X^{(k)}), \\ s' &= (s_{y_o}, s_{y_1}, \dots, s_{y_n} ; y_i \in X^{(k+1)}). \end{aligned} \tag{3.3}$$

Upon defining

$$(U_i^{(k)})_{ss'} = \prod_{\substack{j=0 \\ j\neq i}}^{n} \delta(s_j,s_j')\ \omega(s_i,s_{i+1},s_i',s_{i-1}) \quad (3.4)$$

for i=1,...,n-1,and

$$V_j^{(k)} = U_\partial^{(k)}\ U_{n-1}^{(k)}\ U_{n-2}^{(k)} \cdots U_j^{(k)} \quad (3.5)$$

where $U_\partial^{(k)}$ is a matrix designed to take into account the boundary conditions (and indeed irrelevant in view of the thermodynamic limit),one can write

$$A^{(k)} = V_2^{(k)}\ V_3^{(k)} \cdots V_n^{(k)} \quad (3.6)$$

Thus Z(L') is indeed a trace over a long string (word) of factors $U_i^{(k)}$. It is easy to check from the very definition (3.4) that the latter have the commutation relations

$$U_i^{(k)}\ U_j^{(k)} = U_j^{(k)}\ U_i^{(k)} \quad \text{if } |i-j| \geqslant 2, \text{ all } k \quad (3.7)$$

Moreover the requirement of commutativity of the whole family of transfer matrices,implying the integrability of the system (i.e. the existence of an infinite number of conservation laws) leads to the "star-triangle" relations /2,13/

$$U_{i+1}^{(k)}(\theta)\ U_i^{(k)}(\theta+\theta')\ U_{i+1}^{(k)}(\theta') = U_i^{(k)}(\theta')\ U_{i+1}^{(k)}(\theta+\theta')\ U_i^{(k)}(\theta) \quad (3.8)$$

for k=1,..,4 and i=1,...,n-2;where the argument θ of $U_i^{(k)}$ is a reminder of Baxter's parametrization of the ω's in terms of Jacobi elliptic functions.

One immediately notices that (3.7) together with (3.8) for $\theta = \theta' = 0$ are but the defining relations (2.9) of the braid group B_n.

The (3.8) define then a fibration by B_n over the abelian one-dimensional manifold W of θ (elliptic curve).

Since the braid group af an algebraic function is an invariant of the function,but not of the corresponding covering,once more the partition function can be expected to be an automorphic function of the group G' of bundle homeomorphisms whose action are those covering translations (displacements of half period) commuting with the braid group. G' can be represented as the double divisor of the principal polarization of W /14/.

4. The Euclidean Algebra

The concluding remarks of previous section are far reaching. In order to make them more rigorous,let Ell(R_f), $R_f = \mathbf{Z}+fR$, $f \in \mathbf{Z}$, $f \geqslant 1$, where R is a ring of integers in the field $K = \text{End } W \otimes \mathbf{Q}$, be the set of classes of elliptic curves with endomorphism ring R_f.

Let j(W) be the invariants of W associated with R_f,and Cl(R_f) the group of projective modules of rank 1 over R_f.

The Artin map takes elements of Cl to elements of the Galois group, which permutes the j(W)'s,thus Cl acts on Ell by translations.

In other words the elliptic curves with endomorphism ring R_f correspond in one-to-one way with the class group Cl(R_f),and,conversely, given R_f the corresponding elliptic curves have complex multiplication.

Of course $G' \subset Cl$. However one can define a generalized abelian variety A isomorphic to the q-fold product of W, where q is the dimension of the irreducible root system Ω generated by solving the conjugacy problem for the braid group; $A \sim W \times \Omega$.

The natural line bundle over A is the orbit space B' of the group of affine transformations generated by the lattice of Ω and the Weyl group G_w (which replaces G'). The set Γ^k of theta functions of degree k - which are the holomorphic sections of B' - on the lattice of affine linear functions which take integral values on Ω, form in a natural way a Γ^o -graded algebra Γ ;

$$\Gamma^r \; \Gamma^s \subseteq \Gamma^{r+s} \quad ; \quad \Gamma = \bigoplus_{s=0}^{\infty} \Gamma^s \tag{4.1}$$

in that Γ^o has a natural ring structure /15/.

Γ is isomorphic, as $\mathbf{Z}$-graded algebra, to the principal subalgebra of an affine Lie algebra.

The stucture of affine Lie algebras, which form an important subclass of Kac-Moody algebras /16/, implies a remarkable set of combinatorial identities, which are an extreme generalization of the denominator and of Weyl's character formulas /17/. In particular the Rogers-Ramanujan identities /18/ turn out to have a canonical interpretation in terms of the infinite dimensional Lie algebra A_1^1. The latter, which is the simplest affine Lie algebra, is of special interest for physics, in that it possesses a principal Heisenberg subalgebra whose stucture allows to construct A_1^1 in terms of creation and annihilation operators and a vertex operator /19/ acting on the basic module, identified with a Fock space; or as an algebra of operators on a mixed boson-fermion Fock space /20/.

It is not surprising that in the original solution by Baxter of some vertex models the Rogers-Ramanujan identities play a crucial role /2/. Also, the possibilities of constructing new models - possibly superintegrable - are striking.

As a concluding, though yet very speculative, remark, let's notice how the affine algebra description neatly applies to the vertex models and their generalizations in that these are implicitly tackled in the thermodynamic limit. However, even in the finite case, such as that we dealt with for the Ising model, there appears to emerge a similar structure: extraspecial groups /21/ - which play a fundamental role in simple finite group theory /3/ - , i.e. finite p-groups (p prime) whose commutator subgroup, Frattini subgroup and center all coincide and have order p, can be thought of as finite analog of the Heisenberg group.

References

/1/ M.Rasetti, in "Group Theoretical Methods in Physics", M.Serdaroglu and E.Inönü eds., Springer Verlag, Berlin 1983, page 513

M.Rasetti, in "Selected Topics in Statistical Mechanics", N.N.Bogolubov and V.N.Plechko eds., J.I.N.R. Publ., Dubna 1981, p.181

/2/ R.J.Baxter, "Exactly Solved Models in Statistical Mechanics", Academic Press, London 1982

G.E.Andrews, Proc.Natl.Acad.Sci.(U.S.A.) 78, 5290(1981)

/3/ R.L.Griess, jr., Inventiones Math. 69, 1(1982)

/4/ M.Rasetti and G.D'Ariano, in "Differential Geometric Methos in The-

oretical Physics",D.H.Doebner ed.,Springer Verlag,in press
/5/ M.E.Fisher,J.Math.Phys. 7,1776(1966)
/6/ F.Lund,M.Rasetti and T.Regge,Commun.Math.Phys. 51,15(1976)
M.Rasetti and T.Regge,Rivista Nuovo Cimento 4,1(1981)
/7/ P.W.Kasteleyn,J.Math.Phys. 4,287(1963)
/8/ A.Kerber,"Representations of Permutation Groups,I",Springer Verlag, Berlin 1971
/9/ A.Hatcher and W.Thurston,"A Presentation for the Mapping Class Group of a Closed Orientable Surface",to be published
/10/H.S.M.Coxeter and W.O.Moser,"Generators and Relations for Discrete Groups",Springer Verlag,Berlin 1965
/11/E.Artin,Ann.Math. 48,101(1947)
/12/G.Burde,Abh.Math.Sem.Univ.Hamburg 27,97(1964)
/13/A.B.Zamolodchikov,Commun.Math.Phys. 79,489(1981)
/14/I.V.Cherednik,Dokl.Akad.Nauk USSR 249,1095(1979)
/15/H.Garland,J.Algebra 53,480(1978)
/16/V.G.Kac,Izv.Akad.Nauk USSSR 32,1323(1968)
R.V.Moody,J.Algebra 10,211(1968);Canad.J.Math. 21,1432(1969)
/17/J.Lepowsky and S.Milne,Advances in Math. 29,15(1978)
/18/G.E.Andrews,"The Theory of Partitions",Addison Wesley,Reading 1976
/19/J.Lepowsky and R.L.Wilson,Commun.Math.Phys. 62,43(1978)
I.B.Frenkel and V.G.Kac,Invent.Math. 62,23(1980)
/20/J.Lepowsky and R.L.Wilson,Proc.Natl.Acad.Sci.(U.S.A.) 78,7254(1981)
/21/P.Hall and G.Higman,Proc.London Math.Soc. 7,1(1956)

PHASE COEXISTENCE IN MANY-FERMION SYSTEMS

Allan I. Solomon
Faculty of Mathematics, Open University, Milton Keynes, England
and
Joseph L. Birman
Physics Department, City College, CUNY, New York 10031, U.S.A.

We summarize work that we have carried out recently on the application of dynamical groups to the problem of many-fermion systems capable of simultaneously exhibiting more than one condensed phase. The classical example of a single condensed phase is the superconductor; the methods we employ in our analyses are group theoretical analogues of the celebrated treatment of Bardeen, Cooper and Schrieffer(1). A somewhat less familiar example of a single fermion condensed phase is the structural transition observed in the crystalline lattices of intermetallic compounds of the form A_3B, known as β-tungstens. This phase transition, referred to as a Peierls or martensitic transition, is characterized by the presence of charge-density waves; we shall refer to the condensed state as the CDW state. Since such compounds may also be superconducting, the exciting possibility exists of observing the simultaneous occurrence of both states, SC and CDW, in a given sample; and this has been done, for example in the Raman scattering experiments of Sooryakumar and Klein(2). Similarly, the coexistence of magnetism and superconductivity has been the subject of much theoretical interest(3). Due to the rather simple form of the hamiltonians used in conventional treatments - essentially pair-reduced, mean field models - one may readily present a unified group theoretical approach to the general coexistence problem of many-fermion systems. We outline such an approach in the following note.

We may write the general many-fermion interacting Hamiltonian H as

$$H = \sum_{k\alpha} \epsilon_k a^+_{k\alpha} a_{k\alpha} + \frac{1}{2} \sum V_{kk'k''k'''} a^+_{k\alpha} a^+_{k'\beta} a_{k''\beta} a_{k'''\alpha} \delta(k + k' - k'' - k''')$$

where the creation and annihilation operators for fermions of momentum k and spin σ satisfy the anticommutation relations

$$[a_{k\sigma}, a^+_{k'\sigma'}]_+ = \delta_{kk'}\delta_{\sigma\sigma'}$$

All the physics of the interaction is contained in the potential V; clearly no progress can be made unless the interaction is specified and perhaps some further reducing approximation is made. Fortunately, nature has conspired with group theorists to ensure that in a large variety of physical phenomena the dominant terms in the interaction are pairing terms - terms which couple operators having some fixed total momentum Q and perhaps spin. Retaining only such dominating terms, our hamiltonian reduces to

$$H^{red} = \sum_{k\alpha} \epsilon_k a^+_{k\alpha} a_{k\alpha} + \frac{1}{2}\sum V_{k,k',Q} a^+_{k\alpha} a^+_{-k+Q\beta} a_{k'\beta} a_{-k'+Q\alpha}$$

where in the interaction term we may sum over a set of n values of Q, $Q_1, Q_2, \ldots, Q_n$ (as well as k, k', α, β). For example, the BCS reduced hamiltonian has Q = 0, spin $\alpha + \beta = 0$. Such pairing reduced hamiltonians are amenable to a Lie algebraic treatment in the following way: Consider the 4n operators

$$\{A_1, A_2, \ldots, A_{4n}\} = \{a_{k_1\uparrow}, a_{k_1\downarrow}, a_{k_2\uparrow}, a_{k_2\downarrow}, \ldots, a_{k_n\downarrow}; a^+_{-k_1\downarrow}, a^+_{-k_1\uparrow}, \ldots, a^+_{-k_n\uparrow}\}.$$

with $\qquad k_i = k - Q_i.$

These obey $\qquad [A_i, A^+_j]_+ = \delta_{ij} \qquad (i,j = 1,2,\ldots,4n)$

These $(4n)^2$ operators $X_{ij} = A^+_i A_j$ satisfy the commutation relations $[X_{ij}, X_{k\ell}] = \delta_{jk} X_{i\ell} - \delta_{i\ell} X_{kj}$ of $g\ell(4n)$; hermitian combinations of pairs of A_i, such as occur in the hamiltonian, lead to the algebra u(4n). Such operators X_{ij} incorporate only pairing terms of total momentum $Q_i - Q_j$, and so H^{red} may be rewritten as a bilinear form in X_{ij}. Taking all possible pairs of A_i and A^+_i leads to the larger algebra O(8n), as Lipkin has shown[(4)]; this algebra goes beyond the simple fixed-momentum pairing terms included here. In fact, we shall generally only consider spin-singlet pairing, (except in the case of superfluid helium three, for which the spin-triplet pairing is known to be important[(5)]) which reduces the number of A_i by half

$$\{A_1, A_2, \ldots, A_{2n}\} = a_{k_1\uparrow}, a_{k_2\uparrow}, \ldots, a_{k_n\uparrow}; a^+_{-k_1\downarrow}, a^+_{-k_2\downarrow}, \ldots, a^+_{-k_n\downarrow}\}$$

leading to the algebra u(2n). Further, the present physical phenomena that we are interested in involve only two total momentum states Q = 0 for superconductivity, and $Q = 2k_F$, where k_F is the wave-vector of the Fermi surface, for CDW states. Thus n = 2, and the relevant Lie algebra is u(4)(or a subalgebra of u(4), depending on the precise form of the hamiltonian). The u(2) subalgebras of u(4) are essentially the spectrum generating algebras of the individual SC, CDW and - it transpires - anti-ferro-magnetic hamiltonians (AF).

By the above process we may write the general reduced hamiltonian in the form bilinear in the X_{ij}

$$H^{red} = \sum_{i,k} \epsilon_i(k) X_{ii}(k) + \sum_{i,j,k,k'} m_{ij}(k,k') X_{ij}(k) X_{ij}^{+}(k').$$

We need to further approximate in order to express our model as an element of a Lie algebra. The mean-field approximation H^{mf} to H^{red} is given by

$$H^{mf} = \sum_k H(k)$$

where

$$H(k) = \sum_i \epsilon_i(k) X_{ii}(k) + \Big(\sum_{i,j} m_{ij}(k) X_{ij}(k) + \text{h.c.} \Big)$$

with

$$m_{ij}(k) = \langle\langle \sum m_{ij}(k,k') X_{ij}^{+}(k') \rangle\rangle.$$

The complex numbers $m_{ij}(k)$ are self-consistently determined from the thermodynamic expectation $\langle\langle\ \rangle\rangle$, defined by

$$\langle\langle A \rangle\rangle = \frac{\text{trace}\{\exp(-\beta H^{mf}) A\}}{\text{trace}\{\exp(-\beta H^{mf})\}}$$

where β is the inverse temperature, $\beta = 1/k_{Boltzmann}T$. This final stage has explicitly expressed our hamiltonian as a sum of elements X_{ij} of the Lie algebra $u(4)$ (in the case we have selected). More precisely, each $H(k)$ is an element of $su(4)_{(k)}$, since the kinetic energy term $\sum_i \epsilon_i(k) = 0$ leads to a traceless algebra; and the spectrum generating algebra associated with H^{mf} is $\bigoplus_k su(4)_{(k)}$.

To proceed further with explicit calculations of energy gaps, coexistence domains, etc., we must take a basis for our $u(4)$ algebra. A convenient one is the set

$$\{\tau_\mu \otimes \tau_\nu : \mu,\nu = 0,1,2,3\}$$

of direct products of the Pauli matrices. For calculations including spin-triplet effects, a basis for the algebra $u(8)$ is required, and this is conveniently provided by

$$\{\tau_\mu \otimes \tau_\nu \otimes \tau_\rho : \mu,\nu,\rho = 0,\ldots,3\}.$$

This approach is implicit in the work of Horovitz[6], and Gutfreund and Little[7].

We review the results obtained and obtainable by the preceding methods. The energy spectrum is the most immediate, depending as it does only on an implicit diagonalization of the hamiltonian H^{mf}. Since this calculation may be performed in any representation, it amounts to the diagonalization of a 4 × 4 matrix in our case. The resulting spectrum depends on the 3 Casimir invariants of su(4), and we therefore expect to see 3 energy gaps if the three phases (SC, CDW and AF) are simultaneously present. These would manifest themselves, for example, as peaks in Raman scattering experiments such as those of reference (2). (The general result for an su(n) model would be n-1 gaps.) We may next calculate domains of coexistence of the phases, for example at zero temperature in terms of the interaction strengths (coupling constants). The presence of a given phase is detected by the non-vanishing of the respective order parameter; these order parameters are also elements of the Lie algebra, which again makes the calculation of their expectations purely algebraic. This program has been carried out in the absence of antiferromagnetism (and for real CDW coupling) for which the su(4) ~ so(6) algebra reduces to so(5) - a rank-2 algebra with correspondingly simpler structure and less laborious calculation - leading to the two-gap picture observed in the experiments of reference (2).

We summarize the group chain of this note as follows:

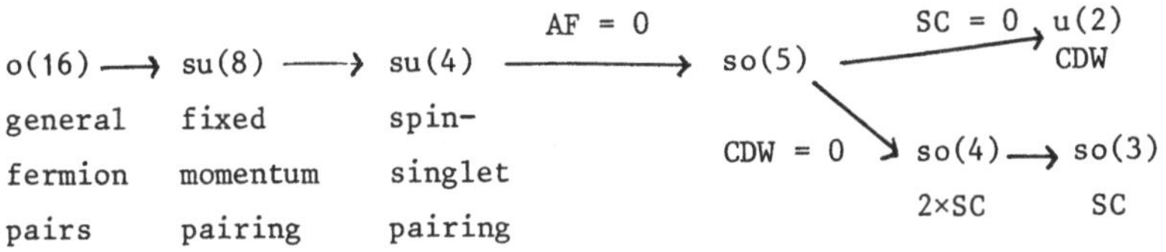

References

(1) J. Bardeen, L.N. Cooper and J. R. Schrieffer, Phys.Rev. 108, 1175(1957)

(2) R. Sooryakumar and M. V. Klein, Phys.Rev.Lett. 45, 660(1980)

(3) M. Ishikawa, Contemp.Phys. 23, 443 (1982)

(4) H. J. Lipkin, "Lie Groups for Pedestrians" (North-Holland, 1965), Chapter 5.

(5) B. Horovitz, Solid State Comm. 18, 445 (1976)

(6) H. Gutfreund and W. A. Little, Rice University Studies 56 (1980)

(7) J. L. Birman and A. I. Solomon, Phys. Rev. Lett. 49, 230 (1982)

MEAN FIELD RENORMALIZATION GROUP APPROACH TO LATTICE MODELS

A.L. Stella
Dipartimento di Fisica ed Unità GNSM del CNR
Università di Padova, Italy

Even if not able to determine critical exponents correctly, mean field and other classical theories of phase transitions are generally much easier to apply than renormalization group methods [1], and often provide a very useful first insight into the phase diagrams of complicated models.

Recently the possibility has been shown of embodying classical ideas within a modern renormalization group strategy [2]. This led to the development of a very simple, flexible and efficient approach to the static, as well as dynamic critical properties of lattice models.

The renormalization group deals with the scaling symmetry of critical behaviour. This statistical symmetry arises when an infinite number of random variables becomes strongly dependent in a probabilistic sense [3]. To be concrete, let us consider an Ising model, with nearest neighbour reduced exchange coupling K and magnetic field h, on a d-dimensional hypercubic lattice. Near criticality ($K \sim K_C$, $h \sim 0$) the canonically averaged magnetization per site, m, scales like

$$m(k,h) = \ell^{-d+y_H}\, m(K_C + \ell^{y_T}(K-K_C),\, \ell^{y_H} h), \tag{1}$$

as far as its leading singular part is concerned. In eq. (1) y_T and y_H are the thermal and magnetic exponents, respectively, and ℓ is a rescaling factor.

A renormalization group strategy amounts to construct a suitable regular mapping, $K' = K'(K,h)$, $h' = h'(K,h)$, such that, for a given value of ℓ,

$$m(K,h) = \ell^{-d+y_H}\, m(K',h'). \tag{2}$$

Eq. (2) is expected to hold for $h \sim 0$ (thus, by symmetry, also $h' = \lambda_H(K)h \sim 0$), and, for $K \to K_C$, applies to the leading singular part of m. The assumed regularity immediately implies $K'(K_C,0) = K_C$ and $h'(K_C,0) = 0$. Moreover, by comparison with eq. (1), one gets $\ell^{y_T} = \partial K'/\partial K\big|_{K_C,0}$ and $\ell^{y_H} = \partial h'/\partial h\big|_{K_C,0}$. So, from the properties of a regular mapping, information is obtained about the location and nature of critical singularities.

A simple strategy of approximating the above mapping makes use of the effective field idea of classical mean field or cluster approaches [2]. In these approximations one considers a cluster with one (mean field), or more interacting spins feeling the action of the surronding spins in the lattice in an effective sense; in other words the surronding spins $S_i (S_i = \pm 1)$ are replaced by m in the effective cluster hamiltonian. The average magnetization in the cluster is then given by a function $f(K,h,m)$. Equating this function to m itself gives a classical equation of state. The regularity in K and h, which is obviously implied by such cluster calculation of the equation of state, leads unavoidably to classical exponents (e.g. $\beta = (d - y_H)/y_T = \frac{1}{2}$).

In the mean field renormalization approach, however, the above regularity is properly employed for the construction of the renormalization mapping. According to the modern picture of criticality, the parameter ℓ in eq. (1) can be interpreted as a length rescaling. The idea is thus to replace the usual self-consistency condition of classical approximations by the new requirement that the magnetization should scale according to eq. (1), when one looks at it on different length scales.

In the cluster calculations described above the length scale is given by the linear size of the cluster, L. So, for a d-dimensional cubic cluster containing L^d spins, the average magnetization should be denoted by $f_L(K,h,m)$. An approximate self-consistent realization of eq. (2) is then obtained by imposing

$$f_L(K,h,m) = \ell^{-d+y_H} f_{L'}(K',h',\ell^{d-y_H} m) \tag{3}$$

with $\ell = \frac{L}{L'} (>1)$. Developing eq. (3) for small m and h, one obtains the renormalization mapping in the neighborhood of $h = 0$.

The above scheme has been successfully applied to a variety of classical and quantum spin models. The results generally improve with both L and L' increasing, and are already satisfactory with the smallest L and L' (e.g. $L = 2$, $L' = 1$) [2].

A very promising direction of improvement turns out to be that of embodying reaction field corrections in the calculation of the functions f above [4]. Corrections are derived in a spirit similar to the one inspiring the, so called, Onsager's reaction field correction to the mean field equation of state [5].

The effective field provided by the spins surronding a given cluster is thus made dependent on the internal configuration of the cluster itself, in a way consistent with m being the average magnetization. Along these lines one can e.g. obtain K_C, y_T and y_H for a d = 2 Ising model, all within ∿10% of the exact results, with calculations involving no more than 4 spins. A similar agreement with expected results is obtained in d = 3 with calculations involving just up to 8 interacting spins. Thus the method seems to exploit in an optimal way the information contained in the short range fluctuations of the order parameter. It would be interesting, in the future, to combine it with Monte Carlo or transfer matrix techniques, in order to increase substantially cluster sizes and, hopefully, accuracy.

Another appealing feature of mean field renormalization is its wide range of applicability, which essentially is the same as for classical approximations. Besides the static properties of ordered spin systems, also disordered spin models [6] (e.g. dilute Ising model) and geometrical critical phenomena [7-8], like percolation (both directed and undirected) and SAW's, have been treated. For SAW's, in particular, the critical properties are extracted quite accurately already by relatively small cell calculations, without using reaction field corrections [7]. In the case of directed percolation the method has been extended to large cells and combined with a suitable extrapolation technique, yielding results comparable with those of series expansions and finite size scaling methods [8].

Very promising applications have been performed recently also to the critical dynamics of kinetic spin models evolving according to a master equation [9]. In this context progress of real space renormalization is very much hindered by the necessity of controlling memory effects [10] and by the possible dangerous consequences of arbitrary truncations in the equations of motion [11]. Due to its phenomenological and self-consistent character, the dynamical extension of the mean field renormalization method does not deal with these problems [4].

Slightly out of equilibrium and close to criticality, the time dependent magnetization, m(K,h,t), scales, for long times, like in eq. (1), with , in addition, t replaced by $\ell^{-z}t$ on the right hand side. The z exponent is the dynamical index, and $\Delta = z/y_T$ is the exponent of the temperature divergence of the relaxation time. Along lines similar to those illustrated in the static case, the combination of different dynamical cluster approximations [12] allows to compute z in addition to the static quantities [4]. For the Glauber model a reaction field dynamical calculation involving up to 4 spins yields $\Delta = 2.24$ in d = 2, and one involving up to

8 spins yields $\Delta = 1.46$ in $d = 3$. These values are remarkably close, especially in $d = 2$, to present estimates by Monte Carlo or series expansions methods ($\Delta = 2.2 \pm 0.1$ in $d = 2$ [13], $\Delta = 1.32 \pm 0.03$ in $d = 3$ [14]).

REFERENCES

1. Recent reviews of real space renormalization methods and applications are contained in "Real-Space Renormalization", ed. by T.W. Burkhardt, J.M.J. van Leeuwen, Topics in Current Physics (Springer Berlin, Heidelberg, New York 1982).
2. J.O. Indekeu, A. Maritan, A.L. Stella: J. Phys. A 15, L291 (1982).
3. See e.g. the talk by G. Jona-Lasinio, this conference.
4. J.O. Indekeu, A.L. Stella, L. Zhang: to be published, preprint KUL-TF-83/20 (1983).
5. L. Onsager: J. Am. Chem. Soc. 58, 1486 (1936). See also R. Dekeyser, F. Halzen: Phys. Rev. 181, 949 (1969).
6. M. Droz, A. Maritan, A.L. Stella: Phys. Lett. 92A, 287 (1982).
7. K. De' Bell: J. Phys. A16, 1279 (1983).
8. K. De' Bell, T. Lookmann: Dalhousie University preprint (1983).
9. R.J. Glauber: J. Math. Phys. 4, 294 (1963).
10. G.F. Mazenko, O.T. Valls: ref. [1], pag. 87. This review and ref. [4] contain references on related dynamical real space renormalization work.
11. J.O. Indekeu, A.L. Stella: Phys. Lett. 78A, 160 (1980).
12. M. Suzuki, R. Kubo: J. Phys. Soc. Japan 24, 51 (1968).
13. N. Jan, D. Stauffer: Phys. Lett. 93A, 39 (1982).
14. Z. Racz, M.F. Collins: Phys. Rev. B13, 3074 (1976).

LINEAR-ANTILINEAR REPRESENTATIONS OF MAGNETIC LINE GROUPS

Milan Vujičić and Milan Damnjanović

Department of Physics, Faculty of Science, POB 550, 11001 Belgrade, Yugoslavia

Line groups describe[1] the symmetry of systems that are translationally periodical in one direction (e.g. quasi-one-dimensional solids and stereoregular polymers). They can be applied also in investigations of highly anisotropic three-dimensional crystals (e.g. ferromagnetics and ferroelectrics). When spin subsystems are considered one needs magnetic line groups[2]. Linear-antilinear matrix representations[3] of magnetic line groups are necessary in order to predict degeneracies of energy bands of magnetic systems, as well as in the symmetry approach to spin-wave dispersion relations.

Each magnetic line group $L(L')$ has a line group L' as an index-two subgroup, and therefore it can be written in a coset form

$$L(L') = L' + g\theta L' , \tag{1}$$

where $g\theta$ is a coset representative (θ is the time reversal, and g is an element of the Euclidean group such that $L' + gL'$ is a line group again). If g belongs to L', then $L(L') = L' \otimes \{E,\theta\}$ is a grey group, otherwise it is a black-and-white one.

In quantum mechanics elements of L' are linear operators, in contrast to the elements of the coset $g\theta L'$ which are antilinear ones. In the co-representation theory[4] all these operators are represented by matrices which are linear operators in the space C^n of number columns. Therefore a co-representation is not a homomorphic mapping. In order to achieve a homomorphism the elements of the coset should be represented by antilinear operators in C^n, i.e. by antimatrices[3].

To construct all irreducible linear-antilinear representations of $L(L')$ one uses the irreducible representations[5] of L'. For each irreducible representation of L' the complex-g-conjugate representation $\bar{d}^{*}(L') = \{ d^{*}(g^{-1}hg) \mid h \in L' \}$ is found. Then, from the character of $d(L')$ one evaluates $X \equiv (|L'|^{-1}) \sum_{h\in L'} \chi\left[(gh)^2\right]$ which must be $+1$, -1 or 0 ($\chi\left[(gh)^2\right]$ is the trace of $d\left[(gh)^2\right]$). If $X=+1$, then there exists unitary matrix Z such that $\bar{d}^{*}(L') = Z^{-1}d(L')Z$, and $ZZ^{*} = d(g^2)$. In this case the corresponding linear-antilinear irreducible representation of $L(L')$ is

$$d(h) = d(h) , \ d_a(g\theta h) = ZK_o d(h) , \ h \in L' , \tag{2}$$

where K_o is the complex conjugation in the space of number columns and index a denotes atimatrices. To obtain Z the specific structure of the irreducible representations of L' is made use of to simplify the general method[3]. In both other cases ($X = -1$ and $X = 0$) the $*$-induction method is applied, yielding the corresponding linear-antilinear irreducible representation of $L(L')$:

$$D(h) = \begin{bmatrix} d(h) & 0 \\ 0 & \bar{d}^*(h) \end{bmatrix}, \quad D_a(g\theta h) = \begin{bmatrix} 0 & d(g^2) \\ I & 0 \end{bmatrix} K_o D(h), \ \forall h \in L' . \tag{3}$$

All the linear-antilinear irreducible representations have been derived[6], whereas those of grey groups have been already published[7].

References

1. M. Vujičić, I.B.Božović and F.Herbut, J.Phys.A 10,1271 (1977)
2. M. Damnjanović and M.Vujičić, Phys.Rev. B 25,6987 (1982)
3. F. Herbut, M.Vujičić and Z.Papadopolos, J.Phys. A 13,2577 (1980)
4. C.J. Bradley and A.P.Cracknell, The Mathematical Theory of Symmetry in Solids, Clarendon, Oxford (1972)
5. I.B. Božović, M.Vujičić and F.Herbut, J.Phys. A 11,2133 (1978)
   I.B. Božović and M.Vujičić, J.Phys. A 14,777 (1981)
6. M. Damnjanović, Ph.D. Thesis, Belgrade (1981) (in Serbo-Croat)
7. I.B. Božović and N.Božović, J.Phys. A 14, 1825 (1981)

ANDERSON TRANSITION AND NONLINEAR σ-MODEL

Franz Wegner
Institut für Theoretische Physik, Ruprecht-Karls-Universität
D-6900 Heidelberg, Fed. Rep. of Germany

1. Anderson Transition

A particle (e.g. an electron) moving in a random one-particle potential may have localized and extended eigenstates depending on the energy of the particle. The energy E_c which separates the localized states from the extended states is called the mobility edge. Extended states can carry a direct current whereas localized states are bound to a certain region and can move only with the assistance of other mechanisms (e.g. phonon-assisted hopping). Thus the residual conductivity is expected to vanish for Fermi energies E in the region of localized states, and to be nonzero for E in the region of extended states. This transition from an insulating behaviour to a metallic one is called Anderson transition.

It will be shown that this problem can be mapped onto a field theory of interacting matrices. The critical behaviour near the mobility edge will be discussed. The theory has a G(m,m) symmetry which for finite frequency breaks to a G(m) x G(m) symmetry. Depending on the potential G stands for the unitary, orthogonal and symplectic group. Due to the replica trick m equals 0. The replica trick can be circumvented by using fields composed of commuting and anticommuting components. Then one deals with unitary graded and unitary orthosymplectic symmetries.

I refer to lectures given in Les Houches /1/ and in Sanda-Shi /2/, where, however, the graded groups have not yet been used. Most of the material presented here can be found in the original papers /3,4,5/.

2. Mapping on a Static Problem and Continuous Symmetry

Consider a one-particle tight-binding model

$$H = \sum_{rr'} f_{rr'} |r\rangle\langle r'| \tag{1}$$

where the ket $|r\rangle$ stands for an atomic orbital at site r of a d dimensional Bravais lattice. Then the Green's function

$$\mathcal{G}(r,r',z_p) = \langle r| \, (z_p - H)^{-1} \, |r'\rangle \tag{2}$$

can be expressed as expectation value over the field ϕ

$$\mathcal{G}(r,r',z_p) = s_p \langle \phi^*_{pa}(r') \, \phi_{pa}(r) \rangle \tag{3}$$

with respect to the "density"

$$\prod_p \{\det (z_p - f)\}^m \, e^{-\mathcal{H}} \tag{4}$$

where

$$\mathcal{H} = -c \, \mathrm{tr} \, (\phi \, s \, \phi^+ f) + c \, \mathrm{tr} \, (\phi \, s \, z \, \phi^+). \tag{5}$$

The field ϕ is written as a 2m x N matrix where the columns are labelled by the energy index $p = 1,2$ and the replica index $a = 1,2,...m$, the rows by the N lattice points r. $c = 1$. s and z are 2m x 2m diagonal matrices with diagonal elements $-i,+i,-i,+i,...$ for s and $z_1,z_2,z_1,z_2,...$ for z. We assume $z_p = E - i \, s_p \, \omega/2$, E real, $\mathrm{Im}\,\omega > 0$. The factors s guarantee the convergency of the integrals.

In order to get rid of the determinant in (4) we may formally choose $m = 0$. This is called the replica trick. Although this means literally that no degrees of freedom are left and (3) becomes meaningless, one can in practice do the calculation for general m and finally set $m = 0$. Diagrammatic expansions are well-defined for $m = 0$.

A mathematical and conceptual clean way is to add anticommuting components to ϕ /6,5/. Then the integral over the anticommuting components yields the determinant. Thus we may choose ϕ to be a 4 x N matrix with N rows

$$\phi(r) = (S_1(r), \, S_2(r), \, \zeta_1(r), \, \zeta_2(r)) \tag{6}$$

where S_p are complex, ζ_p anticommuting components. We denote the set of graded matrices

$$X = \begin{pmatrix} a & \zeta \\ \eta & b \end{pmatrix} \tag{7}$$

by $M(n_1,m_1,n_2,m_2)$ where the blocks a,ζ,η,b are $n_1 \times m_1$, $n_1 \times m_2$, $n_2 \times m_1$, $n_2 \times m_2$ matrices and a,b are even, ζ,η odd elements of the graded algebra. Thus $\phi \in M(N,0,2,2)$, $s,z \in M(2,2,2,2)$, $f \in M(N,0,N,0)$. For an elementary introduction to graded matrices and groups see /7/. Working with these matrices all traces have to be read as graded traces.

$\mathcal{H}$ is invariant under linear transformations $\phi \to \phi U$ with $U \in U(m) \times U(m)$ and $U \in UPL$

(1,1) x UPL(1,1), resp. In the limit $\omega \to 0$ the symmetry group is U(m,m) or the pseudo-unitary graded subgroup of UPL(2,2) obeying $UsU^{+} = s$. Thus the contribution proportional to ω, $\mathrm{tr}\,(\phi\phi^{+})$ breaks this larger symmetry. The expectation value of the symmetry breaking term

$$\sum_{p} < \phi^{*}_{pa}(r)\,\phi_{pa}(r) > = i(\mathcal{G}(r,r,E + \omega/2) - \mathcal{G}(r,r,E - \omega/2)) \tag{8}$$

is proportional to the density of states ρ in the limit $\omega \to 0$, thus playing the role of the order parameter.

3. Composite Variables and Nonlinear σ-Model

Now let us consider the ensemble average over the random potentials H. Suppose the matrix elements f are Gaussian distributed with

$$\overline{f_{rr'}} = 0,\ \overline{f_{rr'}f_{r''r'''}} = \delta_{rr'''}\delta_{r'r''}\,M_{r-r'}. \tag{9}$$

This model is called local-gauge invariant since the distribution of the Hamiltonians H is invariant under gauge transformations

$$|r> \to \exp(i\Phi_{r})|r>. \tag{10}$$

Since these transformations are unitary we call it a unitary ensemble. Accordingly the only nonvanishing one-particle Green's function is

$$\overline{\mathcal{G}(r,r',z_{p})} = \delta_{rr'}G(z_{p}) = s_{p}\ <\phi^{*}_{pa}(r')\phi_{pa}(r)> \tag{11}$$

and the only nonvanishing two-particle Green's functions $K(z_1,z_2)$ are

$$K(r,r',z_1,z_2) = \overline{<r|(z_1 - H)^{-1}|r'>\ <r'|(z_2 - H)^{-1}|r>} = <\phi^{*}_{1a}(r')\phi_{1a}(r)\phi^{*}_{2b}(r)\phi_{2b}(r')> \tag{12}$$

$$K''(r,r',z_1,z_2) = \overline{<r|(z_1 - H)^{-1}|r>\ <r'|(z_2 - H)^{-1}|r'>} = <\phi^{*}_{1a}(r)\phi_{1a}(r)\phi^{*}_{2b}(r')\phi_{2b}(r')> \tag{13}$$

where the average is taken with respect to

$$\overline{\exp(-\mathcal{H})} = \exp\{-\ c\ \mathrm{tr}(\phi\ s\ z\ \phi^{+}) + c/2 \sum_{rr'} \mathrm{tr}(\phi(r)\ s\ \phi^{+}(r')\phi(r')\ s\ \phi^{+}(r))\} \tag{14}$$

By means of a Gaussian transformation we can transform to composite variables Q (effectively $Q(r) \sim \sqrt{s}\,\phi^{+}(r)\phi(r)\,\sqrt{s}$) where Q(r) is a 2m x 2m matrix and $Q(r) \in M(2,2,2,2)$, resp., so that

$$G(z_p) \sim \langle Q^{pp}_{aa}(r) \rangle \tag{15}$$

$$K(r,r',z_1,z_2) \sim \langle Q^{12}_{ab}(r)\, Q^{21}_{ba}(r') \rangle \tag{16}$$

$$K''(r,r',z_1,z_2) \sim \langle Q^{11}_{aa}(r)\, Q^{22}_{bb}(r') \rangle \tag{17}$$

which is averaged with respect to exp(-L) with

$$L = 1/2\; c\; \Sigma\, w_{rr'}\, \mathrm{tr}(Q(r)\, Q(r')) + c \sum_r \mathrm{tr}\, \ln(z - Q(r)), \qquad \sum_{r'} w_{r-r'} M_{r'-r''} = \delta_{rr''}. \tag{18}$$

Even if local-gauge invariance is absent, then a similar transformation to matrices Q can be performed /12/.

The saddlepoints of L are matrices Q with eigenvalues

$$\lambda_p = 2E_o^{-2}\{z_p - s_p(E_o^{2} - z_p^{2})^{1/2}\}, \qquad E_o^{2} = 4 \sum_r M_r \tag{19}$$

If instead of the Hamiltonian (1) we choose a model Hamiltonian /8/

$$H = n^{-1/2} \sum_{r\alpha, r'\beta} f_{r\alpha, r'\beta}\; |r\alpha\rangle \langle r'\beta| \tag{20}$$

with n orbitals per lattice site $\alpha = 1,2,\ldots n$, and

$$\overline{f_{r\alpha, r'\beta}} = 0 \qquad \overline{f_{r\alpha r'\beta} f_{r''\gamma r'''\delta}} = \delta_{rr'''}\delta_{r'r''}\delta_{\alpha\delta}\delta_{\beta\gamma}\, M_{r-r'} \tag{21}$$

then we obtain again the Lagrangian (18) but with c = n. In the limit $n \to \infty$ the saddlepoints become very sharp and

$$G(r\alpha, r'\beta, z_p) = \delta_{rr'}\delta_{\alpha\beta}\lambda_p. \tag{22}$$

This is the d dimensional generalization of a model by Wigner /9/. The eigenstates are in the interval $(-E_o, +E_o)$ obeying a semicircle law.

(If Q contains only commuting components, then the saddlepoint argument is correct. If Q contains also anticommuting components, then the UPL-symmetry is used additionally). Starting from this limit one can perform a systematic expansion in powers of n^{-1} which was useful to obtain information on the mobility edge behaviour /10/ and which can also be used to investigate interacting systems /11/.

The contour of the integrals over Q can be made to run through the noncompact saddle-point manifold obtained from the diagonal matrix with eigenvalues λ_p obtained by pseudo-unitary transformations. If we assume that the fluctuations of the eigenvalues of Q are irrelevant similar to the fluctuations of the length of the vector in the n-vector model, then Q obeys

$$(Q - \lambda_1)(Q - \lambda_2) = 0. \tag{23}$$

From $Q^{pp} \approx \lambda_p$ for small Q^{12}, Q^{21} one obtains from (23)

$$Q^{11} = 1/2(\lambda_1 + \lambda_2) + \{(1/2(\lambda_1 - \lambda_2))^2 - Q^{12}Q^{21}\}^{1/2}, \tag{24}$$

and similarly for Q^{22}. Thus the "longitudinal" components Q^{11}_{ab}, Q^{22}_{ab} can be expressed in terms of the "transversal" components Q^{12}_{ab}, Q^{21}_{ab}. The matrix elements Q^{12}_{ab} are independent. The Taylor expansion of (24)

$$Q^{11}_{ab} = \lambda_1\delta_{ab} - \frac{1}{\lambda_1-\lambda_2} \sum_c Q^{12}_{ac}\, Q^{21}_{cb} + \dots \tag{25}$$

yields

$$< Q^{11}_{ab} > = \lambda_1\delta_{ab} \tag{26}$$

since the summation runs over m = 0 equal contributions in the case of the replica trick, and $< Q^{11}_{a2}\, Q^{11}_{2a} > = - < Q^{11}_{a1}\, Q^{11}_{1a} >$ due to the UPL(1,1) symmetry. This tells us that

$$G(r,r,z_p) = \lambda_p \tag{27}$$

holds. The transverse fluctuations do not affect the averaged one-particle Green's function which thus shows no critical behaviour at the mobility edge. This is a particular feature of the m = 0 component problem.

We may construct an effective interaction L_{eff} for the matrices Q. Any local potential of Q obeying the fall pseudo U(m,m) symmetry and UPL(2,2) symmetry is a function of λ_1 and λ_2 only and thus a constant. Any interaction containing one gradient can be expressed as a surface integral. The most simple interaction containing two gradients is

$$L^o{}_{eff} = 1/4\, \tilde{K} \int d^dr\, \mathrm{tr}(\nabla Q(r) \nabla Q(r)). \tag{28}$$

It is the only interaction containing two derivatives ∇ obeying rotational and inflection symmetry. If inflection symmetry is violated, then a term proportional

$$\int d^d r \; tr(Q(r) \, [\partial_x Q(r), \partial_y Q(r)] \,) \tag{29}$$

may appear. Levine, Libby and Pruisken /13/ have recently emphasized that this term should be important to explain the Hall current for the quantized Hall effect. To (28) the symmetry breaking term has to be added

$$L_{eff} = L^{o}_{eff} + \frac{i\omega}{4v} \quad d^d r \; tr(s \; Q(r)) \tag{30}$$

where v is the volume per lattice site.Harmonic approximation which is exact in the limit $n \to \infty$ yields for (30) the two-particle Green's function

$$K(q.z_1,z_2): = \sum_{r,\beta} e^{iqr} \; K(0\alpha, r\beta, z_1, z_2) \sim \frac{\rho(E)}{-i\omega + Dq^2} \tag{31}$$

in the hydrodynamic limit, and thus a diffusion type behaviour. Therefore the Goldstone mode due to the continuous symmetry of U(m,m)/U(m) x U(m) corresponds to the diffusion mode in the random potential.

4. Mobility Edge Behaviour

In order to obtain the power laws near the mobility edge we now use the well-known results from conventional critical phenomena. The analogies with an isotropic ferromagnet at temperature T, magnetic field h, correlation length ξ, and with magnetization m, and transverse and longitudinal susceptibilities $\chi_{\perp}$ and $\chi_{\parallel}$, are listed in the first two rows below

$E - E_c$	$\hat{=}$	$T - T_c$	1
$-i\omega$	$\hat{=}$	h	$\beta+\gamma = d\nu$
q	$\hat{=}$	q	ν
ξ	$\hat{=}$	ξ	$-\nu$
ρ	$\hat{=}$	m	$\beta = 0$
K	$\hat{=}$	$\chi_{\perp}$	$-\gamma = -d\nu$
K''	$\hat{=}$	$\chi_{\parallel}$	$-\gamma = -d\nu$

The localization length ξ indicates the range over which the wavefunction decays. The third column shows how these quantities scale, that is, if $E - E_c$, $-i\omega$, and q are multiplied by factors b^1, $b^{\beta+\gamma}$, b^{ν}, then ξ, ρ, K, K'' are rescaled by factors $b^{-\nu}$, $b^{-\gamma}$, resp. Since $\beta = 0$, and the scaling law $2\beta + \gamma = d\nu$ holds, all exponents can be expressed by ν. Thus, e.g.,

$$K(b^{\nu}q,\ b^{d\nu}\omega,\ b(E - E_c)) = b^{-d\nu}\ K(q,\omega,E - E_c). \tag{32}$$

The conductivity σ can be expressed by the diffusion constant D and by K as

$$\sigma \sim \rho D \sim \rho^2 \partial K^{-1}/\partial q^2 \tag{33}$$

thus it scales like

$$\sigma \sim b^s \sim (E - E_c)^s \tag{34}$$

with

$$s = 2\beta + \gamma - d\nu = (d - 2)\nu. \tag{35}$$

All the scaling laws agree with the prediction from real-space renormalization for the homogeneous fixed point ensemble /14/.

The model (28),(30) is very similar to the nonlinear σ-model on matrices with unitary symmetry $U(2m)/U(m) \times U(m)$ for $m = 0$. The "pseudo"-symmetries yield the same diagrams; only some overall signs are flipped. The W-function and thus the critical exponent s has been calculated in $d = 2 + \varepsilon$ dimensions /15-17/ yielding

$$s = 1/2 + O(\varepsilon).$$

This holds for systems in which time-reversal invariance is broken but isotropy and inversion symmetry is maintained in the average.

For time-reversal invariant systems two universality classes known may be represented by the model (21) with /8/

$$\overline{f_{r\alpha r'\beta}\ f_{r''\gamma r'''\delta}} = (\delta_{rr'''}\delta_{r'r''}\delta_{\alpha\delta}\delta_{\beta\gamma} + \delta_{rr''}\delta_{r'r'''}\delta_{\alpha\gamma}\delta_{\beta\delta})M_{r-r'} \tag{36}$$

and by a model with a spindependent potential /15/

$$H = n^{-1/2}\ \Sigma f_{r\alpha\sigma r'\beta\sigma'}\ |r\alpha\sigma\rangle\langle r'\beta\sigma'| \tag{37}$$

with $\alpha = 1,2,...n/2$ and

$$\begin{aligned}\overline{f_{r\alpha\sigma r'\beta\sigma'}\ f_{r''\gamma\sigma''r'''\delta\sigma'''}} &= (\delta_{rr'''}\delta_{r'r''}\delta_{\alpha\delta}\delta_{\beta\gamma}\delta_{\sigma\sigma'''}\delta_{\sigma'\sigma''} \\ &+ \delta_{rr''}\delta_{r'r'''}\delta_{\alpha\gamma}\delta_{\beta\delta}\sigma\sigma'\delta_{\sigma,-\sigma''}\delta_{\sigma',-\sigma'''})M_{r-r'}.\end{aligned} \tag{38}$$

Using the replica trick the models are governed by orthogonal $O(m,m)/O(m) \times O(m)$ and symplectic $Sp(m,m)/Sp(m) \times Sp(m)$ symmetries, resp., with $m = 0$. The orthogonal case yields /10,17,16,18/

$$s = 1 + O(\varepsilon^4). \tag{39}$$

In the symplectic case the W-function does not show a zero up to four-loop order in the physical region.

For two-dimensional systems the d.c. conductivity vanishes at all energies in the orthogonal and unitary case /15,10,19,20/. In the symplectic case there may be a region where the conductivity behaves better than ohmic /21/. In all cases one finds for σ of a square of length L in the quasi-metallic region

$$\sigma = \sigma_0 - \frac{e^2\alpha}{\pi^2 h} \ln(\frac{L}{a}) + \ldots \tag{40}$$

with $\alpha = 1,0,-1/2$ in the orthogonal, unitary and symplectic case, respectively.

The time-reversal invariant systems can also be treated by introducing fields ϕ and Q composed of commuting and anticommuting components. Then the matrices are of the form (7) where the block a contains real elements, b 2 x 2 submatrices of quaternion form, ζ and η pairs of adjoint Grassmann variables. Then the underlying symmetry is the unitary orthosymplectic group. The system is still described by $\mathcal{H}$, eq.(5) and L, eq.(18) with $c = 1/2$, $n/2$ for model (21), (36) and $c = -1/2$, $-n/2$ for model (37), (38). (For details see /5/). Although the paths of integration over the Q-matrices are different the saddlepoints are the same and the 1/n expansion of one system can be obtained from the other by changing the sign of n. This symmetry relation has been first obtained by Oppermann and Jüngling /15/ on a diagrammatic basis, and is related to the fact that manifolds $O(2m_1 + 2m_2)/O(2m_1) \times O(2m_2)$ and $Sp(-m_1-m_2)/Sp(-m_1) \times Sp(-m_2)$ yield the same low temperature expansion /22/.

Moreover, there is a second way to handle the unitary ensemble by introducing $\phi \in M(0,N,2,2)$. It also yields (5) and (18), but with $c = -1$, $-n$. Thus the saddlepoint expansion is invariant under the change of 1/n into -1/n. This has also been observed in /15/ and is related to the formal equivalence of $U(m_1 + m_2)/U(m_1) \times U(m_2)$ and $U(-m_1-m_2)/U(-m_1) \times U(-m_2)$ /22/.

References

1 D.J. Thouless, p.5, E. Abrahams, p.9, F. Wegner, p.15, Phys. Reports 67 (1980)

2 D.J. Thouless, p.2, F.J. Wegner, p.8, S. Hikami, p.15, P. Wölfle and D. Vollhardt, p.26, in Y. Nagaoka, H. Fukuyama (eds.) "Anderson Localization", Springer Series in Solid-State Sciences 39 (1982)

3 F. Wegner, Z. Phys. B 35 (1979) 207

4 L. Schäfer, F. Wegner, Z. Phys. B 38 (1980) 113

5 F. Wegner, Z. Phys. B 49 (1983) 297

6 K.B. Efetov, Zh. Eksp. Teor. Fiz. 82 (1982) 872, JETP 55 (1982) 514

7 V. Rittenberg, M. Scheunert, J. Math. Phys. 19 (1978) 709

8 F.J. Wegner, Phys. Rev. B 19 (1979) 783

9 E. Wigner, Ann. Math. 62 (1955) 548; 67 (1958) 325

10 R. Oppermann, F. Wegner, Z. Phys. B 34 (1979) 327

11 R. Oppermann, Z. Phys. B 49 (1983) 273
R. Oppermann, J. Phys. Soc. Jap. 52 (1983) no. 10
M. Ma, E. Fradkin, preprint

12 A.M. Pruisken, L. Schäfer, Phys. Rev. Lett. 46 (1981) 490; Nucl. Phys. B 200[FS4] (1982) 20

13 H. Levine, S.B. Libby, A.M.M. Pruisken, preprint

14 F.J. Wegner, Z. Phys. B 25 (1976) 327

15 R. Oppermann, K. Jüngling, Phys. Lett. 76 (1980) 449; Z. Phys. B 38 (1980) 93

16 S. Hikami, Prog. Theor. Phys. 64 (1980) 1466; Phys. Rev. B 24 (1981) 2671

17 E. Brézin, S. Hikami, J. Zinn-Justin, Nucl. Phys. B 165 (1980) 528

18 S. Hikami, Nucl. Phys. B 215 [FS7] (1983) 555

19 E. Abrahams, P.W. Anderson, D.C. Licciardello, T.V. Ramakrishnan, Phys. Rev. Lett. 42 (1979) 673

20 L.P. Gorkov, A.I. Larkin, D.E. Khmelnitzkii, Pis. Zh. Eksp. Teor. Fiz. 30 (1979) 248; JETP Lett. 30 (1979) 228

21 S. Hikami, A.I. Larkin, Y. Nagaoka, Prog. Theor. Phys. 63 (1980) 707

22 F.J. Wegner, Nucl. Phys. B 180 [FS2] (1981) 77

DO ENERGY BANDS IN SOLIDS HAVE AN IDENTITY?

J. Zak
Department of Physics
Technion - Israel Institute of Technology
Haifa, Israel

In solids energy levels are grouped into bands. The number of levels in a band equals the number of atoms in the solid and for an infinite solid each band contains an infinite number of energy levels. The question we are asking in the title is whether all the levels belonging to a given energy band in a solid have a common symmetry label, or an identity that follows from the symmetry of the solid. More precisely, we are asking whether an energy band as a whole entity can be specified by the space group of the solid.

It is customary to label energy bands in solids by means of atomic angular momentum quantum numbers s, p, d and so on. Such a specification of energy bands was first used by Bloch[1] and it originates from the tight binding expression of a Bloch function $\psi_{\ell k}(\vec{r})$ by means of atomic orbitals $a_\ell(\vec{r})$

$$\psi_{\ell k}(\vec{r}) = \Omega^{-1/2} \sum_{\vec{R}_m} e^{i\vec{k}\cdot\vec{R}_m} a_\ell(\vec{r}-\vec{R}_m) \tag{1}$$

where $\vec{k}$ is the Bloch momentum and Ω is the volume of the reciprocal lattice unit cell. When $a_s(\vec{r})$ is an s-orbital $(\ell = s)$, $\psi_{sk}(\vec{r})$ with $\vec{k}$ varying in the whole Brillouin Zone are the Bloch functions of what is called an s-band. Correspondingly, when $\ell = p$ there are three atomic orbitals $a_{p_x}(\vec{r})$, $a_{p_y}(\vec{r})$ and $a_{p_z}(\vec{r})$, and Rel. (1) defines three Bloch functions $\psi_{p_xk}(\vec{r})$, $\psi_{p_yk}(\vec{r})$ and $\psi_{p_zk}(\vec{r})$. These Bloch functions with $\vec{k}$ varying over the Brillouin Zone define a p-band. The s-band is simple with one Bloch function for each $\vec{k}$-vector in the Brillouin Zone, while the p-band is composite and for each $\vec{k}$ we have three Bloch functions. In general, when the Bloch functions are composed of different orbitals, say s and p, then the energy band is a composite sp-band. This atomic labelling of energy bands in solids is physically very appealing because atoms are the building stones of the solid. However, from the point of view of symmetry the atomic labels for energy bands in solids have a number of shortcomings. One of them is trivial and is connected with the fact that the atomic angular momentum ℓ is not a conserved quantity in solids. The conservation of ℓ is a consequence of the continuous rotational symmetry of the atom around its nucleus. A solid has only discrete rotational symmetry and the angular momentum ℓ is no longer a conserved quantity. Another shortcoming of atomic labels for energy bands is less trivial and is connected with the concept of symmetry centers in solids. For an atom when one talks about the symmetry center it is unmistakably connected with its nucleus. In solids the situation is different and there is usually more than one symmetry center of a given kind.[2] In Fig. 1 the difference between a symmetry center in an atom and a solid is explained. Fig. 1a shows an atom with its symmetry center at the nucleus. Fig. 1b shows a diatomic molecule with the symmetry center halfway between the atoms. For a triatomic linear molecule (Fig. 1c) the symmetry center is on the central atom. Continuing this way, we find that for 4 atoms the center is between the atoms, for 5 on an atom and so on. This means that for an odd number of atoms

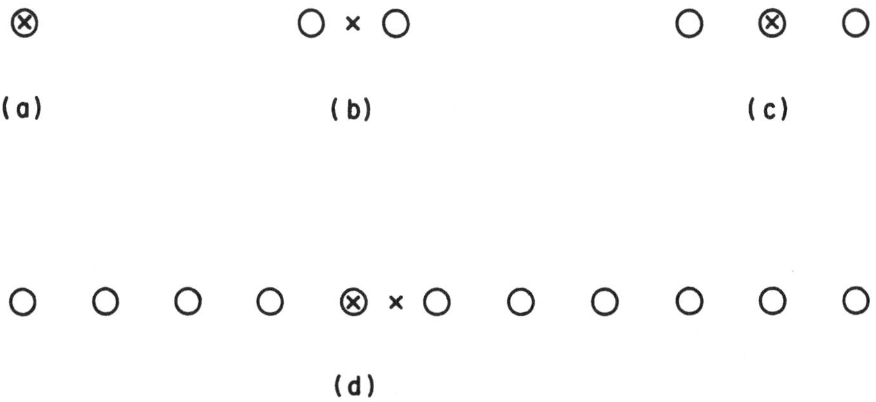

Fig. 1. a) an atom with the inversion symmetry center on the nucleus, b) a diatomic molecule with the inversion center half way between the identical atoms, c) a triatomic molecule with the inversion center or the central atom, d) an infinite chain of atoms representing a one-dimensional solid; it has two inequivalent inversion symmetry centers.

in the molecule the center is on an atom while for an even number of atoms it is in between the atoms. A one-dimensional solid can be represented by an infinite chain of atoms. Since infinity is both even and odd, one should expect that an infinite chain will have symmetry centers on the atoms and in between them. Fig. 1d shows such a chain: there are two inequivalent inversion symmetry centers, one on an atom and one halfway between the atoms. The conclusion is that unlike a finite chain of atoms which has one symmetry center (Figs. 1a - 1c), a solid which is represented by an infinite chain of atoms has two inequivalent symmetry centers. The situation is similar in solids of higher dimensions. Thus, in Fig. 2 we show the inequivalent inversion centers for a two dimensional crystal (there are 4 such inequivalent centers). We see therefore that solids because of their translational symmetry acquire a number of point symmetry centers. For solids with different space group symmetries, those centers are listed in the International X-Ray Crystallography Tables.[3]

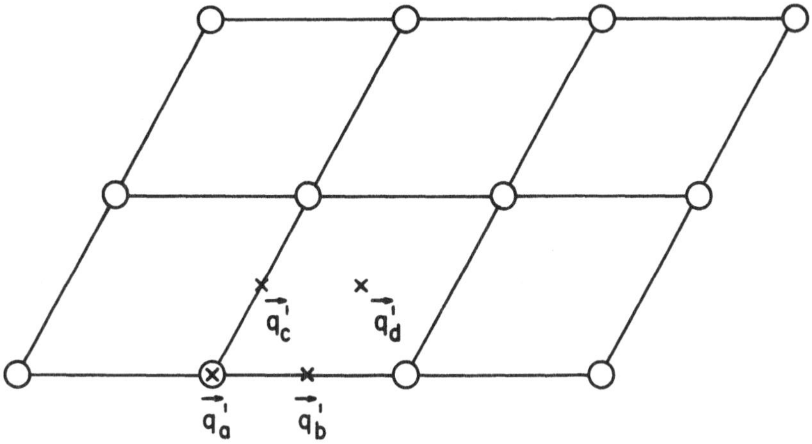

Fig. 2. Inversion centers for a general two dimensional crystal: $\vec{q}_a' = (000)$, $\vec{q}_b' = (\frac{a_1}{2}\ 0)$, $\vec{q}_c' = (0\ \frac{a_2}{2})$, $\vec{q}_d' = (\frac{a_1}{2}\ \frac{a_2}{2})$.

Let us now return to Eq. (1). The idea contained in this equation is that the symmetry of the extended Bloch functions $\psi_{\ell k}(\vec{r})$ (for all k-vectors in the Brillouin Zone!) is defined by the symmetry of the single localized orbital $a_\ell(\vec{r})$. This idea of Eq. (1) can be used in a symmetry specification of energy bands consistent with the space group symmetry of the solid. For this purpose the symmetry of the localized orbitals has to be determined by means of the point group of the solid with respect to a well specified symmetry center.[2] When the localized orbital $a_\ell(\vec{r})$ specified by the symmetry center $\vec{q}'$ and the representation indices ℓ and n (ℓ the representation and n the number of the function) then Eq. (1) will specify the symmetry of the extended functions $\psi_k(\vec{r})$ at all the $\vec{k}$-vectors in the Brillouin Zone. With the full indexation of the functions, Eq. (1) will become

$$\psi_{nk}^{(\vec{q}',\ell)}(\vec{r}) = \Omega^{-\frac{1}{2}} \sum_{R_n} \exp(i\vec{k}\cdot\vec{R}_m)\ a_n^{(\vec{q}',\ell)}(\vec{r} - \vec{R}_m) \qquad (2)$$

One can show that the knowledge of the symmetry of the localized orbital $a_n^{(\vec{q}',\ell)}(\vec{r})$ in Eq. (2) fully defines the symmetry of the Bloch functions $\psi_{nk}^{(\vec{q}',\ell)}(\vec{r})$ at all the symmetry points $\vec{k}$ in the Brillouin Zone. This can best be seen by writing the localized orbital $a(\vec{r})$ in the kq-representation.[4] By denoting the kq-wave function by $C(\vec{k},\vec{q})$ we have

$$C_n^{(\vec{q}',\ell)}(\vec{k},\vec{q}) = \Omega^{-1/2}\sum_{R_m} e^{i\vec{k}\cdot\vec{R}_m} a_n^{(\vec{q}',\ell)}(\vec{q}-\vec{R}_m) \tag{3}$$

From Rels. (2) and (3) it follows that the Bloch function coincides with the kq-transform of the localized orbital

$$\psi_{nk}^{(\vec{q}',\ell)}(\vec{r}) = C_n^{(\vec{q}',\ell)}(\vec{k},\vec{r}) \tag{4}$$

It is obvious that if we know the behavior of the localized orbitals on the right hand side of Eq. (4) under the operation of a space group element $(\alpha|\vec{t})$ (α is a point group element and $\vec{t}$ is a translation) this fully defines the transformation of the extended functions on the left hand side of Eq. (4). Thus, if the localized orbitals transform according to an irreducible representation $D^{(\ell)}$ then also the extended function transform in the same way

$$\alpha\, \psi_{nk}^{(\vec{q}',\ell)}(\vec{r}) = \sum_{n'} D_{n'n}^{(\ell)}(\alpha)\, \psi_{n'k}^{(\vec{q}',\ell)}(\vec{r}) \tag{5}$$

It should be pointed out that in Eq. (5) both $\vec{k}$ and $\vec{r}$ are variables on which the space group element $(\alpha|\vec{t})$ operates. However, when $(\beta|\vec{b})$ is a symmetry element of the groups of $\vec{k}$, G_k[5], then

$$(\beta|\vec{b})\vec{k} = \vec{k} + \vec{K} \tag{6}$$

where $\vec{K}$ is a vector of the reciprocal lattice. Since by definition[6] for any element $(\alpha|\vec{t})$

$$(\alpha|\vec{t})C(\vec{k},\vec{q}) = C(\alpha^{-1}\vec{k}\,,\ (\alpha|\vec{t})^{-1}\vec{q}) \tag{7}$$

and since $C(\vec{k},\vec{q})$ satisfies the following periodicity conditions

$$C(\vec{k},\vec{q}) = C(\vec{k}+\vec{K},\ \vec{q}) = e^{-i\vec{k}\cdot\vec{R}_m} C(\vec{k},\vec{q}+\vec{R}_m) \tag{8}$$

it follows from (4), (5) and (6) that for any symmetry point $\vec{k}$ in the Brillouin Zone

$$\beta\, \psi_{nk}^{(\vec{q}',\ell)}(\vec{r}) =$$

$$= \psi_{nk}^{(\vec{q}',\ell)}(\beta^{-1}\vec{r}) = \sum_{n'} D_{n'n}^{(\ell)}(\beta)\, \psi_{n'n}^{(\vec{q}',\ell)}(\vec{r}) \tag{9}$$

In Eq. (9) the space group elements β of G_k operate only on the spatial coordinate $\vec{r}$. What this means is that the knowledge of the symmetry behavior of the localized orbitals (the knowledge of D) fully defines the transformation properties of the extended Bloch functions $\psi_{nk}(\vec{r})$ at all the symmetry points in the Brillouin Zone.We shall say that we know the symmetry of an energy band (or its symmetry identity) if we know the symmetry of all the Bloch functions for this band at all the symmetry points in the Brillouin Zone. From Eqs. (4), (5) and (9) it follows that the symmetry label of the localized orbital $C_n^{(\vec{q}',\ell)}$ defines the symmetry label of the corresponding energy bands as a whole entity.

The localized orbitals $C_n^{(\ell,q')}$ form a basis for a band representation of the space group.[6] By definition, the f square integrable functions $C_n^{(\vec{q}',\ell)}(\vec{k},\vec{q})$, n = 1,2,...,f belong to a band representation if they transform as follows

$$(\alpha|\vec{t})\; C_n^{(\vec{q}',\ell)}(\vec{k},\vec{q}) = \sum_{n'=1}^{f} D_{n'n}^{(q',\ell)}[(\alpha|\vec{t}),\vec{k}]\; C_{n'}^{(\vec{q}',\ell)}(\vec{k},\vec{q}) \qquad (10)$$

The matrix D in (10) is k-dependent and non-singular at each $\vec{k}$ in the Brillouin Zone. In the kq-representation the band representation in Rel. (10) appears with an f-dimensional matrix D(k). However, since k is a variable the band representation is actually infinite-dimensional. This can be seen by writing Rel. (10) in the r-representation. By using Eq. (3) we have

$$(\alpha|t)\psi_i(r) = \Omega^{-\frac{1}{2}} \sum_{i',m} D_{i'i}[(\alpha|\vec{t}),\vec{R}_m]\psi_{i'}(\vec{r}-\vec{R}_m) \qquad (11)$$

where the matrix $D[(\alpha|\vec{t}),\vec{R}_m]$ gives the Fourier coefficients of the matrix $D[(\alpha|\vec{t}),\vec{k}]$ in (10). Rel. (11) shows that a band representation in an infinite crystal is infinite-dimensional. The apparent finite dimensionality that appears in the kq-representation expresses the fact that the band is built out of a finite number of localized functions. The definition of a band representation can also be written as a correspondence between the elements of the group $(\alpha|t)$ and the k-dependent matrices

$$(\alpha|\vec{t}) : \quad D[(\alpha|\vec{t}),\vec{k}] \qquad (12)$$

By applying to Rel. (10) another element $(\beta|\vec{\mu})$ we find the correspondence

$$(\beta|\vec{\mu})(\alpha|\vec{t}) : \; D[(\beta|\vec{\mu}),\vec{k}]\; D[(\alpha|\vec{t}),\beta^{-1}\vec{k}] \qquad (13)$$

where in the second matrix $\vec{k}$ is replaced by $\beta^{-1}\vec{k}$. The latter result follows from the transformation of a kq-function, $C(\vec{k},\vec{q})$, under a space group element, Rel. (7).

The multiplication rule (13) influences the whole algebra of band representations. In particular, it influences the definition of equivalent band representations. Thus, if the functions $C'(\vec{k},\vec{q})$ form a new basis connected by the matrix $T(\vec{k})$ to the old basis then the equivalent band representation D' will be given by the matrices

$$D'[(\alpha|\vec{t}),\vec{k}] = T^{-1}(\vec{k})\; D[(\alpha|\vec{t}),\vec{k}]\; T(\alpha^{-1}\vec{k}) \qquad (14)$$

In the matrix on the right the vector $\vec{k}$ is replaced by $\alpha^{-1}\vec{k}$. With the aid of (14) one can define the concepts of reducible and irreducible band representations. The band representation $D[(\alpha|\vec{t}),\vec{k}]$ is reducible if a matrix $T(\vec{k})$ exists for which all the matrices in (14) assume a quasidiagonal form. This is equivalent to saying that the basis of the band representation can be split into invariant sub-bases. If such a matrix T does not exist then $D[(\alpha|\vec{t}),\vec{k}]$ is called an irreducible band representation. Like in usual group representations, the irreducible band representations play a central role in specifying bands on the basis of space group symmetries. For the construction of band representations the reader is referred to Ref. (6).

As an example of labelling energy bands by the symmetry of the solid, let us consider the two-dimensional space group p2 containing only one point group element, e.g. the inversion I or the rotation by π (Ref. 3, page 58). Let $\vec{a}_1$ and $\vec{a}_2$ be the unit cell vectors of the Bravais lattice and we shall apply the Born von Karman boundary conditions

$$(E|2N\vec{a}_1) = (E|2N\vec{a}_2) = 1 \tag{15}$$

where N is an integer. The group p2 becomes finite with $8N^2$ elements $(E|m_1\vec{a}_1 + m_2\vec{a}_2)$, $(I|m_1\vec{a}_1 + m_2\vec{a}_2)$, where $m_1, m_2 = 0, \pm 1, \pm 2, \ldots, \pm (N-1), N$. We shall denote these elements by (m_1m_2) and $(m_1m_2)'$ correspondingly. There are $2N^2 + 6$ classes: one class is formed by each of the elements (00), (N0), (0N), (NN); one class by each pair of elements with $m_1, m_2 \neq 0$ or N (m_10), $(-m_10)$; $(0m_2)$, $(0-m_2)$; (m_1N), $(-m_1N)$; (Nm_2), $(N,-m_2)$; (m_1m_2), $(-m_1-m_2)$; they form $2N^2 - 2$ classes; one class by each set of N^2 elements $(2m_12m_2)'$ $(2m_1{+}1\ 2m_2{+}1)'$, $(2m_1\ 2m_2{+}1)'$, $(2m_1{+}1\ 2m_2)'$. All irreducible representations of the space group p2 are given in Table 1.

The symmetry centers for the group p2 are shown in Fig. 2. To each of these centers we can assign an orbital $C^{(q',+)}$ or $C^{(q',-)}$ where + or - means even or odd with respect to inversion about the corresponding center $\vec{q}'$. Since there are 4 symmetry centers one can construct 8 different orbitals $C^{(q',\ell)}$ (ℓ denotes the symmetry of inversion) which form bases for 8 band representations of p2 . The characters for these band representations are listed in Table 2.

It is clear that as representations of p2 the band representations are reducible and their reduction is given in Table 3. Each column of this table shows the symmetries of the Bloch functions at the different symmetry points in the Brillouin Zone corresponding to a fixed symmetry of the localized orbital.

The set of the Bloch symmetries Γ_i , X_j , Y_k , R_ℓ in Table 3 is called a continuity chord.[7] The latter defines the symmetry of the energy band. As is seen from Table 3 there is a one-to-one correspondence between a band representation $(\vec{q}',\ell)$ and a continuity chord Γ_i , X_j , Y_k , R_ℓ : given $(\vec{q}',\ell)$ we know uniquely the continuity chord and vice versa, if Γ_i , X_j , Y_k , R_ℓ is given we know uniquely the band representation label. This means that the symmetry of an energy band or its identity can be defined either by the band representation label $(\vec{q}',\ell)$ or by the continuity chord Γ_i , X_j , Y_k , R_ℓ .

The label $(\vec{q}',\ell)$ of the band representation is a symmetry label of the corresponding localized orbital $C^{(q'\ell)}(\vec{k},\vec{q})$: this orbital transforms according to the representation $D^{(\ell)}$ for the point group elements α operating with

$\vec{k}$ \ α	$(m_1 m_2)$	$(2m_1 2m_2)'$	$(2m_1+1\ 2m_2+1)'$	$(2m_1\ 2m_2+1)'$	$(2m_1+1, 2m_2)'$
Γ_1 (00)	1	1	1	1	1
Γ_2 (00)	1	-1	-1	-1	-1
X_1 $(\frac{\pi}{a_1}\ 0)$	$(-1)^{m_1}$	1	-1	1	-1
X_2 $(\frac{\pi}{a_2}\ 0)$	$(-1)^{m_1}$	-1	1	-1	1
Y_1 $(0\ \frac{\pi}{a_2})$	$(-1)^{m_2}$	1	-1	-1	1
Y_2 $(0\ \frac{\pi}{a_2})$	$(-1)^{m_2}$	-1	1	1	-1
R_1 $(\frac{\pi}{a_1}\ \frac{\pi}{a_2})$	$(-1)^{m_1+m_2}$	1	1	-1	-1
R_2 $(\frac{\pi}{a_1}\ \frac{\pi}{a_2})$	$(-1)^{m_1+m_2}$	-1	-1	1	1
G $(k_1\ k_2)$	$2\cos(k_1 m_1 a_1 + k_2 m_2 a_2)$	0	0	0	0

Table 1. Irreducible Representations of the Space Group p2 . The number of general points G in the Brillouin Zone is $4N^2-4$.

respect to the symmetry center $\vec{q}'$ (Eq. (5)). The continuity chord Γ_i, X_j, Y_k, R_ℓ is obtained by reducing the band representation $D^{(q',\ell)}$ (Eq. (10)). The question can be asked whether all possible combinations of Bloch symmetries at different points in the Brillouin Zone can occur as continuity chords of an energy band? The answer to this question is negative. This is well demonstrated on the example of the group p2 , As is seen in Table 3 not all combinations of Bloch symmetries appear in the reduction of all the band representations. In fact, since there are 4 symmetry points in the Brillouin Zone of the space group p2 and at each point there are 2 possible symmetries (even or odd) it is possible to construct $2^4 = 16$ different combinations of Bloch symmetries. As is seen from Table 3 only half of them, namely 8, are continuity chords. For example, the combination Γ_1, X_1, Y_1, R_2 is not a continuity chord. What this actually means is that some combinations of Bloch symmetries are not connected with continuous localized orbitals $C(\vec{k},\vec{q})$. The term continuity chord originates from the continuity of the orbitals that form a basis for a band representation. Only those combinations of Bloch symmetries that correspond to a reduction of a band representation are called continuity chords and only they define the symmetry identity of an energy band.

$\vec{q}'$ \ α	(00)	(m_1m_2)	$(2m_12m_2)'$	$(2m_1+1\ 2m_2+1)'$	$(2m_1\ 2m_2+1)'$	$(2m_1+1,2m_2)'$
$(\vec{q}'_a, +)$	$4N^2$	0	4	0	0	0
$(\vec{q}'_a, -)$	$4N^2$	0	-4	0	0	0
$(\vec{q}'_b, +)$	$4N^2$	0	0	0	0	4
$(\vec{q}'_b, -)$	$4N^2$	0	0	0	0	-4
$(\vec{q}'_c{}' +)$	$4N^2$	0	0	0	4	0
$(\vec{q}'_c, -)$	$4N^2$	0	0	0	-4	0
$(\vec{q}'_d, +)$	$4N^2$	0	0	4	0	0
$(\vec{q}'_d, -)$	$4N^2$	0	0	-4	0	0

Table 2. Band Representations of the Space Group p2.

	$(\vec{q}'_a, +)$	$(\vec{q}'_a, -)$	$(\vec{q}'_b, +)$	$(\vec{q}'_b, -)$	$(\vec{q}'_c, +)$	$(\vec{q}'_c, -)$	$(\vec{q}'_d, +)$	$(\vec{q}'_d, -)$
Γ	Γ_1	Γ_2	Γ_1	Γ_2	Γ_1	Γ_2	Γ_1	Γ_2
X	X_1	X_2	X_2	X_1	X_1	X_2	X_2	X_1
Y	Y_1	Y_2	Y_1	Y_2	Y_2	Y_1	Y_2	Y_1
R	R_1	R_2	R_2	R_1	R_2	R_1	R_1	R_2
G	ΣG_i	ΣG_i	ΣG_i	ΣG_i	ΣG_i	ΣG_i	ΣG_i	ΣG_i

Table 3. Continuity chords or the reduction of the band representations into irreducible representations of the space group p2. ΣG_i is the direct sum over all the $4N^2-4$ general points in the Brillouin Zone.

Having all possible band representations of a space group[6] one can find, by reducing them, all the possible continuity chords and correspondingly, all the possible symmetries of the energy bands. Without the concept of band representations, the question of whether or not a set of Bloch functions with well defined symmetries at different points in the Brillouin Zone can form a continuous energy band is not trivial.[2] The irreducible representations of space groups are defined separately at each $\vec{k}$-vector in the Brillouin Zone.[5] The information we know from them is what are the possible Bloch symmetries at each given $\vec{k}$. However, in order to go from one $\vec{k}$ to another $\vec{k}$ in the Brillouin Zone some kind of continuity has to be invoked. Thus, if one knows the symmetries of Bloch functions at a high symmetry point $\vec{k}$ then in the vicinity of it one can find the Bloch symmetries by using the compatibility relations.[7] The concept of a band representation enables one to find the continuity chords and correspondingly all those Bloch functions that can form an energy band. The knowledge of the continuity chords of a given space group enables one to solve also another problem: Given the symmetry of all the Bloch functions in an energy band (either from experiment or calculations) what is the symmetry of the localized orbitals that form this band. In some sense, this is an inverse symmetry problem and it can be solved by knowing the continuity chords. It was already remarked about Table 3 that for each continuity chord there is a well defined symmetry of a localized orbital. This turns out to be a general result, and it is sufficient to know the symmetries of Bloch functions at a small number of symmetry points in the Brillouin zone in order to be able to determine the symmetry of the corresponding localized orbital.[9]

A central rôle in defining the symmetries of energy bands is played by the symmetry centers $\vec{q}'$ in the Wigner Seitz cell. The symmetry of the localezed orbitals (Rel. (2) or (3)) is determined with respect to these centers. The pair of indeces $(\vec{q}',\ell)$ labels the symmetry of the localized orbitals and, correspondingly, also the symmetry of the band which is formed by these orbitals (Rel. (10)). ℓ is the index of an irreducible representation of the localized orbitals as defined with respect to a given symmetry center $\vec{q}'$. The rôle played by the symmetry centers $\vec{q}'$ in defining band representations is very much the same as the one played by the Bloch momentum $\vec{k}$ in defining irreducible representations of space groups. While the bases of the band representations are localized functions $\psi^{(q',\ell)}(\vec{r})$, the bases for irreducible representations are Bloch functions $\psi_k^{(j)}(\vec{r})$, where j labels the representation for a given $\vec{k}$-vector. The pairs of indeces $(\vec{q}',\ell)$ and $(\vec{k},j)$ play therefore a very similar rôle. As is well known the Bloch momentum $\vec{k}$ is a conserved quantity in translationally invariant systems! One should expect also the symmetry center $\vec{q}'$ to be a conserved quantity in periodic solids. This expectation is based on the simple observation that the symmetry centers $\vec{q}'$ describe well defined positions in the crystal which remain constant during the motion of the electrons. When expressed as quantum mechnical operators these symmetry centers should commute with the Hamiltonian of the problem. For a one-dimensional problem it is quite elementary to construct an operator whose eigenvalues are the symmetry centers q' in the Wigner Seitz unit cell.[10] This operator, called the band center Q has the form

$$Q_{mn} = \exp(i\ q_n \frac{2\pi}{a})\delta_{mn} \tag{16}$$

where q_n is the projection of the coordinate operator x on the band n and a is the lattice constant. q_n can be shown to be equal to the expectation value of x in the state described by a Wannier function for the band n. It assumes the values 0 and $\frac{a}{2}$ corresponding to the symmetry centers of a one-dimensional crystal. One should expect that the symmetry centers $\vec{q}'$ which label energy bands are conserved quantities in crystalline solids. In summary, energy bands in solids have an identity that is given by the symmetry label $(\vec{q}', \ell)$ of the localized orbitals $C^{(\vec{q}', \ell)}$ which span the energy band (Eqs. (2) - (4)). The orbitals $C^{(\vec{q}', \ell)}$ form a basis for a band representation of the space group. The reduction of the band representation defines the continuity chord of the energy band. The latter is a combination of all the Bloch symmetries at different points in the Brillouin zone consistent with the continuity of the band. The symmetry identity of an energy band is fully defined by either the label of a band representation or by a continuity chord.

REFERENCES

1. F. Bloch, Z. Phys. 52, 555 (1928).
2. J. Des Cloizeaux, Phys. Rev. 129, 554 (1963).
3. International Tables of X-Ray Crystallography, Kynoch Press, 1952, Vol. 1.
4. J. Zak, in Solid State Physics, edited by F. Seitz, D. Turnbull and H. Ehrenreich, Academic Press (1972), Vol. 27.
5. G.F. Koster, in Solid State Physics, edited by F. Seitz and D. Turnbull, Academic Press (1957), Vol. 5.
6. J. Zak, Phys. Rev. B 26, 3010 (1982).
7. J. Zak, Phys. Rev. B 25, 1344 (1982).
8. L.P. Bouckaert, R. Smoluchowski and E.P. Wigner, Phys. Rev. 50, 58 (1936).
9. J. Zak, Phys. Rev. Letters 47, 450 (1981).
10. J. Zak, Phys. Rev. Letters 48, 359 (1982).

COUPLING COEFFICIENTS FOR THE SPACE GROUP OF THE HEXAGONAL CLOSE-PACKED STRUCTURE

L. Ziemczonek

Department of Physics, Pedagogical University of Słupsk, Arciszewskiego 22B, 76-200 Słupsk, Poland

M. Suffczyński

Institute of Physics, Polish Academy of Sciences, Lotników 32/46, 02-668 Warsaw, Poland

We have computed the Clebsch-Gordan coefficients of the representations for the space group of the hexagonal close-packed structure for the points: Γ, A, H, K, L, M. We enumerate all arms of the wave vector stars and all wave vector selection rules.

The space group D_{6h}^4 ($P6_3/mmc$) of the hexagonal close-packed structure is the symmetry group of the metals of the second column of the periodic table of the elements and of graphite [1]. Also ice I cristallizes in the h.c.p. structure 2 . Ice I exists in a wide range of temperatures: from about -130°C to 0°C and in a range of pressure: from 0 kbar to about 2 kbar [3]. Ice I has been studied by several methods [3]. The first Brillouin zone for the h.c.p. structure has been shown in several places [6-8].

Up to now the Clebsch-Gordan coefficients (CGc's) for the irreducible representations of the space group with h.c.p. structure have not been published. Some CGcs, which we have computed by Berenson and Birman method [8], are presented explicitly in the Tables II and III.

Birman and Berenson [4] have shown that the elements of the first order scattering tensor are precisely CGc's multiplied by certain constants and the elements of the second order tensor are bilinear sums of CGcs. Also the matrix elements of the effective Hamiltonian are products of appropriate CGcs multiplied by symmetrized tensorial field quantities [5] .

The irreducible representations of the space group D_{6h}^4 and the selection rules for their products are given in 6 . The leading wave vector selection rules are constructed with the help of the table 5 from [7] and are the same as thus in 6

In table I we give the wave vector selection rules and blocks. The canonical wave vectors, numbering of symmetry operations, labels and generators of the irreducible representations are as given in the tables of Miller and Love [9].

In Tables II and III: $a=\sqrt{2}/2$.

In Table II the left margin refers to the representations listed above, the right margin to ones listed below the table.

The matrices of CGc's for $\Gamma_m \otimes \Gamma_m$ where $m = 1^{+}_{-}, 2^{+}_{-}, 3^{+}_{-}, 4^{+}_{-}$ are $\hat{1}$.

Acknowledgement

One of the authors (L.Z.) expresses his gratitude to Dr H. Kunert for fruitful discussions and for his help in the preparation of the manuscript.

References.

[1] Olbrychski, K., Gorzkowski, W., Acta Phys. Pol. A 41 (1972) 575.

[2] Landolt-Börnstein, Numerical Data and Functional Relationships in Science and Technology. New Series, Group III, Vol. 7, Crystal Structure Data of Inorganic Compounds, Part b. Springer-Verlag, Berlin, Heidelberg-New York 1975.

[3] Eisenberg, D., Kauzmann, W., The Structure and Properties of Water, Oxford University Press 1969.

[4] Birman, J., Berenson, R., Phys. Rev. B9 (1974) 4512.

[5] Birman, J., Lee Ting-Kuo, Berenson, R., Phys. Rev. B14 (1976) 318.

[6] Cracknell, A.P., Davies, B.L., Miller, S.C. and Love, W.F., Kronecker Product Tables, vol. 1-4 (IFI/Plenum, New York, Washington, London) 1979.

[7] Davies, B.L., Cracknell, A.P., On the completness of tables of irreducible representations of classical space groups, in Communications to the Royal Society of Edinburgh (Physical Sciences) 8 (1976) 81.

[8] Berenson, R. and Birman, J.L., J. Math. Phys. 16 (1975) 227.

[9] Miller, S.C. and Love, W.F., Tables of Irreducible Representations of Space Groups and Co-representations of Magnetic Space Groups (Pruett Press, Boulder, Colorado, 1967).

Table I

Leading wave vector selection rules (LWVSR), wave vector selection rules and blocks in D_{6h}^4.

	$\underline{k}' + \underline{k}'' = \underline{k}$	$\underline{k}' + \underline{k}'' = \underline{k}$	$\sigma'\ \sigma''\ \sigma$
LWVSR	$\underline{k}_\Gamma + \underline{k}_\Gamma = \underline{k}_\Gamma$	$\underline{k}_A + \underline{k}_A = \underline{k}_\Gamma$	1 1 1
LWVSR	$\underline{k}_H + 2\underline{k}_H = \underline{k}_\Gamma$	$\underline{k}_K + 2\underline{k}_K = \underline{k}_\Gamma$	1 2 1
	$2\underline{k}_H + \underline{k}_H = \underline{k}_\Gamma$	$2\underline{k}_K + \underline{k}_K = \underline{k}_\Gamma$	2 1 1
LWVSR	$2\underline{k}_H + 2\underline{k}_H = \underline{k}_K$	$2\underline{k}_K + 2\underline{k}_K = \underline{k}_K$	2 2 1
	$\underline{k}_H + \underline{k}_H = 2\underline{k}_K$	$\underline{k}_K + \underline{k}_K = 2\underline{k}_K$	1 1 2
LWVSR	$\underline{k}_L + \underline{k}_L = \underline{k}_\Gamma$	$\underline{k}_M + \underline{k}_M = \underline{k}_\Gamma$	1 1 1
	$2\underline{k}_L + 2\underline{k}_L = \underline{k}_\Gamma$	$2\underline{k}_M + 2\underline{k}_M = \underline{k}_\Gamma$	2 2 1
	$3\underline{k}_L + 3\underline{k}_L = \underline{k}_\Gamma$	$3\underline{k}_M + 3\underline{k}_M = \underline{k}_\Gamma$	3 3 1
LWVSR	$2\underline{k}_L + 3\underline{k}_L = \underline{k}_M$	$2\underline{k}_M + 3\underline{k}_M = \underline{k}_M$	2 3 1
	$2\underline{k}_L + \underline{k}_L = 3\underline{k}_M$	$2\underline{k}_M + \underline{k}_M = 3\underline{k}_M$	2 1 3
	$3\underline{k}_L + 2\underline{k}_L = \underline{k}_M$	$3\underline{k}_M + 2\underline{k}_M = \underline{k}_M$	3 2 1
	$3\underline{k}_L + \underline{k}_L = 2\underline{k}_M$	$3\underline{k}_M + \underline{k}_M = 2\underline{k}_M$	3 1 2
	$\underline{k}_L + 3\underline{k}_L = 2\underline{k}_M$	$\underline{k}_M + 3\underline{k}_M = 2\underline{k}_M$	1 3 2
	$\underline{k}_L + 2\underline{k}_L = 3\underline{k}_M$	$\underline{k}_M + 2\underline{k}_M = 3\underline{k}_M$	1 2 3

Table II

C G c's for $\Gamma_n \otimes \Gamma_n$ $(n=5^{+}_{-}, 6^{+}_{-})$ and $K_j \otimes K_j$ $(j=1, 2, 3, 4)$ in D^4_{6h}.

$\Gamma_n \otimes \Gamma_n = \Gamma_{1+} + \Gamma_{2+} + \Gamma_{5+}$

$\alpha' \alpha''$	$\alpha=1$	1	1 2	
1 1	0	0	0 1	1 1 1 1
1 2	a	a	0 0	1 1 1 2
2 1	a	−a	0 0	1 1 2 1
2 2	0	0	1 0	1 1 2 2
	1	1	1 $1=\alpha$	$\alpha' \alpha'' \sigma' \sigma''$
	1		1 $2=\sigma$	

$\Gamma_{1+} + \Gamma_{3-} + K_1 = K_j \otimes K_j$

Table III

C G c's for $A_j \otimes A_j$ $(j=1, 2)$ in D^4_{6h}.

$A_j \otimes A_j = \Gamma_{1+} + \Gamma_{3+} + \Gamma_{2-} + \Gamma_{4-}$

$\alpha' \alpha''$	$\alpha=1$	1	1	1
1 1	0	a	0	a
1 2	a	0	a	0
2 1	a	0	−a	0
2 2	0	a	0	−a

CANONICAL TRANSFORMATIONS

AND

QUANTUM MECHANICS

HARMONIC ANALYSIS ON PHASE SPACE AND BORN'S METRIC FOR SPACE TIME

S. Twareque Ali
Department of Mathematics, Concordia University
Montréal, P.Q., Canada H4B 1R6

1. INTRODUCTION

In the past few years extensive work on formulating relativistic quantum mechanics on stochastic phase space has been carried out (see, for example, [1,2] and the references cited therein). This approach is able, at the single particle level, to avoid, among other things, the violations of causality associated with Hegerfeldt's theorem [3] , the standard instability problems associated with the Zitterbewegung and the difficulty of the non-existence of a conserved probability (as opposed to charge) current in the usual formulation. The key concept in the stochastic phase space approach is that of localization in phase space regions (where position and momentum are simultaneously determined only to an accuracy allowed by the uncertainty principle). The theory yields a covariant and conserved probability current as a consequence of the invariance of the volume element in phase space under Lorentz transformations. We analyze in this note a specific representation of the Poincaré group on phase space, assciated with the stochatic phase space approach. A complete harmonic analysis of this representation connects in an interesting way concepts on reproducing kernel Hilbert spaces, positive operator valued (POV) measures and systems of covariance. For definitions and basic results on these topics we refer to [4,5] and [6] . The decomposition mentioned above is rendered unique by invoking Born's reciprocity principle [7] and introducing the eigenstates of his metric operator for space time.

Let X be a locally compact topological space, $B(X)$ the Borel sets of X, and μ a Borel measure on X. To each point $x \in X$, let us associate a Hilbert space $\mathcal{K}_x$, and let

$$\tilde{\mathcal{H}} = \int_X^{\oplus} \mathcal{K}_x \, d\mu(x) . \tag{1}$$

Let $K(x,y)$ be a reproducing kernel [4,6] on $\tilde{\mathcal{H}}$ and $\mathbb{P}_K$ the associated projection operator,

$$(\mathbb{P}_K \tilde{\Psi})(x) = \int_X K(x,y)\, \tilde{\Psi}(y)\, d\mu(y) , \tag{2}$$

$$\mathbb{P}_K \tilde{\mathcal{H}} = \mathcal{H}_K , \tag{3}$$

for all $\tilde{\Psi} \in \tilde{\mathcal{H}}$. The evaluation map $E_x^K : \tilde{\mathcal{H}} \longrightarrow \mathcal{K}_x$, associated [4] with $K(x,y)$ is then

(4) $$E_x^K(\psi_K) = \int_X K(x,y)\,\psi_K(y)\,d\mu(y),$$

for all $\psi_K \in \mathcal{H}_K$, and

(5) $$K(x,y) = E_x^K E_y^{K*},$$

for all $(x,y) \in X \times X$, with E_y^{K*} being the adjoint of E_y^K. Canonically associated with $K(x,y)$ is the POV-measure $a^K(\Delta)$, $\Delta \in B(X)$ on $\mathcal{H}_K$:

(6) $$a^K(\Delta) = \int_\Delta E_x^{K*} E_x^K\, d\mu(x),$$

having μ-density [4] $x \longrightarrow F(x) \in \mathcal{L}(\mathcal{H}_K)^+$,

(7) $$F(x) = E_x^{K*} E_x^K.$$

Moreover, if $\tilde{E}$ is the projection valued (PV) measure on $\tilde{\mathcal{H}}$,

(8) $$(\tilde{E}(\Delta)\tilde{\psi})(x) = \chi_\Delta(x)\,\tilde{\psi}(x)$$

then,

(9) $$a^K(\Delta) = \mathbb{P}_K\,\tilde{E}(\Delta)\,\mathbb{P}_K,$$

and $\tilde{E}$ is the (minimal) Naimark extension of a^K.

2. A POINCARÉ GROUP REPRESENTATION ON PHASE SPACE

The representation U_o^P of the Poincaré group $\mathcal{P}$ on phase space, that we analyze here is carried [8,9] by the Hilbert space $\mathcal{H}_o^P = L^2(\sigma_o \times \mathcal{V}_m^+, d\Sigma)$, where σ_o is a space like hypersuface with surface measure $d\sigma^\mu(q)$ $(\mu = 0,1,2,3)$, $\mathcal{V}_m^+$ is the forward mass hyperboloid, with Lorentz invariant measure $d\Omega_m(p) = d\underline{p}/2p_o$ and $d\Sigma(q,p) = 2p_\mu d\sigma^\mu(q)d\Omega_m(p)$. The representation U_o^P in question is

(10) $$(U_o^P(g)\phi)(q,p) = \phi(\Lambda^{-1}(q-a), \Lambda^{-1}p)$$

(11) $$\phi(q,p) = (\exp[-\tfrac{i}{\hbar}H_o t]\,\phi)(\underline{q},\underline{p})$$

(12) $$H_o = c[-\hbar^2\nabla_{\underline{q}}^2 + m^2c^2], \qquad g = (a,\Lambda) \in \mathcal{P}$$

where $\phi \in \mathcal{H}_o^P$. This representation is reducible but not globally unitary. However, each irreducible subrepresentation of U_o^P is unitary and is carried by a reproducing kernel Hilbert space $\hat{\mathcal{H}}_o^{\eta}$, with resolution generator η and is equivalent to one of the standard Wigner representations corresponding to mass m and spin j [9]. Denoting by $R(k,\Lambda)$ the usual Wigner rotation, by $Y_{js}(\underline{k})$, $s = -j,\dots,+j$, the spherical harmonics of the angles of $\underline{k}$, and P_j a Legendre polynomial, we have the following spectral analysis of U_o^P.

Theorem 1

Let $\{\hat{e}_n\}_{n=1}^{\infty}$ be any orthonormal basis in $L^2_{\tilde{\mu}}(\mathbb{R}^1)$, where

$$d\tilde{\mu}(p_o) = p_o(p_o^2 - m^2c^2)^{1/2}\, dp_o\,, \qquad mc \le p_o < \infty\,, \tag{13}$$

and, $d\tilde{\mu}(p_o) = 0$ for $p_o < mc$. Then the family of functions

$$\Phi_{nj}(q,p) = \left[\frac{2mc}{(2\pi\hbar)^3}\right]^{1/2} \sum_{s=-j}^{j} \int_{\mathcal{V}_m^+} \exp\left(-\frac{i}{\hbar} k\cdot q\right) \hat{e}_n\left(\frac{p\cdot k}{mc}\right) \times Y_{js}(R(k,\Lambda_p)\underline{p})\, \tilde{\Phi}_s(k)\, d\Omega_m(k)\,, \tag{14}$$

for all $\tilde{\Phi}_s \in L^2(\mathcal{V}_m^+)$, $s = -j,\dots,+j$, is a closed subspace $\hat{\mathcal{H}}_o^{nj}$ of $\mathcal{H}_o^P$. Each $\hat{\mathcal{H}}_o^{nj}$ is left invariant by U_o^P, and there exists an invertible linear mapping θ of a dense set D_θ of vectors in $\mathcal{H}_o^P$ onto a Hilbert space $\bar{\mathcal{H}}_o$, such that the subspaces $\theta\hat{\mathcal{H}}_o^{nj}$ are mutually orthogonal and,

$$\bar{\mathcal{H}}_o = \oplus \sum_{n=1}^{\infty} \sum_{j=0}^{\infty} \theta\, \hat{\mathcal{H}}_o^{nj} \tag{15}$$

Theorem 2

Each of the irreducible subspaces $\hat{\mathcal{H}}_o^{nj}$ is a reproducing kernel Hilbert space, with reproducing kernel

$$\hat{K}_{nj}(q,p;q',p') = \frac{mc(2j+1)}{2\pi\hbar^3} \int_{\mathcal{V}_m^+} \exp\left[\frac{i}{\hbar}(q'-q)\cdot k\right] \times P_j\left(\frac{R(k,p)\underline{p}\cdot R(k,p')\underline{p}'}{|\underline{p}||\underline{p}'|}\right) \hat{e}_n\left(\frac{p\cdot k}{mc}\right) \hat{e}_n^*\left(\frac{p'\cdot k}{mc}\right) d\Omega_m(k) \tag{16}$$

and canonically associated POV-measure

$$\hat{E}_o^{nj}(\Delta) = \int_{\Delta} |\hat{U}_o^{nj}((0,\underline{q}),\Lambda_p)\eta_{nj}\rangle\langle \hat{U}_o^{nj}((0,\underline{q}),\Lambda_p)\eta_{nj}|\, d\underline{q}\, d\underline{p}\,, \tag{17}$$

$\Delta \in \mathcal{B}(\Gamma)$. Furthermore, the carrier space $\mathcal{H}_o^P$ is the minimal extension of $\hat{\mathcal{H}}_o^{nj}$ in the sense of Naimark.

If we restrict ourselves to the Euclidean subgroup of $\mathcal{P}$, then $\{E_o^{nj}, U_o^{nj}\}$ is a system of covariance on $\hat{\mathcal{H}}_o^{nj}$ and can be lifted (minimally) to the system of imprimitivity $\{E_o, U_o^P\}$ on $\mathcal{H}_o^P$.

The decomposition in (13) is arbitrary to the extent that the $\hat{e}_n$'s can be chosen arbitrarily. However, since all spacetime excitons arising from Born's quantum metric operator [7,10] can be used to define extended test particles on phase space, one may use the eigenvectors of the operator

$$D^2 = Q^\mu Q_\mu + P^\mu P_\mu \tag{18}$$

to get a set of $\hat{e}_n$'s which will then fix the decomposition uniquely.

REFERENCES

1. S.T. Ali, Lecture Notes in Physics, Vol. 139, Springer (1981).
2. E. Prugovečki, Stochastic Quantum Mechanics and Quantum Space-time, Reidel, Dordrecht-Boston-London (1983), in press.
3. G.C. Hegerfeldt and S.N.M. Ruijsenaars, Phys. Rev. D22 (1980), 377
4. U. Cattaneo, J. Math. Phys. 23 (1982), 659
5. S.T. Ali, Lecture Notes in Mathematics, Vol. 905, Springer (1982), P. 207
6. S.T. Ali, Harmonic analysis on phase space I: Reproducing kernel Hilbert spaces, POV-measures and systems of covariance, to appear.
7. M. Born, Rev. Mod. Phys. 21 (1949), 463.
8. S.T. Ali, J. Math. Phys. 20 (1979), 1385 and 21 (1980), 818
9. S.T. Ali and E. Prugovečki, Harmonic analysis and systems of covariance for phase space representations of the Poincaré group, Concordia University preprint (1983)
10. J.A. Brooke and E. Prugovečki, Lett. Nuovo Cim. 33 (1982), 171.

GENERALIZED CHEBYSHEV POLYNOMIALS AND CHARACTERS OF GL(N,C) AND SL(N,C)

(fragments of results)

H. BACRY (Marseille,France)

1) Chebyshev polynomials and SL(2,C).

For $A \in SL(2,C)$ and $t = \mathrm{Tr}\, A$, we have $A^2 - tA + I = 0$

Therefore, for $n \in \mathbb{Z}$, we have

$$A^n = f_n(t)\, A - g_n(t)\, I$$

where f_n and g_n are polynomials with integral coefficients. In fact,

$$g_n = f_{n-1}$$

We also define the polynomials $l_n(t) = \mathrm{Tr}(A^n)$.

Both f_n and l_n satisfy the recurrence relation :

$$f_{n+1}(t) = t\, f_n(t) - f_{n-1}(t)$$

with the "initial conditions"

$$f_0 = 0 \quad , \quad f_1 = 1$$
$$l_0 = 2 \quad , \quad l_1 = t$$

The l_n (resp. f_n) are related with the Chebyshev polynomials of the first (resp. second) kind by *

$$l_n(t) = 2\, T_n(\tfrac{t}{2}) \qquad l_{-n}(t) = l_n(t)$$

$$f_n(t) = U_{n-1}(\tfrac{t}{2}) \qquad f_{-n}(t) = f_n(t)$$

By making $t = 2\cos\theta$, one gets $f_n(2\cos\theta) = \frac{\sin n\theta}{\sin\theta}$. One recognize the character of $A \in SU(2)$ in the representation of dimension n.

2) Generalized Chebyshev polynomials.

We take GL(3,C) because the general case of GL(N,C) is obvious. The Cayley-Hamilton equation for $A \in GL(3,C)$ is

$$A^3 - tA^2 + uA - vI = 0 \qquad (1)$$

t (trace), u (?), v (determinant) are the characters of the representations $\square$, $\begin{smallmatrix}\square\\\square\end{smallmatrix}$, $\begin{smallmatrix}\square\\\square\\\square\end{smallmatrix}$, respectively.

---

* The labelling of the f_n's is justified by the fact that f_n divides f_m iff n divides m.

We define

$$l_n = \mathrm{Tr}(A^n)$$

$$A^{n+1} = f_n\, A^2 - g_n\, A + h_n\, I$$

l_n, f_n, g_n, h_n are polynomials in t, u, v, with integral coefficients satisfying the same recurrence relation

$$f_{n+1} = t\, f_n - u\, f_{n-1} + v\, f_{n-2}$$

with the "initial conditions" $f_0 = g_{-1} = g_1 = h_0 = h_1 = 0$, $f_1 = h_{-1} = 1$, $f_{-1} = g_0 = -1$, $l_0 = 3$, $l_1 = l_{-1} = t$. Iff $v = \pm 1$, they are polynomials even with $n < 0$. For $v = 1$, l_n and f_n will be called <u>generalized Chebyshev polynomials</u>.

<u>Generalized Cayley-Hamilton formula</u> :

If A is diagonal (eigenvalues a, b, c), we can define

$$\chi_{n_1 n_2 n_3} = \det \left[\begin{vmatrix} a^{n_1} & a^{n_2} & a^{n_3} \\ b^{n_1} & b^{n_2} & b^{n_3} \\ c^{n_1} & c^{n_2} & c^{n_3} \end{vmatrix} \begin{vmatrix} a^2 & a & 1 \\ b^2 & b & 1 \\ c^2 & c & 1 \end{vmatrix}^{-1} \right]$$

which is the character associated with the Young diagram 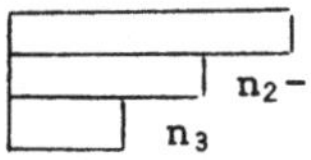

Whenever $n_1 > n_2 > n_3$. We also have $f_n = \chi_{n+1,1,0}$, $g_n = \chi_{2,N+1,0}$, $h_n = \chi_{2,1,n+1}$.

We have the generalized C-H formula

$$\chi_{n_2,n_3,0}\, A^{n_1} - \chi_{n_1,n_3,0}\, A^{n_2} + \chi_{n_1,n_2,0}\, A^{n_3} - \chi_{n_1,n_2,n_3}\, I = 0$$

Moreover χ_{n_1,n_2,n_3} satisfy the same recurrence relation as the f_n's for two fixed indices.

<u>Generating functions for characters</u> :

$$\sum_{n_i=0}^{\infty} \chi_{n_1,n_2,n_3}\, x^{n_1} y^{n_2} z^{n_3} = \frac{(x-y)(x-z)(y-z)}{\det\left[(1-Ax)(1-Ay)(1-Az)\right]}$$

and, for dimensions of the representations

$$\sum_{n_i=0}^{\infty} d_{n_1,n_2,n_3}\, x^{n_1} y^{n_2} z^{n_3} = \frac{(x-y)(x-z)(y-z)}{(1-x)^3(1-y)^3(1-z)^3}$$

The generating functions permit to get the expressions of χ_{n_1,n_2,n_3}.

In particular,

$$f_n(t,u,1) = \sum_{k+2l=n-1} \frac{(k+l)!}{k!\, l!} (-)^l\, t^k\, u^l$$

$$= \sum_{k+2l=n-1} \frac{(k+1)!}{k!\, l!} \frac{(n+2k+l+1)!}{(n-k-2l-1)!\,(3k+3l+2)!} (-)^l (t-3)^k (u-3)^l$$

Differential equations for characters of SL(3,C) (v=1)

$$\text{Define } \Delta_3 = (t^2-3u)\frac{\partial^2}{\partial t^2} + (tu-9)\frac{\partial^2}{\partial t\partial u} + (u^2-3t)\frac{\partial^2}{\partial u^2}$$

$$D_3 = t\frac{\partial}{\partial t} + u\frac{\partial}{\partial u}$$

One has

$$(\Delta_3+4D_3)\,\chi_{n_1,n_2,n_3} = \frac{1}{2}\left[(n_1-n_2)^2+(n_1-n_3)^2+(n_2-n_3)^2\right]\chi_{n_1,n_2,n_3}$$

One has also

$$(\Delta_3+D_3)\,1_n = n^2\,1_n$$

Remark : the determinant of Δ_3, namely $(t^2-3u)(u^2-3t)-(tu-9)^2$ is the discriminant of Eq. (1).

Case of GL(N,C) : The $\chi_{n_1,n_2\ldots}$ are eigenfunctions of an operator $\Delta_N+(N+1)D_N$. The operator Δ_N is associated with a symmetric matrix of dimension N-1, the coefficients of which are polynomials of degree 2 in t, u, v, ... The determinant of this matrix is the discriminant of the C-H equation. The rank of the matrix describes the degree of degeneracy of the eigenvalues (i.e. the number of "=" signs needed to express the equality of roots).

Orthogonality relations : For n n

$$\int \chi_{n_1,n_2,0}(t,\bar{t},1)\,\chi_{n'_1,n'_2,0}(t,\bar{t},1)\,d\mu(t) = \delta_{n_1n'_1}\,\delta_{n_2n'_2}$$

Where $d\mu(t)$ is the measure induced by the Haar measure on the space of conjugacy classes. For SL(3,C) this domain is the interior of a hypocycloid in the complex t-plane.

TENSOR OPERATORS AS AN EXTENSION OF THE UNIVERSAL ENVELOPING ALGEBRA*

L.C. Biedenharn
Department of Physics
Duke University
Durham, North Carolina 27706 U.S.A.

and

Daniel E. Flath
Department of Mathematics
Duke University
Durham, North Carolina 27706 U.S.A.

Abstract

A global, coordinate-free, algebraic formulation of SU3 tensor operator structure is presented. This structure is shown to contain the universal enveloping algebra of SU3 and to resolve the multiplicity problem in conceptual terms.

§1. Operators play a fundamental rôle in the structure of quantum mechanics, and the concept of a *tensor operator* was introduced by Born and Jordan in order to extend to operators the symmetry classification (by the quantal rotation group) used so advantageously in classifying (Hilbert space) state vectors. The concept of a tensor operator was formalized by Michel [1]: *A tensor operator is a set of linear operators* $\{T(\alpha)\}$, *indexed by a group representation label* α, *which obeys the equivariance condition:*

$$U_g T(\alpha) U_g^{-1} = T(g(\alpha)), \tag{1.1}$$

for every $g \in G$ (the symmetry group). (Here U_g is the unitary transformation associated with g and $g(\alpha)$ denotes the action of g on the representation label α.)

The theory of tensor operators has, however, been developed fully only for the symmetry group SU2 (the quantal angular momentum group); it is no exaggeration to say that the resulting theory [2] (the Racah-Wigner algebra) is of fundamental importance in almost all applications of quantum physics. Since the symmetry group $SU3^{color}$ is, like SU2, currently believed to be both an exact physical symmetry and of fundamental importance in hadron physics, it would appear useful to

*Supported in part by the National Science Foundation.

develop an analogous theory of tensor operators for SU3. As is well-known, the principal obstacle in the way of such a development is the problem of multiplicity (to be discussed below). A constructive resolution of this problem was announced [3] at one of our earlier conferences (Nijmegen, 1975), but the construction was computationally based and somewhat complicated. It would clearly be desirable to obtain a global, coordinate-free resolution of the multiplicity-problem in purely conceptual terms.

That is the purpose of the present talk: to analyze SU3 tensor operators conceptually in terms of a beautiful algebraic structure--a structure extending the universal enveloping algebra of SU3 whose very existence was previously quite unsuspected and thus had an element of surprise.

§2. In conceptual terms, the fundamental problem is the *explicit* decomposition of all finite dimensional tensor product representations $V \otimes W$ of the simple Lie algebra su_3. (For the Lie algebra su_2, the famous Wigner-Clebsch-Gordan coefficients provide the complete solution.)

An equivalent, but more formal, statement of the problem is the decomposition of all homomorphisms, linear over $\mathbb{C}$, of the representation V into the representation W, that is, of the space $\mathrm{Hom}_{\mathbb{C}}(V,W)$. This is exactly the space of tensor operators in Michel's formulation cited above.

An irreducible tensor operator is thus an irreducible sub-representation of $\mathrm{Hom}_{\mathbb{C}}(V,W)$. It entails no loss of generality to take V and W irreducible.

Irreducible representations are labelled by their highest weights, but for irreducible tensor operators, it is important to note that *two distinct concepts of weight* enter:

i) the highest weight of the tensor operator as an (irreducible) SU3 representation;

ii) the weight which is the difference between the highest weights of the irreps (irreducible representations) V and W.

Let us call the weight in (ii), the *shift weight* (or, more briefly, *shift*), since it is the "shift in representation labels" effected by the tensor operator.

At first glance, tensor operators with the same highest weights but different shift weights would seem to be quite unrelated (since they map between different spaces). But the fact that every shift weight of a tensor operator is necessarily an actual weight of the tensor operator suggests otherwise. The analogy between these two 'weight concepts' is even closer, as we illustrate now by the multiplicity problem.

Denote by V_λ the (finite dimensional, unitary) irrep of SU3 with highest weight λ.

Lemma 2.1 ([4],[5]): *The multiplicity of* V_λ *in* $\mathrm{Hom}_{\mathbb{C}}(V_\alpha, V_{\alpha+\mu})$ *is dominated by the multiplicity of the weight* μ *in* V_λ*, with equality for generic* α. (Generic means that the irrep V_α has Young frame labels $[M_{13}, M_{23}, 0]$ with the non-negative integers M_{13}, M_{23} sufficiently large and with $(M_{13} - M_{23})$ also sufficiently large.)

This result strongly suggests that the multiplicity of V_λ in $\mathrm{Hom}_{\mathbb{C}}(V_\alpha, V_{\alpha+\mu})$ is to be considered as "independent" of α, in the sense that one considers the tensor operator as defined by its action on the set of *all* irreps V_α, for which the generic behaviour dominates.

§3. Let us remark that this point of view contrasts sharply with the customary (mathematical) view in which the multiplicity of V_λ in $\mathrm{Hom}_{\mathbb{C}}(V_\alpha, V_\beta)$ is viewed as a joint property (the intertwining number) of the three irreps α, β, λ, each playing an (essentially) equivalent rôle in the triple. It is the fundamental structure of quantum mechanics that justifies our assignment of a special rôle to the irrep λ carried by the tensor operator. (Probability amplitudes in quantum mechanics are 'triples' $\langle\alpha|T(\gamma)|\beta\rangle$ composed of initial state $|\beta\rangle$, final state $\langle\alpha|$ and the operator $T(\gamma)$; the operator plays a distinguished rôle in the 'triple'.) Thus one is to view the multiplicity as enumerating different components of *one* tensor operator (each component possessing two weights) with action on the set of all irreps defining (in some way) each component individually.

§4. Consider now the special case for which the shift μ is zero. The set of all such operators constitutes the universal enveloping algebra $\mathcal{U}$ of the group. We may view the algebra $\mathcal{U}$ as a sub-algebra of the algebra of all mappings (linear over $\mathbb{C}$) of $\bigoplus V_\alpha$ into itself, that is of the endomorphisms $\mathcal{E} \equiv \mathrm{End}_{\mathbb{C}}(\bigoplus V_\alpha)$, where the sum includes precisely one finite dimensional irrep of the group from each isomorphism class. Elements of $\mathcal{U}$ map each V_α into itself (since $\mu = 0$).

What we seek is a suitable enlargement--remaining within the algebra $\mathcal{E}$--of the enveloping algebra $\mathcal{U}$ which contains also operators (transformations) mapping each V_α to each V_β. Let us denote this extension of the enveloping algebra by $\mathcal{A}$.

For SU2, one can realize the algebra $\mathcal{A}$ by means of a pair of boson operators, a_1^+ and a_2^+, and the Jordan-Schwinger map [2] (of the Pauli matrices into SU2 generators). The space $\bigoplus V_\alpha$ is then the ring of polynomials in the two 'variables' a_1^+ and a_2^+ (terminated by the vacuum ket $|0\rangle$, defined by $a_i|0\rangle = 0$). These results are well-known, and carrying out the construction poses no difficulties since the two weights label each component of the tensor operator uniquely.

§5. Let us consider next SU3; for this group the multiplicity can be arbitrarily large for large enough irreps. To construct the space $\oplus V_\alpha$, we take two sets of three bosons: a_1, a_2, a_3 and b_1, b_2, b_3 where:

$$[a_i, a_j^+] = [b_i, b_j^+] = \delta_{ij}, \tag{5.1}$$

with all other commutators zero. The SU3 action is again the Jordan map, J:

$$J: \quad (e_{ij}, i, j = 1,2,3) \rightarrow (E_{ij} = a_i^+ a_j - b_j^+ b_i). \tag{5.2}$$

Under the commutator action of the E_{ij}, $\{a_i^+\}$ transforms as the irrep [100] and $\{b_i^+\}$ transforms as [110]. The space of states generated by the polynomial ring over the six bosons $\{a_i^+, b_j^+\}$ (terminated by the vacuum ket) will be denoted by W. The highest weight states in W are generated by the boson operators a_1^+, b_3^+ and the SU3 invariant operator $M^+ = a_1^+b_1^+ + a_2^+b_2^+ + a_3^+b_3^+$. It is easily shown [6] that the sub-space V of W consisting of all vectors in W annihilated by the operator $M = (M^+)^+$ is the desired space $\oplus V_\alpha$. That is, $V \equiv$ kernel (M) of W *consists of a multiplicity free sum of finite dimensional irreps of* SU3, *with precisely one subrepresentation from each equivalence class.*

The algebra A, just as in the SU2 case, will be a sub-algebra of $\mathrm{End}_{\mathbb{C}}(V)$.

Since the weights for the fundamental irreps [100] and [110] are of multiplicity *one*, the nine components of the tensor operators corresponding to each of these irreps are uniquely labelled by the weight and the shift; we denote these fundamental operators by $\begin{pmatrix}\mu\\ \alpha\end{pmatrix}$, μ = shift weight and α = (SU3)-weight. The algebra A is then, by definition, generated by the nine fundamental operators belonging to the tensor operator $\langle 100\rangle$. (Equivalently, one could use the nine components of $\langle 110\rangle$.)

Let us give these generators more explicitly:

$$\begin{pmatrix}001\\100\end{pmatrix} = b_1 \tag{5.3}$$

$$\begin{pmatrix}010\\100\end{pmatrix} = b_3^{+}a_2 - b_2^{+}a_3 \tag{5.4}$$

$$\begin{pmatrix}100\\100\end{pmatrix} = a_1^{+}(2 + \Sigma_i(a_i^{+}a_i + b_i^{+}b_i)) - (\Sigma_i a_i^{+}b_i^{+})b_1. \tag{5.5}$$

The remaining six fundamental operators of $\langle 100\rangle$ are generated by commutation with the SU3 generators (E_{ij}).

It may be verified directly (albeit laboriously) that these operators have the claimed shift and weight properties; a more structural verification is given in [7].

The algebra A generated by the nine $\langle 100\rangle$ can be shown to be precisely of the right 'size' for our purposes:

Lemma 5.6([6]): *Every* T *in* $\mathrm{Hom}_{\mathbb{C}}(V_\alpha, V_\beta)$ *is the restriction of some element of* A; in particular, *the universal enveloping algebra* U *of* SU3 *is contained in* A.

To get a better grasp on the algebra A, let us consider the Lie algebra, generated by commutation, from the nine components of the fundamental tensor operator $\langle 100\rangle$.

Theorem 5.7([6]): *The Lie algebra generated by the nine components of* $\langle 100\rangle$ *is isomorphic to the twenty-eight dimensional (complex) Lie algebra* so(8). *The nine components of* $\langle 100\rangle$, *the nine components of* $\langle 110\rangle$, *the eight generators of* SU3, *and two* SU3-*invariant operators account for the dimension* 28. (The two SU3-invariant operators are easily understood: they are the two operators R and S yielding the number of "a" quanta and the number of "b" quanta, respectively.)

It follows from this theorem that A is isomorphic to a quotient of the universal enveloping algebra of so(8).

Moreover, A under the commutation action of the so(8) generators is itself an so(8) representation. It is shown in [6] that:

Theorem 5.8: *The representation carried by* A, *under commutation with the* so(8) *generators, is a direct sum of* so(8) *irreps of the form* (0p00), *each irrep* (p = 0,1,...) *occurring once and only once.* (We use the notation of [8]; (0000) denotes the identity irrep and (0100) the 28 dimensional adjoint irrep.)

Using the Weyl branching rules for the chain $D4 \supset B3 \supset A3 \supset A2$ this theorem provides an enumeration of all SU3 tensor operators. To confine attention to SU3 components having highest weight, let us define B as the commutant of the raising operators $\{E_{12}, E_{23}\}$ in A. Let $B\binom{\mu}{\beta}$ be the space of all operators T in B such that:

(i) $T(V_\alpha) \subset V_{\alpha+\mu}$, that is, μ is a *shift weight*, and

(ii) T is of highest weight β.

Our principal result is then:

Theorem 5.9([6]): (a) $\mathcal{B}\binom{0}{0}$ *consists of all polynomials in the two* SU3 *invariant operators* R *and* S *(the two "number of quanta" operators);*

(b) $\mathcal{B}\binom{\mu}{\beta}$ *is a free* $\mathcal{B}\binom{0}{0}$ *-module of rank equal to the multiplicity of the weight* μ *in the irrep* V_β*;*

(c) *An explicit* $\mathcal{B}\binom{0}{0}$*-basis for* $\mathcal{B}\binom{\mu}{\beta}$ *can be given.*

This theorem establishes the "α-independent" construction of the irreducible tensor operators in $\mathrm{Hom}_{\mathbb{C}}(V_\alpha, V_{\alpha+\mu})$ for SU3 as discussed, and motivated, by our introductory remarks. From its evident parallel to the Poincaré-Birkhoff-Witt theorem for enveloping algebras, theorem (5.9) validates our assertion that A is an extension of the concept of universal enveloping algebra--but our claim is limited to SU3 (and, of course, SU2).

The algebra A has one further nice property:

Lemma 5.10([6]): *A contains no non-zero proper two-sided ideal.*

§6. What meaning do these results have for physicists interested, as mentioned in §1, in *fully explicit results for specific tensor operators?* (After all, almost every physicist nowadays knows that there are two adjoint SU3-tensor operators, F and D. But which is which in an unlabelled multiplicity space of two dimensions?)

To answer this let us note that the so(8) generators act on the space $V = \oplus V_\alpha$, by juxtaposition (*not* commutation), and it can be shown [6] that V itself is an irreducible infinite dimensional so(8) representation with a highest weight. This representation can be unitarized [9] and as such turns out to be an irrep of a real form of so(8) isomorphic to so(6,2). The inner product on V induces a Hilbert-Schmidt inner product on each irreducible tensor operator in $\mathrm{Hom}_{\mathbb{C}}(V_\alpha, V_{\alpha+\mu})$. (This structure is rather obvious to physicists as the familiar boson operator norm.)

It is now clear that to obtain explicit normalized unit tensor operators (the analog to Wigner-Clebsch-Gordan matrix coefficients) one can simply carry out a Gram-Schmidt process on the bases for $\mathcal{B}\binom{\mu}{\beta}$ asserted in Theorem (5.9). *But a Gram-Schmidt process is far from canonical, there being unlimitedly many free choices involved!* (Only individual authors could be happy with such idiosyncratic tables!) Note that the so(8) decomposition chain of D4 does not, per se, give an adequate resolution (since so(8) distinguishes between an SU3 operator and the same operator multiplied by an SU3 invariant: cf. Theorem (5.9b)).

The resolution of this difficulty lies in the result [7]:

Lemma 6.1: *The* SU3 *tensor operators* $\in A$ *in a given multiplicity set are uniquely distinguished by the order in which they occur for the first time in the ordered sequence of* so(8) *irreps:* (0000), (0100), ..., (0p00),... . (To illustrate, the F operator occurs in (0100) for the first time, whereas the D operator occurs first in (0200).)

Once an order is imposed on the Gram-Schmidt process there are no free choices. This resolves the problem of explicitly determining all SU(3) tensor operators uniquely, despite the occurrence of multiplicity.

What is remarkable is that this global, coordinate-free, formulation of the SU3 tensor operator problem (canonically splitting all multiplicities) agrees in every detail with the canonical labelling based upon the characteristic-null space of the operators [3], and with the labelling induced by the intrinsic zeroes of the projective operator matrix elements [10] ("isoscalar factors"). All of these ways of determining the explicit matrix elements agree, and the resulting special functions (matrix elements) are, themselves, possessed of quite remarkable symmetry properties [11].

§7. Does the theory described here generalize to other simple Lie algebras? This is, in our view, an important open problem, to which we can only remark that already for SU4 the structure is significantly more complex.

References

[1] L. Michel, "Application of Group Theory to Quantum Physics; Algebraic Aspects" in *Lecture Notes in Physics*: Battelle Rencontres (V. Bargmann, Ed.) pp. 36-143, Springer-Verlag, Berlin 1970.

[2] L. C. Biedenharn and J.D. Louck, "Angular Momentum in Quantum Physics", Vol. 8, Encyclopedia of Mathematics and Its Applications, (G.-C. Rota, Ed.), Addison-Wesley Publishing Co. (Reading, MA) 1981.

[3] L.C. Biedenharn, M.A. Lohe, and J.D. Louck, "The Canonical Resolution of the Multiplicity Problem for U(3): An Explicit and Complete Constructive Solution", in the Proceedings of the Fourth International Colloquium in Group Theoretical Methods in Physics, University of Nijmegen, The Netherlands, Springer Verlag (Berlin) 1976.

[4] G.E. Baird and L.C. Biedenharn, "On the Representations of the Semisimple Lie Groups IV. A canonical classification for Tensor Operators in SU_3, J. Math. Phys. 5 (1964), 1730-1747.

[5] A.U. Klimyk, "Decomposition of a Tensor Product of Irreducible Representations of a Semisimple Lie Algebra into a Direct Sum of Irreducible Representations", *Amer. Math. Soc. Translations*, Series 2, Vol. 76, Amer. Math. Soc. (Providence) 1968.

[6] D.E. Flath and L.C. Biedenharn, "Beyond the Enveloping Algebra of $\mathfrak{sl}_3$", preprint, 1982.

[7] L.C. Biedenharn and D.E. Flath, "On the Structure of Tensor Operators in SU3", preprint, 1983.

[8] W.G. McKay and J. Patera, "Tables of Dimensions, Indices and Branching Rules for Representations of Simple Lie Algebras", M. Dekker, New York (1981).

[9] T.J. Enright, R. Howe, and N.R. Wallach, "A Classification of Unitary Highest Weight Modules", preprint, 1981.

[10] L.C. Biedenharn, A. Giovannini, and J.D. Louck, "Canonical Definition of Wigner Coefficients in U_n", J. Math. Phys. 8, 691-700 (1967).

[11] L.C. Biedenharn, R.A. Gustafson, M.A. Lohe, J.D. Louck, and S.C. Milne, "Special Functions and Group Theory in Theoretical Physics", (A joint report based on three invited papers presented at the Mathematisches Forschungsinstitut Oberwolfach, 13-19 March 1983), to be published by Reidel (Dordrecht).

A GROUP-THEORETICAL CRITERION FOR AN EINSTEIN-PODOLSKY-ROSEN STATE

F. Herbut and M. Vujičić
Department of Physics, Faculty of Science, University of Belgrade, 11001 BEOGRAD, P.O.B. 550, Yugoslavia

In Bohm's well-known [1] total-spin-zero state vector of two distant particles

$$\Psi_{12} \equiv |\vec{s}_1+\vec{s}_2=0\rangle = 2^{-1/2}(|s_z=1/2\rangle_1|s_z=-1/2\rangle_2-|s_z=-1/2\rangle_1|s_z=1/2\rangle_2)$$

one has the <u>reduced statistical operators</u> (RSO's)

$$\rho_1 \equiv Tr_2|\Psi_{12}\rangle\langle\Psi_{12}| = (1/2)(|s_z=1/2\rangle_1\langle s_z=1/2|_1+|s_z=-1/2\rangle_1\langle s_z=-1/2|_1)$$
$$= 1/2 ,$$

and symmetrically $\rho_2 \equiv Tr_1|\Psi_{12}\rangle\langle\Psi_{12}|$. The antiunitary <u>correlation operator</u> [2] U_a is determined as

$$|s_z=-1/2\rangle_2 = U_a|s_z=1/2\rangle_1, \quad |s_z=1/2\rangle_2 = U_a|s_z=-1/2\rangle_1 .$$

Whereas U_a is uniquely defined by Ψ_{12}, the latter decomposes nonuniquely, e.g. in the basis $\{|s_x=1/2\rangle_1, |s_x=-1/2\rangle_1\}$ it has the same form as in the above basis $\{|s_z=1/2\rangle_1, |s_z=-1/2\rangle_1\}$.

The single-particle spin-projection operators satisfy

$$s_{z,1}\Psi_{12} = -s_{z,2}\Psi_{12} , \qquad s_{x,1}\Psi_{12} = -s_{x,2}\Psi_{12} .$$

Hence, $\{s_{z,1},-s_{z,2}\}$ and $\{s_{x,1},-s_{x,2}\}$ are pairs of <u>twin observables</u>.[2] Further, they are <u>incompatible on</u> Ψ_{12} :

$$[s_{z,1},s_{x,1}]\Psi_{12} \neq 0 ,$$
$$[s_{z,2},s_{x,2}]\Psi_{12} \neq 0 ,$$

making Ψ_{12} an <u>Einstein-Podolsky-Rosen</u> (EPR) <u>state</u>.

<u>In general</u>, any state vector Φ_{12} describing two distant particles, for an arbitrary spectral decomposition of its first-particle RSO into ray projectors

$$\rho_1 = \sum_m r_m|\varphi_m\rangle_1\langle\varphi_m|_1 ,$$

has its <u>Schmidt canonical form</u>

$$\Phi_{12} = \sum_m r_m^{1/2} \varphi_m \otimes (U_a \varphi_m) ,$$

and

$$\rho_2 = \sum_m r_m U_a |\varphi_m\rangle\langle\varphi_m| U_a^\dagger$$

is the corresponding spectral form of ρ_2 .

The equation

$$A_1 \Phi_{12} = A_2 \Phi_{12} ,$$

or equivalently the equations

$$[A_1, \rho_1] = 0 , \qquad A_2 Q_2 = U_a A_1 U_a^{-1} Q_2 \tag{1a,b}$$

(Q_2 being the range projector of ρ_2), define pairs of single-particle twin observables $\{A_1, A_2\}$. Their respective predictive measurements give <u>equal results</u> (prediction and change of state) on Φ_{12}. Moreover, the direct measurement of A_1 is ipso facto the <u>distant measurement</u> [2] of A_2 (and vice versa).

If there exist two incompatible pairs of twin observables $\{A_1, A_2\}$, $\{B_1, B_2\}$ on Φ_{12}, i.e. such that

$$[A_1, B_1] \Phi_{12} \neq 0 ,$$

or equivalently

$$[A_1 Q_1, B_1 Q_1] \neq 0 \tag{2}$$

(Q_1 being the range projector of ρ_1), then either of the incompatible observables A_2 and B_2 can be measured distantly, i.e. measured without interaction of either the apparatus or the first particle with the second one. Hence, if there exist twin observables A_1 and B_1 satisfying (1a) and (2), then Φ_{12} is an <u>EPR state</u>.

If one takes the detectable parts $A_1 Q_1$ of all observables A_1 compatible with ρ_1, i.e. satisfying (1a), then one has a <u>Lie algebra</u> L_1^c with "i/ħ times the commutator" as the Lie product. (For Bohm's state Ψ_{12} L_1^c contains all observables.) Thus, Φ_{12} is an EPR state <u>if and only if</u> its L_1^c is <u>non-Abelian</u>.

Approaching the same problem via symmetry transformations, one can say that for a given state Φ_{12} two unitary operators U_1 and U_2 are <u>twins</u> if

$$U_1 U_2 \Phi_{12} = \Phi_{12} ,$$

or equivalently if

$$[U_1, \rho_1] = 0 \;, \qquad U_2 Q_2 = U_a U_1 U_a^{-1} Q_2 \;.$$

If one takes all unitary operators U_1^c commuting with ρ_1 that are canonical (regarding Φ_{12}), i.e. of the form

$$U_1^c = U_1^c Q_1 + (1 - Q_1) \;,$$

then one obtains a <u>Lie group</u> G_1^c. (For Bohm's Ψ_{12} $G_1^c = U(2)$.) The equation

$$U_1^c = \exp[iA_1 Q_1]$$

makes it clear that L_1^c is the Lie algebra of the Lie group G_1^c. Hence, one can also say that Φ_{12} is an EPR state <u>if and only if</u> G_1^c is <u>non-Abelian</u>.

[1] D. Bohm, Quantum Theory (New York, Prentice-Hall, 1951).

[2] F. Herbut and M. Vujičić, Ann. Phys. <u>96</u> (1976) 382 ,
M. Vujičić and F. Herbut, A Quantum Mechanical Theory of Distant Correlations, 1983, preprint.

GROUP THEORY ALGEBRAS AND BOSONIZATION

Andrés J. Kálnay and Ricardo A. Tello-Llanos

Centro de Física, IVIC, Apdo. 1827,
Caracas 1010A, Venezuela

ABSTRACT

Taking into account the well known relevance of group theory to physics and the relation between groups and algebras we discuss the application of the method of bosonization to a wide class of algebras as well as some physical implications. Seemingly different physical and mathematical structures are unified within the bosonization scheme.

1. MATHEMATICAL FORMALISM

Let us denote bosonization of a physical or mathematical formalism a realization of it in terms of Bose creation and annihilation operators. There are several bosonizations, e.g. that by Okubo[1] for the Fermi algebra. Much of the work on bosonization was done for group theory: to quote a few examples we mention some of the papers by Baird, Bargmann, Biedenharn, Doebner, Hassan, Melsheimer, Moshinsky, Salam and Strathdee.[2-6] We shall concern ourselves with the bosonization of algebras, research of obvious relevance for group theory. This bosonization is that shown in Ref. 7, some examples being advanced in Ref. 8. This bosonization is the natural extension to more general associative algebras of the bosonization of Fermi algebra discussed in Ref. 9. We must state that this bosonization has roots on the work of several researchers on bosonizations of Fermi, angular momentum and other algebras: see for example the papers by Jordan, Kademova and Schwinger.[10-12] Our research concerns general finite associative algebras, over the real or the complex fields, in the hope that, under suitable restrictions the essential results could have generalization for infinite dimensional algebras. Let $G_1, G_2, \ldots G_L$, be the generators of an algebra of dimension N, the algebra being specified by M algebraic equations

$$P_a(G_1, G_2, \ldots, G_L) = 0 \qquad a=1,2,\ldots,M, \tag{1.1}$$

Associate G_j of the original algebra with the Bose constructed operator

$$G_j(b,b^+) \equiv \sum_{r,s} M_{jrs} b_r^+ b_s, \quad j=1,2,\ldots,L, \tag{1.2}$$

where the M_{jrs} are complex numbers. Then we have proven:[7]

Theorem 1.1. A unique (up to equivalence) set of coefficients M_{irs} exists, such that the relations that follow are true,

$$P_a\left[G_1(b,b^+),\ G_2(b,b^+), \ldots,\ G_L(b,b^+)\right] \mathcal{B}_1 = 0 \tag{1.3}$$

where $\mathcal{B}_1$ is the one-boson subspace of the Bose state vector space $\mathcal{B}$.

Theorem 1.2. When having $\mathcal{B}_1$ as domain, the operators (1.2) are the generators of an associative finite dimensional operator algebra $A^{\mathcal{B}_1}$ which puts forward a faithful representation of the algebra A.

The algebra $A^{\mathcal{B}_1}$ is the bosonization of the original algebra A.

Let us now call $\mathcal{B}_p$ that subspace of $\mathcal{B}$ with p and only p bosons. It is obvious that the operators (1.2) transform $\mathcal{B}_p$ into a subspace of itself (eventually equal to $\mathcal{B}_p$). Let us call $A^{\mathcal{B}_p}$ the algebra generated by the operators (1.2) but now having $\mathcal{B}_p$ as domain. When A is the Fermi algebra, then we have shown that $A^{\mathcal{B}_p}$ is an irreducible representation of the parafermi algebra of order p of parastatistics.[11,9] Therefore, we shall call B-para-algebra of order p the abstract algebra isomorphic to $A^{B}P$. Only in exceptional cases a B-para-algebra is isomorphic to A. Recalling that in standard algebraic usage the word "non" means "not necessarily" ("not" being reserved for an absolute negation), we can deduce a result that extends one of Ref. 7:

Theorem 1.3. Given two B-para-algebras or different orders, both constructed starting from the same algebra A, then these B-para-algebras are non isomorphic among themselves.

Proof: Follows from eq. (1.2) and standard properties of the Bose algebra.

This result is the generalization to general finite dimensional algebras of the well known fact that parafermi algebras of different order of parastatistics are not isomorphic. In particular, they are not isomorphic to the Fermi algebra, the same as a B-para-algebra $A^{B}P$ of order greater than one is non isomorphic to A. Each of the inequivalent $A^{B}P$ is characterized by a set of identities other than (1.1), the same as the parafermi algebras of different order of parastatistics are characterized by inequivalent identities. An important consequence is that the bosonization correlates inequivalent algebras: all the $A^{B}P$ obtained starting from the same A may look (as abstract algebras) as non correlated algebraic structures, but the bosonization shows in a natural way that they belong to a single family of algebras. (The same as all parafermi algebras of different order of parastatistics are known to be correlated by the fact that they belong to the family of the general parafermi algebras.)

The inequivalent parafermi algebras (of different order p or parastatistics) are known to be characterized by relations of the form

$$f_i f_j^{+} |0\rangle = p\delta_{ij} |0\rangle \qquad (1.4)$$

(Greenberg-Messiah Theorem).[13] The same, the $A^{B}P$ obey double commutation relations that generalize parafermi double commutation relations, and, under certain restrictions the $A^{B}P$ satisfy a natural generalization of eq. (1.4).[14] Surprisingly, the hint for an easy discovery of the restrictions is obtained when one recognizes that the underlying c-number formalism is Nambu's generalized mechanics.[15] To be more specific what is needed is the generalization for Nambu's mechanics of the notion of canonical momenta.[16]

Finally, let us mention that the B-para-algebras obtained from Grassmann algebras are the so called paragrassmann algebras.[17]

2. PHYSICS

Since bosonization associates in a natural family a set of algebras (Sec. 1) it will also associate in a natural family a set of physical structures when there are algebraic structures. An example previously discussed is that of the parafermi systems. From the work by Schwinger,[12] reworded in our language, it results that the B-para-algebra of order p obtained from the bosonization of spin 1/2 algebra is the algebra of spin p/2: an obviously natural family is again obtained. Moreover, the bosonization of spin 1/2 algebra has Nambu's mechanics as the c-number underlying formalism,[18] which shows a remarkable selfconsistency of the theory. In fact, the angular momentum of a classical rotator is the leading example for Nambu's equations of motion.[15] Finally, it was shown in Ref. 8 that the bosonization of Dirac's wave equation leads to a unique equation of motion for all free fields (irreducible representations of Poincaré group).

Therefore, bosonization has an unifying power, in addition to its pragmatic interest as a computational tool and its conceptual interest for the search of the elementary quantum statistics.[19]

REFERENCES

1. S. Okubo, Phys. Rev. C 10, 2048 (1974).
2. C. E. Baird and L. C. Biedenharn, J. Math. Phys. 4, 1449 (1963).
3. V. Bargmann and M. Moshinsky, Nucl. Phys. 18, 697 (1960).
4. H. D. Doebner and O. Melsheimer, J. Math. Phys. 9, 1638 (1968).
5. M. Hage Hassan, J. Phys. A 12, 1633 (1979).
6. Abdus Salam and J. Strathdee, Phys. Rev. 148, 1352 (1966).
7. C. A. González-Bernardo, A. J. Kálnay and R. A. Tello-Llanos, Bose Realizations of Finite Associative Algebras and Physical Applications (to be published).
8. C. A. González-Bernardo, A. J. Kálnay and R. A. Tello-Llanos, Lett. N. Cimento 33, 74 (1982).
9. A. J. Kálnay, Prog. Theor. Phys. 54, 1848 (1975).
10. P. Jordan, Zeitschrift fuer Physik, 94, 531 (1935).
11. K. Kademova, Int. J. Theor. Phys. 3, 109 (1970).
12. J. Schwinger, On Angular Momentum, p. 229 in: L. C. Biedenharn and H. van Dam, Quantum Theory of Angular Momentum, Academic Press, (N. York, 1965).
13. O. W. Greenberg and A. M. L. Messiah, Phys. Rev. 138, B 1155 (1965).
14. C. A. González-Bernardo, A. J. Kálnay and R. A. Tello-Llanos (work in progress).
15. Y. Nambu, Phys. Rev. D 7, 2405 (1973).
16. A. J. Kálnay and R. Tascón, Int. J. Theor. Phys. 16, 635 (1977).
17. A. J. Kálnay, Rep. Math. Phys. 9, 9 (1976).
18. A. J. Kálnay and R. A. Tello-Llanos (work in progress).
19. C. A. González-Bernardo and A. J. Kálnay, Search of Elementary Quantum Statistics, Int. J. Theor. Phys. (in press).

SO(3) COMMUTATORS FOR ANGULAR MOMENTUM AND ROTATION OBSERVABLES

J. Krause

Universidad Católica de Chile, Casilla 114-D, Santiago, Chile.

The modern trend in elementary particles raises the question of extending the canonical commutators to non-Abelian dynamical variables. In this work we exemplify the issue, since we quantize SO(3) by means of a geometric approach which stems from three essential features. 1) We use (in a systematic way) the group law of multiplication of the parameters of ordinary rotations and the affine structure induced by this law on the group manifold. 2) We adopt the 3-dimensional sphere S_3 (embedded in E_4) as a representative of the group manifold. 3) We define position operators on S_3 and study their quantum kinematics.

To this end, let us label the elements of SO(3) in the $(\phi,\hat{n})$ parametrization [1], and define the following embedding:

$$q^0 = \cos\phi,\ q^1 = \sin\phi\,\sin\theta\,\cos\rho,\ q^2 = \sin\phi\,\sin\theta\,\sin\rho,\ q^3 = \sin\phi\,\cos\theta,$$

where, clearly, $\hat{n} = (\theta,\rho)$ and $0 \leq \phi \leq \pi$, $0 \leq \theta \leq \pi$, $0 \leq \rho \leq 2\pi$. Then the group multiplication rule for finite rotations becomes:

$$q''^{\mu} = \sigma^{\mu}_{\nu\lambda} q'^{\nu} q^{\lambda},\quad (\mu,\nu,\lambda,\ldots = 0,1,2,3),$$

with

$$\sigma^0_{00} = 1,\quad \sigma^0_{0j} = \sigma^0_{j0} = 0,\quad \sigma^0_{jk} = -\delta_{jk},$$

$$\sigma^i_{00} = 0,\quad \sigma^i_{0j} = \sigma^i_{j0} = \delta^i_j,\quad \sigma^i_{jk} = e_{ijk},\quad (i,j,k,\ldots = 1,2,3).$$

In this manner, the Hurwitz measure comes out in the form:

$$d\tau(q) = u_0 \sin^2\phi\,\sin\theta\,d\phi\,d\theta\,d\rho,$$

as expected.

Next, let us consider an irreducible linear vector representation of SO(3) on a Hilbert space $\mathcal{H}$ of some thought out physical system:

$q \to U(q)$. Then, for any chosen fixed vector $|e\rangle = |1,\vec{0}\rangle \in \mathcal{H}$, we define the vectors $|q\rangle = U(q)|e\rangle$. These vectors behave transitively on $\mathcal{H}$ upon the action of the (irreducible) unitary group $U\{SO(3)\}$, and they provide us with a complete orthogonal basis on $\mathcal{H}$. So we define the following position operators on S_3:

$$\cos\Phi = \int_{S_3} d\tau(q)\,|q\rangle \cos\phi \langle q| \ ,$$

$$N_j \sin\Phi = \int_{S_3} d\tau(q)\,|q\rangle n_j \sin\phi \langle q| \ .$$

Then $U(e + \delta q) = I - 2i\delta\phi\hat{n}\cdot\vec{J}$ (the factor 2 comes out automatically, and $\vec{J}$ is the angular momentum, i.e., $[J_j, J_k] = i\epsilon_{jkl}J_l$, $\hbar = 1$), and thus we obtain the following angular momentum-angle commutators:

$$[\cos 2\Phi, J_j] = -iN_j \sin 2\Phi \ ,$$

$$[\sin 2\Phi, J_j] = \ iN_j \cos 2\Phi \ .$$

Also, for the axis-of-rotation observable, we get:

$$N_j = J_j - e^{2i\Phi} J_j e^{-2i\Phi} \ ,$$

from where the commutators $[N_j, J_k]$ immediately follow. Of course, the spectrum of 2Φ is now: $0 \leq 2\phi \leq 2\pi$, as it must be. The elementary commutators $[\cos\Phi, J_3] = i\sin\Phi$, and $[\sin\Phi, J_3] = -i\cos\Phi$, for the group $O(2)$ (i.e., when $\hat{n} = \hat{k}$ remains frozen and $0 \leq \phi \leq 2\pi$) are well known in the literature since long time. Our results for SO(3) are a direct generalization thereof and may be of interest for studying minimum uncertainty rotation-states and coherent isospin-states [2]. Details will be published elsewhere [3].

This work was supported by DIUC, through Grant N° 46/83.

References

1) See, for instance, L.C. Biedenharn and J.D. Louck, Angular Momentum in Quantum Physics (Addison-Wesley, Reading, Mass., 1981).
2) K. Yamada, Phys. Rev. D 25, 3256 (1982).
3) J. Krause, "SU(2) quantum kinematics", preprint, PUCCH, 1983.

INTEGRALS OF MOTION OF NONSTATIONARY QUANTUM SYSTEMS.

V.I. Man'ko

P.N. Lebedev Institute of Physics, Moscow, USSR.

The aim of this article is to review the problem of quantum integrals of motion for the nonstationary systems with n degrees of freedom accooding to the result of [1],[2]. There are two sorts of quantum integrals of motion the operators and the quantities. We will discuss both. These last integrals of motion are closely connected with integral invariants by Poincaré of classical mechanics and have been found in ref.[2]. The operator integrals of motion are now of special interest because namely they are planned to be used in the experiments on gravitational wave detection in order to avoid the quantum restrictions of sensitivity of gravitational wave detector [3-5].

Let a quantum system be described by a time-dependent Hamiltonian $\hat{H}(t)$. In Schrödinger picture we will say that an operator $\hat{I}(t)$ is the quantum integral of motion if

$$\frac{d}{dt} \operatorname{Tr} \hat{\rho}(t) \hat{I}(t) = 0 , \tag{1}$$

where $\hat{\rho}(t)$ is the density matrix of the system which can be expressed in the pure states $|\psi(t)\rangle$ of the system as $\hat{\rho}(t)=|\psi(t)\rangle\langle\psi(t)|$. According to ref.[1] there exist 2n independent integrals of motion of the system with n degrees of freedom which can be considered as the initial coordinates operators $\underset{\sim}{x}_0, \underset{\sim}{p}_0$ in the phase space of the system. Any other invariant can be expressed as a function of these 2n invariants. These integrals of motion have the following properties [1]:

a) Any function of integrals of motion is the invariant too.

b) If we have a solution $|\psi\rangle$ to Schrödinger equation and act on it by an integral of motion $\hat{I}(t)$ we obtain a new state vector $|\varphi\rangle=\hat{I}(t)|\psi\rangle$ which also satisfies the Schrödinger equation.

c) Any integral of motion $\hat{I}(t)$ can be expressed in terms of evolution operator $\hat{U}(t)$ of the system $(|\psi(t)\rangle=\hat{U}(t)|\psi(0)\rangle)$ in the form

$$\hat{I}(t) = \hat{U}(t)\, \hat{I}(0)\, \hat{U}^{-1}(t) . \tag{2}$$

d) If we have the observaible $\hat{I}_H(t)$ in Heisenberg picture such that $\hat{I}_H(0)=\hat{I}(0)$ then integral of the motion $\hat{I}(t)$ is connected with this observaible by the formula [1]

$$\hat{I}(t) = \hat{U}^{2}(t)\, \hat{I}_H(t)\, U^{-2}(t) . \tag{3}$$

From the property a) it follows that for any integral of motion $\hat{I}(t)$ all the momenta characterising the distribution function of this observaible (like the mean value $\langle\hat{I}(t)\rangle$, the variance $\sigma_I = \langle\hat{I}^2(t)\rangle - \langle\hat{I}(t)\rangle^2$, etc) are conserved numbers i.e. the distribution function for conserved variable $\hat{I}(t)$ does not change in time.

It is interesting also that the quantity of independent integrals of motion of the system does not dependent on the symmetry of its potential well. So, for the 3-dimensional harmonic oscillators with Hamiltonians

$$\hat{H}_1 = \hat{\underset{\sim}{p}}^2/2m + m\omega^2\hat{\underset{\sim}{r}}^2/2 , \tag{4}$$

$$\hat{H}_2 = \hat{\underset{\sim}{p}}^2/2m + m\Omega^2(\hat{x}^2+\hat{y}^2)/2 + m\omega^2\hat{z}^2/2 , \tag{5}$$

$$\hat{H}_3 = \hat{\underset{\sim}{p}}^2/2m + m(\omega_1^2\hat{x}^2 + \omega_2^2\hat{y}^2 + \omega_3^2\hat{z}^2)/2 , \tag{6}$$

this quantity is equal to 6 in all cases though in the first case we have the spherical symmetry and angular momentum conservation ($\hat{L}_x, \hat{L}_y, \hat{L}_z$ conserve), in the second case we have the axial symmetry only and the angular momentum projection conservation ($\hat{L}_z$ conserves), in the third case we have no symmetry and no one angular momentum projection conserves. The explanation of this mystery is contained in the fact that instead of disappearing integrals of motion which are time-independent there arise new invariants which obligely are time-dependent ones. It is interesting to note that if for discussed harmonic oscillators all these 6 independent invariants are found explicitely (see [1]) for the hydrogen atom one knows in explicit form only 5 independent invariants.

If we will define the symmetry of Schrödinger equation according to the definition of symmetry of any equation given in reference [6] as the set of the operators which acting an any solution to Schrödinger equation create a new solution, than we can conclude that the set of integrals of motion for any quantum system coincides with the set of the symmetry operators of the same system. The symmetry of quantum system is not obligely (according to such a definition) the group. It can be Lie algebra, superalgebra or something else. Among all the operators of symmetry of the system one can find, of course, the operators which belong to a group. But this is only a small part of symmetry operators of the quantum system. So, it is worthy, may be, to distinguish two notions - the symmetry of the quantum system and the symmetry group of the quantum system . Integrals of motion of the quantum system are simultaneously the symmetry operators of the system.

As an example of nonstationary system we will demonstrate 2 independent integrals of motion found in ref. [7] for one-dimensional forced harmonic oscillator with time-dependent frequancy described by the Hamiltonian

$$\hat{H} = \hat{p}^2/2 + \omega^2(t)\hat{x}^2/2 - f(t)\hat{x}, \quad \hbar = m = 1 . \tag{7}$$

Introducing the compex function $\varepsilon(t)$ which satisfies the equation $\ddot{\varepsilon} + \omega^2\varepsilon = 0$, and the function $\delta(t) = -i2^{-1/2}\int_0^t f(\tau)\varepsilon(\tau)d\tau$ one can find the nonhermitian invariant.

$$\hat{a} = i2^{1/2}[\varepsilon(t)\hat{p} - \dot{\varepsilon}(t)\hat{x}] + \delta(t) . \tag{8}$$

The hermitian and antihermitian parts of this invariant are just 2 independent integrals of motion whose mean values determine the initial coordinates of the oscillator in its phase space. Measuring the invariant $\hat{a}^+(t)\hat{a}(t)$ in the experiments with parametric gravitational wave detector which is described by the Hamiltonian (7) gives the possibility to improve essentially the sensitivity of the detector [5]. Let us discuss now the invariants of the quantum systems which are universal ones and do not depend on the Hamiltonians like the volume in the phase space of classical system does not change in time for any Hamiltonian system being one of the integral Poincaré invariants. Following to ref. [2] we consider N operators $\hat{Z}_1\hat{Z}_2\ldots\hat{Z}_n$ which obey the commutation relations of a Lie algebra

$$[\hat{Z}_\alpha, \hat{Z}_\beta] = i\hbar c^\gamma_{\alpha\beta}\hat{Z}_\gamma ; \quad \alpha, \beta = 1, 2, \ldots, N, \tag{9}$$

and the system with the Hamiltonian

$$\hat{H}(t) = b^\gamma(t)\hat{Z}_\gamma \tag{10}$$

which is arbitrary linear form of this Lie algebra generators with the time dependent coefficients. The Heisenberg operators $\hat{Z}_\alpha(t)$ can be found in the form

$$Z_\alpha(t) = \Lambda_{\alpha\beta}(t)\, Z_\beta , \tag{11}$$

where the matrix $\Lambda(t)$ satisfies the equation

$$\dot{\Lambda}_{\alpha\beta} = c^\delta_{\alpha\gamma}\, b^\gamma(t)\Lambda_{\delta\beta} , \tag{12}$$

and the relation

$$\Lambda g \tilde{\Lambda} = g , \qquad g_{\alpha\beta} = c^\rho_{\alpha\delta}\, c^\delta_{\beta\rho} \tag{13}$$

$\tilde{\Lambda}$ means transposed matrix Λ. Let us introduce the matrix Z

$$Z_{\alpha\beta} = \frac{1}{2} < \hat{z}_\alpha(t)\hat{z}_\beta(t) + \hat{z}_\beta(t)\hat{z}_\alpha(t) > - < \hat{z}_\alpha(t) > < \hat{z}_\beta(t) > . \tag{14}$$

Then it can be cheked that if $\det g \neq 0$ the following determinant (μ is a complex variable) does not depend on time

$$\det[Z(t) - \mu g] = \sum_{m=0}^{N} D_m \mu^m . \tag{15}$$

This determinant is a polinomial in respect to variable μ . So, the coefficients of this polinomial $\mathcal{D}_m$ are the integrals of motion of the system with the Hamiltonian (10). These invariants $\mathcal{D}_m$ are c-numbers. For the quantum systems with the hamiltonians which are arbitrary nonstationary quadratic forms in respect to coordinates and momenta the analogous invariants can be constructed. These invariants turned out to be for such systems the functions of all corresponding integral Poincaré invariants of classical systems with the same Hamiltonians. So, for such a one-dimensional quadratic system this invariant Δ is equal to

$$\Delta = \sigma_p \sigma_q - \sigma_{pq}^2 = inv , \tag{16}$$

where

$$\sigma_p = \langle \hat{p}^2 \rangle - \langle p \rangle^2 , \quad \sigma_q = \langle \hat{q}^2 \rangle - \langle \hat{q} \rangle^2 , \quad \sigma_{pq} = \frac{1}{2} \langle \hat{p}\hat{q} + \hat{q}\hat{p} \rangle - \langle \hat{p} \rangle \langle \hat{q} \rangle . \tag{17}$$

This invariant $\Delta^{1/2}$ is equal to the volume in the phase space of the system if we describe the states of the system in terms of Wigner function.

REFERENCES

1. I.A. Malkin and V.I. Man'ko, "Dynamical symmetric and coherent states of quantum systems", Moscow, publisher "Nauka", (1979).
2. V.V. Dodonov and V.I. Man'ko , "Group Theoretical Methods in Physics", Proceedings of the International Seminar, Zvenigorod, 24-26 November 1982, edited by M.A. Markov, publisher Nauka, Vol. 2, pp. 11-34, (1983).
3. V.B. Braginsky, Y.I. Vorontsov and K.S. Thorne, Science 209, p. 547,(1980).
4. C.M. Caves, K.S. Thorne and R.W.P. Drever , Rev. Mod.Phys. 52 ,341,(1980).
5. V.V. Dodonov, V.I. Man'ko and V.N. Rudenko, P.N. lebedev Institute of Physics Reporters, Trudy, publisher Nauka, Moscow, Vol. 152, p.12,(1983).
6. I.A. Malkin and V.I. Man'ko, ZETP Letters, 2, 230 (1965).
7. I.A. Malkin and V.I. Man'ko, Physics Letters, 32A , 243, (1970).

GEOMETRIC PROPERTIES OF THE LOWEST ENERGY STATE FOR A POLYNOMIAL HAMILTONIAN *

W. Schweizer and P. Kramer
Institut für Theoretische Physik
der Universität Tübingen
Auf der Morgenstelle 14
D - 7400 Tübingen

Consider a Hamiltonian H which is an element of the enveloping algebra of a Lie algebra L. For simplicity we choose L as SU(2) or SU(1,1) and construct a polynomial Hamiltonian H which is at most quadratic in the basis of the Lie algebra. Without loss of generality we choose the quadratic part entirely symmetric. This allows us to represent the Hamiltonian H by irreducible tensor operators J^k_q of maximum rank two[1], where k denotes the rank and q the components.

(1) $$H = \sum_{k=0}^{2} \sum_{q=-k}^{+k} \alpha_{kq} (-1)^q J^k_{-q} \qquad \text{with}$$

$$\overline{\alpha_{kq}} = (-1)^q \alpha_{k-q} \qquad J^{k+}_{-q} = (-1)^q J^k_q \qquad \text{for SU(2) and}$$

$$\overline{\alpha_{kq}} = \alpha_{k-q} \qquad J^{k+}_{-q} = J^k_q \qquad \text{for SU(1,1) .}$$

As trial states we choose the coherent state $|z\rangle$ of the group SU(2) respectively SU(1,1)[2] and define for an arbitrary operator A

(2) $$A(z',\bar{z}) := \langle z'|A|z\rangle / \langle z'|z\rangle = \langle A\rangle \ .$$

The tensor operators J^k_q are composed of the basis operators J_i of the Lie algebra by $[J^{k-1}\, J^1]^k_q$ and therefore J^2_q by $\langle J_i \cdot J_j\rangle$. To get J^k_q by the basis functions J_i of the Lie algebra we need the connection between the product of the expectation values and the expectation value of the operator products. Therefore, we define the irreducible tensors $(\cdot J^k_q)$ by $(\cdot J^k_q) = [J^{k-1} J^1]^k_q$ and get

(3) Proposition[3]: ∀ irreducible tensor operators J^k_r $k>0$:

$$J^k_r = (2j)^{-k} (\cdot J^k_r) \cdot \prod_{l=0}^{k-1} (2j \pm l) \qquad (\text{- for SU(2), + for SU(1,1)}).$$

With this proposition we get $H = H(J_1 J_2 J_3)$.

Now consider a space with axes labelled by J_1, J_2, J_3. Then H may be interpreted as a two dimensional surface in this 3-dimensional space. Because of $J(z,\bar{z})^2 = \sum_i g_{ii} J_i(z,\bar{z})^2$

*Work supported by the Deutsche Forschungsgemeinschaft

$= j^2$ = const. (g_{ij} the metric tensor of the Lie algebra) $J_q(z,\bar{z})$ is the restriction of the coordinate functions to the sphere of radius j, respectively to the hyperboloid with main axes j. The phase spaces are hypersurfaces given by constant values of the Casimir observables J^2, on the dual space L^* to the Lie algebra, hence L^* will be Euklidean for SU(2) and Minkowskean for SU(1,1). The generalized Poisson bracket $\{\cdot,\cdot\}$ [2] is defined by

(4) $$\{F,G\} = -\frac{(1\pm z\bar{z})^2}{2j}\left(\frac{\partial F}{\partial z}\frac{\partial G}{\partial \bar{z}} - \frac{\partial F}{\partial \bar{z}}\frac{\partial G}{\partial z}\right)$$ (+ for SU(2), - for SU(1,1)

for arbitrary functions $F(z,\bar{z})$ and $G(z,\bar{z})$.

(5) Proposition[3]: $\forall$ coherent states $|z\rangle$, A arbitrary operator, J_q basis operators of the Lie algebra:

$$\{J_q,A\}(z,\bar{z}) = \langle z|[J_q,A]|z\rangle/\langle z|z\rangle \ .$$

Note that this would not be true for arbitrary operators $\{B,A\}$.

Now we are able to look for those states $|sc\rangle := |z\rangle/\langle z|z\rangle^{1/2}$ which solve the variational equation $\delta H(z,\bar{z}) = 0$.

(6) Proposition[3]: $\lim_{z'\to z} H(z',\bar{z})$ = Extremum $\Longleftrightarrow \langle z|[J_q,H]|z\rangle = 0$

$\forall\ q\epsilon\{1,2,3\}$; and with (5) $\{J_q,H\}(z,\bar{z}) = 0$.

To give a geometrical picture for the extremal condition above, we define a selfconsistent linear Hamiltonian H_{sc} which generalizes the well-known Hartree-Fock Hamiltonian by

(7) $\langle sc|H_{sc}|sc\rangle = \langle sc|H|sc\rangle$ and $\langle sc|[J_q,H_{sc}]|sc\rangle = \langle sc|[J_q,H]|sc\rangle$

$\forall\ q\epsilon\{1,2,3\}$ and this is a nonlinear equation like the Hartree-Fock equations. The classical equations corresponding to this quantum mechanical conditions are

(8) $H_{sc}(z,\bar{z}) = H(z,\bar{z})$ and $\{J_q,H_{sc}\}(z,\bar{z}) = \{J_q,H\}(z,\bar{z})$.

Note that these two equations have only a local validity.

Because H_{sc} is a linear operator, H_{sc} is linear too, and therefore a plane in the space introduced above. From (6) and (8) we get

(9) $\{J_q,H_{sc}\} = 0$,

which leads to the geometrical interpretation: The extremal state $|sc\rangle$ corresponds to the point of contact of the energy surface with the sphere, respectively the hyperboloid J^2, while the linearized Hamiltonian H_{sc} corresponds to the plane tangent to the energy surface at this point of contact.

This interpretation enables us to construct an iteration procedure: Draw the energy surface and the J^2-surface cutting each other. Now, in the first iteration procedure, construct the tangential plane to the energy surface at an arbitrary point of intersection. By parallel shifting of this plane, construct a tangential plane to the J^2-surface. Alter[3] the energy surface till it intersects the J^2-surface at this point and construct again the plane tangent to the energy surface at this point. Repeat the whole procedure till the point of contact J^2-surface - energy surface is constructed. The second iteration procedure differs from that one above in starting with a plane tangent to the J^2-surface. Because only the energy can be varied this has some geometrical consequences. But under consideration of those, the whole procedure is similar to the first one. For calculations, the two graphical iteration procedures can be transcribed into numerical ones which are tested at some examples in[3].

References:

1) Edmonds A.R. 1957 Angular Momentum in Quantum Mechanics, Princeton University Press, Princeton.
2) Kramer p. and Saraceno M. 1981 Lecture Notes in Physics 140, Springer Verlag, Berlin.
3) Schweizer W. and Kramer P. 1983 submitted for publication

GROUPES DIFFERENTIELS ET PHYSIQUE MATHEMATIQUE

Jean-Marie Souriau

Beaucoup de théories physiques font jouer un rôle essentiel à un certain groupe (le "groupe des symétries" de la théorie).

Très souvent, il s'agit d'un groupe de Lie; mais il y a d'autres exemples importants où interviennent des groupes de dimension infinie:

- les principes de la mécanique classique - et ceux de la mécanique quantique dans la formulation de Dirac - utilisnt la symétrie par le groupe des "transformations canoniques" (difféomorphismes symplectiques);
- la théorie des particules élémentaires est aujourd'hui fondée sur les "groupes de jauge" ou "groupes de courants" (ensembles des applications différentiables d'une variété dans un groupe de Lie);
- la théorie de la gravitation (relativité générale) est une "théorie de jauge" d'un type particulier, construite sur le groupe des difféomorphismes de la variété espace-temps;
- on rencontre aussi des associations (produits semi-directs) de groupes de jauge et de groupes de difféomorphismes: dans l'électrodynamique relativiste, dans les théories de type Kaluza-Klein;
- la physique des solides (dans le cas des structures incommensurables) fait intervenir d'autres groupes qui ne sont plus de dimension infinie, mais qu'on considère généralement comme pathologiques (des quotients d'un groupe de Lie par un sous-groupe non fermé).

Rappelons d'autre part, pour mémoire, les principales structures mathématiques associées aux groupes de Lie qui interviennent en physique:

- les espaces vectoriels tangent et cotangent - munis des représentations adjointe et coadjointe;
- la 3-forme de structure, qui confère à l'espace tangent sa structure d'algèbre de Lie et à l'espace cotangent sa structure de Poisson;
- l'application exponentielle;
- les structures homologiques, topologiques et homotopiques; en particulier l'existence (pour tout groupe de Lie connexe) d'un revêtement simplement connexe, possédant des propriétés universelles, joue un rôle fondamental dans plusieurs branches de la physique mathématique;
- enfin l'étude des représentations unitaires des groupes de Lie (analyse harmonique non commutative) constitue un chapitre essentiel des mathématiques comme de la physique théorique.

Ce double inventaire suggère la question suivante: est-il possible d'étendre les propriétés mathématiques "utiles" des groupes de Lie à une catégorie plus vaste - catégorie qui engloberait les divers groupes que rencontre le physicien?

Ce projet peut se réaliser simplement: il suffit de "faire sauter un axiome". Voici comment:

On sait que la définition des groupes de Lie fait intervenir la stucture de groupe, la structure de variété, et un axiome de compatibilité.

Ici, c'est la structure de variété (ou "difféologie") que nous allons élargir, par une axiomatique où ne figure pas l'existence de cartes. Les objets munis d'une telle structure - ou "espaces différentiels" - constituent une catégorie particulièrement stable par rapport aux constructions ensemblistes (sommes, produits, quotients, etc.).

On obtient donc les "groupes différentiels" en remplaçant dans la définition des groupes de Lie la structure de variété par celle d'espace différentiel.

Or cette catégorie beaucoup plus large conserve la plupart des propriétés élémentaires des groupes de Lie.

Pour établir ces résultats, plusieurs changements de point de vue sont nécessaires: par exemple la théorie de l'homotopie se développe sans faire intervenir de topologie; la topologie canonique d'un groupe différentiel G et son espace tangent $\mathfrak{g}$ s'obtiennent à partir de l'analyse harmonique; $\mathfrak{g}$ est un espace vectoriel topologique localement convexe - mais il n'est pas nécessairement un modèle local de G , parce que l'application exponentielle n'est généralement définie que sur une partie étoilée de $\mathfrak{g}$; etc.

Ceci explique pourquoi les groupes différentiels ne peuvent pas s'atteindre en choisissant un espace-type pour les modeler; au contraire, ce sont les groupes différentiels eux-mêmes qui permettent de définir globalement les espaces utiles.

Faute de place, nous ne pouvons pas donner ici un exposé détaillé. On le trouvera dans la référence suivante:

Colloque "Géométrie Symplectique et de Contact, Feuilletages et Quantification Géométrique", Lyon. P. Dazord et N. Desolneux-Moulis éditeurs (1984)

avec le sommaire suivant:

I - Groupes différentiels

- Difféologies. Espaces différentiels, applications différentiables, difféomorphismes.
- Finesse des difféologies. Images d'une difféologie, submersions. Quotients d'un espace différentiel. Exemple du quotient irrationnel du tore. Image réciproque d'une difféologie. Sous-espaces. Sommes et produits d'espaces différentiels.
- Groupes différentiels. Sous-groupes et groupes quotients. Exemples des groupes de difféomorphismes et des groupes de jauge.
- D-morphismes. D-actions. Espaces de Klein et espaces homogènes.
- Homologie des groupes différentiels: D-morphismes stricts, suites D-exactes, lemme des "9".
- Homotopie des groupes différentiels et des espaces homogènes: groupes connexes et simplement connexes; revêtement universel et groupe d'homotopie d'un groupe différentiel (resp. d'un espace différentiel homogène).
- Rayons, étoiles. Exemple des champs de vecteurs. Difféologie forte.
- Etats d'un groupe; subordination, harmonies; topologies harmoniques.
- Topologie canonique d'un groupe différentiel; réduction séparée.
- Espaces tangent et cotangent d'un groupe différentiel. Topologie localement convexe de l'espace tangent. Représentations adjointe et coadjointe.
- Application exponentielle.
- 3-forme de structure d'un groupe différentiel. Algèbre de Lie. Structure

symplectique des orbites coadjointes.

- Spectre associé à un état et un rayon. Relations d'incertitude.
- Groupe "statistique" d'une variété X ; observables; interprétation des lois de probabilités sur X .

II - Quantification géométrique.

- Structure des systèmes dynamiques classiques: espace d'évolution, espace des mouvements.
- Mécanique statistique classique, formulation par le groupe statistique de l'espace des mouvements.
- Structure symplectique et de contact des systèmes dynamiques: 2-forme de Lagrange; difféologie hamiltonienne. 1-forme préquantique. Groupes des symplectomorphismes et des quantomorphismes; groupes dynamiques et quantodynamiques. Groupe quantique.
- Groupes infinitésimalement proches; cas du groupe statistique et du groupe quantique.
- Axiomatique des états quantiques; spectres; formulation hilbertienne; quantification géométrique d'un observable classique par un self-adjoint. Axiome harmonique; états mélangés. Axiome de fonctionnalité. Exemples.

GAUGE INVARIANCE AND CANONICAL TRANSFORMATIONS IN DIRAC GENERALIZED MECHANICS

Ricardo A. Tello Llanos

Centro de Fisica, IVIC, Aptdo. 1827,
Caracas 1010A, Venezuela

There are several transformations allowed by the classical theory in the description of a given dynamical system. Does such transformations lead to different physical images when the theory is quantized?

We are concerned with a particular case of this general question, case wich was pointed out by Kâlnay and Ruggeri[1]: Can two quantum systems whose classical Lagrangians differ by a total time derivative may be physically inequivalent? Of course, in normal cases they can not. But when the classical Lagrangian is a singular one ($\det(\partial L/\partial \dot{q}_i \partial \dot{q}_j)=0$) the answer seemed to be affirmative.

In the reference quoted above it was shown that the presence of second class constraints (terminology of Dirac[2]) leads to a change in the functional form of the Dirac brackets when a total time derivative is added to the Lagrangian. Dirac brackets are necessary for quantization instead the usual Poisson brackets. Therefore, it results a change in the commutation relations between the quantum canonical variables and unitary equivalence can not be reached. Consequently, one must to recognize that canonical quantization leads to different physical situations at least in this sufficient general case when a gauge transformation is applied (following Levy-Leblond[3] we call the addition of a time derivative a gauge transformation. See also paper by Anderson and Bergmann[4] about field theories with gauge covariance.)

Nevertheless, our own aim is to reestablish the unicity of the quantization procedure, at least in the reffered case. For it, we consider without a serious loss of generality a Lagrangian with linear dependence on the velocities taken from a paper by Newman and Bergmann[5] and limit ourselves for simplicity to the case of a finite number of degrees of freedom:

$$L_g(q,\dot{q}) = f^i(q)\dot{q}_i - Q(q) + dg/dt \quad , \quad (q) = (q_1,q_2,\dots,q_n) \tag{1}$$

(summation convention is assumed.)

Following Kâlnay and Ruggeri we ask ourselves if there is some consequence of the presence of the gauge function $g(q)$ and, if there is any, can be found a unitary transformation to the case when the gauge function is taken equal to zero?

We will sketch the proof that there are not consequences and that the function $g(q)$ do not affect the transformation properties of the quantum description. A more complete argumentation will be published elsewhere[6]. In fact what we do is to achieve a gauge invariant canonical quantization by means of a canonical transformation at the classical level.

Applying the variational principle suitable for such Lagrangians, we obtain the Hamiltonian $H = Q + u_i Y^i$ up to a linear combination of primary constraints $Y^i = p^i - f^i - \partial^i g$ with indetermined multipliers u_i. At this step arise the following consistency conditions

$$\partial^i Q + f^{ij} u_j = 0 , \quad f^{ij} = \partial^i f^j - \partial^j f^i , \quad \partial^i f = \partial f / \partial q^i , \quad (2)$$

where the f^{ij} are gauge independent and equal to the Poisson brackets between the primary constraints.

The following argumentation depends of the regularity or not of the matrix f^{ij}. In the former case we can use the inverse matrix f_{ij} to solve the equations (2) respect to the u_i unknowns. As a result, all the constraints will be second class and we can use them to eliminate from the theory a number n of irrelevant canonical variables. If we eliminate the momenta (the only gauge dependent variables) we arrive to a gauge independent canonical formalism which quantization will be also gauge invariant.

In the more involved case with a degenerated matrix f^{ij} there exists a number $n - r$ of null vectors v_i^A , A runnig from one to $n - r$, where r is the rank of f^{ij}. Then, from eq. (2) follow the conditions

$$v_i^A \partial^i Q = 0 \quad (3)$$

Suppose that eq. (3) are in fact identities. In this case the number of second class constraints is equal to r and arise $n - r$ first class constraints $Y^A = v_i^A Y^i$. Introducing the Dirac brackets we can eliminate only $r < n$ irrelevant variables (momenta) and arrive to a theory with a number of gauge dependent variables. After relabeling they will be q_a, q_A, p^A, where a runs from one to r and A from $r + 1$ to n. Now we can perform a canonical transformation such that $q'_A = 0$. The remaining variables q_a satisfy the gauge invariance condition $\{ q_a , Y^A \} = 0$, have gauge invariant Dirac brackets and the Hamiltonian take a gauge independent form $H = Q(q_a, q_A = 0)$.

If equations (3) are not identities, further conditions (secondary constraints) may appear and repeated use of the outlined procedure leads finally to a gauge invariant description.

The Lagrangian $L = \frac{1}{2}(\dot{q}_1)^2 q_2 - \frac{1}{2}(q_1)^2 q_2$ used by Kálnay and Ruggeri to illustrate their result, which we has shown is wrong, can be replaced by the equivalent first order Lagrangian $L = q_2 q_3 \dot{q}_1 - \frac{1}{2}(q_1)^2 q_2 - \frac{1}{2}(q_3)^2 q_2$. The develope of this example according to our general procedure allows to see clearly the mistake.

REFERENCES

1. Kálnay, A.J. and Ruggeri, G.J. (1973). International Journal of Theoretical Physics, 8,189.

2. Dirac, P.A.M. (1964). Lectures on Quantum Mechanics, Belfer Graduate School of Sciences Monograph Series No. 2. Yeshiva University, New York.

3. Levy-Leblond, J.M. (1969). Communications in Mathematical Physics,12,64.

4. Anderson, J.L. and Bergmann, P.G. (1951). Physical Review, 83,1018.

5. Newman, E. and Bergmann, P.G. (1955). Physical Review, 99, 587.

6. Tello-Llanos, R.A. Work in progress.

AUTHOR INDEX

Lecture Notes in Physics

Vol. 173: Stochastic Processes in Quantum Theory and Statistical Physics. Proceedings, 1981. Edited by S. Albeverio, Ph. Combe, and M. Sirugue-Collin. VIII, 337 pages. 1982.

Vol. 174: A. Kadić, D.G.B. Edelen, A Gauge Theory of Dislocations and Disclinations. VII, 290 pages. 1983.

Vol. 175: Defect Complexes in Semiconductor Structures. Proceedings, 1982. Edited by J. Giber, F. Beleznay, J. C. Szép, and J. László. VI, 308 pages. 1983.

Vol. 176: Gauge Theory and Gravitation. Proceedings, 1982. Edited by K. Kikkawa, N. Nakanishi, and H. Nariai. X, 316 pages. 1983.

Vol. 177: Application of High Magnetic Fields in Semiconductor Physics. Proceedings, 1982. Edited by G. Landwehr. XII, 552 pages. 1983.

Vol. 178: Detectors in Heavy-Ion Reactions. Proceedings, 1982. Edited by W. von Oertzen. VIII, 258 pages. 1983.

Vol. 179: Dynamical Systems and Chaos. Proceedings, 1982. Edited by L. Garrido. XIV, 298 pages. 1983.

Vol. 180: Group Theoretical Methods in Physics. Proceedings, 1982. Edited by M. Serdaroğlu and E. İnönü. XI, 569 pages. 1983.

Vol. 181: Gauge Theories of the Eighties. Proceedings, 1982. Edited by R. Raitio and J. Lindfors. V, 644 pages. 1983.

Vol. 182: Laser Physics. Proceedings, 1983. Edited by J.D. Harvey and D.F. Walls. V, 263 pages. 1983.

Vol. 183: J.D. Gunton, M. Droz, Introduction to the Theory of Metastable and Unstable States. VI, 140 pages. 1983.

Vol. 184: Stochastic Processes – Formalism and Applications. Proceedings, 1982. Edited by G.S. Agarwal and S. Dattagupta. VI, 324 pages. 1983.

Vol. 185: H.N. Shirer, R. Wells, Mathematical Structure of the Singularities at the Transitions between Steady States in Hydrodynamic Systems. XI, 276 pages. 1983.

Vol. 186: Critical Phenomena. Proceedings, 1982. Edited by F.J.W. Hahne. VII, 353 pages. 1983.

Vol. 187: Density Functional Theory. Edited by J. Keller and J.L. Gázquez. V, 301 pages. 1983.

Vol. 188: A.P. Balachandran, G. Marmo, B.-S. Skagerstam, A. Stern, Gauge Symmetries and Fibre Bundles. IV, 140 pages. 1983.

Vol. 189: Nonlinear Phenomena. Proceedings, 1982. Edited by K.B. Wolf. XII, 453 pages. 1983.

Vol. 190: K. Kraus, States, Effects, and Operations. Edited by A. Böhm, J.W. Dollard and W.H. Wootters. IX, 151 pages. 1983.

Vol. 191: Photon Photon Collisions. Proceedings, 1983. Edited by Ch. Berger. V, 417 pages. 1983.

Vol. 192: Heidelberg Colloquium on Spin Glasses. Proceedings, 1983. Edited by J.L. van Hemmen and I. Morgenstern. VII, 356 pages. 1983.

Vol. 193: Cool Stars, Stellar Systems, and the Sun. Proceedings, 1983. Edited by S.L. Balliunas and L. Hartmann. VII, 364 pages. 1984.

Vol. 194: P. Pascual, R. Tarrach, QCD: Renormalization for the Practitioner. V, 277 pages. 1984.

Vol. 195: Trends and Applications of Pure Mathematics to Mechanics. Proceedings, 1983. Edited by P.G. Ciarlet and M. Roseau. V, 422 pages. 1984.

Vol. 196: WOPPLOT 83. Parallel Processing: Logic, Organization and Technology. Proceedings, 1983. Edited by J. Becker and I. Eisele. V, 189 pages. 1984.

Vol. 197: Quarks and Nuclear Structure. Proceedings, 1983. Edited by K. Bleuler. VIII, 414 pages. 1984.

Vol. 198: Recent Progress in Many-Body Theories. Proceedings, 1983. Edited by H. Kümmel and M. L. Ristig. IX, 422 pages. 1984.

Vol. 199: Recent Developments in Nonequilibrium Thermodynamics. Proceedings, 1983. Edited by J. Casas-Vázquez, D. Jou and G. Lebon. XIII, 485 pages. 1984.

Vol. 200: H.D. Zeh, Die Physik der Zeitrichtung. V, 86 Seiten. 1984.

Vol. 201: Group Theoretical Methods in Physics. Proceedings, 1983. Edited by G. Denardo, G. Ghirardi and T. Weber. XXXVII, 518 pages. 1984.

Selected Issues from

Lecture Notes in Mathematics

Vol. 836: Differential Geometrical Methods in Mathematical Physics. Proceedings, 1979. Edited by P. L. García, A. Pérez-Rendón, and J. M. Souriau. XII, 538 pages. 1980.

Vol. 837: J. Meixner, F. W. Schäfke and G. Wolf, Mathieu Functions and Spheroidal Functions and their Mathematical Foundations Further Studies. VII, 126 pages. 1980.

Vol. 838: Global Differential Geometry and Global Analysis. Proceedings 1979. Edited by D. Ferus et al. XI, 299 pages. 1981.

Vol. 840: D. Henry, Geometric Theory of Semilinear Parabolic Equations. IV, 348 pages. 1981.

Vol. 841: A. Haraux, Nonlinear Evolution Equations- Global Behaviour of Solutions. XII, 313 pages. 1981.

Vol. 842: Séminaire Bourbaki vol. 1979/80. Exposés 543–560. IV, 317 pages. 1981.

Vol. 843: Functional Analysis, Holomorphy, and Approximation Theory. Proceedings. Edited by S. Machado. VI, 636 pages. 1981.

Vol. 845: A. Tannenbaum, Invariance and System Theory: Algebraic and Geometric Aspects. X, 161 pages. 1981.

Vol. 849: P. Major, Multiple Wiener-Itô Integrals. VII, 127 pages. 1981.

Vol. 851: Stochastic Integrals. Proceedings, 1980. Edited by D. Williams. IX, 540 pages. 1981.

Vol. 852: L. Schwartz, Geometry and Probability in Banach Spaces. X, 101 pages. 1981.

Vol. 856: R. Lascar, Propagation des Singularités des Solutions d'Equations Pseudo-Différentielles à Caractéristiques de Multiplicités Variables. VIII, 237 pages. 1981.

Vol. 858: E. A. Coddington, H. S. V. de Snoo: Regular Boundary Value Problems Associated with Pairs of Ordinary Differential Expressions. V, 225 pages. 1981.

Vol. 861: Analytical Methods in Probability Theory. Proceedings 1980. Edited by D. Dugué, E. Lukacs, V. K. Rohatgi. X, 183 pages. 1981.

Vol. 866: J.-M. Bismut, Mécanique Aléatoire. XVI, 563 pages. 1981.

Vol. 878: Numerical Solution of Nonlinear Equations. Proceedings, 1980. Edited by E. L. Allgower, K. Glashoff, and H.-O. Peitgen. XIV, 440 pages. 1981.

Vol. 881: R. Lutz, M. Goze, Nonstandard Analysis. XIV, 261 pages. 1981.

Vol. 888: Padé Approximation and its Applications. Proceedings, 1980. Edited by M. G. de Bruin and H. van Rossum. VI, 383 pages. 1981.

Vol. 898: Dynamical Systems and Turbulence, Warwick, 1980. Proceedings. Edited by D. Rand and L.-S. Young. VI, 390 pages. 1981.

Vol. 901: Séminaire Bourbaki vol. 1980/81 Exposés 561–578. III, 299 pages. 1981.

Vol. 904: K. Donner, Extension of Positive Operators and Korovkin Theorems. XII, 182 pages. 1982.

Vol. 905: Differential Geometric Methods in Mathematical Physics. Proceedings, 1980. Edited by S.J. Andersson, H.-D. Doebner, and. H.R. Petry. VI, 309 pages. 1982.

Vol. 909: Numerical Analysis. Proceedings, 1981. Edited by J.P. Hennart. VII, 247 pages. 1982.

Vol. 912: Numerical Analysis. Proceedings, 1981. Edited by G. A. Watson. XIII, 245 pages. 1982.

Vol. 920: Séminaire de Probabilités XVI, 1980/81. Proceedings. Edité par J. Azéma et M. Yor. V, 622 pages. 1982.

Vol. 921: Séminaire de Probabilités XVI, 1980–81 Supplément: Géométrie Différentielle Stochastique. Proceedings. Edité par J. Azéma et M. Yor. III, 285 pages. 1982.

Vol. 922: B. Dacorogna, Weak Continuity and Weak Lower Semicontinuity of Non-Linear Functionals. V, 120 pages. 1982.

Vol. 923: Functional Analysis in Markov Processes. Proceedings, 1981. Edited by M. Fukushima. V, 307 pages. 1982.

Vol. 926: Geometric Techniques in Gauge Theories. Proceedings, 1981. Edited by R. Martini and E.M.de Jager. IX 219 pages. 1982.

Vol. 927: Y. Z. Flicker, The Trace Formula and Base Change for GL (3). XII, 204 pages. 1982.

Vol. 928: Probability Measures on Groups. Proceedings 1981. Edited by H. Heyer. X, 477 pages. 1982.

Vol. 929: Ecole d'Eté de Probabilités de Saint-Flour X – 1980. Proceedings, 1980. Edited by P.L. Hennequin. X, 313 pages. 1982.

Vol. 930: P. Berthelot, L. Breen, et W. Messing, Théorie de Dieudonné Cristalline II. XI, 261 pages. 1982.

Vol. 931: D.M. Arnold, Finite Rank Torsion Free Abelian Groups and Rings. VII, 191 pages. 1982.

Vol. 932: Analytic Theory of Continued Fractions. Proceedings, 1981. Edited by W.B. Jones, W.J. Thron, and H. Waadeland. VI, 240 pages. 1982.

Vol. 934: M. Sakai, Quadrature Domains. IV, 133 pages. 1982.

Vol. 935: R. Sot, Simple Morphisms in Algebraic Geometry. IV, 146 pages. 1982.

Vol. 936: S.M. Khaleelulla, Counterexamples in Topological Vector Spaces. XXI, 179 pages. 1982.

Vol. 937: E. Combet, Intégrales Exponentielles. VIII, 114 pages. 1982.

Vol. 938: Number Theory. Proceedings, 1981. Edited by K. Alladi. IX, 177 pages. 1982.

Vol. 942: Theory and Applications of Singular Perturbations. Proceedings, 1981. Edited by W. Eckhaus and E.M. de Jager. V, 363 pages. 1982.

Vol. 953: Iterative Solution of Nonlinear Systems of Equations. Proceedings, 1982. Edited by R. Ansorge, Th. Meis, and W. Törnig. VII, 202 pages. 1982.

Vol. 956: Group Actions and Vector Fields. Proceedings, 1981. Edited by J.B. Carrell. V, 144 pages. 1982.

Vol. 957: Differential Equations. Proceedings, 1981. Edited by D.G. de Figueiredo. VIII, 301 pages. 1982.

Vol. 963: R. Nottrot, Optimal Processes on Manifolds. VI, 124 pages. 1982.

Vol. 964: Ordinary and Partial Differential Equations. Proceedings, 1982. Edited by W.N. Everitt and B.D. Sleeman. XVIII, 726 pages. 1982.

Vol. 968: Numerical Integration of Differential Equations and Large Linear Systems. Proceedings, 1980. Edited by J. Hinze. VI, 412 pages. 1982.

Vol. 970: Twistor Geometry and Non-Linear Systems. Proceedings, 1980. Edited by H.-D. Doebner and T.D. Palev. V, 216 pages. 1982.

Vol. 972: Nonlinear Filtering and Stochastic Control. Proceedings, 1981. Edited by S.K. Mitter and A. Moro. VIII, 297 pages. 1983.

Vol. 978: J. Ławrynowicz, J. Krzyż, Quasiconformal Mappings in the Plane. VI, 177 pages. 1983.

Vol. 979: Mathematical Theories of Optimization. Proceedings, 1981. Edited by J.P. Cecconi and T. Zolezzi. V, 268 pages. 1983.